Lecture Notes in Computer Science   3892

Commenced Publication in 1973
Founding and Former Series Editors:
Gerhard Goos, Juris Hartmanis, and Jan van Leeuwen

Alessandra Carbone   Niles A. Pierce (Eds.)

DNA Computing

11th International Workshop on DNA Computing, DNA11
London, ON, Canada, June 6-9, 2005
Revised Selected Papers

 Springer

Volume Editors

Alessandra Carbone
Université Pierre et Marie Curie
Department of Computer Science
INSERM U511, 91 Boulevard de L'Hôpital, 75013 Paris, France
E-mail: Alessandra.Carbone@lip6.fr

Niles A. Pierce
California Institute of Technology
Applied and Computational Mathematics, Bioengineering
MC 114-96, Pasadena, CA 91125, USA
E-mail: niles@caltech.edu

Library of Congress Control Number: 2006925104

CR Subject Classification (1998): F.1, F.2.2, I.2.9, J.3

LNCS Sublibrary: SL 1 – Theoretical Computer Science and General Issues

ISSN   0302-9743
ISBN-10   3-540-34161-7 Springer Berlin Heidelberg New York
ISBN-13   978-3-540-34161-1 Springer Berlin Heidelberg New York

Springer is a part of Springer Science+Business Media

springer.com

© Springer-Verlag Berlin Heidelberg 2006
Printed in Germany

Typesetting: Camera-ready by author, data conversion by Scientific Publishing Services, Chennai, India
Printed on acid-free paper   SPIN: 11753681   06/3142   5 4 3 2 1 0

Preface

Biomolecular computing has emerged as an interdisciplinary field that draws on chemistry, computer science, mathematics, molecular biology, and physics. The International Meeting on DNA Computing (formerly DNA Based Computers) is a forum where scientists with different backgrounds, yet sharing common interests in biomolecular computing and DNA nanotechnology, meet and present their latest results. Continuing this tradition, the 11th International Meeting on DNA Computing was held June 6–9, 2005 at the University of Western Ontario in London, Ontario, Canada. For the first time, the meeting was organized under the auspices of the newly founded International Society for Nanoscale Science, Computation and Engineering (ISNSCE). The DNA11 Program Committee received 79 submissions, of which 23 were presented orally and 47 were presented as posters. The meeting was attended by 131 registered participants from 15 countries.

The meeting began with tutorials on computer science for life science researchers by Mark Daley (University of Western Ontario) and on molecular biology for computer scientists by Junghuei Chen (University of Delaware). Ned Seeman (New York University) concluded the first day with a survey on DNA nanotechnology. The remaining three days included contributed oral and poster presentations as well as invited lectures by James Gimzewski (University of California, Los Angeles) on nanomechanical probes of biosystems, Pehr Harbury (Stanford University) on DNA display, Eshel Ben-Jacob (Tel Aviv University) on bacterial intelligence, Erik Klavins (University of Washington) on robotic self-organization, and Dipankar Sen (Simon Fraser University) on DNA biosensors.

This volume contains 34 papers selected from the contributed oral and poster presentations. The wide-ranging topics include in vitro and in vivo biomolecular computation, algorithmic self-assembly, DNA device design, DNA coding theory, and membrane computing. The style of the contributions varies from theoretical molecular algorithms and complexity results, to experimental demonstrations of DNA computing and nanotechnology, to computational tools for simulation and design.

We wish to express our gratitude to the Program Committee members and external reviewers for evaluating the manuscripts. We also appreciate the efforts of the International Steering Committee chaired by Grzegorz Rozenberg in providing intellectual continuity for the meeting series. Profound thanks are due to Lila Kari, Mark Daley and all the members of the Organizing Committee for their substantial efforts in preparing for and running the meeting. We are grateful to all of the organizations that provided financial support for the meeting. Finally, we wish to thank all the members of the research community who contributed to the vitality of DNA11.

December 2005

Alessandra Carbone
Niles A. Pierce

Organization

Program Committee

Alessandra Carbone (Co-chair)	Université Pierre et Marie Curie, France
Niles A. Pierce (Co-chair)	California Institute of Technology, USA
Yaakov Benenson	Weizmann Institute, Israel
Junghuei Chen	Delaware Biotechnology Institute, USA
Robert Corn	University of California, Irvine, USA
Ashish Goel	Stanford University, USA
Lila Kari	University of Western Ontario, Canada
Chengde Mao	Purdue University, USA
Giancarlo Mauri	University of Milan, Bicocca, Italy
Gheorghe Păun	Romanian Academy, Bucharest & Sevilla University, Spain
John Rose	University of Tokyo, Japan
Paul Rothemund	California Institute of Technology, USA
William Shih	Harvard University, USA
Milan Stojanovic	Columbia University, USA
Masayuki Yamamura	Tokyo Institute of Technology, Japan
Hao Yan	Arizona State University, USA
Takashi Yokomori	Waseda University, Japan
Byoung-Tak Zhang	Seoul National University, South Korea

Steering Committee

Grzegorz Rozenberg	University of Leiden, Netherlands (Chair)
Leonard Adleman	University of Southern California, USA (Honorary Member)
Anne Condon	University of British Columbia, Canada
Masami Hagiya	University of Tokyo, Japan
Lila Kari	University of Western Ontario (UWO), Canada
Laura Landweber	Princeton University, USA
Richard Lipton	Georgia Institute of Technology, USA
Giancarlo Mauri	University of Milan, Bicocca, Italy
John Reif	Duke University, USA
Harvey Rubin	University of Pennsylvania, USA
Nadrian Seeman	New York University, USA
Erik Winfree	Caltech, USA

Organizing Committee
at the University of Western Ontario

Lila Kari (Chair)   Argyrios Margaritis
Mark Daley (Vice-chair)   Kathleen Hill
Michael Bauer   Susanne Kohalmi
Meg Borthwick   Elena Losseva
Fernando Sancho Caparrini   Kalpana Mahalingam
Gang Du   Shiva Singh
Greg Gloor

Sponsors

The Fields Institute of Research in Mathematical Sciences
MITACS (The Mathematics of Information Technology and Complex Systems
Institute)
Biomar Inc.
University of Western Ontario
Faculty of Science at the University of Western Ontario

Table of Contents

Self-correcting Self-assembly: Growth Models and the Hammersley Process

Yuliy Baryshnikov[1], Ed Coffman[2], Nadrian Seeman[3], and Teddy Yimwadsana[2]

[1] Bell Labs, Lucent Technologies, Murray Hill, NJ 07974
ymb@research.bell-labs.com
[2] Department of Electrical Engineering, Columbia University, NY 10027
{egc, teddy}@ee.columbia.edu
[3] Chemistry Dept., New York University, New York 10003
ned.seeman@nyu.edu

Abstract. This paper extends the stochastic analysis of self assembly in DNA-based computation. The new analysis models an error-correcting technique called *pulsing* which is analogous to checkpointing in computer operation. The model is couched in terms of the well-known tiling models of DNA-based computation and focuses on the calculation of computation times, in particular the times to self assemble rectangular structures. Explicit asymptotic results are found for small error rates q, and exploit the connection between these times and the classical Hammersley process. Specifically, it is found that the expected number of pulsing stages needed to complete the self assembly of an $N \times N$ square lattice is asymptotically $2N\sqrt{q}$ as $N \to \infty$ within a suitable scaling. Simulation studies are presented which yield performance under more general assumptions.

1 Introduction

In many respects, the current state of DNA-based computing resembles the state of standard, electronic computing a half century ago: a fascinating prospect is slow to develop owing to inflexible interfaces and unacceptably low reliability of the computational processes. We concentrate in this paper on the latter aspect, specifically addressing the interplay between the reliability and speed of DNA computing.

While DNA-based computational devices are known to be extremely energy efficient, their reliability is seen as a major obstacle to becoming a viable computing environment. As DNA based computing becomes more fully developed, the speed of self assembly will become a crucial factor; but as of now, little is known concerning the fundamental question of computation times. We emphasize the intrinsic connection between the two problems of reliability and speed, because of the unavoidable trade-off that exists between them. A clear understanding of the limitations of self-assembly reliability and speed, specifically that of DNA-based computing, and the interplay between these properties, will be paramount in determining the full potential of the paradigm. Our past work,

A. Carbone and N.A. Pierce (Eds.): DNA11, LNCS 3892, pp. 1–11, 2006.

briefly reviewed later, analyzed for a given function the time required to determine its value on given inputs, and therefore established theoretical limits on the performance of DNA-based computers. In the simplest instance, the analysis of computation times has surprising connections with interacting particle systems and variational problems, as shown in [1], and as further developed here. The critical new dimension of this paper is that of error correction; the new contributions lie in (a) a novel approach to dramatic improvements in the reliability of computations and (b) in the analysis of the inevitable performance losses of reliable computations.

The early theoretical work on DNA-based computation focused chiefly on various measures of complexity, in particular, program-size and time complexity [2,3,4]. However, Adleman et al [2,5] also investigated interesting combinatorial questions such as the minimum number of tile types needed for universality, and stochastic optimization questions such as the choice of concentrations that leads to minimum expected assembly times. Apart from these works, the mathematical foundations of computational speed in a stochastic context appear to be restricted to the work of Adleman et al [6] and Baryshnikov et al [1,7,8,9]. The former work studies random self assembly in one dimension. In a problem called $n-linear$ $polymerization$, elementary particles or monomers combine to form progressively larger polymers. The research of Baryshnikov et al [8] on linear self assembly has resulted in exact results for dimer self assembly, which reduces to an interesting maximal matching problem.

Any implementation of DNA computing is constrained fundamentally by the fact that all basic interactions have known and fixed energy thresholds, and these thresholds are much lower than those in electronic devices. This means that any realistic computational device based on organic structures like DNA is forced to operate at signal-to-noise ratios several orders of magnitude lower than those in electronic computing. Therefore, error correction at the computation stage becomes a necessity. Not surprisingly, recent research on error correction has concentrated on approaches that are analogous, at some level, to the repetition coding of information theory, or the concurrent execution of the same computational algorithm with subsequent comparison of results. Note also that biological computations achieve redundancy at little or no extra cost, as by the inherent virtues of the process, many copies of it are run independently. The work within this circle of ideas will be reviewed shortly. However, in our view, the notion of $pulsing$ should be introduced into this paradigm, a concept analogous to checkpointing in computer operation[1]. In our context, described in more detail below, the temperature (or some other parameter) of a self assembly process is pulsed periodically, a method in wide use to grow crystals; the pulse effectively rescues the most recent fault-free state (the current state if there have been no errors since the previous pulse). Clearly, the cost of washing out defective subassemblies must be balanced by a higher speed of these checkpointed computations.

---

[1] Checkpointing techniques have been in use since the early days of computing; see, e.g., [10].

Fig. 1. The set of four tiles on the left implements the operation $\oplus$. A tile glues to other tiles only if their corner labels match. On the right, operation $0 \oplus 1 = 1$ is performed. The input tiles are preassembled and the correct output tile simply attaches itself to the pattern, effectively obtaining the value of the output bit.

Formally, the now standard *tiling system* introduced and validated as a universal computational framework in [11,12] will be our abstract model, and follows the adaptation to elementary logic units of DNA computing described by Winfree and Rothemund [13,3]. The tile is modeled as shown in Figure 1 as a marked or labeled square. Briefly, in the simplest version, the label values are 0 or 1, and they break down into two input labels on one edge and two output labels on the opposite edge. As illustrated in Figure 1, a computational step consists of one tile bonding to others according to given rules that match input labels of one tile to the output labels of one or two adjacent tiles. Successive bonding of tiles in a self assembly process performs a computation.

Currently, in a typical implementation of this scheme, the tiles are DNA-based molecular structures moving randomly, in solution, and capable of functioning independently and in parallel in a self assembly process. This process results in a crystal-like construct modeled as a two dimensional array whose allowable structural variations correspond to the possible results of a given computation. We emphasize the contrast with classical computing paradigms: the random phenomena of self assembly create a randomness in the time required to perform a given computation.

2   Growth Models

The tiling self assemblies of the last section are growth processes. Through the abstraction described next, the times to grow constructs or patterns can be related to classical theories of particle processes; growth in the latter processes is again subject to rules analogous to those governing the self assembly process of the previous section. An initial set of tiles (the input) is placed along the coordinate axes, and growth proceeds outward in the positive quadrant; the placement of a new tile is allowed only if there are already tiles to the left and below the new tile's position which match labels as before. The left-and-below constraint is equivalent to requiring a newly added tile to bond at both of its input labels. The completion of the computation occurs at the attachment of the upper-right corner tile which can happen only after all other tiles are in place.

The fundamental quantity of interest is the computation time, or equivalently, the time until the final square (represented by the upper right corner square in position (M, N)) is in place. Let $T_{i,j}$ be the time it takes for a tile to land at position (i, j) once the conditions are favorable; that is, once both positions $(i, j-1)$ and $(i-1, j)$ are tiled. In a reference theory for self assembly, it is natural to take the $T_{i,j}$'s as independent exponential random variables with unit means. Let $C_{i,j}$ be the time until the square (i, j) becomes occupied, so that the random completion time of interest is given by $C_{M,N}$.

On discovering the isomorphic relationship between the self assembly process and the totally asymmetric simple exclusion process (TASEP), Baryshnikov et al [1] exploited the results on TASEP behavior in the hydrodynamic limit to show that, as N, M grow to infinity such that M/N tends to a positive constant, one has [14, p. 412]

$$C_{M,N}/(\sqrt{M} + \sqrt{N})^2 \sim 1, \tag{1}$$

a formula quantifying the degree of parallelism in the computation. One can generalize this formula to schemes where the tiles can depart as well (like the schemes described in [15]) with the rates of departures $\rho < 1$, and also to more general shapes D than mere squares. In this model, growth is clearly not monotonic, but still can be mapped to a generalization of TASEP for which similar results are known. Baryshnikov et al [1] proved the following:

Theorem 1. *The time $E_{\lambda D, \rho}$ required to complete computation on a DNA computer of shape λD with tiles arriving at rate 1 and departing at rate ρ is given by*

$$\lim_{\lambda \to \infty} \lambda^{-1} E_{\lambda D} = \frac{1}{1-\rho} \sup_\gamma \int \left(\sqrt{\frac{d\xi}{dz}} + \sqrt{\frac{d\eta}{dz}} \right)^2 dz. \tag{2}$$

3   Error Correction

Since the early days of computing, various methods have been used to deal with error prone computers, parity checking being a standard one. Checkpointing was a popular method implemented in operating systems, and it is still being used today in high-performance systems. This method creates milestones, or checkpoints, at different locations in the process. All the required information is stored at a checkpoint so that the process can restart from that location without having to perform the work done before that checkpoint. Typically a control mechanism periodically creates checkpoints. When the controlled process fails, it is rolled back to the most recent checkpoint and resumes from that location.

Current developments in tile self-assembly resemble developments in the early computing era, after a Hegelian development cycle. We expect the checkpointing method, being a simple but elegant error-correction technique to become a viable tool in the area, at least until a dramatic change in the underlying chemical technology takes place. The narrow question we address below is how to apply it to DNA tile self-assembly. We briefly review the literature before turning to our new approach.

Alternative approaches. The two most frequent errors in DNA self-assembly are growth errors and nucleation errors. Growth errors occur when a wrong type of tile, an *error* tile, attaches to the lattice; a sublattice that forms with the error tile at its origin will then be corrupt. A nucleation error occurs when only one side of a tile attaches to the lattice, and hence at a wrong position. Thermodynamic controls that slow down growth can be introduced to help ensure the relatively early separation of error tiles.

A tile can also be designed to have its own error-correction capability, or a new type of tile that assists the self-assembly process in lowering error rate can be introduced. Several methods for this have been proposed. For example, Winfree and Bekbolatov's Proofreading Tile Set [15] shows that the error rate can be reduced significantly by creating an original Wang Tile using four or nine smaller tiles (2×2 or 3×3) in order to ensure that the small incorrect tiles will fall off before they are assembled to form an incorrect Wang tile. Chen and Goel's Snake Tile Set [16] improves the Proofreading Tile Set by ensuring that the smaller tiles can be assembled only in certain directions.

Reif et al [17] use pads to perform error checking when a new tile is attached to the lattice. Each pad acts as a kind of adhesive, connecting two Wang tiles together, whereas in the original approach the Wang tiles attach to each other. This method allows for redundancy: a single pad mismatch between a tile and its immediate neighbor forces at least one further pad mismatch between a pair of adjacent tiles. This padding method can be extended further to increase the level of redundancy.

Chen et al's Invadable Tile Set [18] applies the invading capability of the DNA strand to emulate the invasion of a tile. In this model, the tiles are designed so that the correct tile can invade any incorrect tile during the lattice growth process. Fujibayashi and Murata's Layered Tile Model [19] significantly reduces the error rate by using two layers of tiles: the Wang tile layer and the protective tile layer. The protective layer does not allow tiles to attach to the lattice incorrectly. When the attachment is correct, the protective tile releases the rule tile, to which the next tile attaches itself. As one must expect, all methods have one or more shortcomings or costs associated with them, such as prolonged self-assembly times, enlarged lattices, potential instabilities, and failure to deal effectively with both error types.

Modeling checkpointing in DNA self-assembly. We now return to periodic temperature pulsing, whereby pulses remove the defective parts of a crystal; in particular, the hydrogen bonds between improperly attached DNA tiles are broken so that defective substructures can separate from the lattice, thus restarting growth at an earlier fault-free structure. Parameters other than temperature can also be considered in the pulsing approach. Pulsing applied to the DNA tile self-assembly model removes the incorrectly attached tiles from the assembly at a higher rate than the correct ones. More targeted pulsing systems can employ enzymatic or conformational ways to shift the binding energy.

In our model of self-assembly with checkpointing, we consider a lattice of size $N \times N$. (While our results are valid for general shapes, the square lattice

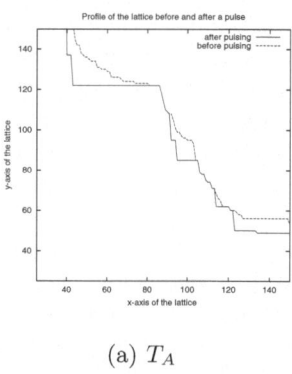

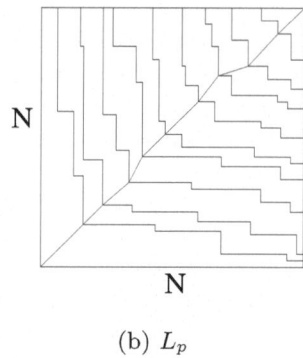

(a) T_A (b) L_p

Fig. 2. (a) - the profile of DNA tile self-assembly process before and after a pulse. (b) the relationship between the number of layers and the longest increasing subsequence for a Poisson Point Process on the plane. Crystal size: 500×500.

helps focus discussion.) We study the standard growth process described earlier, with the modification that there are two competing populations of tiles to be attached, correct tiles and erroneous tiles. With an appropriate rescaling, the waiting time until a tile attaches at a vacant position is taken to be exponential with mean 1 (all attachment times are independent). Attached tiles are erroneous with probability q.

We call an error tile that attaches to a valid structure a *seed error tile*. Any tile attached to the northeast of the seed error tile is automatically an error tile as it participates in an incorrect computation. (See Figure 2(a)). In our initial analysis we assume that a pulse succeeds in washing out all defective regions of the structure. (See Figure 2(a).)

A growth *stage* consists of a growth period of duration P between consecutive pulses. At the end of one such stage, the locations of the seed error tiles define the boundary of the lattice for the next growth period, on which more tiles will be attached. A growth *layer* is the region of tiles that attach to the lattice during one growth step. The number of stages required to complete the $N \times N$ lattice is the number of pulses or layers required to complete the lattice.

The profiles at the beginning of each growth stage (which we will call the *rectified profiles*, as they are the result of the removal of all erroneous tiles, including the seed tiles) form a Markov process which is clearly significantly more complicated than the growth process without pulsing. Moreover, it is easily seen that these processes cannot be mapped onto 1-dimensional particle processes with local interaction. Hence to evaluate performance, we are forced to resort to asymptotic analysis and simulation studies.

There are several remarks we can make before starting our analysis. Denote the pulsing times by $t_1 < t_2 < \ldots$, and the corresponding rectified profiles, that is the profiles after pulses, by $\mathcal{S}_i$. Clearly, these profiles, from one to the next, are nondecreasing. In fact, one can describe the evolutions of these profiles using the growth models as in Section 2: the only modification is that if the tile (k, l) is an error seed, then the completion time for this tile becomes

$$C_{k,l} = \min\{t_i > \max(C_{k-1,l}, C_{k,l-1})\}.$$

Using this nondecreasing property, and the standard (in the analysis of algorithms) "principle of deferred decisions", it is not difficult to verify that the collection of the *seed error tiles* over all growth process, form an independent q-Bernoulli sample from the $N \times N$ lattice points.

Small q asymptotics. When q is small, the sample from the planar domain $N \times N$, each dimension rescaled by $\sqrt{q}$ (so that the expected number of seed error tiles in any part of the domain of unit area is 1), approaches, in distribution, the Poisson sample.

The number of pulses required to complete a crystal can be approximated using Hammersley's Process. In our version of this process, we have an underlying Poisson process in two dimensions with samples Ω taken from the square $S = [0, a] \times [0, a]$, with $a = \sqrt{q}N$.

For each $z = (x, y) \in S$ let $n(z)$ be the length of the longest *monotone subsequence* in Ω between $(0, 0)$ and z, that is the maximal length ℓ of a chain of points $(x_1, y_1) < (x_2, y_2) < \ldots < (x_\ell, y_\ell) < (x, y)$, where $(u, v) < (u', v')$ iff $u < u'$ and $v < v'$. See Figure 2(b): The right-hand picture shows layers of points in Ω which can be reached via monotone paths in Ω of length 0 (these are the points adjoining a coordinate axis), 1 (next layer) and so on. A longest path connecting the origin and the point (a, a) is shown as well.

The problem of finding this length is closely related to the famous problem of finding the longest increasing subsequence in a random permutation (Ulam's problem). It turns out that the expected value $\mathbb{E}\ell \sim 2a$, as $a \to \infty$, see, e.g., [20].

This implies immediately information about the asymptotic scaling of the number L_p of pulses necessary to achieve a correct assembly when the interpulse time P is large compared to $q^{-1/2}$. In the limit of small q and large N, this yields a very precise description of this number. Indeed, we have

Theorem 2. *The following assertions hold:*

1. *For any given sample of seed error tiles, L_p is at least the length of the longest increasing subsequence in the sample.*
2. *If $qP^2 \to \infty$, then for samples of seed error tiles, L_p is asymptotically the length of a longest increasing subsequence.*
3. *If $N, qP^2 \to \infty$, and $q \to 0$, then L_p grows as*

$$L_p \sim 2N\sqrt{q}.$$

Sketch of the proof. The first assertion follows immediately from the fact that along any increasing sequence of seed error tiles, the higher error tile is corrected by a pulse coming later than the pulse correcting the previous error tile. Further, the second assertion follows from the fact that if the time P is large enough, the rectified profiles $\mathcal{S}_i$ coincide with the layers of points reached by the increasing subsequences in Ω having constant length (this can be proved by induction). The last assertion is simply the limit corresponding to the Poisson approximation introduced earlier.

Remarks.

1. Using this result, we can adjust the value of P so as to obtain an estimate of the minimal time required to complete the lattice for any given N, q, and average time p_s taken by a pulse to remove error tiles. More details can be found in [21].
2. The Layered Tile Set [19] can be seen as a variant of our method where P is 1 time unit. When the value of P is very small, the total number of pulses becomes very large because the process pulses once per time unit. As a result, when the value of P is small, the completion time for the formation of the crystal is inversely large. Furthermore, in the case of a high p_s, a very low value of P will not be suitable for the process because of the length of time required during the checkpointing process. Therefore, if P is adjusted appropriately, our checkpointing method will be better than the Layered Tile Set technique.
3. The total growth time in the model is the sum of the interpulse time (plus a constant pulse setup time) times the number of pulses. In the regime described in part 3 of of Theorem 2, this gives the growth rate $\sqrt{q}PN$, which can be arbitrarily close to the obvious lower bound $\Omega(N)$.

Simulation analysis. The total crystal completion time, C, consists of the total time required by tile attachment, C_A, pulsing setup time and the pulsing overhead, C_p. Our simulations determine the effect of P and q on C_A and C_p. The simulation of a 500×500 lattice yielded C_A and C_p for various P and q. The total pulsing overhead time, C_p, is given by $C_p = p_s L_p$ (recall that p_s is the average time taken by a pulse to remove all erroneous tiles).

Our self assembly simulations created more than a million tiles. Developing the simulator was a challenge in itself, given current limits on computer memory. We designed our simulator so that it contains only the information of the crystal, which for our purposes will suffice without having to assign memory space for each tile. Implementation details can be found in [21].

Figure 3 shows the effects of P and q on the performance of self-assembly with pulsing. Since the total time C required to complete the crystal is $C_p + C_A$, we see that C in general has an optimal point for given p_s.

For example, Figure 4(a) shows the total-time surface plot as a function of (P, q). For simplicity, we assume that the time required for each pulse, p_s, is linearly proportional to the growth time, $p_s = 0.2P + 2$, to show how p_s can affect the total time, C. For a given value of q, there is an optimal P that minimizes the total time to complete the self-assembly. Figure 4(b) shows the total time for different values of P with the error probability $q = 0.05$. The figure shows that one obtains the highest over-all lattice growth rate when P is approximately 9 time units.

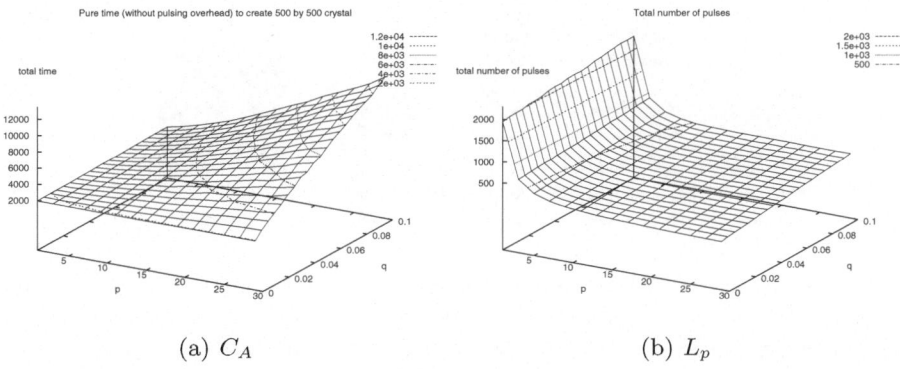

(a) C_A (b) L_p

Fig. 3. The performance of pulsing for various P and q. Crystal size: 500×500.

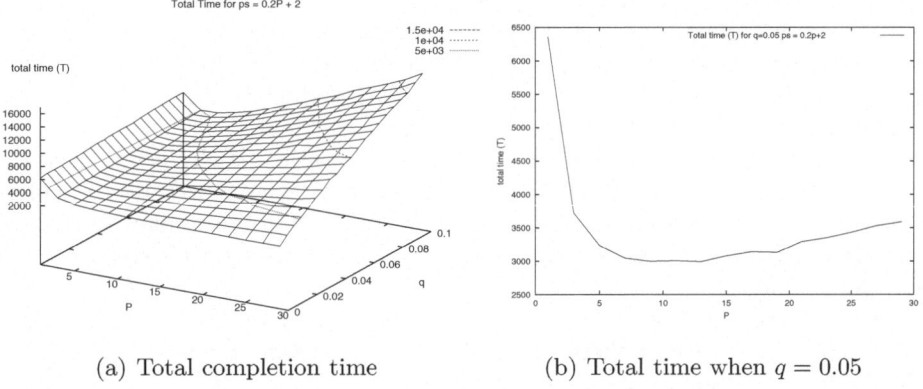

(a) Total completion time   (b) Total time when $q = 0.05$

Fig. 4. (a) total-time pulsing performance, various (P,q). (b) cross section of (a), q=0.05.

4   Future Directions

We have introduced and analyzed the performance of an error-correcting self assembly pulsing (checkpointing) technique. This comprises the modeling and analysis component of an over-all project that will include the essential experimental
component. To fully verify the validity of the method, we propose an experimental setup that produces a simple periodic system capable of errors, as a testbed for
examining our proposed error-correcting techniques. It is possible to 'tag' particular elements in 2D crystals; the addition of a DNA hairpin that sticks up from the
plane of the array is the simplest tag. Thus, rather than using a single motif to act
as a 'tile' in forming the crystal, we can use two or more different tiles, say, an A-tile
and a B-tile. An example of this approach is shown for DX molecules in Figure 5.

The experiments we propose here are based on this notion of two differently
marked tiles. The idea is to make a 2-D array with two tiles whose sticky ends
are distinct, as they are in the DX example above. It is unlikely that individual

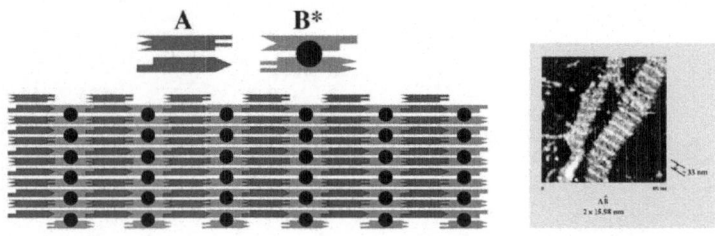

Fig. 5. A Schematic DX Array with 2 Components, A and B*; B* contains a hairpin (the black circle) that protrudes from the plane. Sticky ends are shown geometrically. Note that the A and B* components tile the plane. Their dimensions in this projection are 4 × 16 nm, so a 32 nm stripe is seen on the AFM image to the right.

4 nm x 16 nm DX tiles can be recognized, but many motifs with large dimensions in both directions exist (see e.g., [22]). The uniqueness of the sticky ends is central to the robust formation of the pattern shown above. Were the sticky ends of the two molecules identical, a random array of protruding features would result, rather than the well-formed stripes shown. It is clear that there must exist some middle ground between unique sticky ends and identical sticky ends. Each tile contains four sticky ends, each of six or so nucleotides. We propose to explore steps on the way from uniqueness to identity (starting from unique) so that we can get a set of tiles that produce an array with a well-defined error rate, low but detectable.

Once we have such a system, it will then be possible to do prototype rescue operations. The basis of these operations will be thermodynamic pulsing, but the study of techniques based on recognition of structural differences in the 2D array, or some combination of the two basic approaches, will be pursued as well.

References

1. Baryshnikov, Y., Coffman, E., Momčilović, P.: DNA-based computation times. In: Proc. of the Tenth International Meeting on DNA Computing, Milan, Italy (2004)
2. Adleman, L., Cheng, Q., Goel, A., Huang, M.D.: Running time and program size for self-assembled squares. In: Proc. ACM Symp. Th. Comput. (2001) 740–748
3. Rothemund, P., Winfree, E.: The program-size complexity of self-assembled squares. In: Proc. ACM Symp. Th. Comput. (2001) 459–468
4. Winfree, E.: Complexity of restricted and unrestricted models of molecular computation. In Lipton, R., Baum, E., eds.: DNA Based Computing. Am. Math. Soc., Providence, RI (1996) 187–198
5. Adleman, L., Cheng, Q., Goel, A., Huang, M.D., Kempe, D., de Espanés, P.M., Rothemund, P.: Combinatorial optimization problems in self-assembly. In: Proc. ACM Symp. Th. Comput., Montreal, Canada (2002) 23–32
6. Adleman, L., Cheng, Q., Goel, A., Huang, M.D., Wasserman, H.: Linear self-assemblies: Equilibria, entropy, and convergence rates. In Elaydi, Ladas, Aulbach, eds.: New progress in difference equations, Taylor and Francis, London (2004)

7. Baryshnikov, Y., Coffman, E., Momčilović, P.: Incremental self-assembly in the fluid limit. In: Proc. 38th Ann. Conf. Inf. Sys. Sci., Princeton, NJ (2004)
8. Baryshnikov, Y., Coffman, E., Winkler, P.: Linear self-assembly and random disjoint edge selection. Technical Report 03-1000, Electrical Engineering Dept., Columbia University (2004)
9. Baryshnikov, Y., Coffman, E., Momčilović, P.: Phase transitions and control in self assembly. In: Proc. Foundations of Nanoscience: Self-Assembled Architectures and Devices, Snowbird, UT (2004)
10. E. G. Coffman, J., Flatto, L., Wright, P.E.: A stochastic checkpoint optimization problem. SIAM J. Comput. **22** (1993) 650–659
11. Wang, H.: Dominoes and AEA case of the decision problem. In: Proc. of the Symposium in the Mathematical Theory of Automata, Polytechnic Press, Brooklyn, NY (1963)
12. Berger, R.: The undecidability of the domino problem. In: Memoirs of the American Mathematical Society. Volume 66. (1966)
13. Winfree, E.: Algorithmic Self-Assembly of DNA. PhD thesis, California Institute of Technology, Pasadena, CA (1998)
14. Liggett, T.M.: Interacting Particle Systems. Springer-Verlag, New York (1985)
15. Winfree, E., Bekbolatov, R.: Proofreading tile sets: Error correction for algorithmic self-assembly. In: Proceedings of the Ninth International Meeting on DNA Based Computers, Lecture Notes in Computer Science. Volume 2943. (2004) 126–144
16. Chen, H.L., Goel, A.: Error free self-assembly with error prone tiles. In: Proceedings of the Tenth International Meeting on DNA Based Computers, Milan, Italy (2004)
17. Reif, J.H., Sahu, S., Yin, P.: Compact error-resilient computational dna tiling assemblies. In: Proceedings, Tenth International Meeting on DNA Based Computers, Lecture Notes in Computer Science, Springer-Verlag, New York (2004) 293–307
18. Chen, H.L., Cheng, Q., Goel, A., Huang, M.D., de Espanes, P.M.: Invadable self-assembly: Combining robustness with efficiency. In: ACM-SIAM Symposium on Discrete Algorithms. (2004)
19. Fujibayashi, K., Murata, S.: A method of error suppression for self-assembling DNA tiles. In: Proceedings of the Tenth International Meeting on DNA Based Computers: Lecture Notes in Computer Science, Springer Verlag, New York (2004) 284–293
20. Aldous, D., Diaconis, P.: Hammersley's interacting particle process and longest increasing subsequences. Probab. Th. Rel. Fields **103** (1995) 199–213
21. Baryshnikov, Y., Coffman, E., Yimwadsana, T.: Analysis of self-correcting self-assembly growth models. Technical Report 03-1001, Electrical Engineering Dept., Columbia University (2005)
22. Mao, C., Sun, W., Seeman, N.: Designed two-dimensional DNA Holliday Junction Arrays visualized by atomic force microscopy. J. Am. Chem. Soc. **121** (1999) 5437–5443

Recognizing DNA Splicing

Matteo Cavaliere[1], Nataša Jonoska[2], and Peter Leupold[3]

[1] Department of Computer Science and Artificial Intelligence,
University of Sevilla,
Avda. Reina Mercedes s/n, 41012 Sevilla, Spain
martew@inwind.it
[2] Department of Mathematics,
University of South Florida, Tampa, FL 33620, USA
jonoska@math.usf.edu
[3] Research Group on Mathematical Linguistics,
Rovira i Virgili University,
Pl. Imperial Tàrraco 1, 43005 Tarragona, Spain
klauspeter.leupold@estudiants.urv.es

Abstract. Motivated by recent techniques developed for observing evolutionary dynamics of a single DNA molecule, we introduce a formal model for accepting an observed behavior of a splicing system. The main idea is to input a marked DNA strand into a test tube together with certain restriction enzymes and, possibly, with other DNA strands. Under the action of the enzymes, the marked DNA strand starts to evolve by splicing with other DNA strands. The evolution of the marked DNA strand is "observed" by an outside observer and the input DNA strand is "accepted" if its (observed) evolution follows a certain expected pattern. We prove that using finite splicing system (finite set of rules and finite set of axioms), universal computation is attainable with simple observing and accepting devices made of finite state automata.

1 Introduction: (Bio)Accepting Devices

Recently, several techniques for observing the dynamics of a single DNA molecule and in general of a single biomolecule have been developed. Some of these come from the study of protein dynamics and interactions in living cells. For instance, a well established methodology is the *FRAP*, fluorescent recovery after photobleaching, [13]; other known methodologies are *FRET*, [11], fluorescence resonance energy transfer and *FCS*, [19], fluorescent correlation spectroscopy. A survey on the techniques to observe dynamics of biomolecules, with their advantages and disadvantages, can be found in [14]. Usually these techniques can be used to observe only three different colors in fluorescent microscope, but it is possible to obtain more colors by *multiplexing*, as suggested by [12].

A totally new way to mark (and then, to observe) single DNA molecules is represented by *quantum dots*; by using this technique it is possible to tag individual DNA molecules; in other words they can be used like fluorescent biological labels, as suggested by [3], [8]. A very recent review on the use of quantum dots in vivo imaging can be found in [16].

A. Carbone and N.A. Pierce (Eds.): DNA11, LNCS 3892, pp. 12–26, 2006.
© Springer-Verlag Berlin Heidelberg 2006

In many techniques presented in [14], studying of the dynamics of DNA strands is divided in two separate phases: the registration of the dynamics (on a special support like channels of data) and then the investigation of the collected data. Hence, the model that is introduced in this paper uses "observer" and "decider" as two independent devices.

The theoretical model is used to construct accepting devices using DNA operations. The evolution/observation strategy was initially introduced in a formal computing model inspired by the functioning of living cells, known as membrane systems [5].

Since then, the evolution/observation idea has been [2], [4], [6]. considered in different formal models of biological systems. In all these developments, the underlying idea is that a *generative* device is constructed by using two systems: a mathematical model of a biological system that "lives" (evolves) and an observer that watches the entire evolution of this system and translates it into a readable output.

Thus the main idea of this approach is that the computation is made by observing the entire life of a biological system. Differently from the previously mentioned works, in [7] the evolution/observation strategy has been used to construct an *accepting* device. There, it has been suggested that it is possible to imagine any biological system as an accepting device. This is achieved by taking a model of a biological system, introducing an input to such a system and observing its evolution. If the evolution of the system is of an expected type, (for example follows a regular predetermined pattern) the input is accepted by the (bio)system, otherwise it can be considered rejected.

An external *observer* is fundamental in extracting a more abstract, formal behavior from the evolution of the biological system. A *decider* is the machine that checks whether the behavior of the biological system is of the expected type.

Splicing systems belong to a formal model of recombination of double stranded DNA molecules (for simplicity we call them DNA strands) under the action of a ligase and restriction enzymes (endonucleases), [10]. The main purpose of this paper is to illustrate the accepting strategy of oberver/decider to splicing systems. For the motivations and background on splicing systems we refer to the original paper [10] or to the corresponding chapter in [18].

In [4] an observer was associated to splicing systems to construct a *generative* device. Here we construct an *accepting device* by joining a *decider* to the observer of the splicing system. We call such a system *Splicing Recognizer* (in short, SR). A schematic view of the model is depicted in Figure 1.

The SR works in the following way. An input *marked* DNA strand (represented by a string w) is inserted in a test tube. Due to the presence of restriction enzymes, the input strand changes, as it starts to recombine with other DNA strands present in the test tube. A sequence of *intermediate* marked DNA strands is generated. This constitutes the evolution of the input marked DNA strand. Schematically this is presented with the sequence of w, w', w'', w''' in Figure 1.

The external observer associates to each intermediate marked strand a certain label taken from a finite set of possible labels. It writes these labels onto an output tape in their chronological order. In Figure 1 this corresponds to the string $a_1a_2a_3a_4$. This string represents a code of the obtained evolution. When the marked strand becomes of a certain predetermined "type" the observation stops.

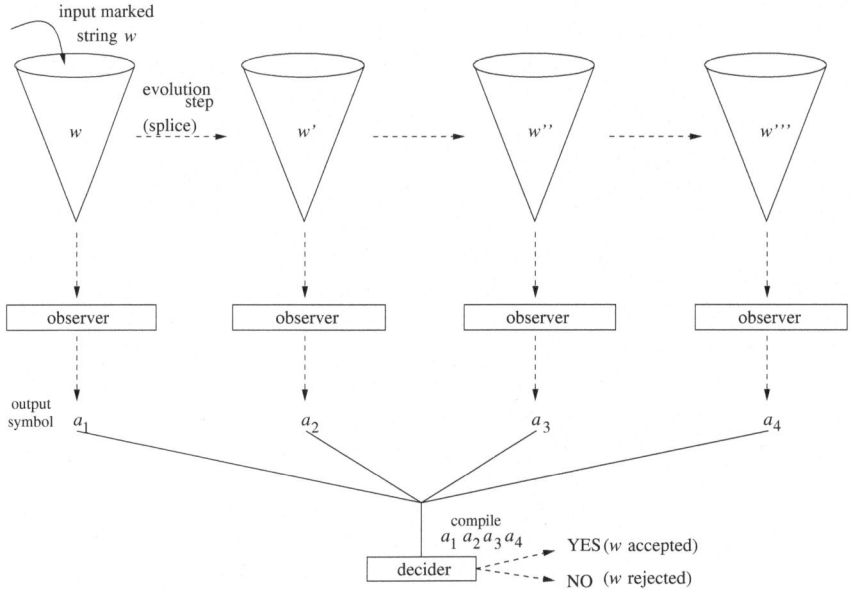

Fig. 1. The splicing/observer architecture

At this point the decider checks if the entire evolution of the input marked DNA strand described by the string $a_1a_2a_3a_4$ has followed a certain pattern, i.e. if it is in a certain language. If this is true, the input string w is accepted by the SR; otherwise it is considered to be rejected.

This paper shows that using this strategy, it is possible to obtain very powerful accepting systems even when very simple components are used.

For instance, we show that having just a finite state automaton as observer of the evolution of a finite splicing system (with a finite set of splicing rules) is already enough to simulate a Turing machine. This is a remarkable jump in acceptance power since it is well known that a finite splicing system by itself can generate only a subclass of the class of regular languages. The results are not surprising, since by putting extra control with the decider, the computational power of the whole system increases. Similar results, but in the generative sense, were obtained without the decider in [4] but these required a special observation of a right-most evolution, which is not the case with the results presented here.

2   Splicing Recognizer: Definition

In what follows we use basic concepts from formal language theory. For more details on this subject the reader should consult the standard books in the area, for instance, [20], [21].

Briefly, we fix the notations used here. We denote a finite set (the alphabet) by V, the set of words over V by V^*. By REG, CF, CS, and RE we denote the classes of languages generated by regular, context-free, context-sensitive, and unrestricted grammars respectively.

2.1   Splicing with a Marked String

As underlying biological system we consider a splicing system (more precisely an H scheme, following the terminology used in [18]). As discussed in the Introduction, the splicing system used has the particular feature that, at any time, exactly one string of the produced language is marked.

First we recall some basic notions concerning splicing systems. However, in what follows, we suppose the reader is already familiar with this subject, as for instance, presented in [18].

Consider an alphabet V (splicing alphabet) and two special symbols $\#$ and $\$$ not in V. A splicing rule (over V) is a string of the form $u_1\#u_2\$u_3\#u_4$, where $u_1, u_2, u_3, u_4 \in V^*$.

For a splicing rule $r = u_1\#u_2\$u_3\#u_4$ and strings $x, y, z_1, z_2 \in V^*$ we write $(x, y) \Longrightarrow_r (z_1, z_2)$ iff $x = x_1u_1u_2x_2$, $y = y_1u_3u_4y_2$, $z_1 = x_1u_1u_4y_2$, $z_2 = y_1u_3u_2x_2$. We refer to z_1 (z_2) as the first (second) string obtained by applying the splicing rule r.

An H scheme is a pair $\sigma = (V, R)$ where V is an alphabet, and $R \subseteq V^*\#V^*\$V^*\#V^*$ is a set of splicing rules. For a given H scheme $\sigma = (V, R)$ and a language $L \subseteq V^*$ we define $\sigma(L) = \{z_1, z_2 \in V^* \mid (x, y) \Longrightarrow_r (z_1, z_2),$ for some $x, y \in L, r \in R\}$.

When restriction enzymes (and a ligase) are present in a test tube, they do not stop acting after one cut and paste operation, but they act iteratively.

Given a *initial language* $L \subseteq V^*$ and an H scheme $\sigma = (V, R)$ we define the iterated splicing as: $\sigma^0(L) = L$, $\sigma^{i+1}(L) = \sigma^i(L) \cup \sigma(\sigma^i(L)), i \geq 0$.

In this work, as previously discussed, we are interested in observing the evolution of a specific marked string introduced, at the beginning, in the initial language L and called *input marked string*.

Given an initial language L, an *input marked string* $w \in L$, a *target marked language* L_t and an H scheme σ, the scheme defines a sequence of marked strings that represents the evolution of the input marked string w, according to the splicing rules defined in σ (for simplicity we suppose $w \notin L_t$). The sequence of marked strings, $\langle w_0 = w, w_1, \cdots, w_k \rangle$, for $k \geq 1$ and $w_k \in L_t$, is constructed in the following iterative way (w_i is the marked string associated to the set $\sigma^i(L), 0 \leq i \leq k$).

Each new marked string is obtained by splicing the old marked string, until a marked string w_k from the target marked language L_t is reached or the marked string cannot be spliced.

The first string of the sequence is the input marked string, $w_0 = w$.

If $w_i \in L_t$, $i \geq 1$, then the sequence ends (the marked string is among the ones of the target marked language).

If there is no $x \in \sigma^i(L)$, $i \geq 0$, such that $(w_i, x) \Longrightarrow_r (z_1, z_2)$ or $(x, w_i) \Longrightarrow_r (z_1, z_2)$ for some $r \in R$, then the sequence ends (the marked string cannot be spliced).

If $x, y \in \sigma^i(L)$, $i \geq 0$, with $w_i = x$ (or $w_i = y$) and there exists a rule $r \in R$ such that $(x, y) \Longrightarrow_r (z_1, z_2)$, then $w_{i+1} = z_1$. In this case, if the marked string can be subject to more than one splicing rule, producing different strings, the choice of the next marked string is done in a non-deterministic way. Notice that we always consider the first string produced as the new marked one.

Because the update of a marked string is made in a non-deterministic way, given an input marked string w, an initial language L, a target marked language L_t, and an H scheme σ, it is possible to get different sequences of intermediate marked strings. The collection of all these sequences is denoted by $\sigma(w, L, L_t)$.

For a splicing rule $r = u_1 \# u_2 \$ u_3 \# u_4$ we denote by $rad(r)$ the length of the longest string u_1, u_2, u_3, u_4; we say that this is the *radius* of r. The radius of an H scheme is the maximal radius of its rules.

In what follows, we denote by FIN_{H_k} the class of H schemes with radius at most k and using finite set of splicing rules.

2.2 Observer

For the observer as described in the Introduction we need a device mapping arbitrarily long strings, into just one singular symbol. As in earlier work [6] we use a special variant of finite automata with some feature known from Moore machines: the set of states is labelled with the symbols of an output alphabet Σ. Any computation of the automaton produces as output the label of the state it halts in (we are not interested in accepting / not accepting computations and therefore also not interested in the presence of final states); because the observation of a certain string should always lead to a fixed result, we consider here only deterministic and complete automata.

Formalizing this, a *monadic transducer* is a tuple $O = (Z, V, \Sigma, z_0, \delta, l)$ with state set Z, input alphabet V, initial state $z_0 \in Z$, and a complete deterministic transition function δ as known from conventional finite automata; further there is the output alphabet Σ and a labelling function $l : Z \mapsto \Sigma$. The output of the monadic transducer is the label of the state it stops in. For a string $w \in V^*$ and a transducer O we then write $O(w)$ for this output; for a sequence $\langle w_1, \ldots, w_n \rangle$ of $n \geq 1$ strings over V^* we write $O(w_1, \ldots, w_n)$ for the string $O(w_1) \cdots O(w_n)$.

For simplicity, in what follows, we present only the mappings that the observers define, without giving detailed implementations for them.

2.3 Decider

As *deciders* we require devices accepting a certain language over the output alphabet Σ of the corresponding observer as just introduced. For this we do not need any new type of device but can rely on conventional finite automata with input alphabet Σ. The output of the decider D, for a word $w \in \Sigma^*$ in input, is denoted by $D(w)$. It consists of a simple yes or no.

2.4 Splicing Recognizer

Putting together the components just defined in the way informally described in the Introduction, a *splicing recognizer* (in short SR) is a quintuple $\Omega = (\sigma, O, D, L, L_t)$; $\sigma = (V, R)$ is an H scheme, O is an observer $(Z, V, \Sigma, z_0, \delta, l)$, D is a decider with input alphabet Σ, L and L_t are finite languages, respectively, the initial and the target marked language for σ.

The *language accepted by SR* Ω is the set of all words $w \in V^*$ for which there exists a sequence $s \in \sigma(w, L, L_t)$ such that $D(O(s)) = $ yes; formally

$$L(\Omega) := \{w \in V^* \mid \exists s \in \sigma(w, L, L_t)[D(O(s)) = \text{ yes}]\}.$$

3 A Short Example

It is well-known in the splicing literature that the family of languages generated by splicing systems using only a finite set of splicing rules and a finite initial language is strictly included in the family of regular languages [18]. In the following example we show that an SR composed by such an H scheme with a finite set of rules, finite initial language, finite target marked language and finite state automata as observer and decider, can recognize non regular languages. This example is just a hint towards the fact that the combination splicing system-observer-decider can be powerful even when the single components are simple.

In particular, we construct an SR recognizing the language $\{o_l a^n b^n o_r \mid n \geq 0\}$ that is known to be non-regular. The SR $\Omega = (\sigma, O, D, L, L_t)$ is defined as follows: the H scheme is $\sigma = (V, R)$, with $V = \{o_l, o_r, a, b, a', b', X_1, Y_1, X_2, Y_2\}$ and $R = \{r_1 : \#bo_r\$X_2\#b'o_r, \ r_2 : o_l a'\#Y_2\$o_l a\#, \ r_3 : \#b'o_r\$X_1\#o_r, \ r_4 : o_l\#Y_1\$o_l a'\#\}$. The initial language is $L = \{X_2 b'o_r, o_l a'Y_2, X_1 o_r, Y_1 o_l\}$. The target marked language is $L_t = \{o_l o_r\}$.

The observer O has input alphabet V and output alphabet $\Sigma = \{l_0, l_1, l_2, l_3, \perp\}$. The mapping it implements is:

$$O(w) = \begin{cases} l_0 & \text{if } w \in o_l(a^*b^*)o_r, \\ l_1 & \text{if } w \in o_l(a^*b^*b')o_r, \\ l_2 & \text{if } w \in o_l(a'a^*b^*b')o_r, \\ l_3 & \text{if } w \in o_l(a'a^*b^*)o_r, \\ \perp & \text{else.} \end{cases}$$

The decider D is a finite state automaton, with input alphabet Σ, that gives a positive answer exactly if a word belongs to the regular language $l_0(l_1 l_2 l_3 l_0)^*$.

The observer checks that the splicing rules are applied in the order r_1, r_2, r_3, r_4, and this corresponds to remove, in an alternating way, a b from the right and an a from the left of the input marked string. In this way, at least one of the evolutions of the input marked string is of the kind accepted by the decider if, and only if, the input marked string is in the language $\{o_l a^n b^n o_r \mid n \geq 0\}$. Notice that, at each step, the marked string present is spliced only with one of the strings present in the initial language.

To clarify the working of the SR Ω we show the acceptance of the input marked string $w_0 = o_l aabbo_r$. For simplicity, we only show the evolution of the input marked string and the output of the observer, step by step.

- Step 0: input marked string $w_0 = o_l aabbo_r$; $O(w_0) = l_0$;
- Step 1: apply rule r_1; new marked string $w_1 = o_l aabb'o_r$; $O(w_1) = l_1$;
- Step 2: apply rule r_2; new marked string $w_2 = o_l a'abb'o_r$; $O(w_2) = l_2$;
- Step 3: apply rule r_3; new marked string $w_3 = o_l a'abo_r$; $O(w_3) = l_3$;
- Step 4: apply rule r_4; new marked string $w_4 = o_l abo_r$; $O(w_4) = l_0$;
- Step 5: apply rule r_1; new marked string $w_5 = o_l ab'o_r$; $O(w_5) = l_1$;
- Step 6: apply rule r_2; new marked string $w_6 = o_l a'b'o_r$; $O(w_6) = l_2$;
- Step 7: apply rule r_3; new marked string $w_7 = o_l a'o_r$; $O(w_7) = l_3$;
- Step 8: apply rule r_4; new marked string (in the target marked language) $w_8 = o_l o_r$; $O(w_8) = l_0$.

Obviously the entire observed evolution $l_0 l_1 l_2 l_3 l_0 l_1 l_2 l_3 l_0$ is of the kind accepted by the decider D, so the string w_0 is accepted by the SR Ω.

4   Preliminary Results

An SR can accept even non context-free languages as stated in the following theorem. The trick used here consists in the rotation of the input marked string, during its evolution. The regular observer can control that this kind of rotation is done in a correct way.

Theorem 1. *There is a SR Ω such that $L(\Omega)$ is a non context-free, context-sensitive language. Moreover, the splicing scheme of Ω can be taken to be of radius ≤ 3.*

Proof. We construct an SR Ω accepting the non context-free language $\{o_l w o_r \mid w \in \{a, b, c\}^+, \#_a(w) = \#_b(w) = \#_c(w)\}$.

The SR $\Omega = (\sigma, O, D, L, L_t)$ is defined as follows: the H scheme is $\sigma = (V, R)$, with $V = \{a, b, c, o_l, o_r, X_1, X_2, X_3, X_4, X_5, X_6, X_a, X'_a, X_b, X'_b, X_c, X'_c\}$. The set of splicing rules of R is divided in two groups, according to their use.

The first group consists of the rules used to rotate the marked string.

$r_1 : \{d\#o_r\$X_1\#X_a o_r \mid d \in \{a, b, c\}\}$,
$r_2 : \{\#dX_e o_r\$X_2\#X_d X_e o_r, \mid e, d \in \{a, b, c\}, e \neq d, \}$
$r_3 : \{o_l X'_e\#X_3\$o_l\#d, \mid e, d \in \{a, b, c\}\}$
$r_4 : \{\#X_d X_e o_r\$X_4\#X_e o_r, \mid e, d \in \{a, b, c\}, e \neq d\}$
$r_5 : \{o_l e\#X_5\$o_l X'_e\# \mid e \in \{a, b, c\}\}$.

The second group of splicing rules is used to remove a symbol a, b, or c from the marked string.

$r_6: \#aX_ao_r\$X_6\#X_bo_r,$

$r_7: \#bX_bo_r\$X_6\#X_co_r,$

$r_8: \#cX_co_r\$X_6\#X_ao_r.$

The initial language of the SR is $L = \{X_1X_eo_r, o_lX'_eX_3, X_4X_eo_r, o_leX_5 \mid e \in \{a, b, c\}\} \cup \{X_2X_dX_eo_r \mid d, e \in \{a, b, c\}, e \neq d\} \cup \{X_6X_bo_r, X_6X_cor, X_6X_ao_r\}$. Notice the language is finite. The target marked language is $L_t = \{o_lX_ao_r\}$.

The observer O has input alphabet V and output alphabet $\Sigma = \{l_0, \perp\} \cup \{l_{e,1}, l_{e,2}, l_{e,3}, l_{e,4} \mid e \in \{a, b, c\}\}$.

The mapping implemented by the observer is

$$O(w) = \begin{cases} l_0 & \text{if } w \in o_l\{a, b, c\}^+o_r, \\ l_{e,1} & \text{if } w \in o_l\{a, b, c\}^+X_eo_r, \ e \in \{a, b, c\} \\ l_{e,2} & \text{if } w \in o_l\{a, b, c\}^*X_dX_eo_r, \ e, d \in \{a, b, c\} \\ l_{e,3} & \text{if } w \in o_lX'_d\{a, b, c\}^*X_dX_eo_r, \ e, d \in \{a, b, c\} \\ l_{e,4} & \text{if } w \in o_lX'_d\{a, b, c\}^*X_eo_r, \ e, d \in \{a, b, c\} \\ \lambda & \text{if } w \in \{o_lX_ao_r\} \\ \perp & \text{else.} \end{cases}$$

The decider D is a finite state automaton, with input alphabet Σ, that gives a positive answer exactly if and only if, a word belongs to the regular language $l_0(l_{a,1}(l_{a,2}l_{a,3}l_{a,4}l_{a,1})^*l_{b,1}(l_{b,2}l_{b,3}l_{b,4}l_{b,1})^*l_{c,1}(l_{c,2}l_{c,3}l_{c,4}l_{c,1})^*)^+$.

At the beginning of the computation the input marked string is of the kind $o_l\{a, b, c\}^+o_r$ and it is mapped by the observer to l_0. If the input marked string is not of this type, then the observer outputs something different from l_0, and the entire evolution is not accepted by the decider D. In the first step, the splicing rule $d\#o_r\$X_1\#X_ao_r$ from r_1 is used, and in this way a new marked string of the type $o_l\{a, b, c\}^+X_ao_r$ is obtained and mapped by the observer to $l_{a,1}$. The introduced symbol X_a indicates that we want to search (and then to remove) a symbol a from the obtained marked string. This searching is done by rotating the marked string, until a symbol a becomes the symbol immediate to the left of X_a. The rotation of the string is done by using the splicing rules given in the first group.

A rotation of the string consists in moving the symbol immediately to the left of X_a, to the right of o_l; one rotation is done by applying, in a consecutive way, a rule from r_2, from r_3, from r_4 and finally from r_5 (the precise rules to apply depend on the symbol to move during the rotation). The sequence of marked strings obtained during a rotation is mapped by the observer to the string $l_{a,2}l_{a,3}l_{a,4}l_{a,1}$. The $*$ present in the regular expression describing the decider language, indicates the possibility to have 0, or more consecutive rotations before a symbol a comes to be the symbol immediately to the left of X_a.

The observer checks that each rotation is made in a correct way; that is, the symbol removed from the left of X_a by using a rule from r_4, is exactly the same symbol introduced to the right of o_l, by using the corresponding rule in r_3. This condition is checked in the fourth line of the observer mapping; if this regular condition is not respected, then the observer outputs $\perp$ and the entire evolution of the input marked string is not accepted by the decider D.

Once a symbol a becomes the symbol immediately to the left of X_a, and the rotations can stop, then it is deleted by using the splicing rule r_6. When rule r_6 is applied, the new marked string obtained is of the kind $o_l\{a, b, c\}^+ X_b o_r$ that is mapped by the observer to $l_{b,1}$; the inserted symbol X_b, indicates that now we search the symbol b.

In an analogous way, by using consecutive rotations, a symbol b is placed immediately to the left of X_b and then is removed by using rule r_7. In this case, the sequence of marked strings obtained during each rotation is mapped by the observer to $l_{b,2}l_{b,3}l_{b,4}l_{b,1}$. Once rule r_7 is applied, the new marked string obtained is of the kind $o_l\{a, b, c\}^+ X_c o_r$ and is mapped by the observer to $l_{c,1}$.

Again analogously, the symbol c is searched for and then deleted by using rule r_8; in this case, the sequence of marked strings obtained during each rotation is mapped by the observer to the string $l_{c,2}l_{c,3}l_{c,4}l_{c,1}$. At this point the entire process can be iterated. By searching and removing a new symbol a, and then again a b, and again a c, until the marked string $o_l X_a o_r$, from the target language is reached (the string obtained when all symbols a, b and c, have been deleted from the input marked string). Notice that at each step the current marked string is spliced with a string from the initial language.

This explanation shows that all strings from the language $\{o_l\{a, b, c\}^+ o_r \mid \#_a = \#_b = \#_c\}$ can indeed be accepted by Ω. The fact that only such strings can be accepted is guaranteed by the particular form of sequences accepted by the decider in combination with the very specific form of the observed strings leading to such a sequence.   □

5   Universality

Following the idea used in the proof of Theorem 1, it is possible to prove that SRs are universal. In informal words this means that *it is possible to simulate an accepting Turing machine by observing, with a very simple observer, the evolution of a very simple splicing system.*

The universality is not unexpected since, H systems with observer and decider are similar to splicing systems with regular target languages, known to be universal, [17].

Theorem 2. *For each RE language L over the alphabet A there exists an SR Ω using a splicing scheme $\sigma \in FIN_{H_4}$, such that Ω accepts the language $\{o_l' w o_r' \mid w \in L\}$, with $o_l', o_r' \notin A$.*

Proof. Any SR of the specified type can be simulated by a Turing machine. Thus we only show that, for any Turing machine, there can be constructed an equivalent SR system Ω composed of a splicing system using a finite set of rules, a finite initial language and target marked language and by an observer and a decider that are finite state machines. In this proof we use off-line Turing machines with only a single combined input/working tape. The set δ of transitions is composed of elements of the form $Q \times A \to Q \times A \times \{+, -\}$, where Q is the set of states, A the tape alphabet, and $+$ or $-$ denotes a move to the right or left, respectively.

An input word is accepted, if and only if, the Turing machine stops in a state that belongs to $F \subset Q$ of final states. Without loss of generality, we suppose that the machine M accepts the input, if and only if it reaches a configuration where the tape is entirely empty, and M is in a state that belongs to F. The initial state of M is $q_0 \in Q$. The special letter $\square \in A$ denotes an empty tape cell.

We construct an SR Ω simulating M. Before giving the formal details, we outline the basic idea of the proof. The input string to the Turing machine is inserted as input marked string to the SR Ω, delimited by two external markers o'_l, o'_r. This does not restrict the generality of the theorem, because these two symbols could be added to any input string in two initalizing steps by the SR. However, we want to spare ourselves the technical details of this.

Initially, an arbitrary number of empty tape cells $\square$ is added to the left and to the right of the input marked string. When this phase is terminated, some new markers o_l and o_r are added to the left and right of the produced marked string; starting from this step, the transitions of the Turing machine M are simulated on the current marked string; the marked string contains, at any time, the content of the tape of M, the current state and the position of the head of M over the tape. To read the entire tape of M the marked string is rotated using a procedure very similar to the one described in the proof of Theorem 1; like there, the observer can check that the rotations are done in a correct way. The computation of Ω stops when the target marked string is reached, that is when a marked string representing an empty tape is reached.

Formally, the SR $\Omega = (\sigma, O, D, L, L_t)$ is constructed in the following way.

The H scheme $\sigma = (V, R)$ has alphabet $V = \{o_r, o_l, o'_r, o'_l, X_1, X_2, \cdots, X_{12}\} \cup A' \cup \{X_e, X'_e \mid e \in A'\}$ where $A' = A \cup (A \times Q)$.

The splicing rules present in R are divided in groups, according to their use.

Initialization
$r_1 : \{o'_l(a, q_0)\#X_1\$o'_l a\#, \ a \in (A - \{\square\})\};$
$r_2 : \{\#o'_r\$X_2\#\square o'_r\};$
$r_3 : \{o'_l\square\#X_3\$o'_l\#\};$
$r_4 : \{\#o'_r\$X_4\#o_r\};$
$r_5 : \{o_l\#X_5\$o'_l\#\};$

Rotations
$r_6 : \{a\#eo_r\$X_6\#X_e o_r, \ e \in A', a \in A\};$
$r_7 : \{o_l X'_e\#X_7\$o_l\#f, \ e, f \in A'\};$
$r_8 : \{a\#X_e o_r\$X_8\#o_r, \ e \in A', a \in A\};$
$r_9 : \{o_l e\#X_9\$o_l X'_e\#f, \ e, f \in A'\};$

Transitions
$r_{10} : \{\#(a, q_1)bo_r\$X_{10}\#c(b, q_2)o_r,$
$q_1, q_2 \in Q, a, b, c \in A, \ (q_1, a) \rightarrow (q_2, c, +) \in \delta \};$
$r_{11} : \{\#b(a, q_1)do_r\$X_{11}\#(b, q_2)cdo_r,$
$q_1, q_2 \in Q, a, b, c, d \in A, \ (q_1, a) \rightarrow (q_2, c, -) \in \delta\};$

Halting phase

$r_{12}: \{o_l\#\$X_{12}\#o_r\}.$

The initial language L is the finite language containing the strings used by the mentioned splicing rules; in particular, $L = \{o_l'(a, q_0)X_1 \mid a \in (A - \{\square\})\} \cup \{X_2 o_r', o_l'\square X_3, X_4 o_r, o_l X_5, X_8 o_r, X_{12} o_r\} \cup \{X_6 X_e o_r, o_l X_e' X_7, o_l e X_9 \mid e \in A'\} \cup \{X_{10} c(b, q_2) o_r \mid q_2 \in Q, c, b \in A\} \cup \{X_{11}(b, q_2) cdo_r \mid b, c, d \in A, q_2 \in Q\}.$

The target marked language is $L_t = \{o_l o_r\}$. The observer has input alphabet V and output alphabet $\Sigma = \{l_0, l_1, \cdots, l_8, l_f, \bot\}.$

The mapping implemented by the observer is

$$O(w) = \begin{cases} l_0 & \text{if } w \in o_l'(A - \{\square\})^+ o_r', \\ l_1 & \text{if } w \in o_l'(a, q_0)(A - \{\square\})^* o_r', \ a \in (A - \{\square\}), \\ l_2 & \text{if } w \in o_l'(A' - \{\square\})^+(\square)^+ o_r', \\ l_3 & \text{if } w \in o_l'(\square)^+(A' - \{\square\})^+(\square)^+ o_r', \\ l_4 & \text{if } w \in \{o_l' w' o_r \mid w' \in (\square)^*(A' - \{\square\})^+(\square)^*, length(w') \geq 3\}, \\ l_5 & \text{if } w \in (o_l(A')^+ o_r - \{w \mid w \in E\}), \\ l_6 & \text{if } w \in o_l(A')^* X_e o_r, \ e \in A', \\ l_7 & \text{if } w \in o_l X_e'(A')^* X_e o_r, \ e \in A', \\ l_8 & \text{if } w \in o_l X_e'(A')^* o_r, \ e \in A', \\ l_f & \text{if } w \in E, \\ \bot & \text{else.} \end{cases}$$

where $E = o_l(\square)^*(\square, q)(\square)^+ o_r \cup o_l(\square)^+(\square, q)(\square)^* o_r \cup o_l(\square)^+(\square, q)(\square)^+ o_r,$ $q \in Q.$

The decider is a finite state automaton, with input alphabet Σ that accepts the regular language $E_1 \cup E_2$, where $E_1 = l_0 l_1 (l_2)^+ (l_3)^* l_4 (l_5 \cup l_5 l_5)(l_6 l_7 l_8 (l_5 \cup l_5 l_5))^* l_f$ and $E_2 = l_0 l_1 l_4 (l_5 \cup l_5 l_5)(l_6 l_7 l_8 (l_5 \cup l_5 l_5))^* l_f.$

The main point of the proof is to show that, given an input marked string w, at least one of its (observed) evolutions is of the type accepted by the decider if, and only if, the string w is accepted by the Turing machine M.

We now describe the (observed) evolution of a correct input marked string; from this, we believe it will be clear that non correct strings will not have an evolution of the kind accepted by the decider, and, therefore will not be accepted by the SR Ω. The reader can compare the observed evolution of the input marked string with the language accepted by the decider.

Actually we introduce in the system Ω not the string w but a string of the type $o_l' w o_r'$ where o_l', o_r' are left and right delimiters. In general the input marked string will be of the type $o_l'(A - \{\square\})^+ o_r'$ and is mapped by the observer to l_0. The pairs in $Q \times A$ are used to indicate in the string the state and the position of the head of M. Initially the head is positioned on the leftmost symbol of the input marked string, starting in state q_0 (by using a rule in r_1); the obtained marked string is of the kind $o_l'(a, q_0)(A - \{\square\})^* o_r', \ a \in (A - \{\square\})$ mapped to l_1 by the observer.

Then empty cells $\square$ are added to the right and to the left of the marked string using rules in r_2 and in r_3, respectively. The marked string obtained at the end of this phase will be of the kind $o'_l(\square)^+(A' - \{\square\})^+(\square)^+o'_r$ mapped to l_3 by the observer. This phase is optional, and therefore the language of the decider is described by the union of E_1 where the adding of spaces is used and E_2, where no spaces are added, i.e., l_2 and l_3 are missing.

Then, by using rules in r_4 and in r_5 the delimiters o'_l and o'_r are changed into o_l and o_r, respectively. When a rule in r_4 is applied, the marked string obtained is of the kind $o'_l w' o_r, w' \in (\square)^*(A' - \{\square\})^+(\square)^*$ mapped to l_4 if the size of the string w' (possibly, including empty cells) is at least of 3 symbols; this condition is useful during the following phases of rotations and does not imply a loss of generality.

When a rule in r_5 is applied, also o'_l is removed and the marked string obtained is mapped to l_5 by the observer. This means that the symbol indicating the head of M, (a, q_1), is exactly one symbol away from o_r, then a splicing rule in r_{10} or in r_{11} is applied. The one symbol left between the symbol representing the head and the delimiter o_r is useful in case of the simulation of a right-moving transition. The rule sets r_{10} and r_{11} correspond to transitions moving right and left, respectively.

Once a transition is simulated, the obtained marked string is again of the type mapped to l_5 by the observer (this is why it is possible to have in the language of the decider the substring $l_5 l_5$). At any rate it is not possible to have immediately another transition after a transition, because the symbol corresponding to the head of M is moved. At least one rotation must be first executed.

In case the symbol representing the head of M is not exactly one symbol away from o_r, then the marked string is rotated until this condition is not true any more. The rotation of one symbol in the string (i.e., moving the symbol present to the left of o_r, to the immediate right of o_l) is done by applying, in this order, splicing rules from r_6, r_7, r_8 and from r_9. The marked strings obtained during this phase are mapped by the observer to l_6, l_7, l_8 and finally l_5. At the end of a rotation a transition can be simulated; more consecutive rotations can be done until the necessary condition to simulate a transition is reached. This explains why $(l_6 l_7 l_8 (l_5 \cup l_5 l_5))^*$ forms part of the decider language.

When, after a transition, the marked string obtained represents the empty tape of M, then the computation of the SR stops. The marked strings representing an empty tape are the ones in the language E and they are mapped by the observer to l_f. After the observer has output l_f, the splicing rule in r_{12} can be applied and the unique string in the target marked language $o_l o_r$ can be reached. If the rule in r_{12} is applied before the observer outputs l_f, then the entire evolution is not accepted by the decider. Notice that during the entire computation the marked string can be spliced only with a string from the initial language.

From the above explanation, it follows that an input marked string written in the form $o'_l w o'_r$ is accepted by Ω, if and only if, w is accepted by the Turing machine M. $\qquad\square$

6   Concluding Remarks

We have presented another approach to compute by using DNA molecules (and in general, biological systems), using the idea of evolution and observation.

The paper shows that observing an evolution of only one marked DNA strand by means of a simple observer and decider can be a powerful tool which theoretically is sufficient to simulate a Turing machine. The components involved are rather simple (finite splicing and finite state automata), that the computational power seems to stem mainly from the ability to observe, in real-time, the changes (the dynamics) of a particular (marked) DNA strand, under the action of restriction enzymes.

The proposed approach suggests several problems, if this were to be implemented in practice.

For instance, the process of observation as defined here is non-deterministic; meaning, the marked DNA strand inputed is accepted if, at least one of its observed evolution follows an expected pattern, while there might be several possible evolutions of this DNA strand since there might be several different ways to splice the strand. From a practical point of view this would require several copies of the same input DNA strand, each copy marked with a different "color". The observer should follow, separately, the evolution of each one of these strands. This theoretically requires an unbounded number of copies of DNA strands, each one marked with a different color. In practice, however, using many marked copies may increase the chance to obtain the needed evolutions.

A possible way to implement this might be the use of the multiplexing technique introduced in [12] used to mark several molecules, each one with a different "color". Another way may be marking the strands with quantum dots, [3]. However, none of these techniques have been used for observing splicing and the problems that may arise during the implementation may be numerous.

Further theoretical investigations may provide better solutions if it can be shown that by increasing the complexity of the observer and the decider a ("more") deterministic way of generating the splicing evolutions can be employed. We recall that in the model presented here the observers and deciders are with very low computational power, i.e. finite state automata.

Another problem that needs to be taken care of if implementing an SR is the real-time observation: in the model presented here it is supposed that the observer is able to catch, in the molecular soup, every single change of the marked DNA strand. In practice, it is very questionable whether every step of the evolution can be observed. It should be assumed that only some particular types of changes, within a certain time-interval can be observed (see [14]). Therefore another variant of SR needs to be, at least theoretically, investigated in which an observer with "realistic" limitations on the ability of observation is considered. For instance the observer might be able to watch only a window or a scattered subword of the entire evolution.

On the other hand, universal computational power has been obtained here by using an H scheme of radius 4. We conjecture that it is possible to decrease the radius, hence the question arises of what is the minimum radius that provides universal computation.

It remains also to investigate SRs using simpler and more restricted variants of H schemes, like the ones with simple splicing, [15] and semi-simple splicing rules, [9]. Notice that from a pure theoretical point of view, observer and decider could be joined in an unique finite state automaton, which may provide a better framework for theoretical investigation. In this paper we prefer to leave the two "devices" of observer and decider separated since this situation can be envisioned to be closer to reality.

Moreover, we can interpret a given H scheme with an observer as a device computing a function, by considering as input the input marked string, and as output its (observed) evolution. What kind of functions can be computed in this way?

These are only a few of the possible directions of investigation that the presented approach suggests. We believe that some of these directions will provide useful results for using recombinant DNA for computing.

Acknowledgments

The authors want to thank Peter R. Cook for providing extremely useful references. M. Cavaliere and P. Leupold are supported by the FPU grant of the Spanish Ministry of Science and Education. N. Jonoska has been supported in part by NSF Grants CCF #0432009 and EIA#0086015.

References

1. L.M. Adleman, Molecular Computation of Solutions to Combinatorial Problems, *Science* 226, 1994, pp. 1021–1024.
2. A. Alhazov, M. Cavaliere, Computing by Observing Bio-Systems: the Case of Sticker Systems, *Proceedings of DNA 10 - Tenth International Meeting on DNA Computing*, Lecture Notes in Computer Science 3384 (C. Ferretti, G. Mauri, C. Zandron eds.), Springer, 2005, pp. 1–13.
3. M. Bruchez, M. Moronne, P. Gin, S. Weiss, A.P. Alavisatos, Semiconductor Nanocrystals as Fluorescent Biological Labels, *Science*, 281, 1998, pp. 2013-2016.
4. M. Cavaliere, N. Jonoska, (Computing by) Observing Splicing Systems. Manuscript 2004.
5. M. Cavaliere, P. Leupold, Evolution and Observation – A New Way to Look at Membrane Systems, *Membrane Computing*, Lecture Notes in Computer Science 2933 (C. Martín-Vide, G. Mauri, Gh. Păun, G. Rozenberg, A. Salomaa eds.), Springer, 2004, pp. 70–88.
6. M. Cavaliere, P. Leupold, Evolution and Observation — A Non-Standard Way to Generate Formal Languages, *Theoretical Computer Science* 321, 2004, pp. 233-248.
7. M. Cavaliere, P. Leupold, Evolution and Observation — A Non-Standard Way to Accept Formal Languages. *Proceedings of MCU 2004, Machines, Computations and Universality*, Lecture Notes in Computer Science 3354 (M. Margenstern ed.), Springer, 2005, pp. 152–162.

8. W.C.W. Chan, S. Nie, Quantum Dot Bioconjugates for Ultrasensitive Nonisotopic Detection, *Science* 281, 1998, pp. 2016-2018.

9. E. Goode, D. Pixton, Semi-Simple Splicing Systems, *Where Mathematics, Computer Science, Linguistics and Biology Meet,* (C. Martin-Víde, V. Mitrana eds.), Kluwer Academic Publisher, 2001, pp. 343 – 352.

10. T. Head, Formal Language Theory and DNA: An Analysis of the Generative Capacity of Specific Recombinant Behaviors, *Bulletin of Mathematical Biology* 49, 1987, pp. 737-759.

11. T.M. Jovin, D.J. Arndt-Jovin, in *Cell Structure and Function by Microspectrofluorimetry,* (E. Kohen, J.S. Ploem, J.G. Hirschberg, eds.), Academic, Orlando, Florida, pp. 99–117.

12. J.M. Levsky, S.M. Shenoy, R.C. Pezo, R.H. Singer, Single-Cell Gene Expression Profiling, *Science* 297, 2002, pp. 836–40.

13. J. Lippincott-Schwartz et al., in *Green Fluorescent Proteins,* (K. Sullivan, S. Kay, eds.), Academic, San Diego, 1999, pp. 261-291.

14. J. Lippincott-Schwartz, E. Snapp, A. Kenworthy, Studying Protein Dynamics in Living Cells, *Nature Rev. Mol. Cell. Biol.,* 2, 2001, pp. 444–456.

15. A. Mateescu, Gh. Păun, G. Rozenberg, A. Salomaa, Simple Splicing Systems, *Discrete Applied Mathematics,* 84, 1998, pp. 145–163.

16. X. Michalet, F.F. Pinaud, L.A. Bentolila, J.M. Tsay, S. Doose, J.J. Li, G. Sundaresan, A.M. Wu, S.S. Gambhir, S. Weiss, Quantum Dots for Live Cells, in Vivo Imaging and Diagnostic, *Science,* 307, 2005, *www.sciencemag.org.*

17. Gh. Păun, Splicing systems with targets are computationally universal, *Information Processing Letters,* 59 (1996), pp. 129-133.

18. Gh. Păun, G. Rozenberg, A. Salomaa, *DNA Computing - New Computing Paradigms,* Springer-Verlag, Berlin, 1998.

19. R. Rigler, E.S. Elson, *Fluorescent Correlation Spectroscopy,* Springer, New-York, 2001.

20. G. Rozenberg, A. Salomaa (eds.), *Handbook of Formal Languages.* Springer-Verlag, Berlin, 1997.

21. A. Salomaa, *Formal Languages,* Academic Press, New York, 1973.

On Computational Properties of Template-Guided DNA Recombination*

Mark Daley[1,2] and Ian McQuillan[2]

[1] Department of Computer Science and Department of Biology,
University of Western Ontario,
London, Ontario, N6A 5B7, Canada
daley@csd.uwo.ca
[2] Department of Computer Science,
University of Saskatchewan,
Saskatoon, Saskatchewan, S7N 5A9, Canada
imcquill@csd.uwo.ca

Abstract. The stichotrichous ciliates have attracted the attention of both biologists and computer scientists due to the unique genetic mechanism of gene descrambling. It has been suggested that it would perhaps be possible to co-opt this genetic process and use it to perform arbitrary computations *in vivo*. Motivated by this idea, we study here some basic properties and the computational power of a formalization inspired by the template-guided recombination model of gene descrambling proposed by Ehrenfeucht, Prescott and Rozenberg. We demonstrate that the computational power of a system based on template-guided recombination is quite limited. We then extend template-guided recombination systems with the addition of "deletion contexts" and show that such systems have strictly greater computational power than splicing systems [1, 2].

1 Introduction

The stichotrichous ciliates are a family of single-celled organisms that have come to be studied by both biologists and computer scientists due to the curious mechanism of gene scrambling. Every stichotrichous ciliate has both a functional macronucleus, which performs the "day-to-day" genetic chores of the cell, and an inert micronucleus. Although stichotrichs reproduce asexually, they do also conjugate to exchange genetic material. This hopefully increases the genetic diversity and strength of both organisms involved in conjugation.

The micronucleus contains germline DNA which becomes important during the process of conjugation between two cells. Specifically, when two ciliate cells conjugate, they destroy their macronuclei and exchange haploid micronuclear genomes. Each cell then builds a new functional macronucleus from the genetic material stored in the micronucleus.

* This research was funded in part by institutional grants of the University of Saskatchewan and the University of Western Ontario, the SHARCNET Research Chairs programme, the Natural Sciences and Engineering Research Council of Canada and the National Science Foundation of the United States.

A. Carbone and N.A. Pierce (Eds.): DNA11, LNCS 3892, pp. 27–37, 2006.
© Springer-Verlag Berlin Heidelberg 2006

The interest in this process from a computational point of view comes from the fact that the genes in the micronucleus are stored in a scrambled order. Specifically, the micronuclear gene consists of fragments of the macronuclear gene in some permuted order. That is, if we denote a functional macronuclear gene with the string "1-2-3-4-5", then the equivalent gene in the micronucleus may appear as "2-ϵ_1-4- ϵ_2-1-ϵ_3-3-ϵ_4-5", where the ϵ's represent so-called internally eliminated sequences (or IES's) which are removed from the macronuclear version of the gene. Each sequence, 1 through 5, is referred to as a macronuclear destined sequence (or MDS).

The cell must thus have some mechanism to de-scramble these fragments in order to create a functional gene which is capable of generating a protein. For more information on the biological process of gene de-scrambling, we refer to [3].

Several models for how this de-scrambling process takes place have been proposed in the literature. There are two primary theoretical models which have been investigated: the Kari-Landweber model [4, 5] which consists of a binary inter- and intra-molecular recombination operation and the Ehrenfeucht, Harju, Petre, Prescott and Rozenberg model [6, 7, 8] which consists of three unary operations inspired by intramolecular DNA recombination.

Recently, a new model has been proposed by Prescott, Ehrenfeucht and Rozenberg [9] based on the recombination of DNA strands guided by templates.

The basic action of the model is to take two DNA segments and splice them together via a template intermediary, if the form of the segments matches the form of the template. Consider DNA segments of the form $u\alpha\beta d$ and $e\beta\gamma v$ where $u, v, \alpha, \beta, d, e, \gamma$ are subsequences of a DNA strand. If we wish to splice these two strands together, we require a template of the form $\bar{\alpha}\bar{\beta}_1\bar{\beta}_2\bar{\gamma}$ where $\bar{\alpha}$ denotes a DNA sequence which is complementary to α and $\beta = \beta_1\beta_2$. Specifically, the $\bar{\alpha}\bar{\beta}_1$ in the template will bind to the $\alpha\beta_1$ in the first strand and $\bar{\beta}_2\bar{\gamma}$ will bind to the $\beta_2\gamma$ in the second strand. The molecules then recombine according to the biochemistry of DNA and we are left with d and e being cleaved and removed, a new copy of the template $\bar{\alpha}\bar{\beta}\bar{\gamma}$ and the product of our recombination: $u\alpha\beta\gamma v$. For more details on this operation, we refer to [9].

It has been suggested that the *in vivo* computational process of gene descrambling may be able to be controlled in such a way that it would be possible to perform an arbitrary computation with a ciliate. Taking this as our motivation, in this paper we present a generalized version of the template-guided recombination operation and study the basic properties and computational power of both non-iterated and iterated versions. We conclude that, even in the iterated case, the computational power is quite limited and propose a straightforward extension to a model which is strictly more computationally powerful than splicing systems.

The paper is organized as follows; Section 2 of the paper will present formal language theoretic prerequisites and notation. In section 3 we consider the basic closure properties and the computational power of the template-guided recombination operation. We then contrast this by recalling results on an iterated version of this operation. The limited computational power of both the iterated and non-iterated versions leads us to study a context-aware extension of the operation, which proves strictly more powerful than splicing systems, in Section 4. We present our conclusions in section 5.

2 Preliminaries

We refer to [10] for language theory preliminaries. Let Σ be a finite alphabet. We denote, by Σ^* and Σ^+, the sets of all words and non-empty words, respectively, over Σ and the empty word by λ. A language L is any subset of Σ^*. Let $x \in \Sigma^*$. We let $|x|$ denote the length of x. For $n \in \mathbb{N}_0$, let $\Sigma^{\leq n} = \{w \in \Sigma^* \mid |w| \leq n\}$, $\Sigma^{\geq n} = \{w \in \Sigma^* \mid |w| \geq n\}$ and $\Sigma^n = \{w \in \Sigma^* \mid |w| = n\}$. A homomorphism $h : X^* \rightarrow Y^*$ is termed a coding if $|h(a)| = 1$ for each $a \in X$ and h is termed a weak coding if $|h(a)| \leq 1$ for each $a \in X$. Let $L, R \subseteq \Sigma^*$. We denote by $R^{-1}L = \{z \in \Sigma^* \mid yz \in L \text{ for some } y \in R\}$ and $LR^{-1} = \{z \in \Sigma^* \mid zy \in L \text{ for some } y \in R\}$.

We denote the family of finite languages by **FIN**, regular languages by **REG**, linear languages by **LIN**, context-free languages by **CF**, context-sensitive languages by **CS** and recursively enumerable languages by **RE**.

A *trio* is a non-trivial language family closed under λ-free homomorphism, inverse homomorphism and intersection with regular sets. It is known that every trio is closed under λ-free a-transductions[1] and inverse gsm mappings. An *AFL* is a trio closed under arbitrary union, concatenation and $+$. A *full trio*[2] is a trio closed under arbitrary homomorphism. It is known that every full trio is closed under arbitrary a-transductions and hence arbitrary gsm mappings. A *full semi-AFL* is a full trio closed under union. A *full AFL* is a full trio closed under arbitrary union, concatenation and Kleene $*$. It can be seen that **REG**, **CF** and **RE** are full AFL's, **LIN** is a full semi-AFL not closed under concatenation or $*$, and **CS** is an AFL not closed under arbitrary homomorphism. We refer to [11, 12] for the theory of AFL's.

3 Template-Guided Recombination

We will first formally define the template-guided recombination operation as it appears in [13, 14].

Definition 1. *A template-guided recombination system (or TGR system) is a four tuple $\varrho = (T, \Sigma, n_1, n_2)$ where Σ is a finite alphabet, $T \subseteq \Sigma^*$ is the template language, $n_1 \in \mathbb{N}$ is the minimum MDS length and $n_2 \in \mathbb{N}$ is the minimum pointer length.*

For a TGR system $\varrho = (T, \Sigma, n_1, n_2)$ and a language $L \subseteq \Sigma^*$, we define $\varrho(L) = \{w \in \Sigma^* \mid (x, y) \vdash_t w \text{ for some } x, y \in L, t \in T\}$ where $(x, y) \vdash_t w$ if and only if $x = u\alpha\beta d, y = e\beta\gamma v, t = \alpha\beta\gamma, w = u\alpha\beta\gamma v, u, v, d, e \in \Sigma^*, \alpha, \gamma \in \Sigma^{\geq n_1}, \beta \in \Sigma^{\geq n_2}$.

---

[1] An a-transducer is also referred to as a rational transducer.

[2] A full trio is also referred to as a cone.

Let $\mathcal{L}_1, \mathcal{L}_2$ be language families and $n_1, n_2 \in \mathbb{N}$. We write $\hbar(\mathcal{L}_1, \mathcal{L}_2, n_1, n_2) = \{\varrho(L) \mid L \in \mathcal{L}_1, \varrho = (T, \Sigma, n_1, n_2)$ a TGR system, $T \in \mathcal{L}_2\}$ and $\hbar(\mathcal{L}_1, \mathcal{L}_2) = \bigcup_{n_1, n_2 \in \mathbb{N}} \hbar(\mathcal{L}_1, \mathcal{L}_2, n_1, n_2)$

We remark here that while the operation of template-guided recombination bears a superficial resemblance to the splicing operation introduced by Head [1] and extended by Paun, et. al. [2], the operations are, in fact, distinct. While TGR systems are in most cases less computationally powerful than comparable splicing systems, they are often more succinct in terms of the descriptional complexity of a system generating a particular language. Moreover, we will show in this paper that a contextual extension of TGR systems is strictly more computationally powerful than the inherently contextual splicing systems. Further details on the relationship between splicing systems and TGR systems can be found in [13].

Remark 1. In [9], a constant C is defined such that $|\alpha|, |\gamma| > C$ in order to ensure the formation of sufficiently strong chemical bonds. Likewise, [9] also defines constants D and E such that $D < |\beta| < E$. The definition, as above, and also the results in this paper, are general enough to cover any such D and C. In addition, the constant E, as defined above, is shown to be irrelevant in the next proposition. It was noted in [9] that the smallest pointer sequence known was of length 3, although recently, a pointer sequence was discovered experimentally which was only of length one [15]. Also, we believe that the smallest MDS sequence discovered to date is nine nucleotides long [16]. In any case, the notation above is general enough to work for any such constants. We also note that the notation above will work when the two operands x and y in Definition 1 are either the same or when they are not. It has been seen experimentally that two MDS's can be on two different loci but still recombine successfully.

The following proposition, from [13] states that we can always assume that the β subword of a template is of the minimum length, n_2.

Proposition 1. Let $\varrho = (T, \Sigma, n_1, n_2)$ be a TGR system and let $x, y \in \Sigma^*$ and $t \in T$. Then $(x, y) \vdash_t w$ if and only if $x = u\alpha\beta d, y = e\beta\gamma v, t = \alpha\beta\gamma, w = u\alpha\beta\gamma v, u, v, d, e \in \Sigma^*, \alpha, \gamma \in \Sigma^{\geq n_1}, \beta \in \Sigma^{n_2}$.

In the sequel, we shall thus assume, without loss of generality, that β is of length n_2.

We now consider new results regarding the power of template-guided recombination when restricted to a single application of the operation. This is important to the basic theoretical understanding of how the operation functions relative to traditional theoretical computer science. We omit proofs here due to space considerations.

First, we show that, under some weak restrictions, closure under intersection follows from closure under template-guided recombination.

Lemma 1. Let $\mathcal{L}_1$ be a language family closed under left and right concatenation and quotient with a single symbol and under union with singleton languages and let $\mathcal{L}_2$ be a language family closed under left and right concatenation with a single symbol such that $\hbar(\mathcal{L}_1, \mathcal{L}_2, n_1, n_2) \subseteq \mathcal{L}_1$ for some $n_1, n_2 \in \mathbb{N}$. The intersection of a language from $\mathcal{L}_1$ with a language from $\mathcal{L}_2$ belongs to $\mathcal{L}_1$.

Since Σ^* is in every language family containing **REG** and $\Sigma^* \cap T = T$, we obtain:

Corollary 1. *Let $\mathcal{L}_1$ be a language family such that* **REG** $\subseteq \mathcal{L}_1$, *$\mathcal{L}_1$ is closed under left and right concatenation and quotient with a symbol and union with singleton languages and let $\mathcal{L}_2$ be a language family closed under left and right concatenation with a symbol such that $\hbar(\mathcal{L}_1, \mathcal{L}_2, n_1, n_2) \subseteq \mathcal{L}_1$ for some $n_1, n_2 \in \mathbb{N}$. Then $\mathcal{L}_2 \subseteq \mathcal{L}_1$.*

We now continue the characterization of template-guided recombination in terms of AFL theory. We see that, under some restrictions, closure under concatenation follows from closure under template-guided recombination.

Lemma 2. *Let $\mathcal{L}_1$ be a language family closed under limited erasing homomorphism, union, left and right concatenation by a symbol and let $\mathcal{L}_2$ be a language family containing the singleton languages such that $\hbar(\mathcal{L}_1, \mathcal{L}_2, n_1, n_2) \subseteq \mathcal{L}_1$ for some $n_1, n_2 \in \mathbb{N}$. Then $\mathcal{L}_1$ is closed under concatenation.*

We now show that we can simulate template-guided recombination with a few standard operations.

Lemma 3. *Let $\mathcal{L}_1$ be closed under marked concatenation[3], intersection with regular languages and inverse gsm mappings. Let $\mathcal{L}_2$ be closed under inverse gsm mappings and intersection with regular languages. Let $L \in \mathcal{L}_1, T \in \mathcal{L}_2$ and let $\varrho = (T, \Sigma, n_1, n_2)$ be a TGR system. Then there exists $L' \in \mathcal{L}_1, T' \in \mathcal{L}_2$ and a weak coding homomorphism h such that $\varrho(L) = h(L' \cap T')$.*

Since every trio is closed under inverse gsm mappings, we get the following:

Corollary 2. *Let $\mathcal{L}_1$ be a concatenation closed full trio and let $\mathcal{L}_2$ be either a trio or $\mathcal{L}_2 \subseteq$ **REG***. *If $\mathcal{L}_1$ is closed under intersection with $\mathcal{L}_2$ then $\hbar(\mathcal{L}_1, \mathcal{L}_2) \subseteq \mathcal{L}_1$.*

We combine the lemmata above to obtain the following result:

Proposition 2. *Let $\mathcal{L}_1$ be a full semi-AFL and $\mathcal{L}_2$ be a trio or $\mathcal{L}_2 =$ **FIN***. *Then $\hbar(\mathcal{L}_1, \mathcal{L}_2) \subseteq \mathcal{L}_1$ if and only if $\mathcal{L}_1$ is closed under concatenation and $\mathcal{L}_1$ is closed under intersection with $\mathcal{L}_2$.*

Since every full semi-AFL is closed under intersection with regular languages, it now follows that for a full semi-AFL, closure under catenation is necessary and sufficient to show closure under template-guided recombination with regular and finite languages.

Corollary 3. *For every full semi-AFL $\mathcal{L}$, $\hbar(\mathcal{L}, \mathbf{REG}) \subseteq \mathcal{L}$ and $\hbar(\mathcal{L}, \mathbf{FIN}) \subseteq \mathcal{L}$ if and only if it is closed under concatenation.*

---

[3] The marked concatenation of L_1, L_2 is $L_1 a L_2$ where a is a new symbol.

Likewise, the next result concerning the closure of intersection-closed full semi-AFLs now follows since every intersection-closed full semi-AFL is closed under concatenation.

Corollary 4. *For every intersection-closed full semi-AFL $\mathcal{L}$, $\text{th}(\mathcal{L}, \mathcal{L}) \subseteq \mathcal{L}$.*

The above results are sufficient to characterize the closure properties of the families of finite, regular, linear and context-free families. We now show that the family of context-sensitive languages is not even closed under template-guided recombination with singleton languages.

Proposition 3. $\text{th}(\mathbf{CS}, \mathbf{FIN}, n_1, n_2) \not\subseteq \mathbf{CS}$ *for any $n_1, n_2 \in \mathbb{N}$.*

Now, we can completely fill in a table (see Table 1) with the families of languages in the Chomsky hierarchy, the finite languages and the linear languages. A $\sqrt{}$ represents closure of $\mathcal{L}_1$ under template-guided recombination with templates from $\mathcal{L}_2$ and a blank represents non-closure. The results hold for any minimum pointer and MDS length.

Table 1. $\text{th}(\mathcal{L}_1, \mathcal{L}_2) \subseteq \mathcal{L}_1$?

$\mathcal{L}_1 \mid \mathcal{L}_2$	**FIN**	**REG**	**LIN**	**CF**	**CS**	**RE**
FIN	$\sqrt{}$	$\sqrt{}$	$\sqrt{}$	$\sqrt{}$	$\sqrt{}$	$\sqrt{}$
REG	$\sqrt{}$	$\sqrt{}$				
LIN						
CF	$\sqrt{}$	$\sqrt{}$				
CS						
RE	$\sqrt{}$	$\sqrt{}$	$\sqrt{}$	$\sqrt{}$	$\sqrt{}$	$\sqrt{}$

In a biological system it is natural to investigate iterated application of operations as bio-operations are the product of the stochastic biochemical reactions of enzymes, catalysts and substrates in solution. We now recall results on an iterated version of the template-guided recombination operation.

We begin with the definition iterated template-guided recombination from [14]:

Let $\varrho = (T, \Sigma, n_1, n_2)$ be a TGR system and let $L \subseteq \Sigma^*$. Then we generalize ϱ to an iterated operation $\varrho^*(L)$ as follows:

$$\varrho^0(L) = L,$$
$$\varrho^{n+1}(L) = \varrho^n(L) \cup \varrho(\varrho^n(L)), n \geq 0$$
$$\varrho^*(L) = \bigcup_{n=0}^{\infty} \varrho^n(L).$$

Let $\mathcal{L}_1, \mathcal{L}_2$ be language families and $n_1, n_2 \in \mathbb{N}$. We define $\text{th}^*(\mathcal{L}_1, \mathcal{L}_2, n_1, n_2) = \{\varrho^*(L) \mid L \in \mathcal{L}_1, \varrho = (T, \Sigma, n_1, n_2) \text{ a TGR system}, T \in \mathcal{L}_2\}$ and let $\text{th}^*(\mathcal{L}_1, \mathcal{L}_2) = \bigcup_{n_1, n_2 \in \mathbb{N}} \text{th}^*(\mathcal{L}_1, \mathcal{L}_2, n_1, n_2)$.

We now give a short example of an iterated template-guided recombination system.

The following result characterizes all Sd-sortable permutations.

Theorem 2. *Let π be a unsigned permutation. Then π is Sd-sortable if and only if there exists a partition $\{1, 2, \ldots, n\} = M \cup U$, such that the following conditions are satisfied:*

(i) $\pi|_U$ is sorted;
(ii) Nodes of M induce an acyclic dependency subgraph;
(iii) If $k \to l$ is a dependency of π and $l \in M$, then $k \in M$;
(iv) For any $k \in M$, $(k-1)(k+1) \leq_s \pi$;
(v) For any $k \in M$, $(k-1), (k+1) \in U$.

Example 5. Consider permutation $\pi = 1\,3\,8\,10\,5\,7\,2\,9\,11\,4\,6\,12$. Its dependency graph is shown in Fig. 7. Based only on this graph and using Theorem 2 we deduce a sorting strategy for π.

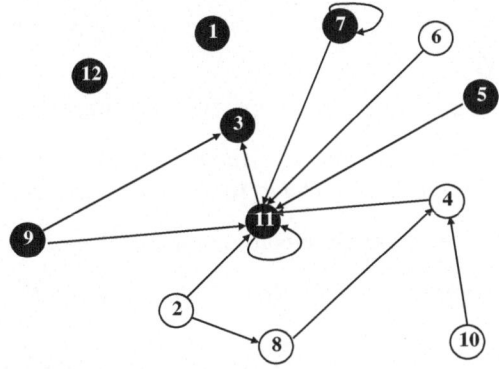

Fig. 7. The dependency graph associated to $\pi = 1\,3\,8\,10\,5\,7\,2\,9\,11\,4\,6\,12$

It follows from Theorem 2(ii) that $7, 11 \in U$. Then it follows from Theorem 2(iii) that $3 \in U$ and from Theorem 2(v) that $1, 12 \in U$. Since $1, 3 \in U$, it follows from Theorem 2(i) that $2 \in M$. Also, since $3, 7, 11 \in U$, it follows from Theorem 2(i) that $4, 6, 8, 10 \in M$ and so, by Theorem 2(v), $5, 9 \in U$. We have now a complete labelling of G_π:

$$M = \{2, 4, 6, 8, 10\}, \ U = \{1, 3, 5, 7, 9, 11, 12\}$$

Permutation π may be sorted now by a composition of operations sd_i with $i \in M$.

The dependency graph imposes the following order of operations: sd_4 after sd_8 and sd_{10}, sd_8 after sd_2. The other operations can be applied in any order. For instance, we can sort π in the following way:

$$(\mathsf{sd}_4 \circ \mathsf{sd}_8 \circ \mathsf{sd}_2 \circ \mathsf{sd}_{10} \circ \mathsf{sd}_6)(\pi) = 1\,2\,3\,4\,5\,6\,7\,8\,9\,10\,11\,12,$$

but also,

$$(\mathsf{sd}_6 \circ \mathsf{sd}_4 \circ \mathsf{sd}_8 \circ \mathsf{sd}_2 \circ \mathsf{sd}_{10})(\pi) == 1\,2\,3\,4\,5\,6\,7\,8\,9\,10\,11\,12.$$

7 {Sd, Sh}-Sortable Permutations

We characterize in this section all signed permutations that can be sorted using our operations. First we give some examples.

Example 6. (i) Signed permutations $\pi_1 = 2\,1\,4\,\overline{3}\,5$ and $\pi_2 = 1\,5\,\overline{2}\,4\,3\,6$ are not {Sd, Sh}-sortable. Indeed, just sh_3 can be applied to π_1, but it does not sort it, and no operation can be applied to π_2.

(ii) Signed permutations $\pi_3 = 9\,2\,\overline{10}\,\overline{11}\,1\,5\,3\,7\,\overline{4}\,6\,8$ and $\pi_4 = 5\,4\,\overline{3}\,\overline{8}\,2\,1\,\overline{9}\,7\,\overline{6}$ are {Sd, Sh}-sortable:

$$(sh_{11} \circ sh_{10} \circ sd_2 \circ sd_5 \circ sh_4 \circ sd_7)(\pi_3) = 9\,10\,11\,1\,2\,3\,4\,5\,6\,7\,8$$

and

$$(sh_4 \circ sh_2 \circ sh_3 \circ sh_3 \circ sd_{\overline{8}} \circ sh_6)(\pi_4) = \overline{5}\,\overline{4}\,\overline{3}\,\overline{2}\,\overline{1}\,\overline{9}\,\overline{8}\,\overline{7}\,\overline{6}.$$

Theorem 3. *No permutation π can be sorted both to an orthodox permutation and to an inverted one.*

The following result gives a duality property of sorting signed permutations.

Lemma 3. *A signed permutation π is sortable to an orthodox permutation π_o if and only if its inversion $\overline{\pi}$ is sortable to the inverted permutation $\overline{\pi_o}$.*

The following result is an immediate consequence of Theorem 3 and of Lemma 3.

Corollary 1. *A permutation π is sortable if and only if either π or $\overline{\pi}$ is sortable to an orthodox permutation.*

Consider in the following only permutations π that are sortable to an orthodox form. Let H be the set of all signed letters in π and let Φ_H be a composition of sh-operations applied on all integers in H. Let $D \subseteq \{1, 2, \ldots, n\} \setminus H$ and Φ_D a composition of sd-operations applied on all integers in D. The *dependency graph* $\Gamma_{\pi, \Phi_H, \Phi_D}$ (or just Γ_{Φ_H, Φ_D} when there is no risk of confusion) generated by Φ_H, Φ_D is the following:

- If $j \in D$ (sd_j is in Φ_D) and $(j-1)i(j+1) \leq_s \|\pi\|$, then edge $(\|i\|, j) \in \Gamma_{\Phi_H, \Phi_D}$. Also, if $(j - 1) \in H$, then edge $(j - 1, j) \in \Gamma_{\Phi_H, \Phi_D}$ and if $j + 1 \in H$, then edge $(j + 1, j) \in \Gamma_{\Phi_H, \Phi_D}$.
- If $i \in H$ (sh_i is in Φ_H), then we have the following two cases:
  - If sh_i is of the form $(i - 1)\overline{i} \to (i - 1)i$, then $(i - 1)i \leq_s \|\pi\|$. For any j, if $(i - 1)ji \leq_s \|\pi\|$, then $(\|j\|, i) \in \Gamma_{\Phi_H, \Phi_D}$;
  - If sh_i is of the form $\overline{i}(i + 1) \to i(i + 1)$, then $i(i + 1) \leq_s \|\pi\|$. For any j, if $ij(i + 1) \leq_s \|\pi\|$, then $(\|j\|, i) \in \Gamma_{\Phi_H, \Phi_D}$.

Example 7. Consider $\pi = 6\,8\,10\,1\,9\,\overline{3}\,7\,4\,2\,\overline{5}$. Clearly, $H = \{3, 5\}$. Assume we apply Sd operations on 2, 7 and 9, thus $D = \{2, 7, 9\}$. Let us build the dependency graph $G = \Gamma_{\pi, \Phi_H, \Phi_D}$, shown in Fig. 8.

We mark by dashed the nodes in H, by white the nodes in D and we mark by black the rest of vertices. For each vertex i from G we have the following edges (j, i):

- Node 1: we do not have edges $(j, 1)$, since $1 \notin H$ and $1 \notin D$;
- Node 2: $2 \in D$, $3 \in H$, thus $(3, 2) \in G$. Since $1\,9\,3 \leq_s \pi$, we have also $(9, 2) \in G$;
- Node 3: $3 \in H$, $3\,7\,4 \leq_s \|\pi\|$, thus $(7, 3) \in G$;
- Node 4: $4 \notin H$ and $4 \notin D$, thus we have no edges $(j, 4)$;
- Node 5: $5 \in H$ and $4\,2\,5 \leq_s \|\pi\|$, thus $(2, 5) \in G$;
- Node 6: $6 \notin H$ and $6 \notin D$, thus we have no edges $(j, 6)$;
- Node 7: $7 \in D$, $6\,8 \leq \pi$, thus we have no edges $(j, 7)$;
- Node 8: $8 \notin H$ and $8 \notin D$, thus we have no edges $(j, 8)$;
- Node 9: $9 \in D$, $8\,10 \leq \pi$, thus we have no edges $(j, 9)$;
- Node 10: $10 \notin H$ and $10 \notin D$, thus we have no edges $(j, 10)$.

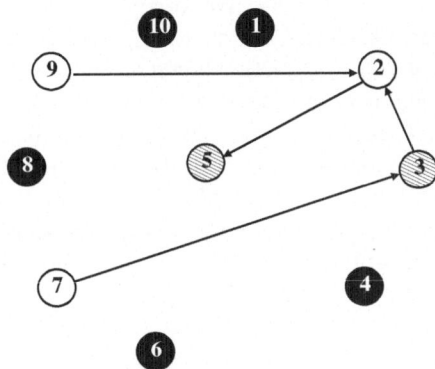

Fig. 8. The dependency graph associated to $\pi = 6\,8\,10\,1\,9\,\overline{3}\,7\,4\,2\,\overline{5}$

Lemma 4. *Let π be an $\mathsf{Sh} \cup \mathsf{Sd}$-sortable permutation over Σ_n and Φ a sorting strategy for π. Let Γ_Φ be the dependency graph associated to π and Φ. Let $\phi_i = \mathsf{sd}_i$ if i is unsigned in π and $\phi_i = \mathsf{sh}_i$ if i is signed in π, for $i \in \Sigma_n$. Then we have the following properties:*

(i) If there is a path from i to j in Γ_Φ and Φ_j is used in Φ, then ϕ_i is applied before ϕ_j in strategy Φ.

(ii) The dependency graph Γ_Φ is acyclic.

The following theorem gives the main result of this section.

Theorem 4. *A permutation π is $\{\mathsf{Sh}, \mathsf{Sd}\}$-sortable to an orthodox form if and only if there is a partition $\{1, 2, \ldots, n\} = D \cup H \cup U$ such that the following conditions are satisfied:*

(i) H is the set of all signed letters in π;

(ii) H sorts $\pi \mid_{H \cup U}$ to an orthodox form with a strategy Φ_H;

(iii) D sorts $\|\pi\|$ with a strategy Φ_D;

(iv) The subgraph of Γ_{Φ_H, Φ_D} induced by $H \cup D$ is acyclic.

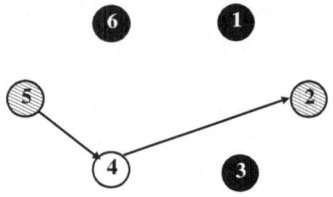

Fig. 9. The dependency graph associated to $\pi = \overline{2}\,4\,3\,\overline{5}\,6\,1$

Example 8. Let $\pi = \overline{2}\,4\,3\,\overline{5}\,6\,1$. We build a sorting strategy for π based on Theorem 4. Consider $H = \{2, 5\}$. Clearly, $\|\pi\| = 2\,4\,3\,5\,6\,1$ is sorted by applying sd_4. Then let $D = \{4\}$ and $U = \{1, 3, 6\}$. We verify now conditions of Theorem 4. Consider $\pi\,|_{H\cup U} = \overline{2}\,3\,\overline{5}\,6\,1$. Then $\mathsf{sh}_2(\mathsf{sh}_5(\pi\,|_{H\cup U})) = 2\,3\,5\,6\,1$, a (circularly) sorted string. The graph $\Gamma_{\mathsf{sh}_2 \circ \mathsf{sh}_5, \mathsf{sd}_4}$ is shown in Fig. 9, where nodes in H are marked by dashed, nodes in D are marked by white and nodes in U are marked by black. Clearly, $H \cup D$ induces an acyclic subgraph in $\Gamma_{\mathsf{sh}_2 \circ \mathsf{sh}_5, \mathsf{sd}_4}$. Thus, by Theorem 4, π is sortable and a sorting strategy should be obtained by combining $\mathsf{sh}_2 \circ \mathsf{sh}_5$ and sd_4 as indicated by the graph. Since $(4, 2)$ is an edge in the graph, it follows that sd_4 must be applied before sh_2. Also, since $(5, 4)$ is an edge, it follows that sh_5 must be applied before sd_4. Consequently, $\mathsf{sh}_2 \circ \mathsf{sd}_4 \circ \mathsf{sh}_5$ must be a sorting strategy for π. Indeed, $\mathsf{sh}_2(\mathsf{sd}_4(\mathsf{sh}_5(\pi))) = 2\,3\,4\,5\,6\,1$, a (circularly) sorted permutation.

Example 9. Let $\pi = 2\,1\,4\,3\,7\,\overline{5}\,9\,6\,8\,\overline{10}\,11$. We build a sorting strategy for π based on Theorem 4. Clearly, $H = \{5, 10\}$. The unsigned permutation $\|\pi\| = 2\,1\,4\,3\,7\,5\,9\,6\,8\,10\,11$ can be sorted by $\mathsf{sd}_2 \circ \mathsf{sd}_4 \circ \mathsf{sd}_9 \circ \mathsf{sd}_7$, thus $D = \{2, 4, 7, 9\}$. Set $U = \{1, 3, 6, 8, 11\}$. The dependency graph G associated to π and $H \cup U$ is shown in Fig. 10. Clearly, permutation $\pi|_{H\cup U} = 1\,3\,\overline{5}\,6\,8\,\overline{10}\,11$ can be sorted to cyclically sorted permutation $1\,3\,5\,6\,8\,10\,11$ by applying sh_5 and sh_{10}. Also, $H \cup D$

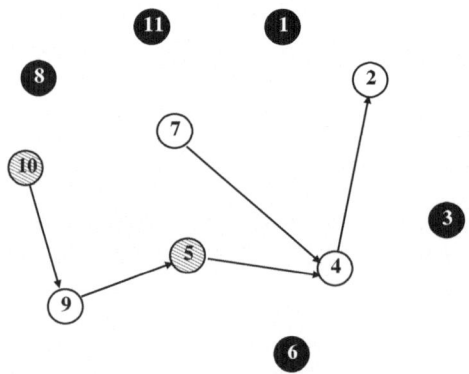

Fig. 10. The dependency graph associated to $\pi = 2\,1\,4\,3\,7\,\overline{5}\,9\,6\,8\,\overline{10}\,11$

induces an acyclic subgraph in G. It follows then that π is sortable. Indeed, a sorting strategy, as suggested by G, is $\mathsf{sd}_2 \circ \mathsf{sd}_4 \circ \mathsf{sd}_7 \circ \mathsf{sh}_5 \circ \mathsf{sd}_9 \circ \mathsf{sh}_{10}$. Another sorting strategy is $\mathsf{sd}_2 \circ \mathsf{sd}_4 \circ \mathsf{sh}_5 \circ \mathsf{sd}_9 \circ \mathsf{sd}_7 \circ \mathsf{sh}_{10}$.

8   Discussion

We consider in this paper a mathematical model for the so called *simple operations* for gene assembly in ciliates. The model we consider here is in terms of *signed permutations*, but the model can also be expressed in terms of *signed double-occurrence strings*, see [14].

Modelling in terms of signed permutations is possible by ignoring the molecular operation Ld that combines two consecutive gene blocks into a bigger block. In this way, the process of combining the sequence of successive coding blocks into one assembled gene becomes the process of sorting the initial sequence of blocks.

It is important to note now that in the molecular model we discus in this paper, each operation affects one single gene block that gets incorporated into a bigger block together with one (in case of Sh) or two (in case of Sd) other blocks. In our mathematical model however, a gene block that was already assembled from several initial blocks is represented as a sorted substring. For that reason, although the molecular operations only displace one block, our model should allow the moving of longer sorted substrings. A mathematical theory in this sense looks challenging. We consider in this paper the simplified variant where our formal operations can only move one block (one letter of the alphabet) at a time. Note however that the general case may in fact be reduced to this simpler variant in the following way: in each step of the sorting, we map our alphabet into a smaller one by denoting each sorted substring by a single letter such that the new string has no sorted substrings of length at least two (this mimics of course the molecular operation Ld).

Deciding whether a given permutation is $\mathsf{Sh} \cup \mathsf{Sd}$-sortable is of course trivial: simply test all possible sorting strategies. The problem of doing this efficiently, perhaps based on Theorems 2 and 4 remains open.

Acknowledgments. The authors were supported by the European Union project MolCoNet, IST-2001-32008. T. Harju gratefully acknowledges support by Academy of Finland, project 39802. I. Petre gratefully acknowledges support by Academy of Finland, projects 203667 and 108421, V. Rogojin gratefully acknowledges support by Academy of Finland, project 203667. G. Rozenberg gratefully acknowledges support by NSF grant 0121422.

References

1. Berman, P., and Hannenhalli, S., Fast sorting by reversals. *Combinatorial Pattern Matching, Lecture Notes in Comput. Sci.* **1072** (1996) 168–185.
2. Caprara, A., Sorting by reversals is difficult. In S. Istrail, P. Pevzner and M. Waterman (eds.) *Proceedings of the 1st Annual International Conference on Computational Molecular Biology* (1997) pp. 75–83.

3. Ehrenfeucht, A., Harju, T., Petre, I., Prescott, D. M., and Rozenberg, G., Formal systems for gene assembly in ciliates. *Theoret. Comput. Sci.* **292** (2003) 199–219.

4. Ehrenfeucht, A., Harju, T., Petre, I., and Rozenberg, G., Characterizing the micronuclear gene patterns in ciliates. *Theory of Comput. Syst.* **35** (2002) 501–519.

5. Ehrenfeucht, A., Harju, T., Petre, I., Prescott, D. M., and Rozenberg, G., *Computation in Living Cells: Gene Assembly in Ciliates*, Springer (2003).

6. Ehrenfeucht, A., Petre, I., Prescott, D. M., and Rozenberg, G., Universal and simple operations for gene assembly in ciliates. In: V. Mitrana and C. Martin-Vide (eds.) *Words, Sequences, Languages: Where Computer Science, Biology and Linguistics Meet*, Kluwer Academic, Dortrecht, (2001) pp. 329–342.

7. Ehrenfeucht, A., Petre, I., Prescott, D. M., and Rozenberg, G., String and graph reduction systems for gene assembly in ciliates. *Math. Structures Comput. Sci.* **12** (2001) 113–134.

8. Ehrenfeucht, A., Petre, I., Prescott, D. M., and Rozenberg, G., Circularity and other invariants of gene assembly in cliates. In: M. Ito, Gh. Păun and S. Yu (eds.) *Words, semigroups, and transductions*, World Scientific, Singapore, (2001) pp. 81–97.

9. Ehrenfeucht, A., Prescott, D. M., and Rozenberg, G., Computational aspects of gene (un)scrambling in ciliates. In: L. F. Landweber, E. Winfree (eds.) *Evolution as Computation*, Springer, Berlin, Heidelberg, New York (2001) pp. 216–256.

10. Hannenhalli, S., and Pevzner, P. A., Transforming cabbage into turnip (Polynomial algorithm for sorting signed permutations by reversals). In: *Proceedings of the 27th Annual ACM Symposium on Theory of Computing* (1995) pp. 178–189.

11. Harju, T., Petre, I., Li, C. and Rozenberg, G., Parallelism in gene assembly. In: *Proceedings of DNA-based computers 10*, Springer, to appear, 2005.

12. Harju, T., Petre, I., and Rozenberg, G., Gene assembly in ciliates: molecular operations. In: G.Paun, G. Rozenberg, A.Salomaa (Eds.) *Current Trends in Theoretical Computer Science*, (2004).

13. Harju, T., Petre, I., and Rozenberg, G., Gene assembly in ciliates: formal frameworks. In: G.Paun, G. Rozenberg, A.Salomaa (Eds.) *Current Trends in Theoretical Computer Science*, (2004).

14. Harju, T., Petre, I., and Rozenberg, G., Modelling simple operations for gene assembly, submitted, (2005). Also as a TUCS technical report TR697, http://www.tucs.fi.

15. Jahn, C. L., and Klobutcher, L. A., Genome remodeilng in ciliated protozoa. *Ann. Rev. Microbiol.* **56** (2000), 489–520.

16. Kaplan, H., Shamir, R., and Tarjan, R. E., A faster and simpler algorithm for sorting signed permutations by reversals. *SIAM J. Comput.* **29** (1999) 880–892.

17. Kari, L., and Landweber, L. F., Computational power of gene rearrangement. In: E. Winfree and D. K. Gifford (eds.) *Proceedings of DNA Bases Computers, V* American Mathematical Society (1999) pp. 207–216.

18. Landweber, L. F., and Kari, L., The evolution of cellular computing: Nature's solution to a computational problem. In: *Proceedings of the 4th DIMACS Meeting on DNA-Based Computers*, Philadelphia, PA (1998) pp. 3–15.

19. Landweber, L. F., and Kari, L., Universal molecular computation in ciliates. In: L. F. Landweber and E. Winfree (eds.) *Evolution as Computation*, Springer, Berlin Heidelberg New York (2002).

20. Prescott, D. M., *Cells: Principles of Molecular Structure and Function*, Jones and Barlett, Boston (1988).

21. Prescott, D. M., Cutting, splicing, reordering, and elimination of DNA sequences in hypotrichous ciliates. *BioEssays* **14** (1992) 317–324.

22. Prescott, D. M., The unusual organization and processing of genomic DNA in hypotrichous ciliates. *Trends in Genet.* **8** (1992) 439–445.
23. Prescott, D. M., The DNA of ciliated protozoa. *Microbiol. Rev.* **58**(2) (1994) 233–267.
24. Prescott, D. M., The evolutionary scrambling and developmental unscabling of germlike genes in hypotrichous ciliates. *Nucl. Acids Res.* **27** (1999), 1243 – 1250.
25. Prescott, D. M., Genome gymnastics: unique modes of DNA evolution and processing in ciliates. *Nat. Rev. Genet.* 1(3) (2000) 191–198.
26. Prescott, D. M., and DuBois, M., Internal eliminated segments (IESs) of Oxytrichidae. *J. Eukariot. Microbiol.* **43** (1996) 432–441.
27. Prescott, D. M., Ehrenfeucht, A., and Rozenberg, G., Molecular operations for DNA processing in hypotrichous ciliates. *Europ. J. Protistology* **37** (2001) 241–260.
28. Prescott, D. M., and Rozenberg, G., How ciliates manipulate their own DNA – A splendid example of natural computing. *Natural Computing* **1** (2002) 165–183.
29. Prescott, D. M., and Rozenberg, G., Encrypted genes and their reassembly in ciliates. In: M. Amos (ed.) *Cellular Computing*, Oxford University Press, Oxford (2003).

Counting Time in Computing with Cells

Oscar H. Ibarra[1] and Andrei Păun[2]

[1] Department of Computer Science,
University of California - Santa Barbara, Santa Barbara, CA 93106
ibarra@cs.ucsb.edu
[2] Department of Computer Science/IfM, Louisiana Tech University,
P.O. Box 10348, Ruston, LA 71272
apaun@latech.edu

Abstract. We consider models of P systems using time either as the output of a computation or as a means of synchronizing the hugely complex processes that take place in a cell. In the first part of the paper, we introduce and study the properties of "timed symport/antiport systems". In the second part we introduce several new features for P systems: the association/deassociation of molecules (modeling for example the protein-protein interactions), ion channel rules and gene activation rules. We show that such timed systems are universal. We also prove several properties concerning these systems.

1 Introduction

We continue the work on symport/antiport P systems [7], [8], [15] using a new paradigm: time as the output of a computation. In recent years, several approaches have been undertaken considering time as part of a biological system's way of "computing". We can mention here the work of W. Maass: he considered a new way to compute with spiking neurons [12]. His model was based on the observation that if considering only the frequency of the neuron's firing signal as a computational framework for the brain, then the brain itself would be very slow computing using only 2-3 spikes per neuron in 150 ms. It is clear that other information is transmitted through these spikes of the neurons. Maass considered a new idea (that seems to be supported experimentally): that also the temporal pattern of the spikes emitted by a neuron is important for the actual message sent. Another feature studied by Maass is that during the "computation", the actual state of a neuron depends on the previous states that the neuron has passed through; maybe even from the birth of the organism. In the current paper we will define significant "configurations" for the system, using similar ideas as in [12] in the sense that for a cell it is important whether it passes through a few "important" configurations.

We also note that in the last year two papers considered the properties of systems in which the rules can take different amounts of time to be completed. Following this idea, in [5] and [14], the authors study the case when the time required for "executing" rules in a system may change sometimes even unpredictably. In such a setting it is an interesting question whether the cell can behave

A. Carbone and N.A. Pierce (Eds.): DNA11, LNCS 3892, pp. 112–128, 2006.
© Springer-Verlag Berlin Heidelberg 2006

in a similar fashion for different times associated with rules; then several such systems with timed rules are studied. In other words, the authors considered the "time-free" devices; such systems will perform the same steps irrespective of the different time lengths associated with each rule in the system. We will follow a similar idea of timing each rule in such a system and propose a new model of P systems in Section 3, but we devise the new model to be closer to cell biology and useful from a cell-simulation point of view: our definition will consider the case when each rule has a specified duration for the reaction they model (which can be determined experimentally).

In this paper, we consider another way of outputting the result of the computation of a P system. The idea originates in [18] as Problem W; the novelty is that instead of the "standard" way to output, like the multiplicities of objects found at the end at the computation in a distinguished membrane as it was defined in the model from [15], it seems more "natural" to consider certain *events* (i.e., configurations) that may occur during a computation and to relate the output of such a computation with the time interval between such distinguished configurations. Our system will compute a set of numbers like in the case of "normal" symport/antiport [15], but the benefit of the current setting is that the computation and the observance of the output are now close to the biology and to the tools used for cell biology. The model of the "timed" P system that we investigate here is the symport/antiport P system. This has been a popular model that has been accepted by the research community immediately after its introduction. Symport/antiport systems are now a very successful model for P systems due to their simplicity and the fact that they observe the basic physical law of matter conservation (the system computes by communication: the objects are not created nor destroyed, but rather only moved in the system). Here, we are studying another way of viewing the output of such a system; the motivation comes from the fact that cells can become fluorescent if, for example, some types of proteins with fluorescence properties are present in the cells. Such a fluorescent "configuration" of a cell will be the configuration that starts the clock used for the output. Even more interesting (making our definition a very natural way of viewing the output of a system) is the fact that there are tools currently used by researchers in cell biology that can detect the fluorescence of each cell individually. The procedure is performed by a device which can check one cell at a time for fluorescence and can automatically decide to put the cell in either the test-tube containing the fluorescent cells or in the test-tube containing the non-fluorescent cells. The procedure does not destroy the cells, meaning that the same process can be performed repeatedly for a given cell computation. Such an automated technique for viewing the output of a computation using cells is highly desirable since it holds the promise of fast readouts of the computations (in contrast with manual "readouts" that could take several days, see for example the well-known Adleman's experiment, [1]).

The main idea of the new definition is that one has a colony of cells; each cell in the colony having the same (nondeterministic) program that performs a computation. There is a configuration of the system that gives the cell a fluorescence, property that can be detected by devices such as *fluorescence acti-vated cell sorters*, in short FACS. Such a device takes the input test-tube (that contains our colony of cells) and splits it into cells that are not fluorescent yet and cells that are fluorescent. The cells that become fluorescent at some time t (i.e. are detected to be fluorescent for the first time) will have a timer associated to their test-tube that starts "ticking", and will be continuously checked whether they are still in the fluorescent state or not. We will consider the moment that a cell is no longer fluorescent as the moment when we receive the "stop clock" signal, and the system outputs the value computed by the cell to be the time interval when the cell was fluorescent and the instant when it is no longer fluores-cent. In this way, by using a FACS one can obtain the output of the computation of such a P system automatically (we consider that it is a easy task to design a system which feeds back the fluorescent test-tube(s) into the FACS incrementing a counter/timer at each feedback, and writing on some medium the content of the counter if a cell was detected to be no longer fluorescent). In other words, we will "output" the duration in the number of "clock cycles" during which the cell was fluorescent. Such a system could output the computation of an entire colony of cells, not only the computation of a single cell. This gives another order of parallelism to our setting which is another strongly desirable feature.

2   Timed Symport/Antiport Systems

We will use a modified definition than the one in [15]; instead of specifying the output region where the result of the computation will be "stored" in a halting computation, we specify two relations C_{start} and C_{stop} (which are computable by multicounter machines) that need to be satisfied by the multisets of objects in the membrane structure at two different times during the computation.

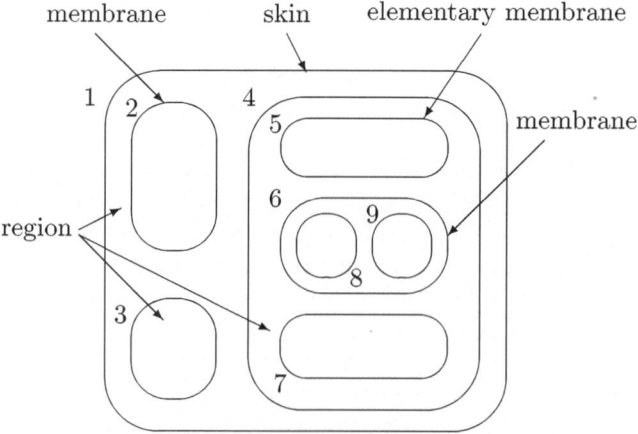

Fig. 1. A membrane structure

An important observation is the fact that we will not require the cell to "stop working" when reaching the result; i.e. we will not require the strong restriction that the system reach a halting configuration for a computation to have a result.

Before progressing further we give some basic notions used in the remainder of the paper; the language theory notions used but not described are standard, and can be found in any of the many monographs available, for instance, in [19].

A membrane structure is pictorially represented by a Venn diagram (like the one in Figure 1), and it will be represented here by a string of matching parentheses. For instance, the membrane structure from Figure 1 can be represented by $[_1[_2\ ]_2[_3\ ]_3[_4[_5\ ]_5[_6[_8\ ]_8[_9\ ]_9]_6[_7\ ]_7]_4]_1$.

A multiset over a set X is a mapping $M : X \longrightarrow \mathbf{N}$. Here we always use multisets over finite sets X (that is, X will be an alphabet). A multiset with a finite support can be represented by a string over X; the number of occurrences of a symbol $a \in X$ in a string $x \in X^*$ represents the multiplicity of a in the multiset represented by x. Clearly, all permutations of a string represent the same multiset, and the empty multiset is represented by the empty string, λ.

We will use symport/antiport rules[1]; mathematically, we can capture the idea of symport by considering rules of the form (ab, in) and (ab, out) associated with a membrane, and stating that the objects a, b can enter, respectively, exit the membrane together. For antiport we consider rules of the form $(a, out; b, in)$, stating that a exits and at the same time b enters the membrane.

Based on rules of this types, we modify the definition from [15] to introduce the model of a *timed symport/antiport system* as a construct,

$$\Pi = (V, \mu, w_1, \ldots, w_m, E, R_1, \ldots, R_m, C_{start}, C_{stop}), \ where :$$

- $V = \{a_1, ..., a_k\}$ is an alphabet (its elements are called *objects*);
- μ is a membrane structure consisting of m membranes, with the membranes (and hence the regions) bijectively labeled with $1, 2, \ldots, m$; m is called the *degree* of Π;
- $w_i, 1 \leq i \leq m$, are strings over V representing multisets of objects associated with the regions $1, 2, \ldots, m$ of μ, present in the system at the beginning of a computation;
- $E \subseteq V$ is the set of objects that are continuously available in the environment in arbitrarily many copies;
- $R_1, \ldots, R_m$ are finite sets of symport and antiport rules over the alphabet V associated with the membranes $1, 2, \ldots, m$ of μ;
- At any time during the computation, a configuration of Π can be represented by a tuple α in N^{mk}, where the (i, j) component corresponds to the multiplicity of symbol a_j in membrane i.
- C_{start} and C_{stop} are recursive subsets of N^{mk} (i.e., they are Turing machine computable or, equivalently, computable by multicounter machines).

---

[1] The definitions have their roots in the biological observation that many times two chemicals pass at the same time through a membrane, with the help of each other, either in the same direction, or in opposite directions; in the first case we say that we have a *symport*, in the second case we have an *antiport* (we refer to [2] for details).

For a symport rule (x, in) or (x, out), we say that $|x|$ is the *weight* of the rule. The *weight* of an antiport rule $(x, out; y, in)$ is $\max\{|x|, |y|\}$. The rules from a set R_i are used with respect to membrane i as explained above. In the case of (x, in), the multiset of objects x enters the region defined by the membrane, from the surrounding region, which is the environment when the rule is associated with the skin membrane. In the case of (x, out), the objects specified by x are sent out of membrane i, into the surrounding region; in the case of the skin membrane, this is the environment. The use of a rule $(x, out; y, in)$ means expelling the objects specified by x from membrane i at the same time with bringing the objects specified by y into membrane i. The objects from E (in the environment) are supposed to appear in arbitrarily many copies since we only move objects from a membrane to another membrane and do not create new objects in the system, we need a supply of objects in order to compute with arbitrarily large multisets. The rules are used in the non-deterministic maximally parallel manner specific to P systems with symbol objects: in each step, a maximally parallel multiset of rules is used.

In this way, we obtain transitions between the configurations of the system. A configuration is described by the m-tuple of multisets of objects present in the m regions of the system, as well as the multiset of objects from $V - E$ which were sent out of the system during the computation. It is important to note that such objects appear only in a finite number of copies in the initial configuration and can enter the system again (knowing the initial configuration and the current configuration in the membrane system, one can know precisely what "extra" objects are present in the environment). On the other hand, it is not necessary to take care of the objects from E which leave the system because they appear in arbitrarily many copies in the environment as defined before (the environment is supposed to be inexhaustible, irrespective how many copies of an object from E are introduced into the system, still arbitrarily many remain in the environment). The initial configuration is $\alpha_0 = (w_1, \ldots, w_m)$. Note that $(w_1, ..., w_m) = (s_{11}, s_{12}, ..., s_{1k}, ... s_{m1}, s_{m2}, ... s_{mk})$, i.e., a tuple in N^{mk}. A sequence of transitions is called a computation.

Let us now describe the way this systems "outputs" the result of its computation: when the system enters some configuration α satisfying C_{start}, we start a counter t that is incremented each time the simport/antiport rules are applied in the nondeterministic parallel manner. At some point, when the system enters some configuration β satisfying C_{stop}, we stop incrementing t, and the value of t represents the output of the computation[2]. If the system never reaches a configuration in C_{start} or in C_{stop}, then we consider the computation unsuccessful, no output is associated with the computation of the system in that case. The set of all such t's (computed as described) is denoted by $N(\Pi)$. The family of all sets $N(\Pi)$ computed by systems Π of degree at most $m \geq 1$, using symport rules of weight at most p and antiport rules of weight at most q, is denoted by $NTP_m(sym_p, anti_q)$ (we use here similar notations with the definitions from [15]).

---

[2] By convention, in the case when a configuration α is reached that satisfies both C_{start} and C_{stop}, then we consider that the system has computed the value 0.

We emphasize the implicit fact in the definition of Π, that we assume that C_{start} and C_{stop} are recursive. Some interesting cases are when C_{start} and C_{stop}:

 - are exactly N^{mk} (i.e., they are *trivial*, as they consist of all the tuples);
 - are computable by deterministic multicounter machines (i.e., recursive);
 - are Presburger relations.

Details about P systems with symport/antiport rules can be found in [15]; a complete formalization of the syntax and the semantics of these systems is provided in [17] where reachability of symport/antiport configurations was discussed.

After defining the new way of considering the time of a computation performed by such a system it is a natural question to ask whether the new definition is powerful enough so that it can carry out universal computation. First we mention a recent result for a very special case where C_{start} consists only of the initial configuration of the system, and C_{stop} consists of all halting configurations. For this case, it was shown in [6] that the set of all times (i.e., intervals between the initial configuration and halting configurations) is recursive making such systems not universal. In light of this result, the proof of universality for our system becomes an interesting contrast. We note that the result in [6] is also in surprising contrast to various results in the literature, where most of the P systems defined so far have been proven to be powerful enough to be universal. We expected to obtain a universality result using similar techniques used for proving the universality of "regular" symport/antiport P systems, and indeed, as the following theorem shows, we are able to prove the universality of timed symport/antiport systems with 3 membranes and symport/antiport rules of minimal weight, as well as universality of one membrane and antiport of weight 2.

Theorem 1. *Using <u>minimal restrictions</u>[3] on the multiplicities of the objects for the C_{start}, C_{stop} rules we have $NRE = NTP_m(sym_r, anti_t)$, for $(m, r, t) \in \{(1, 0, 2), (3, 1, 1)\}$.*

Proof. We use a modification of the construction described in [20] and used for proving $NRE' = NOP_3(sym_1, anti_1)$. It is worth noting that the result from [20] is computing $NRE' = NRE - \{0, 1, 2, 3, 4\}$, thus the best result for regular symport/antiport is still considered to be the one from [9] where it was shown that $NRE = NOP_4(sym_1, anti_1)$. With the help of the timed symport/antiport we will prove that three membranes and minimal symport/antiport are universal by computing exactly NRE.

We will give in the following our construction following the ideas from [20] and [8]. We consider a counter automaton [13] a construct $M = (Q, C, R, q_0, f)$

---

[3] By minimal restrictions on multiplicities we mean the fact that for each object and each membrane, the C_{start}, C_{stop} rules will impose either a fixed multiplicity (for example 5) or not impose any restrictions for the object. For example C_{start} containing $s_{2,3} = 5$ means that the third object has to appear in exactly 5 copies in membrane 2. On the other hand, by $s_{6,2} = i$, $i \geq 0$ we mean that the second object can have any multiplicity in the membrane 6.

where C is the set of $n+1$ counters denoted by $c_0, \ldots, c_n$ and c_0 is the output counter of the automaton. We will construct a timed symport/antiport system Π that generates the set of numbers $L(M)$ as follows:

$$\Pi = (V, [_1[_2[_3 \ ]_3]_2]_1, w_1, w_2, w_3, E, R_1, R_2, R_3, C_{start}, C_{stop}), \quad \text{where:}$$

$$V = \{I_1, \overline{I_1}, I_2, I_3, I_4, \infty_1, \infty_2, \infty_3, \infty_4, P, t, t', t'', \overline{t}, \overline{t}', \overline{t}''\} \cup \{c_i, 0_i \mid$$
$$0 \le i \le n\} \cup \{qr, qr' \mid (q \to r, X) \in R \text{ for } X \in \{i+, i-, i = 0, \lambda\}\},$$
$$w_1 = I_1 \overline{I_1}, I_2 I_3 \infty_1 \infty_2 \infty_4^2, \quad w_2 = \infty_1 \infty_2 \infty_3^2 \overline{t} P 0_0 0_1 \ldots 0_n, \quad w_3 = t',$$
$$E = \{qr, qr' \mid (q \to r, X) \in R \text{ for } X \in \{i+, i-, = 0, \lambda\}\} \cup \{I_4, \overline{t}', \overline{t}''\}$$
$$\cup \{c_i \mid 0 \le i \le n\},$$
$$R_i = R_i^{ini} \cup R_i^{sim} \cup R_i^{timer}, \text{ for } 1 \le i \le 3 \text{ where } R_i^\tau,$$
$$\text{where } \tau \in \{ini, \ sim, \ timer\} \text{ are defined as in the following:}$$
$$R_1^{ini} = \{(qr', in; I_1, out) \mid (q \to r, X) \in R, \ X \in \{i+, i-, = 0, \lambda\}\} \cup \{(I_1, in)\}$$
$$\cup \{(c_j, in; \overline{I_1}, out) \mid 0 \le j \le n\} \cup \{(I_4, in; I_1, out), (\overline{I_1}, in), (I_3, out)\},$$
$$R_2^{ini} = \{((qr', in; I_2, out) \mid (q \to r, X) \in R, \text{ for an operation } X\} \cup \{(I_2, in),$$
$$(I_3, in; \infty_2, out), (I_4, in; I_3, out), (I_1, in; I_4, out), (\infty_4, in; \infty_4, out),$$
$$(\overline{I_1}, in; I_5, out), (\infty_4, in; I_4, out), (\infty_1, in; \infty_1, out), (\infty_2, in; \infty_2, out)\},$$
$$R_3^{ini} = \{(qr', in; I_3, out) \mid (q \to r, X) \in R, \text{ for some } X\} \cup \{(I_3, in),$$
$$(\infty_3, in; I_3, out), (I_1, in; I_5, out), (I_2, in; I_1, out), (\infty_3, in; \infty_3, out)\},$$
$$R_1^{sim} = \{(q_0 q, in; I_5, out), (rs, in; qr', out) \mid q_0, q, r, s \in Q \text{ and } (q_0 \to q, X),$$
$$(q \to r, Y), (r \to s, Z) \in R \text{ for some counter operations } X, Y, Z\},$$
$$R_2^{sim} = \{(rs, in), (c_i, in; rs', out) \mid \text{for } (r \to s, i+) \in R\} \cup \{(rs, in; c_i, out),$$
$$(rs', out) \mid (r \to s, i-) \in R\} \cup \{(rs, in), (rs', out) \mid (r \to s, \lambda) \in R\}$$
$$\cup \{(rs, in; 0_i, out), (0_i, in; c_i, out), (0_i, in; rs', out) \mid (r \to s, i = 0) \in R\},$$
$$R_3^{sim} = \{(rs, in; rs', out) \mid (r \to s, X) \in R, \text{ for a counter operation } X\},$$
$$R_1^{timer} = \{(t, in; qf', out) \mid \text{for } f \text{ the final state}\}$$
$$\cup \{(\overline{t}', in; \overline{t}, out), (t'', in; t', out), (\overline{t}'', in; t'', out)\},$$
$$R_2^{timer} = \{(t, in; \overline{t}, out), (\overline{t}', in; t', out), (\overline{t}'', in)\},$$
$$R_3^{timer} = \{(t, in; t', out), (\overline{t}', in), (\overline{t}'', in), (P, in; \overline{t}', out), (c_0, in; \overline{t}', out),$$
$$(c_0, in; \overline{t}'', out)\}.$$

We will give the rules in C_{start} and C_{stop} in the following format: for each membrane i we will give a rule r_i^α, $\alpha \in \{start, \ stop\}$ associated with that membrane as a sequence of letters each having their multiplicity expressed as the exponent; they will give the exact multiplicities of the objects needed to reach the respective configuration. If an object is not "mentioned" for such a rule associated with a membrane, then we assume that there is no restriction on the multiplicity of that object in that membrane to satisfy the rule C_α. If an object has no exponent, then we assume it has to appear in the configuration exactly one time (i.e. it is considered to have exponent 1). We are now ready to give the *start* and *stop* rules for configurations:

$$C_{start} : r_1^{start} = \infty_1 \infty_2^2 \infty_4^2; \ r_2^{start} = \infty_1 \infty_3^2 \overline{t}''; \ r_3^{start} = c_0^0 t P \overline{t}'.$$

$$C_{stop} : r_1^{stop} = r_3^{stop} = \lambda; \ r_2^{stop} = c_0^0.$$

We will explain briefly the work of the system: from the initial configuration Π will go through three phases of the computation; in the first phase (*Initialization*) the system will bring in from the environment an arbitrary number of objects qr'^4 and c_j that will be used later in the simulation phase. Most of the rules are defined for the initialization phase; such a big number of rules was needed to ensure that only if the system is following a "correct" path in the computation a result is produced. The next phase is the actual *simulation* of the counter automaton (with the use of the objects brought in the system in the previous phase); the two major phases mentioned before are similar to the proofs from [3], [11], [4], [9], [20], the reader can see a detailed explanation of the usage of the rules from R_i^{ini} and R_i^{sim} in [20]. We now pass to describing the final stage of the simulation, the actual output of the contents for the counter c_0 using the time between a configuration α satisfying the rule C_{start} and another configuration β satisfying the rules from C_{stop}.

It is easy to see (for more details we refer to [20]) that, only in the case of a successful simulation, the repartition of the objects in the system will satisfy the rule $C_{end\_sim} : r1 = \infty_1 \infty_2^2 \infty_4^2 F, \ r_2 = \infty_1 \infty_3^2 \overline{t} t P, \ r_3 = t'$ where $F \in \{qf' \mid q \in Q\}$, configuration that has also the property that in membrane 2 there are i copies of the object c_0 where i is the actual result of the computation.

It is clear that we need to go from C_{start} to C_{stop} in exactly i steps. To do this, we perform a few steps of "pre-work" by bringing in the second membrane $\overline{t}'$ and then also $\overline{t}''$, that will move the objects c_0 into membrane 3 one copy for each maximally parallel application of the rules.

Let us describe the "flow" of time in the system starting the end of the simulation phase (the object qf' reaches membrane 1). In that moment qf' is replaced with t in the membrane 1 by the rule $(t, in; qf', out) \in R_1$ and then t enters membrane 2 and sends out in membrane 1 $\overline{t}$: $(t, in; \overline{t}, out) \in R_2$. At the next step two rules can be applied: $(\overline{t}', in; \overline{t}, out)$ in membrane 1 and $(t, in; t', out)$ in membrane 3. Next $\overline{t}'$ moves into membrane 2 while t' reaches membrane 1: $(\overline{t}', in; t', out) \in R_2$ so that at the next step $\overline{t}'$ finally arrives in membrane 3 by $(\overline{t}', in) \in R_3$. During this last step t' is replaced by the rule t'': $(t'', in; t', out) \in R_1$ applicable in membrane 1. Since $\overline{t}'$ is used to move the c_0 objects from membrane 2 into membrane 3, it can start doing this by using the rule $(c_0, in; \overline{t}', out) \in R_3$, but this would mean that the system would never reach a configuration satisfying C_{start} since at least one copy of c_0 would be present in membrane 3 and would never be removed. Instead, we can use the rule $(P, in; \overline{t}', out) \in R_3$ and bring in membrane 3 the symbol P^5 so that $\overline{t}''$ can

---

[4] We note that qr' is a single object "keeping track" of two different states $q, r \in Q$.

[5] The symbol P is used as a padding symbol.

finally come in membrane 2 and start the timer (the rule C_{start} is satisfied by the current configuration). One can note that both $\bar{t}'$ and $\bar{t}''$ can move exactly one copy of c_0 from membrane 2 into membrane 3, and then re-enter membrane 3 so that they can perform this work once more at the next step. Since one copy is moved for each step, the number of steps from C_{start} and C_{stop} is exactly the multiplicity of the symbol c_0 in membrane 2. This in fact means that we proved that $NRE = NTP_3(sym_1, anti_1)$. In the following we prove the second part of the theorem, that one membrane and antiport of size two and no symport are enough for universality:

To prove that $NRE = NTP_1(sym_0, anti_2)$, we follow the constructions from [8] or [7] where a similar result for "regular" simport/antiport P systems was obtained. The unique membrane will start with the start state as its only object in the initial configuration, and the work of the counter automaton can be simulated using the antiport rules in the following way:

For a rule $(p \rightarrow q, \lambda) \in R$ we will have in our timed P system the rule $(q, in; p, out)$; for an increment instruction on the counter c_i: $(p \rightarrow q, i+)$ we will add the following antiport rule to R_1: $(qc_i, in; p, out)$. The decrement instruction can only be applied if the counter is non zero, $(p \rightarrow q, i-)$ is simulated by $(q, in; pc_i, out)$. Finally, $(p \rightarrow q, i = 0)$ is simulated by the rules $(q'\bar{i}, in; p, out)$; $(\infty, in; \bar{i}c_i, out)$, $(q'', in; q', out)$, and $(q, in; q''\bar{i}, out)$ in three steps: first we replace p by q' and $\bar{i}$, then $\bar{i}$ checks whether the register i is empty or not; if nonempty, the special marker ∞ is brought in and the computation cannot continue; but in the case when the register was empty the computation can continue by expelling the two symbols q'' and $\bar{i}$ together to bring in the next state q.

It is clear now that the register machine is simulated in this way only by using antiport rules of weight 2^6. When the final state appears as the current state of the simulation it is time to start "counting" the result; the rule C_{start} can be defined as $r_1^{start} = \infty^0 f$. The rule $(f, in; fc_0, out)$ will expel one symbol c_0 at a time, thus if we define the rule C_{stop} ro be $r_1^{stop} = fc_0^0$ we will have exactly i steps between C_{start} and C_{stop}, where i is the multiplicity of the symbol c_0 (i.e. the contents of the output register) in the system. Following the previous discussion the equality $NRE = NTP_1(sym_0, anti_2)$ was shown, which completes the proof.   □

We will consider next other properties of the newly defined model of P systems with time, such as the possibility of simulating timed systems by using normal symport/antiport systems. We will also consider the case when the start/stop configurations do not have any constrains on the object multiplicities, etc.

2.1 Other Results for Timed Simport/Antiport Systems

The following result shows that a timed system can be simulated by a "time-less" system of the same type.

---

[6] The result can be strengthened in the following way: the construction works even if we only use antiport rules of dimensions $(1, 2)$ or $(2, 1)$ by adding to the only two rules of dimension $(1, 1)$ some padding symbols. For example the rule $(q'', in; q', out)$ can be padded with the extra symbol P in this way $(q''P, in; q', out)$.

Theorem 2. *For every timed symport/antiport system Π we can effectively construct a (regular) symport/antiport system Π' which computes $N(\Pi)$.*

Proof. We start with a timed symport/antiport system Π and proceed to show how to compute $N(\Pi)$. Assume that Π has m membranes and an alphabet V with k symbols. Let α_0 be the initial configuration of Π. Let M_{start} and M_{stop} be deterministic multicounter machines which compute the relations C_{start} and C_{stop}, respectively.

Our procedure will be implemented on a nondeterministic multicounter (or register) machine M. M will have a set C of mk counters. There is a special counter, called T, which will be the timer. We also need counters to simulate M_{start} and M_{stop}. In addition, there are other counters, called D counters, whose use will be explained later. Initially, all the counters are zero.

1. M starts by storing the initial configuration α_0 in the set of counters C. (Since the initial configuration is fixed, this can be incorporated in the finite-state control of M).
2. M nondeterministically selects a maximally parallel multiset of rules applicable to the configuration represented in C, collectively storing the "changes" in the multiplicities of the symbols in the different membranes resulting from the application of the rules in the auxiliary set of counters D. We will explain the details of how this is done later.
3. Using D, M updates C, and resets the counters in D to zeros.
4. M nondeterministically guesses to either go back to step 2 or proceed to the next step.
5. (When M enters this step, it is guessing that the configuration α represented in the C counters is in C_{start}). M simulates M_{start} and checks that α is indeed in C_{start}. If so, M proceeds to the next step; otherwise, M halts in a non-accepting state.
6. As in step 2, M nondeterministically selects a maximally parallel multiset of rules applicable to the configuration represented in C, storing the changes in the multiplicities of the symbols in the different membranes resulting from the step in in the auxiliary set of counters D.
7. Using D, M updates C, resets the counters in D to zeros, and increments the counter T by 1.
8. M nondeterministically guesses to either go back to step 6 or proceed to the next step.
9. M simulates M_{stop} to check that the configuration β represented in the C counters is in C_{stop}. If so, M halts in an accepting state; else it halts in a non-accepting state. (Clearly, when M halts in an accepting state, the value of counter T is the number of steps Π took to reach configuration β from α).

Since steps 6-9 are similar to steps 2-5, we just describe the details of how M carries out step 2.

For every membrane i and the (unique) membrane j enclosing it, define two sets of counters: The first set consists of counters $d_{(i,j,a_1)}, ..., d_{(i,j,a_k)}$, and they will keep track of the multiplicities of the objects that are moved from membrane

i to membrane j as a result of the application of the rules in membrane i (as described below). The other set of counters $d_{(j,i,a_1)}, ..., d_{(j,i,a_k)}$ will keep track of the multiplicities of objects that are moved from membrane j to membrane i. These sets of counters will be called D counters. At the start of step 2, the D counters are reset to zero. Let Q be the set of all rules in the membrane structure.

(a) M nondeterministically picks a rule r in Q. Note that r belongs to a unique membrane, say i. First assume that $i \neq 1$ (i.e., not the skin membrane).
(b) Clearly, r is of the form $(u, out; v, in)$, where u or v, but not both, can be λ (the null string).
(c) Let j be the membrane directly enclosing membrane i. M checks if r is applicable by examining the contents of the counters in C corresponding to the symbols in membrane i and the contents of the counters corresponding to the symbols in membrane j, decrementing these counters appropriately, and then updating the D counters for the pair (i, j) as a result of the application of rule r. M then goes to step (a).

If r is not applicable, then M deletes r from Q. If Q is not empty, M goes to the step (a); otherwise, M has applied a maximal set of rules, and the counters in D can now be used to update the values of the counters in C.

For the case $i = 1$, the enclosing membrane is the environment, which has an abundance of each symbol and, hence, M does not have to keep track of the multiplicities of the symbols in the environment. Note also that multiplicity of each symbol in $V - E$ is bounded and its distribution in the membranes and the environment (although is changing during the computation) can be recorded in the finite-state control of M.

From the discussion above and the fact that a multicounter machine can effectively be simulated by a symport/antiport system, the theorem follows. □

For the trivial case when there are no constraints on the multiplicities of the objects in the membranes, we have:

Theorem 3. *For every timed symport/antiport system Π, with $C_{start} = C_{stop}$ $= N^{mk}$, $N(\Pi)$ is recursive.*

Proof. The idea is the following. Given n, to determine if n is in $N(\Pi)$, simulate all computation paths of Π starting from its initial configuration (note that, in general, there may be several paths because the system is nondeterministic). Use a separate counter for each path to count the number of maximally parallel steps in each path. If there is a path with n steps, then n is in $N(\Pi)$. If each path leads to a halting configuration before n steps, then n is not in $N(\Pi)$. □

Now, from Theorem 1, $N(\Pi)$ is recursive for a timed symport/antiport system even when C_{start} and C_{stop} are very simple cases of Presburger relations. However, from Theorem 3, if C_{start} and C_{stop} are the trivial relations, $N(\Pi)$ is recursive. It follows that the only cases when $N(\Pi)$ would be recursive is

with multiset wo^n, it simulates Z and has a halting computation if and only if Z halts on input n. Moreover, the rules of M are of the form $u \to v$, where $|u| = |v| = 1$ or 2.

It is convenient to use an intermediate P system, called SCPS, which is a restricted version of the the CPS (communicating P system) introduced in [18]. A CPS has multiple membranes, with the outermost one called the skin membrane. The rules in the membranes are of the form:

1. $a \to a_x$,
2. $ab \to a_x b_y$,
3. $ab \to a_x b_y c_{come}$,

where a, b, c are objects, x, y (which indicate the directions of movements of a and b) can be *here*, *out*, or in_j. The designation *here* means that the object remains in the membrane containing it, *out* means that the object is transported to the membrane directly enclosing the membrane that contains the object (or to the environment if the object is in the skin membrane). The designation in_j means that the object is moved into the membrane, labeled j, that is directly enclosed by the membrane that contains the object. A rule of the form (3) can only appear in the skin membrane. When such a rule is applied, c is imported through the skin membrane from the environment (i.e., outer space) and will become an element in the skin membrane. In one step, all rules are applied in a maximally parallel manner. For notational convenience, when the target designation is not specified, we assume that the symbol remains in the membrane containing the rule.

Let V be the set of all objects (i.e., symbols) that can appear in the system, and o be a distinguished object (called the *input symbol*). A CPS M has m membranes, with a distinguished *input membrane*. We assume that only the symbol o can enter and exit the skin membrane (thus, all other symbols remain in the system during the computation). We say that M accepts o^n if M, when started with o^n in the input membrane initially (with no o's in the other membranes), eventually halts. Note that objects in $V - \{o\}$ have fixed numbers and their distributions in the different membranes are fixed initially. Moreover, their multiplicities remain the same during the computation, although their distributions among the membranes may change at each step. The language accepted by M is $L(M) = \{o^n \mid o^n$ is accepted by $M\}$.

It is known that a language $L \subseteq o^*$ is accepted by a deterministic (nondeterministic) CPS if and only if it is accepted by a deterministic (nondeterministic) multicounter machine. (Again, define the language accepted by a multicounter machine Z to be $L = \{o^n \mid Z$ when given n has a halting computation $\}$). The "if" part was shown in [18]. The "only if" part is easily verified. Hence, every unary recursively enumerable language can be accepted by a deterministic CPS (hence, also by a nondeterministic CPS).

An SCPS ('S' for simple) is a restricted CPS which has only rules of the form $a \to a_x$ or $ab \to a_x b_y$. Moreover, if the skin membrane has these types of rules, then $x, y \neq out$ (i.e., no objects are transported to the environment).

Lemma 1. *If a language $L \subseteq o^*$ is accepted by a deterministic (nondeterministic) linear-bounded multicounter machine Z, then it is accepted by a deterministic (nondeterministic) SCPS M.*

Proof. We only prove the case when Z is deterministic, the nondeterministic case being similar. The construction of M is a simple modification of the construction in [18]. Assume Z has m counters $C_1, ..., C_m$. M has the same membrane structure as in [18]. In particular, the skin membrane contains membranes $E_1, ..., E_m$ to simulate the counters, where the multiplicity of the distinguished (input) symbol o in membrane E_i represents the value of counter C_i. There are other membranes within the skin membrane that are used to simulate the instructions of Z (see [18]). All the sets of rules $R_1, ...,$ are the same as in [18], except the instruction

p : If $C_i \neq 0$, decrement C_i by 1, increment C_j by 1, and goto l else goto k

of Z is simulated as in [18], but the symbol o is not thrown out (from the skin membrane) into the environment but added to membrane E_j. It follows from the construction in [18] that M will not have any instruction of the form $ab \rightarrow a_x b_y c_{come}$ and if instructions of the form $a \rightarrow a_x$ or $ab \rightarrow a_x b_y$ appear in the skin membrane, then $x, y \neq out$. Hence, M is a deterministic SCPS. □

Lemma 2. *If a language $L \subseteq o^*$ is accepted by a deterministic (nondeterministic) linear-bounded multicounter machine, then it is accepted by a deterministic (nondeterministic) bounded S/A system.*

Proof. We show how to convert the multi-membrane SCPS M of Lemma 1 to a (one-membrane) bounded S/A system M'. The construction is similar to the one given in [3]. Suppose that M has membranes $1, ..., m$. For each object a in V, M' will have symbols $a_1, ..., a_m$. In particular, for the distinguished input symbol o in V, M' will have $o_1, ..., o_m$. Hence the distinguished input symbol in M' is o_{i_0}, where i_0 is the index of the input membrane in M. We can convert M to a bounded S/A system M' as follows:

1. If $a \rightarrow a_x$ is a rule in membrane i of M, then $(a_i, out; a_j, in)$ is a rule in M', where j is the index of the membrane into which a is transported to, as specified by x.
2. If $ab \rightarrow a_x a_y$ is a rule in membrane i of M, then $(a_i b_i, out; a_j b_k, in)$ is a rule in M', where j and k are the indices of the membranes into which a and b are transported to, as specified by x and y.

Thus, corresponding to the initial configuration wo^n of M, where o^n is in the input membrane i_0 and w represents the configuration denoting all the other symbols (different from o) in the other membranes, M' will have initial configuration $w'o_{i_0}^n$, where w' are symbols in w renamed to identify their locations in M.

Clearly, M' accepts $o_{i_0}^n$ if and only if M accepts o^n, and M' is a deterministic (nondeterministic) bounded S/A system. □

We will prove the converse of Lemma 2 indirectly. A k-head two-way finite automaton (k-2FA) is a finite automaton with k two-way read-only heads operating on an input (with left and right end markers) [8]. A multihead 2FA is a k-2FA for some k.

Lemma 3. *If M is a deterministic (nondeterministic) bounded S/A system with an alphabet V of m symbols (note that V contains the distinguished input symbol o), then M can be simulated by a deterministic (nondeterministic) $m(m+1)$-2FA Z.*

Proof. Suppose M is a deterministic bounded S/A system accepting a language $L(M) \subseteq o^*$. Assume that its alphabet is $V = \{a_1, ..., a_m\}$, where $a_1 = o$ (the input symbol). We construct a deterministic multihead FA Z to accept $L(G)$. The input to Z (not including the left and right end markers) is o^n for some n. We will need the following heads to keep track of the multiplicities of the symbols in the membrane during the computation (note that the bounded S/A system M is given wo^n initially):

1. K_i for $1 \leq i \leq m$. Head K_i will keep track of the current number of a_i's. Initially, K_1 will point to the right end marker (indicating that there are n o's in the input) while all other K_i will point to the appropriate position on the input corresponding to the multiplicity of symbol a_i in the fixed string w.

2. $K_{i,j}$ for $1 \leq i, j \leq m$. These heads keep track of how many a_i's are replaced by a_j's during the next step of M.

One step of M is simulated by a (possibly unbounded) number of steps of Z. At the beginning of the simulation of every step of M, Z resets all $K_{i,j}$'s to the left end marker. To determine the next configuration of M, Z processes the rules as follows:

Let $R_1, R_2, ..., R_s$ be the rules in the membrane. By using $K_1, ..., K_m$ (note each K_i represents the number of a_i's in the membrane), Z applies rule R_1 sequentially a maximal number of times storing the "results" (i.e., the number of a_i's that are converted by the applications of rule R_1) to a_j in head $K_{i,j}$. Thus, each application of R_1 may involve decrementing the K_i's and incrementing some of the $K_{i,j}$'s. (By definition, the sequential application of R_1 has reached its maximum at some point, if further application of the rule is no longer applicable.)

The process just described is repeated for the other rules $R_2, ..., R_s$. When all the rules have been processed, Z updates each head K_j using the values stored in $K_{i,j}$, $1 \leq i \leq m$. This completes the simulation of the unique (because M is deterministic) maximally parallel step of M.

It follows from the above description that a deterministic bounded S/A system can be simulated by a deterministic $m(m+1)$-2FA.

If M is nondeterministic, the construction of the nondeterministic multihead 2FA M is simpler. M just sequentially guesses the rule to apply each time (i.e., any of $R_1, R_2, ..., R_s$) until no more rule is applicable. Note that Z does not need the heads $K_{i,j}$'s. $\square$

For the proof of the next theorem, we need a definition. Define a generalized linear-bounded multicounter machine as follows. As before, at the start of the computation, the input counter is set to a value n (for some n), and all other counters are set to zero. Now we only require that there is a positive integer c such that at any time during the computation, the value of any counter is at most cn. (Thus, we no longer require that the sum of the values of the counters is at most n.) In [3], it was shown that a generalized linear-bounded multicounter machine can be converted to a linear-bounded multicounter machine. For completeness, we describe the construction.

Suppose that Z is a generalized linear-bounded multicounter machine with counters $C_1, ..., C_m$, where C_1 is the input counter. Construct another machine Z' with counters $D, C_1, ..., C_m$, where D is now the input counter. Z' with input n in counter D, first moves n from D to C_1 (by decrementing D and incrementing C_1.) Then Z' simulates Z on counters $C_1, ..., C_m$ (counter D is no longer active).

Let d be any positive integer. We modify Z' to another machine Z'' which uses, for each counter C_i, a buffer of size d in its finite control to simulate Z', and Z'' increments and decrements each counter modulo d. Z'' does not alter the action of Z' on counter D.

By choosing a large enough D, it follows that the computation of Z'' is such that when given input n in counter D and zeros in counters $C_1, ..., C_m$, the sum of the values of counters $D, C_1, ..., C_m$ at any time is at most n. It follows that, given a generalized linear-bounded multicounter, we can construct an equivalent linear-bounded multicounter machine.

The next theorem is similar to a result in [3] concerning BPS.

Theorem 1. *Let $L \subseteq o^*$. Then the following statements are equivalent:*

(1) L is accepted by a bounded S/A system.
(2) L is accepted by a linear-bounded multicounter machine,
(3) L is accepted by a $\log n$ space-bounded Turing machine.
(4) L is accepted by a multihead 2FA

These equivalences hold for both the deterministic and nondeterministic versions.

Proof. The equivalence of (3) and (4) is well known. By Lemmas 2 and 3, we need only show the equivalence of (2) and (4). That a linear-bounded multicounter machine can be simulated by a multihead 2FA is obvious. Thus (2) implies (4). We now show the converse. Let M be a two-way multihead FA M with m heads $H_1, ..., H_m$. From the discussion above, it is sufficient to construct a generalized multicounter machine Z equivalent to M. Z has $2m+1$ counters, $D, C_1, ..., C_m, E_1, ..., E_m$. Z with input n in counter D, and zero in the other counters first decrements D and stores n in counters $C_1, .., C_m$. Then Z simulates the actions of head H_i of M using the counters C_i and E_i. $\square$

Lemmas 2 and 3 and Theorem 1 can be generalized to non-unary inputs, i.e., inputs of the form $a_1^{n_1}...a_k^{n_k}$, where $a_1, ..., a_k$ are distinct symbols. The constructions are straightforward generalizations of the ideas above. Thus, we have:

Corollary 1. *Let $L \subseteq a_1^*...a_k^*$. Then the following statements are equivalent:*

(1) L is accepted by a bounded S/A system.
(2) L is accepted by a linear-bounded multicounter machine,
(3) L is accepted by a $\log n$ space-bounded Turing machine.
(4) L is accepted by a multihead 2FA.

These equivalences hold for both the deterministic and nondeterministic versions.

We now proceed to show that the number of symbols in the alphabet V of a bounded S/A system induces an infinite hierarchy. This is an interesting contrast to a result in [14] that an unbounded S/A system with three objects is universal. The proof follows the ideas in [6], which showed an infinite hierarchy for a variant of SPCS, called RCPS.

We will need the following result from [8]:

Theorem 2. *For every k, there is a unary language L that can be accepted by a $(k+1)$-2FA but not by any k-2FA. The result holds for both deterministic and nondeterministic versions.*

Theorem 3. *For every r, there exist an $s > r$ and a unary language L (i.e., $L \subseteq o^*$) accepted by a bounded S/A system with an alphabet of s symbols that cannot be accepted by any bounded S/A system with an alphabet of r symbols. This result holds for both deterministic and nondeterministic versions.*

Proof. Suppose there is an r such that any unary language language accepted by any bounded S/A system with an arbitrary alphabet can be accepted by a bounded S/A system with an alphabet of r symbols. Let $k = r(r + 1)$. From Theorem 2, there is a unary language L that can be accepted by a $(k + 1)$-2FA but not by any k-2FA. By Theorem 1, this language can be accepted by a bounded S/A system. Then, by hypothesis, L can also be accepted by a bounded S/A system with an alphabet of r symbols. Then, from Lemma 3, we can construct from this bounded S/A system an $r(r + 1)$-2FA accepting L. Hence, L can be accepted by a k-2FA, a contradiction. $\square$

For our next result, we need the following theorem from [17].

Theorem 4. *Nondeterministic and deterministic multihead 2FAs over a unary input alphabet are equivalent if and only if nondeterministic and deterministic linear bounded automata (over an arbitrary input alphabet) are equivalent.*

From Theorems 1 and 4, we have:

Theorem 5. *Nondeterministic and deterministic bounded S/A systems over a unary input alphabet are equivalent if and only if nondeterministic and deterministic linear bounded automata (over an arbitrary input alphabet) are equivalent.*

3   Multi-membrane Special S/A Systems

Let M be a multi-membrane S/A system, which is restricted in that only rules of the form $(u, out; v, in)$, where $|u| = |v| \geq 1$, can appear in the skin membrane. There are no restrictions on the weights of the rules in the other membranes. Clearly, the number of objects in the system at any time during the computation remains the same. We denote by E_t the alphabet of t symbols (for some t) in the environment. There may be other symbols in the membranes that remain in the system during the computation and are not transported to/from the environment, and they are not part of E_t. Note that E_0 means that the environment alphabet is empty (i.e., there are no symbols in the environment at any time). As before, we consider the case where the input alphabet is unary (i.e. $\Sigma = \{o\}$). M's initial configuration contains o^n in the input membrane (for some n) and a *fixed* distribution of some *non-o* symbols in the membranes. The string o^n is accepted if the system eventually halts. We call the system just described a *special S/A system*.

Theorem 6. *Let $L \subseteq o^*$. Then the following statements are equivalent:*

*(1) L is accepted by a multi-membrane special S/A system with **no** symbols in the environment, i.e., has environment alphabet E_0 (= empty set).*

(2) L is accepted by a bounded S/A system.

(3) L is accepted by a linear-bounded multicounter machine.

(4) L is accepted by a log n space-bounded Turing machine.
(5) L is accepted by a multihead 2FA.

These equivalences hold for both the deterministic and nondeterministic versions.

Proof. As in Lemma 3, it is easy to show that a deterministic (nondeterministic) m-membrane special S/A system with no symbols in the environment can be simulated by a deterministic (nondeterministic) two-way FA with $2m$ heads.

By Theorem 1, to complete the proof, we need only show that a linear-space bounded multicounter machine can be simulated by a multi-membrane special S/A with no symbols in the environment. For notational convenience, we will assume the multicounter machine is controlled by a program with instructions of the type $l_i : (ADD(r), l_j)$, $l_i : (SUB(r), l_j, l_k)$, and $l_i : (HALT)$ where l_i is the label for the current instruction being executed and r is the counter which is either being incremented or decremented. If the current instruction is an add instruction, the next instruction to execute will be l_j. If the current instruction is a subtract instruction the next instruction depends on the value of r. If $r \neq 0$, the next instruction is denoted by l_j otherwise the next instruction is denoted by l_k.

The special S/A system simulating a linear-space bounded multicounter machine will use one membrane to simulate each counter of the multicounter machine. These membranes will be placed within a 'program' membrane where the current instruction is brought in, implemented, and then expelled. This entire system is enclosed within a dummy membrane (the skin membrane) containing no rules and a single copy of each instruction object along with a a few auxiliary objects. So the overall system uses $m + 2$ membranes. Obviously, if the skin membrane of the special S/A system contains no rules, no object can ever be brought into the system or expelled from the system. Hence, since the system initially contains $|wo^n|$ symbols, the system will continue to contain $|wo^n|$ symbols after each step of the computation.

To show how any linear-space bounded multicounter machine can be simulated, we give a formal transformation to a special S/A system. Our transformation is similar to the transformation in [14] except that our transformation yields a deterministic (nondeterministic) special S/A system if the original linear-space bounded multicounter machine is deterministic (nondeterministic). (The transformation in [14] only produces a nondeterministic S/A system.) The transformation is done as follows. Consider a multicounter machine Z with m counters. Construct a symport / antiport system M which simulates Z as follows:

$$M = (V, H, \mu, w_1, w_2, \cdots, w_{m+2}, E_0, R_1, R_2, \cdots, R_{m+2}, i_o)$$

where $H = \{1, 2, \cdots, m + 2\}$; $\mu = [_1[_2[_3]_3[_4]_4 \cdots [_{m+2}]_{m+2}]_2]_1$; $w_1 =$ one copy of each element in V except o and l_{01} (we assume Z's program begins with the instruction l_0); $w_2 = l_{01}$; $w_3 = o^n$; $w_i = \lambda$, for all $i = 4, \cdots, m + 2$; $E_0 = \emptyset$ (the environment, E_t, is empty because $t = 0$); No need to specify i_0, since our system is an acceptor.

The elements of V are as follows:

1. o — The symbol o is used as the counting object for the system. The multiplicity of o's in each counter membrane signifies the count of that counter.

number of complete K_3 complexes in **P** and r the probability that three connected junctions by two sticky ends will close in a complete K_3 complex. Then the expected number of K_3 complete complexes in the pot is

$$E(X) = mr;$$

moreover

$$\lim_{r \to 1} P(X = m) = 1,$$

where m denotes the amount of junctions in **P** of each type.

Proof. (Sketch, see the appendix for details): Let S be a set of three different junctions, one from each type in **P**, and let X_S be the indicator random variable for the event that junctions from S form a complete K_3. Then $X = \sum_S X_S$ gives the total number of complete K_3's in the pot. Since the described model is static, equal probability is assigned on each sticky end. The unconditional probability of any two sticky ends to connect is $p = 1/m$ and $E(X_S) = P(X_S = 1) = p^2 r$. We can form m^3 sets S from the pot **P** and by the linearity of the expectation follows that

$$E(X) = \sum_S E(X_S) = m^3 p^2 r = mr$$

Looking over all pairs of sets, each with three different junction types from **P**, through computation of the covariances we obtain

$$Var(X) \sim mr(1 - r).$$

When $r \to 1$, $Var(X) = E(X - EX)^2 \to 0$, and since X is a nonnegative random variable it follows that almost surely $X = E(X) = m$. That means almost surely only K_3 complexes are obtained in **P**.

The case when $p < \frac{1}{m}$ would result with incomplete complexes, and in this paper we do not consider this, but certainly we believe that such analysis may provide valuable information for the understanding of the self-assembly process.

To recapitulate, given m, depending on the amount of solution, and r, depending on the molecular dynamics, the expected number of junctions in K_3 cycles is mr, with standard deviation $\sqrt{mr(1 - r)}$, the later being unobservable under contemporary laboratory conditions.

We can generalize the result (obtained for complete K_3) for circular complexes of any length. Consider a pot that contains n 2-branched different junction types uniformly distributed, capable of forming a cycle of length n.

Theorem 4. Let $P = \{j_1, \ldots, j_n\}$ be a pot type which contains n uniformly distributed 2-armed junctions capable of admitting C_n complex. Let X denote the number of complete C_n complexes in **P** and r the probability that a sequence of n connected junctions will close in a cycle C_n. Then the expected number of cycles of length n in the pot is given by

process is considered. We are mainly interested in the input and the output of the process. So we provide some insight to the question: If we have a certain amount of junction molecules, what kind of complex types are there in the outcome?

We start with a special case of obtaining cyclic molecules with three 2-armed junctions. This corresponds to building a triangle. For this purpose, we consider three different types of 2-armed junctions (tiles) $J_3 = \{j_1, j_2, j_3\}$ which contain 3 different types of complementary free sticky ends $H_3 = \{h_1, h_2, h_3, \hat{h}_1, \hat{h}_2, \hat{h}_3\}$. These junctions are uniformly distributed in a pot $P = \{j_1, j_2, j_3\}$, and are capable of admitting a complete K_3 complex, meaning that we have equal amount of junctions from each junction type. We conveniently represent this amount with an integer m. The sticky end types are adequately arranged (see Fig. 2),

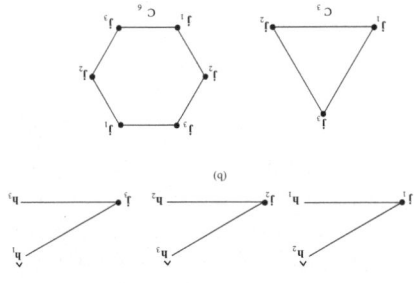
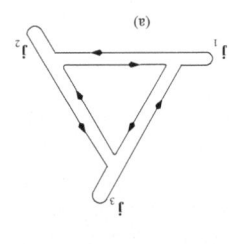

Fig. 2. (a) Three 2-armed junctions (tiles) form a triangle which represents a K_3 complex. (b) Three junction graphs used in a pot to assemble K_3. (c) Complete complexes for this pot will be cycles of length divisible by 3. Cycles K_3 and C_6 are depicted.

$$j_1(h_1) = j_1(\hat{h}_2) = 1,$$

$$j_2(h_2) = j_2(\hat{h}_3) = 1,$$

$$j_3(h_3) = j_3(\hat{h}_1) = 1.$$

With this kind of selection for the junction molecules, complete complexes that are obtained would be cyclic and would involve $3k$ junctions, for some k ($1 \le k \le m$). Junctions from **P** are always assembling according to a specific pattern, $j_1 j_2 j_3$ repeatedly or $j_1 j_3 j_2$ repeatedly, depending on the orientation. We say that a cycle is of *length $3k$*, if it has k junctions from each type adequately arranged, for example $j_1 j_2 j_3$ repeatedly k times, such that the last j_3 junction is glued to the first j_1 junction. We will use notation C_{3k} for cycles of length $3k$, $k \ge 1$ (Note: For cycle of length 3, we will use notation K_3).

The probability of obtaining a certain cyclic complex depends on its size. We employ probabilistic method, often used in random graph theory [5, 19] to obtain the results. We start our analysis by computing the probability of appearance of at least one K_3 complex in the pot described above.

Theorem 3. *Let* $P = \{j_1, j_2, j_3\}$ *be a pot type which contains uniformly distributed 2-armed junctions capable of admitting K_3 complex. Let X denote the*

ends on the junctions/complexes, and not about the underlying graph structure. In order to preserve the information about the graph structure, we define
structure type relation between two complexes. Two complexes are of the same
structure type if there is a graph isomorphism from one to another that preserves
junction types, sticky end types and edges.

We are interested in a collection of junctions of specific types, and potential
complexes that arise by their bonding. A **pot type** is a finite set **P** of junction
types. If **P** is pot type, then a *resolution* of **P**, denoted $C(\mathbf{P})$ is the set of all
possible complete structure types that can be obtained after gluing junctions of
types in **P**.

3 Probabilistic Analysis

Complete complexes are of special interest for the model, since many DNA nanotechnology designs depend on their formation. For example, in [9] it has been
shown that several NP complete problems can be solved if and only if complete
complexes of appropriate size are obtained in the pot. In an experiment one can
expect to obtain many kinds of complete complexes, and not all of them may
represent the designated structures. It can be observed experimentally that a
portion of the DNA material in the pot ends up in incomplete complexes. Also
the appearance of topoisomers of complete complexes and of a complete complex
for a problem that does not have a solution have been reported [16, 12].

At the end of the experiment to solve a given problem there may be some
complete complexes of the desired type, some other complete complexes and
some incomplete complexes. Since our major concern is the construction of complete complexes of certain sizes, we want to explore the density of each complete
complex, for the purpose of evaluating the results of the experiment. We approach this problem by considering a special case, graph assembly of uniformly
distributed junctions (tiles).

For simplicity we avoid thermodynamic properties and consider only the self-
assembly process for which all Watson-Crick connections are equally likely and
no free sticky ends remain after the completion of the experiment. It means that
after self-assembly has occurred only complete complexes are present. We propose a static model (similar to models studied in discrete probability theory)
with uniformly distributed junctions, and Watson-Crick complementary pairs.
Each sticky end has equal probability to connect. We have proved that the probability spaces and distribution for the model exist, but we omit this proves in
the present discussion. This process is not entirely realistic. By observing some
experimental results we amended our model by assuming that in the process
of formation of a complete complex, probability r that the last of the possible
connections within a complex appears after the other connections have been
established is very high (almost certain). Throughout our discussion this probability is denoted with r.

We would like to mention that the evolution of the self-assembly process is not
examined, only distribution of obtained complete complexes at the end of the

junction's arms. Each sticky end is represented with strings over the alphabet $\Delta = \{A, C, G, T\}$, and the set of all strings of the given junction graph G will be the set of all sticky end types, denoted by H_G. The function $\theta : H_G \longrightarrow H_G$ maps every string $\mathbf{h} \in H_G$ to its Watson-Crick complement $\theta(\mathbf{h})$. Then θ is a deranging involution, i.e., for any $\mathbf{h} \in H_G, \theta(\mathbf{h}) \neq \mathbf{h}$, while $\theta(\theta(\mathbf{h})) = \mathbf{h}$. We simplify the notation by writing $\overline{\mathbf{h}}$ for $\theta(\mathbf{h})$.

Definition 2. *A complex $\overline{C}$ is a pair $C = (G, \mathbf{h}) = (J \cup F, E, \mathbf{h})$, where $G = (J \cup F, E)$ is a given junction graph and $\mathbf{h} : F \longrightarrow H_G$ is a map that assigns a free sticky end type to every free vertex. In particular a junction is a complex with $|J| = 1$.*

Self-assembly between complexes is described through a gluing operation as follows. Let $C_1 = (G_1, \mathbf{h}_1) = (J_1 \cup F_1, E_1, \mathbf{h}_1)$ and $C_2 = (G_2, \mathbf{h}_2) = (J_2 \cup F_2, E_2, \mathbf{h}_2)$ be two complexes and let F_1' be a maximal set of sticky end vertices from C_1 that are complementary to a set F_2' of sticky ends from C_2. There will be a bijection $\varphi : F_1' \longrightarrow F_2'$, such that $\theta(\mathbf{h}_1(f)) = \mathbf{h}_2(\varphi(f))$ for every $f \in F_1'$ (F_1' indicates a maximal subset of sticky ends from C_1 that are complementary to sticky ends from C_2). The complex obtained in the self-assembly process by gluing the complexes C_1 and C_2 is the complex $C = (G, \mathbf{h}) = (J \cup F, E, \mathbf{h})$, where

$$J = J_1 \cup J_2$$
$$F = (F_1 \setminus F_1') \cup (F_2 \setminus F_2')$$
$$E = (E_1 \cup E_2)$$
$$\cup \{\{j_1, j_2\} \mid \exists f \in F_1'\{j_1, f\} \in E_1 \text{ and } \{j_2, \varphi(f)\} \in E_2\})$$
$$\setminus \{\{j_1, f\}, \{j_2, \varphi(f)\} \mid f \in F_1'\}$$

$$h(f) = \begin{cases} h_1(f) & f \in F_1 \setminus F_1' \\ h_2(f) & f \in F_2 \setminus F_2'. \end{cases}$$

The set of all possible complexes obtained by gluing C_1 and C_2 is denoted by $C_1 \star C_2$. Intuitively, when two complexes join together, all sticky ends that can connect do connect (there might be a different choice for such connection and the gluing process may not be deterministic, i.e., it does not represent an operation). The vertices that were free before the gluing process begun and have annealed with their complements during the gluing, disappear from the resulting complex. Each connection (annealing) adds an edge in the new complex. Gluing of a 4-armed junction and a 3-armed junction schematically is depicted in Fig. 1 to the right.

For a given complex $C = (G, \mathbf{h})$, we define a *junction type* $\mathbf{j}$ to be a function $\mathbf{j} : H_G \longrightarrow \mathbb{N}$ such that $\mathbf{j}(\mathbf{h})$ will denote the number of sticky ends of type $\mathbf{h} \in H_G$ on the junction vertex $j \in J$. Similarly we define a *complex type* to be a function $C : H_G \longrightarrow \mathbb{N}$, where $C(\mathbf{h})$ is the number of sticky ends of type $\mathbf{h} \in H_G$ in complex C. A complex $C = (G, \mathbf{h})$ is called a *complete complex* if $C(\mathbf{h}) = 0$ for every $\mathbf{h} \in H_G$ i.e., it is a complex without free sticky ends. Since every junction is a complex, every junction type is also a complex type. Both junction and complex types give information only about the free sticky

Example 1. Let $L = \{a^{n_1}a^{n_2}\}$ and let $\varrho = (T, \Sigma, n_1, n_2)$ be a TGR system where $\Sigma = \{a\}$, $T = \{a^{n_1}a^{n_2}a^{n_1}\}$ and $n_1, n_2 \in \mathbb{N}$. Then $\varrho^*(L) = \{a^{n_1+n_2}\} \cup \{a^{2n_1+n_2}a^*\}$. For the case $n_1 = n_2 = 1$, $\varrho(L) = aa^+$.

It is also known from [13] that the closure of a language family under iterated template-guided recombination contains the original language family.

Lemma 4. *Let $\mathcal{L}_1, \mathcal{L}_2$ be language families and let $n_1, n_2 \in \mathbb{N}$. Then $\mathcal{L}_1 \subseteq \pitchfork^*(\mathcal{L}_1, \mathcal{L}_2, n_1, n_2)$.*

We have considered the basic properties of the iterated version of template-guided recombination in [14] and we recall here the definition of a useful template from that paper.

Intuitively, a template word is useful if it can be used as a template to produce any word, not necessarily new. The full formal definition is found in [14]. This notion turns out to be quite important as is shown by the following two results.

We see that every full AFL is closed under iterated template-guided recombination with useful templates from the same full AFL.

Theorem 1. *Let $\mathcal{L}$ be a full AFL, $\varrho = (T, \Sigma, n_1, n_2)$ a TGR system and let $L, T \in \mathcal{L}, L \subseteq \Sigma^*$ and assume that ϱ is useful on L. Then $\varrho^*(L) \in \mathcal{L}$.*

In addition, when the template sets are regular, the useful subset, T_u say, of the template language T on *any* language L has a very simple structure relative to T.

Proposition 4. *Let $\varrho = (T, \Sigma, n_1, n_2)$ be a TGR system, let $L \subseteq \Sigma^*$ and let T_u be the useful subset of T on L. If T is a regular language, it follows that T_u is also regular.*

The language L in the proposition above does not have any restrictions placed on it. It need not even be recursively enumerable. The proof does not, however, provide an effective construction for T_u.

A consequence of Theorem 1, Proposition 4 and the fact that the family of regular languages is the smallest full AFL allows us to show the following key result.

Theorem 2. *Let $\mathcal{L}$ be a full AFL and let $n_1, n_2 \in \mathbb{N}$. Then*

$$\pitchfork^*(\mathcal{L}, \mathbf{REG}, n_1, n_2) = \mathcal{L}.$$

This shows that the operation, as defined, provides very little computational power, regardless of the minimum pointer and MDS length. Indeed, even when we start with regular initial and template languages, we cannot generate any non-regular languages. This is not surprising biologically, however, as one might expect the cell to make use of the least complex computational process to accomplish a given task.

In the next section we show that adding even a small amount of context-sensitiveness to template-guided recombination results in a large increase in computational power.

4   Extension of TGR by Deletion Contexts

As defined above, the operation of template-guided recombination is able to achieve very limited computational power. Even the iterated version is able only to generate regular languages starting from a regular initial language and using a regular set of templates. This is in contrast to the fact that extended splicing systems are able to generate arbitrary recursively enumerable languages starting from regular splicing rules and a finite set of axioms[17]. It is often the case that small alterations to an operation can lead to a huge increase in generative capacity. In this section, we add a feature to this operation in order to achieve more power. It should be stated immediately that it is not clear how realistic this extension is in a biological setting. While the extension presented is certainly not biologically impossible, neither do we have experimental evidence to support it. Despite this, it serves as an aide to the study of what properties should likely be present in order to obtain more general computation.

We begin by defining a more general version of the template-guided recombination operation. Indeed, the new notation allows for extra deletion contexts, beyond the β pointer. The previously studied operation is a special case where all deletion contexts are of length zero. We cannot assume, with this more general notation that the symbol β is always of the minimum pointer length.

Definition 2. *A contextual template-guided recombination system (or shortly, a CTGR system) is a four tuple $\varrho = (T, \Sigma, n_1, n_2)$ where Σ is a finite alphabet, $\#$ is a symbol not in Σ, $T \subseteq \Sigma^* \# \Sigma^* \# \Sigma^*$ is the template language, $n_1 \in \mathbb{N}$ is the minimum MDS length and $n_2 \in \mathbb{N}$ is the minimum pointer length.*

For a CTGR system $\varrho = (T, \Sigma, n_1, n_2)$ and a language $L \subseteq \Sigma^*$, we define $\varrho(L) = \{w \in \Sigma^* \mid (x, y) \vdash_t^c w \text{ for some } x, y \in L, t \in T\}$ where $(x, y) \vdash_t^c w$ if and only if $x = u\alpha\beta d_1 d, y = ee_1\beta\gamma v, t = e_1 \# \alpha\beta\gamma \# d_1, w = u\alpha\beta\gamma v, u, v, d, e \in \Sigma^*, \alpha, \gamma \in \Sigma^{\geq n_1}, \beta \in \Sigma^{\geq n_2}$.

For $k \in \mathbb{N}_0 \cup \{\infty\}$, we denote $\underline{\text{th}}(\mathcal{L}_1, \mathcal{L}_2[k], n_1, n_2) = \{\varrho(L) \mid L \in \mathcal{L}_1, \varrho = (T, \Sigma, n_1, n_2) \text{ a CTGR system}, T \in \mathcal{L}_2, T \subseteq \Sigma^{\leq k} \# \Sigma^* \# \Sigma^{\leq k}\}$ and $\underline{\text{th}}(\mathcal{L}_1, \mathcal{L}_2) = \{\underline{\text{th}}(\mathcal{L}_1, \mathcal{L}_2[\infty], n_1, n_2) \mid n_1, n_2 \in \mathbb{N}\}$.

We then get template-guided recombination as a special case where the contexts are of length zero. Next, we see that if we add in even one symbol of deletion context, we increase the power significantly.

Lemma 5. *Let Σ be an alphabet, $\Sigma_1 = \Sigma \cup \{a_1, a_2, a_3, a_4, a_5\}$, (all new symbols disjoint from Σ), $\mathcal{L}$ a language family closed under left and right concatenation with symbols and $L_1 \cup aL_2 \in \mathcal{L}$ for $L_1, L_2 \in \mathcal{L}$, a a new symbol. Then there exists $L \in \mathcal{L}, T \in \Sigma_1 \# \Sigma_1^* \# \Sigma_1, T \in \mathbf{REG}_0$ and a CTGR system $\varrho = (T, \Sigma_1, 1, 1)$ such that $L_1 \cap L_2 = (a_4 a_1 a_2)^{-1}(\varrho(L))(a_3 a_1 a_5)^{-1}$ and $\varrho(L) \in \underline{\text{th}}(\mathcal{L}, \mathbf{REG}_0[1], 1, 1)$.*

As corollary, we obtain that $\underline{\text{th}}(\mathbf{LIN}, \mathbf{REG}_0[1], 1, 1)$ is equal to $\mathbf{RE}$ after applying an intersection with a regular language and a homomorphism.

Corollary 5. *Let Σ be an alphabet, $L \in \mathbf{RE}, L \subseteq \Sigma^*$. Then there exists an alphabet Σ_1, a homomorphism h from Σ_1^* to Σ^*, languages $R, T \in \mathbf{REG}_0$, a language $L' \in \mathbf{LIN}$ and a CTGR system $\varrho = (T, \Sigma_1, 1, 1)$ such that $h(\varrho(L') \cap R) = L$.*

Thus, even though $\overline{h}(\mathcal{L}_1, \mathbf{REG}) \subseteq S(\mathcal{L}_1, \mathbf{REG})^4$ for every $\mathcal{L}_1$ (see [13]) and $S(\mathcal{L}_1, \mathbf{REG}) \subseteq \mathcal{L}_1$ for every concatenation closed full trio $\mathcal{L}_1$ (see [2]), we see that when we add in even one symbol of deletion contexts, $\overline{h}(\mathcal{L}_1, \mathbf{REG}) \not\subseteq S(\mathcal{L}_1, \mathbf{REG})$ in many cases, for example when $\mathcal{L}_1$ is the family of context-free languages. Consequently, template-guided recombination with deletion contexts can generate more powerful languages than splicing systems.

Lemma 6. *Let $\mathcal{L}_1, \mathcal{L}_2$ be language families, both closed under inverse gsm mappings and intersection with regular languages, let $L \in \mathcal{L}_1, T \in \mathcal{L}_2$ and let $\varrho = (T, \Sigma, n_1, n_2)$ be a CTGR system. Then there exists $L_1, L_2 \in \mathcal{L}_1, T' \in \mathcal{L}_2$ and a weak coding homomorphism h such that $\varrho(L) = h(L_1 \cap L_2 \cap T')$.*

Corollary 6. *Let $\mathcal{L}_1$ be an intersection-closed full trio closed under intersection with $\mathcal{L}_2$, which is either closed under inverse gsm mappings and intersection with regular languages or $\mathcal{L}_2 \subseteq \mathbf{REG}$. Then $\overline{h}(\mathcal{L}_1, \mathcal{L}_2) \subseteq \mathcal{L}_1$.*

Proposition 5. *Let $\mathcal{L}_1$ be a full semi-AFL and let $\mathbf{REG}_0 \subseteq \mathcal{L}_2$ be closed under inverse gsm mappings and intersection with regular languages. Then $\overline{h}(\mathcal{L}_1, \mathcal{L}_2) \subseteq \mathcal{L}_1$ if and only if $\mathcal{L}_1$ is closed under intersection with $\mathcal{L}_1$ and $\mathcal{L}_2$.*

We would also like to study the iterated version of this more general operation. Let $\varrho = (T, \Sigma, n_1, n_2)$ be a CTGR system and let $L \subseteq \Sigma^*, T \subseteq \Sigma^* \# \Sigma^* \# \Sigma^*$. We generalize ϱ to an iterated operation $\varrho^*(L)$ in the natural way:

$$\varrho^0(L) = L,$$
$$\varrho^{n+1}(L) = \varrho^n(L) \cup \varrho(\varrho^n(L)), n \geq 0$$
$$\varrho^*(L) = \bigcup_{n=0}^{\infty} \varrho^n(L).$$

In the following, we show that we are able to generate arbitrary recursively enumerable languages using iterated contextual template-guided recombination with regular templates and a finite initial language and applying an intersection with a terminal alphabet and a coding. The following proof follows the well known "simulate-rotate" proof technique from splicing systems [2]. We apply the final coding homomorphism in the proof in order to stop the β symbol in the definition from "compressing" small amounts of information in an undesirable fashion. It is not clear if the coding is strictly necessary, however it is very simple: mapping three separate symbols onto one for each symbol. We only require deletion contexts of length two.

Proposition 6. *Let $L' \subseteq \Sigma^*$ be an arbitrary recursively enumerable language. Then there exist alphabets $\overline{\Sigma}, W$, a regular template language T, a CTGR system $\varrho = (T, W, 1, 1)$, a finite language $L \subseteq W^*$, and a coding homomorphism h from $\overline{\Sigma}^*$ to Σ^* such that $h(\varrho^*(L) \cap \overline{\Sigma}^*) = L'$.*

We have thus demonstrated that an arbitrary recursively enumerable language can be generated by a CTGR system with a finite initial language and a

---

[4] Where $S(\mathcal{L}_1, \mathbf{REG})$ denotes non-iterated splicing systems with an initial language in $\mathcal{L}_1$ and splicing rules in $\mathbf{REG}$.

regular template language up to a coding homomorphism and intersection with a terminal alphabet.

5   Conclusions

We have considered the basic properties and computational power of an operation inspired by the template-guided recombination model of gene descrambling in stichotrichous ciliates. Specifically, we began by investigating the properties of a non-iterated version of template-guided recombination systems, contributing to their theoretical understanding. We characterized closure properties of families of languages under template-guided recombination in terms of other basic operations and demonstrated that every intersection-closed full semi-AFL is closed under template-guided recombination with templates from the same full semi-AFL.

We then recalled the properties of iterated template-guided recombination systems. The principal result here shows the limited power of template-guided recombination by demonstrating that every full AFL is closed under iterated template-guided recombination with regular templates. This implies that the computational power of any system based on this operation will be quite limited. Indeed, if one enforces the "reasonable" restriction that template and initial languages must be regular, one does not gain any increase in computational power. This motivates the question of what minimal extension would be required to increase the generative capacity beyond the regular languages while still restricting the initial and template languages to be, at most, regular.

We addressed this question by showing that the tight restriction on computational power can be lifted by adding a small degree of context-awareness to a TGR system. We have demonstrated that we are able to generate arbitrary recursively enumerable languages using contextual iterated template-guided recombination with regular templates, a finite initial language and applying an intersection with a terminal alphabet and a coding.

It may be preferable, from the point of view of biocomputing, to show a result which demonstrates the simple template-guided recombination systems to be capable of universal computation; however, from the point of view of judging the closeness of this formalization to the biological process which it models, the opposite may be true. Given that a cell has access to only finite resources, and has serious constraints on the length of time in which the descrambling process must be completed, it seems reasonable that the process must be relatively computationally simple.

We have also shown in this paper that by adding a small amount of context-awareness to a TGR system, we are able to easily generate arbitrary recursively enumerable languages. While this result may be more theoretically satisfying, we caution that the "deletion contexts" required to derive such a result are not present in the biological model given in [9], though they are not biologically impossible. Too little is currently known about the molecular biology of ciliates to make definitive statements; however, we feel that, in the context of formalizing a biological process, a result indicating limited computational power is perhaps preferable.

The simplicity and elegance of template-guided recombination combined with the ubiquity of template-mediated events in biological systems, shows that the operation warrants further investigation both as a possible model of a biological process and as a purely abstract operation.

Acknowledgments

We thank Grzegorz Rozenberg for helpful discussions.

References

1. Head, T.: Formal language theory and DNA: an analysis of the generative capacity of specific recombinant behaviors. Bulletin of Mathematical Biology **49** (1987)
2. Păun, G., Rozenberg, G., Salomaa, A.: DNA Computing : new computing paradigms. Springer-Verlag, Berlin (1998)
3. Prescott, D.: Genome gymnastics: Unique modes of DNA evolution and processing in ciliates. Nature Reviews Genetics **1** (2000) 191–198
4. Kari, L., Landweber, L.: Computational power of gene rearrangement. In Winfree, E., Gifford, D., eds.: DNA5, DIMACS series in Discrete Mathematics and Theoretical Computer Science. Volume 54. American Mathematical Society (2000) 207–216
5. Landweber, L., Kari, L.: The evolution of cellular computing: Nature's solution to a computational problem. In Kari, L., Rubin, H., Wood, D., eds.: DNA4, BioSystems. Volume 52. Elsevier (1999) 3–13
6. Ehrenfeucht, A., Prescott, D., Rozenberg, G.: Computational aspects of gene (un)scrambling in ciliates. In Landweber, L., Winfree, E., eds.: Evolution as Computation. Springer-Verlag, Berlin, Heidelberg (2001) 45–86
7. Ehrenfeucht, A., Harju, T., Petre, I., Prescott, D., Rozenberg, G.: Computation in Living Cells, Gene Assembly in Ciliates. Springer-Verlag, Berlin (2004)
8. Ehrenfeucht, A., Prescott, D., Rozenberg, G.: Molecular operations for DNA processing in hypotrichous ciliates. European Journal of Protistology **37** (2001) 241–260
9. Prescott, D., Ehrenfeucht, A., Rozenberg, G.: Template-guided recombination for IES elimination and unscrambling of genes in stichotrichous ciliates. Journal of Theoretical Biology **222** (2003) 323–330
10. Salomaa, A.: Formal Languages. Academic Press, New York (1973)
11. Berstel, J.: Transductions and Context-Free Languages. B.B. Teubner, Stuttgart (1979)
12. Ginsburg, S.: Algebraic and Automata-Theoretic Properties of Formal Languages. North-Holland Publishing Company, Amsterdam (1975)
13. Daley, M., McQuillan, I.: Template-guided DNA recombination. Theoretical Computer Science **330** (2005) 237–250
14. Daley, M., McQuillan, I.: Useful templates and template-guided DNA recombination. (to appear in *Theory of Computing Systems*)
15. Doak, T.: (Personal communication)
16. Landweber, L., Kuo, T., Curtis, E.: Evolution and assembly of an extremely scrambled gene. PNAS **97** (2000) 3298–3303
17. Păun, G.: Regular extended H systems are computationally universal. Journal of Automata, Languages and Combinatorics **1** (1996) 27–36

Towards Practical Biomolecular Computers Using Microfluidic Deoxyribozyme Logic Gate Networks

Joseph Farfel and Darko Stefanovic

Department of Computer Science,
University of New Mexico
{jfarfel, darko}@cs.unm.edu

Abstract. We propose a way of implementing a biomolecular computer in the laboratory using deoxyribozyme logic gates inside a microfluidic reaction chamber. We build upon our previous work, which simulated the operation of a flip-flop and an oscillator based on deoxyribozymes in a continuous stirred-tank reactor (CSTR). Unfortunately, using these logic gates in a laboratory-size CSTR is prohibitively expensive, because the reagent quantities needed are too large. This motivated our decision to design a microfluidic system. We would like to use a rotary mixer, so we examine how it operates, show how we have simulated its operation, and discuss how it affects the kinetics of the system. We then show the result of simulating both a flip-flop and an oscillator inside our rotary mixing chamber, and discuss the differences in results from the CSTR setting.

1   Introduction

Deoxyribozymes (nucleic acid enzymes) may be used as logic gates, which transform input signals, denoted by a high concentration of substrate molecules, into output signals, which are represented by product created when the deoxyribozyme gate cleaves a substrate molecule [1]. Using these gates, molecular devices have been created in the laboratory that function as a half-adder [2] and a tic-tac-toe automaton [3]. Furthermore, experiments have demonstrated the linking of the output of certain deoxyribozyme gates to the input of others, which opens the prospect of creating complex logic [4].

These gates have so far only been used in the laboratory in very small quantities, and, quite significantly, only in closed reactors. This is due to the expense that inhibits purchasing large amounts of gate molecules and the substrates that act as their input. Using these gates in closed reactor systems has the major drawback of limiting them to performing one-shot computations. Previously, we have simulated multiple gate operation in an open, continuous-influx stirred tank reactor (CSTR), and have shown designs for a flip-flop and an oscillator in this setting [5]. Unfortunately, no such open reactor experiment has been performed, owing to the attendant costs.

We propose a microfluidic system whereby these open reactor experiments may actually be performed in the laboratory at a modest cost in materials and apparatus. We analyze and simulate a molecular flip-flop and oscillator in a microfluidic setting. The reaction kinetics of the flip-flop and oscillator in the CSTR have already been examined in detail. Our simulation changes these kinetics by making the influx and homogeneity of the system time-dependent, varying according to our simulation of a microfluidic mixer, which doubles as the reaction chamber.

A. Carbone and N.A. Pierce (Eds.): DNA11, LNCS 3892, pp. 38–54, 2006.

The extremely small volume of a microfluidic reaction chamber (ours is 7.54 nL) compared to a CSTR (50 mL or more) means that the same or even substantially greater concentrations of oligonucleotide gates and substrates can be obtained within the chamber even with a vastly smaller amount of gate and substrate molecules. This means that the expense of an open-reactor experiment (mostly determined by the amount of substance used—including the substrates, the products, and the gates) can be reduced by several orders of magnitude, and be made reasonable. The initial cost of building the microfluidic system may be large, but the benefit of being able to run experiments with a very small number of pricey deoxyribozyme molecules far outweighs this initial investment. In addition to reducing expense and thereby enabling real-life open-reactor experiments, this approach has numerous other advantages unique to a microfluidic system, including a vast decrease in the time needed to perform logic operations, the possibility of keeping gates inside a chamber (allowing for pre-fabricated chambers, each implementing a certain type of logic), and the ability to link reaction chambers together with externally-controlled valves. Linking chambers together could allow us to create complex networks of reaction chambers, and channels between chambers could even be designed to mimic capillaries connecting living cells in which computation may be taking place *in vivo* at some point in the future. In fact, we consider this microfluidic setting to be the proving ground for deoxyribozyme logic gate circuits for medical applications.

2   The Chemical Kinetics of Deoxyribozyme Gate Networks

The four chemical components present in our reactor are inputs, gates, substrates, and products. All of these components are oligonucleotides. The gates are deoxyribozyme molecules, and under certain input conditions they are active [1]. When a gate becomes active, it cleaves substrate molecules to create product molecules. In more technical terms, the enzymatic (active) gate is a phosphodiesterase: it catalyzes an oligonucleotide cleavage reaction. Input molecules can either activate or deactivate a gate. The effect that a particular type of input molecule has on a gate defines its function. For instance, a simple inverter, or NOT gate, will be active, and cleave substrate to produce product, until an input molecule binds to it, making it inactive. The concentration of product in the system is the output signal of the gate, where a high concentration of product is read as true and a low concentration is read as false (the same is true for high or low input concentrations). Product molecules fluoresce, while substrate molecules do not, so the concentration of product molecules in the system is determined by the level of emitted fluorescence. For the NOT gate example, the concentration of product in the system becomes high when there is no input and becomes low when input molecules are added, as the input molecules deactivate all of the gate molecules and product is no longer being cleaved from substrate. This example of the NOT gate's operation depends on its being in an open reactor, however—if it is in a closed reactor, the product concentration can never go from high to low, but in an open reactor, product is always being removed from the system as part of the system's efflux.

In order to model the operation of these logic gates, we must be well informed of their basic chemical kinetics. The kinetics of the YES gate have been thoroughly examined [5], and we use those results here. In this examination, it is assumed that the bonding between gate and input molecules is instantaneous and complete, since it is known that the cleavage and separation of the substrate molecules into product molecules is the slowest of the reactions, and thus is the rate-limiting process. The rate at which product is produced by a gate is $\frac{dP}{dt} = \beta S G_A$, where P is the product concentration, β is the reaction rate constant, S is the substrate concentration, and G_A is the concentration of active gates. It has been empirically determined that the reaction rate constant for the YES gate is approximately $\beta = 5 \cdot 10^{-7}\,\mathrm{nM^{-1}s^{-1}}$. This value will be assumed as the reaction rate for all deoxyribozyme gates mentioned herein.

The chemical kinetics of an entire system of gates, substrates, inputs, and products in an open, microfluidic reactor can be modeled with a set of coupled differential equations. An example is the case of the inverter, or NOT gate, where the set of equations is:

$$\frac{dG}{dT} = \frac{G^m(T) - E(T)G(T)}{V} \tag{1}$$

$$\frac{dI}{dT} = \frac{I^m(T) - E(T)I(T)}{V} \tag{2}$$

$$\frac{dP}{dT} = \beta H(T)S(T)\max(0, G(T) - I(T)) - \frac{E(T)P(T)}{V} \tag{3}$$

$$\frac{dS}{dT} = \frac{S^m(T)}{V} - \beta H(T)S(T)\max(0, G(T) - I(T)) - \frac{E(T)S(T)}{V} \tag{4}$$

where I^m, G^m, and S^m are the rates of molar influx of the respective chemical species, V is the volume of the reactor, $E(T)$ is the rate of volume efflux, β is the reaction rate constant, and $H(T)$ is a number representing the volume fraction of the reaction chamber that is homogeneous at time T. The influx and efflux of the reactor are time-dependent, because the reactor must close off its input and output periodically in order to mix its contents (*vide infra*). The variable $H(T)$ is needed because in a microfluidic system we cannot assume that the contents of the reactor are always perfectly mixed. New substrate that comes into the system during the period of influx must be mixed before it may react with the gates in the system. This allows for separate influx streams for new gates and for substrates and input molecules. It also allows for the possibility that new gates never enter or leave the system at all; instead, they could be attached to beads which cannot escape semi-permeable membranes at the entrances and exits to the chamber. In either case, only that portion of the total substrate in the chamber that has been mixed with the solution containing the gates may react. The specifics of how the efflux and the homogeneity of the system are calculated are discussed in the next two sections.

3   Microfluidics

In order to simulate an open microfluidic reaction system, we must first analyze the properties of such a system. First, and most obviously, the size of a microfluidic reaction chamber is dramatically small compared to the size of a more conventional open reaction chamber, such as a CSTR. The volume of the smallest CSTR that can be readily assembled is on the order of 50 mL (our previous work used 500 mL), while the volume of a microfluidic reaction chamber is often on the order of 5 nL—a difference of seven orders of magnitude. The reaction chamber we chose for our simulation has a volume of 7.54 nL. This very small volume allows us to have very high concentrations of gate, substrate, input, and product molecules, while keeping the actual number of molecules in the system low.

Fluid flow in microfluidic channels and reaction chambers is different from the flow in a large-scale system because of the very small volumes involved. Namely, the flow is laminar, i.e., there is no turbulence (the Reynolds number of the flowing liquids is typically well below 100). This presents a peculiar challenge: two fluids flowing side by side in a microfluidic channel do not mix except by diffusion, which is a very slow process, but the fluid already in an open reaction chamber must mix quickly with new fluid flowing into the chamber, which contains new supplies of substrates, inputs, and gates, to allow the reaction to continue. This necessitates the use of an active microfluidic mixer for our reaction chamber, to speed up the mixing of the fluids greatly over normal mixing by diffusion.

We have chosen a microfluidic rotary pump to act as our open reaction chamber [6]. This device is an active mixer, mixing the solution within it by pumping it in a circular loop. The design of the device is shown in Figure 1. It consists of a bottom layer with fluid channels, and a top layer with pneumatic actuation channels. Both layers are fabricated with multilayer soft lithography [7]. One input channel in the bottom layer is used for substrate and input influx, while the other channel is used for gate influx—this

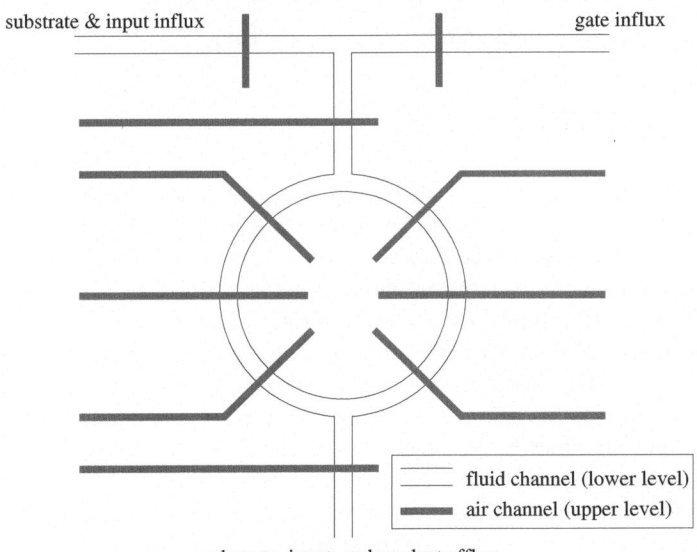

substrate, input, and product efflux

Fig. 1. The rotary mixer. The air channels form microvalves wherever they intersect with the fluid channels.

separation is to keep the substrate and gates from reacting before they have entered the reaction chamber. The pneumatic actuation channels on the top layer form microvalves wherever they intersect with the fluid channels on the bottom layer. A valve is closed when an air channel is pressurized and open when it is not. The actual reaction chamber is the central loop in the diagram. Actuating the valves around the perimeter of the loop in a certain sequence peristaltically pumps the fluid inside either clockwise or counter-clockwise. The frequency of actuation controls the speed at which the fluid rotates.

Continuous-flow mixing is possible with this reaction chamber [6], but it is not feasi-ble for our purposes for two reasons. The first is that the mixer does not completely mix objects with relatively low diffusion constants, such as very large molecules and 1 μm beads, when the flow is continuous. An experiment was performed [6] in which there was a continuous flow through the mixer of one solution containing dye and another solution containing beads. The two solutions entered the mixer side by side in the en-trance channel, flowing laminarly. In the fluid exiting the mixer, the dye was completely mixed, but only one quarter of the beads had crossed over to the other side of the fluid channel. Even if sufficient mixing of oligonucleotide molecules of the size we currently use could be achieved by using a low flow rate or widening and lengthening the mixer loop, this is not conducive to the possibility of attaching gates to beads, so that they may be kept always in the chamber by using semi-permeable membranes. The second problem is that the flow rate required for continuous-flow operation would have to be unreasonably low, in order to allow the gates involved to produce product molecules faster than they are removed from the system. Therefore, our model of the rotary mix-ing chamber uses two discrete, alternating phases: an influx and efflux, or "charging" phase, during which the valves at the chamber entrance and exit are open and the rotary pump is not operating, and a mixing phase, during which the valves at the entrance and exit of the chamber are closed and the pump is operating.

4   Mixing and Diffusion

Through a combination of factors, the rotary pumping in the mixing chamber greatly increases mixing speed compared to spontaneous diffusion. The time it takes to mix fluids is not negligible, however, and so we must examine how it works, and model its operation in our micro-system simulation. The parabolic flow profile present in mi-crofluidic channels (the fluid in the middle moves much faster than the fluid on the very edge, which is stationary) causes interface elongation, which, combined with the shallow channel depth, causes the mixing substances to fold around one another [6]. Where once the two fluids being mixed were completely separated, one in one half of the chamber and the other in the other half, after sufficient mixing time the width of the channel holds many alternating sections ("folds") of the two fluids. The two fluids still mix via diffusion, but folding them around each other greatly reduces the distance across which molecules from one fluid must diffuse into the other.

We can think of a substance as being completely homogeneous in the chamber when enough of that substance has diffused, from the fluid it was in originally, across a char-acteristic distance l, which is the farthest the substance must penetrate into the second fluid. Initially, we have $l_0 = r_0$, where r_0 is half the width of the channel that forms the mixing chamber. This is because we can assume that initially, when there is perhaps one fold in the chamber, the two liquids are side by side, with one liquid filling up half of the channel and the other filling up the other half. In order for a substance to be completely mixed in this situation, it must diffuse from its liquid all the way across half the width of the channel, until it reaches the far edge of the second solution at the chamber's wall. As the mixer continues running, however, the characteristic distance over which the fluids must diffuse to mix is reduced proportionally to the number of rotations, because of the liquids' folding around each other. Specifically, we have $l = l_0/kt$, where k is a constant coefficient determined by the total length of the loop and the pumping speed [6].

Knowing how the maximum characteristic diffusion distance changes over time, it is possible to model the mixing of the system using a diffusion equation. We use an equation which models diffusion of a substance in a fluid that is extended in all dimensions, where the substance is initially confined in one dimension in the region $-h < x < +h$. The regions from $-h$ to $-\infty$ and from $+h$ to $+\infty$ contain fluid with zero initial concentration of the diffusing substance. The substance is free to diffuse in either direction—solutions may be found for negative and positive values of x. The equation is:

$$C(x,t) = \frac{1}{2}C_0\left\{\text{erf}\frac{x-h}{2\sqrt{Dt}} + \text{erf}\frac{x+h}{2\sqrt{Dt}}\right\} \tag{5}$$

where $C(x,t)$ is the concentration of the diffusing substance at location x and time t, C_0 is the concentration initially within the region $-h < x < +h$, D is the diffusion constant of the diffusing substance, and erf is the standard mathematical error function (erf $z = \frac{2}{\sqrt{\pi}}\int_0^z \exp(-\eta^2)d\eta$) [8]. Because the liquids are folding around each other, both h, which bounds the fluid the substance must diffuse out of, and the farthest distance $x = h+l$ to which it must diffuse, are time-dependent. We already know that $l = l_0/kt$, and, since we shall assume that the two fluids have equal-size folds at any given time t, we know that $h = l$.

The only problem with using these equations to model our rotary mixer is that we do not know what the constant k is in the equation for the length of diffusion l. We do know, however, from empirical evidence [6], that at a certain pumping speed it takes 30 seconds to completely mix a solution containing dye with a solution containing $1\,\mu$m beads. We can use this fact to estimate k by noting the value of k for which the concentration of diffusing beads at the maximum mixing distance l is approximately equal to the concentration of beads in the middle of the fluid containing them originally (at $x = 0$) at time $t = 30$ s. Conservatively, we choose to focus on the beads for determining when the fluids are completely mixed because they have a diffusion constant that is much lower than the dye, and thus they diffuse much more slowly. The diffusion constant of the beads is $D = 2.5 \cdot 10^{-9}$ cm^{2}s^{-1}. We find that the concentrations are 97.72% the same when $k = 2$. We do not attempt to get the concentrations to be 100% equivalent, because we realize that the diffusion equation becomes less accurate at the boundary condition at the end of the mixing process, since it assumes that the fluid extends infinitely and substance does not diffuse completely during the duration of the experiment. Also, it is much safer for our purposes to underestimate k than overestimate it, as an underestimate leads to slower mixing, which has the potential to disrupt the kinetics of our chemical

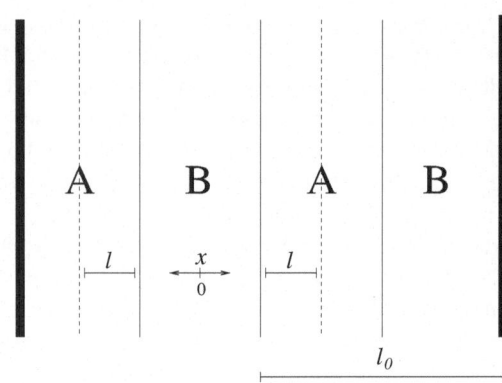

Fig. 2. Folds in a section of the mixer channel

system. We shall see, however, that it does not disrupt it enough to cause the logic that the gates perform to break down.

Using our value of $k = 2$, and the equations for the characteristic length of diffusion and the concentration of a diffusing substance at time t and position x, we can simulate the mixing chamber. There are no beads involved in our experiments; rather, we are only mixing fluids with gate, substrate, and input molecules. So, in accordance with the length of our oligonucleotide strands, we use the diffusion constant for a DNA 50-mer, which is $1.8 \cdot 10^{-7}\,\mathrm{cm^2 s^{-1}}$, in our mixing simulation. The mixing affects the differential equations describing the kinetics of the chemical system within the chamber by way of $H(T)$, which is a function of time (see Section 2). This function returns the fraction of the reaction chamber which is mixed. As noted earlier, during an experiment the rotary mixer alternates spending time in a charging phase, where there is an influx of new substrate, input, and gate molecules and an efflux of homogeneous solution, and a mixing phase, where the influx and efflux valves are closed and the rotary pump is turned on.

5   A Flip-Flop

Now that we can model the microfluidic mixing chamber, we must implement interesting logic in it using networks of deoxyribozyme-based logic gates. Since we are using an open system, we can create circuits which have persistent information that can be accessed and changed over time. The simplest such digital circuit is the *flip-flop*. A flip-flop is a bistable system which represents a single bit of memory. It can be commanded to *set* or *reset* this bit, which causes it to enter its high or low stable state, respectively, or to simply store, or *hold*, the bit in memory, in which case it stays in the state that it was last set or reset to.

We simulated the operation of a biochemical flip-flop within our modeled microfluidic mixing chamber. The flip-flop was implemented as a network, shown in Figure 3, of two deoxyribozyme-based NOT gates connected in a cycle of inhibition [5]. In this system there is no influx of input molecules, only of substrate molecules. We use

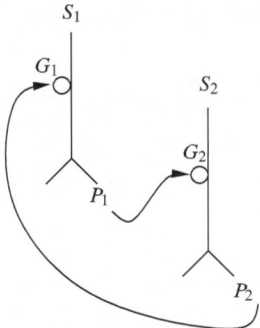

Fig. 3. The flip-flop reaction network

the substrate molecules themselves to control the behavior of the flip-flop. A high concentration of substrate S_2 signifies a set command; a high concentration of substrate S_1 signifies a reset command; and a high concentration of both substrates is used as the hold command. The first gate, G_1, can only cleave substrate S_1, and produces product P_1. The product P_1, in turn, acts as the input molecule for the second NOT gate, G_2, inhibiting its operation. When there is little or no P_1, the second gate G_2 is active, and it cleaves substrate S_2 to produce product P_2, which acts to inhibit the operation of the first gate, G_1. We measure output from the flip-flop in terms of the concentration of the cleaved product P_2, with high or low concentrations corresponding to a logical one or zero, respectively. It is apparent that the commands of set, reset, and hold we mentioned earlier will perform correctly with this inhibition cycle, with certain parameters. If only substrate S_1 is present in the system, only product P_1 and no P_2 will be produced—this corresponds to the reset command. If only S_2 is in the system, only product P_2 will be produced—this corresponds to the set command. However, if both S_1 and S_2 are in the system, we will stay at whatever state we were at previously, because whichever gate was originally producing more product than the other will inhibit the operation of the other gate, and will itself become less inhibited as a result, and thus eventually will become the only operating gate—this corresponds to the hold state. This operation requires that the concentrations of the gates are equal, for symmetry, and also that the efflux of the system is not greater than the rate at which the gates can produce product, so product is not being removed faster than it is being created.

The details of this bistable flip-flop system in a CSTR were examined thoroughly in previous work [5]. In the case of implementing this gate network in a microfluidic rotary mixer, we first define $S_1^m(T)$ and $S_2^m(T)$ to be the variable *molecular* influx of the substrates at time T, with which the flip-flop is controlled. The variable molecular influx of gate molecules, which enter the reactor in a separate stream from the substrate and input molecules, is given by $G_1^m(T)$ and $G_2^m(T)$. The rate of efflux is given by $E(T)$, and is time-dependent, because the system only has influx and efflux during its charging phase, and not during its mixing phase. We define $G_1(T)$, $G_2(T)$, $P_1(T)$, $P_2(T)$, $S_1(T)$, and $S_2(T)$ to be the concentrations within the reactor at time T of gate 1, gate 2, product 1, product 2, substrate 1, and substrate 2, respectively.

We can now represent the dynamics of the flip-flop system with a set of six coupled differential equations:

$$\frac{dG_1}{dT} = \frac{G_1^m(T) - E(T)G_1(T)}{V} \tag{6}$$

$$\frac{dG_2}{dT} = \frac{G_2^m(T) - E(T)G_2(T)}{V} \tag{7}$$

$$\frac{dP_1}{dT} = \beta_1 H(T)S_1(T)\max(0, G_1(T) - P_2(T)) - \frac{E(T)P_1(T)}{V} \tag{8}$$

$$\frac{dP_2}{dT} = \beta_2 H(T)S_2(T)\max(0, G_2(T) - P_1(T)) - \frac{E(T)P_2(T)}{V} \tag{9}$$

$$\frac{dS_1}{dT} = \frac{S_1^m(T)}{V} - \beta_1 H(T)S_1(T)\max(0, G_1(T) - P_2(T)) - \frac{E(T)S_1(T)}{V} \tag{10}$$

$$\frac{dS_2}{dT} = \frac{S_2^m(T)}{V} - \beta_2 H(T)S_2(T)\max(0, G_2(T) - P_1(T)) - \frac{E(T)S_2(T)}{V} \tag{11}$$

where β_1 and β_2 are the reaction rate constants, V is the volume of the reactor, and $H(T)$ is the fraction of the substrate molecules in the chamber which have been mixed (these are the only ones available to react).

In order to achieve flip-flop behavior with this system, we must find appropriate values for the system's efflux, the mixing rate, and the time spent by the system in its mixing phase and charging phase. We fix our mixer's high efflux at $0.12\,\mathrm{nL\,s^{-1}}$. During the charging phase, the mixer has this high efflux value, while during the mixing phase, the efflux is 0. The influx of the mixer is the same as the efflux, to maintain constant volume. We fix the mixing rate based on our empirically determined value for the constant k, which directly controls the mixing speed by determining the number of folds the mixer produces in a given amount of time. This value could be significantly adjusted in reality, as k simply depends on the length of the mixing channel and the speed of the pumping; our value of $k = 2$ reflects what we have determined to be one realistic value. With the efflux and mixing rate fixed, the only variable affecting the operation of the flip-flop is the time the mixing chamber spends in its charging and mixing phases. We find empirically that it works very well to spend 15 seconds in the charging phase and 15 seconds in the mixing phase.

With these parameters, Figure 4 shows the system of equations numerically integrated over a period of $1.2 \cdot 10^4\,\mathrm{s}$. The concentration of each type of gate molecule in the chamber was held steady at 130 nM, with the molecular influx of gates always matching the efflux of gates. We move the system from set, to hold, to reset at $2.5 \cdot 10^3\,\mathrm{s}$ intervals. The rapid, shallow oscillations in product concentration are due to the alternating, discrete sections of charging and mixing the system experiences.

Figure 5 shows the flip-flop switching between the set and reset commands at its maximum rate of speed. This rate was determined in our simulation to be about 900 seconds given to each command. This is over 65 times faster than simulations showed the flip-flop's maximum switching rate to be in the CSTR. We should also note that the

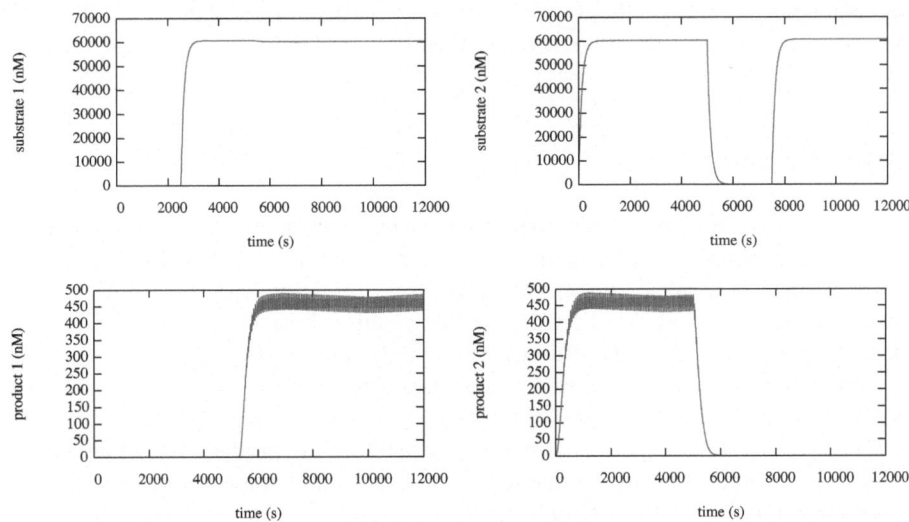

Fig. 4. The flip-flop moved from set, to hold, to reset commands at 2500 s intervals

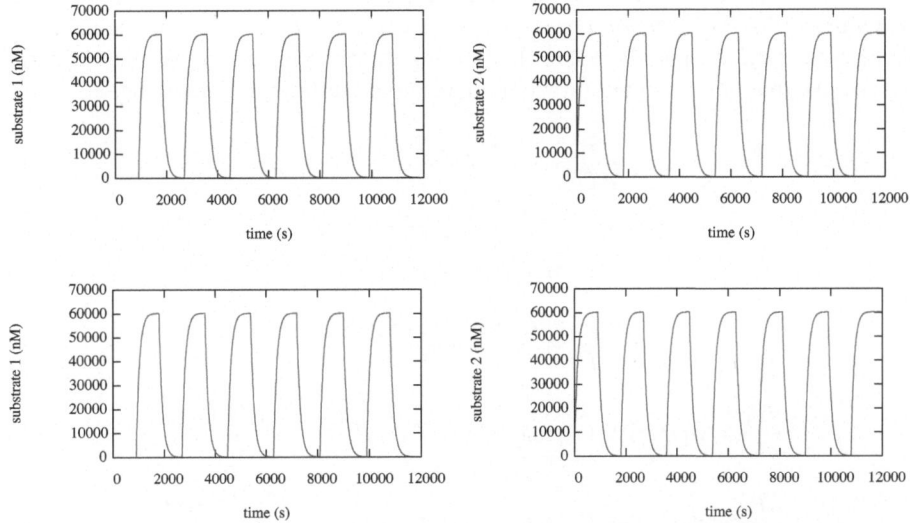

Fig. 5. The flip-flop operating at its maximum switching speed

concentration of substrate within the reaction chamber is a factor of 10 higher than in the CSTR simulation. Because the volume of our mixing chamber is over 7 orders of magnitude smaller than the volume of the CSTR, however, and our flow rate is 5 orders of magnitude lower, the total number of moles of substrate used in the microfluidic simulation is vastly lower than in the CSTR simulation. In fact, the molecular influx of a high substrate signal is only about $7.29\,\mathrm{fmol\,s^{-1}}$. Thus, in the span of a $1.2 \times 10^4\,\mathrm{s}$ experiment (a little over three hours), less than two tenths of a nanomole of substrate is used.

6   An Oscillator

If we increase the number of enzymatic NOT gates in our microfluidic reaction chamber to any odd number greater than one, we can create a biochemical oscillator. We will focus on a network of three NOT gates for simplicity. The three gates are, as before, connected in a cycle of inhibition. We require three different substrates, one matching each gate. Each gate cleaves its substrate into a unique product which inhibits one other gate. Gate G_1 cleaves substrate S_1 to produce product P_1, which acts as input to gate G_2, inhibiting it, while gate G_2 cleaves S_2 to produce P_2, which inhibits gate G_3, and finally gate G_3 cleaves the substrate S_3 to produce P_3, which inhibits gate G_1. As before, there will be one input stream which is a mixed solution containing the three types of substrate molecules, and another stream containing fresh gate molecules. The output of the system will be a solution containing only substrate and product molecules.

We define $G_1(T)$, $G_2(T)$, $G_3(T)$, $S_1(T)$, $S_2(T)$, $S_3(T)$, $P_1(T)$, $P_2(T)$, and $P_3(T)$ to be the concentrations within the reactor at time T of the gates, substrates, and products. We define $G_1^m(T)$, $G_2^m(T)$, $G_3^m(T)$, $S_1^m(T)$, $S_2^m(T)$, and $S_3^m(T)$ to be the molecular influx rate of each species which is replenished during the charging phase. We may describe the system dynamics with the following nine coupled differential equations:

$$\frac{dG_1}{dT} = \frac{G_1^m(T) - E(T)G_1(T)}{V} \tag{12}$$

$$\frac{dG_2}{dT} = \frac{G_2^m(T) - E(T)G_2(T)}{V} \tag{13}$$

$$\frac{dG_3}{dT} = \frac{G_3^m(T) - E(T)G_3(T)}{V} \tag{14}$$

$$\frac{dP_1}{dT} = \beta_1 H(T)S_1(T)\max(0, G_1(T) - P_3(T)) - \frac{E(T)P_1(T)}{V} \tag{15}$$

$$\frac{dP_2}{dT} = \beta_2 H(T)S_2(T)\max(0, G_2(T) - P_1(T)) - \frac{E(T)P_2(T)}{V} \tag{16}$$

$$\frac{dP_3}{dT} = \beta_3 H(T)S_3(T)\max(0, G_3(T) - P_2(T)) - \frac{E(T)P_3(T)}{V} \tag{17}$$

$$\frac{dS_1}{dT} = \frac{S_1^m(T)}{V} - \beta_1 H(T)S_1(T)\max(0, G_1(T) - P_3(T)) - \frac{E(T)S_1(T)}{V} \tag{18}$$

$$\frac{dS_2}{dT} = \frac{S_2^m(T)}{V} - \beta_2 H(T)S_2(T)\max(0, G_2(T) - P_1(T)) - \frac{E(T)S_2(T)}{V} \tag{19}$$

$$\frac{dS_3}{dT} = \frac{S_3^m(T)}{V} - \beta_3 H(T)S_3(T)\max(0, G_3(T) - P_2(T)) - \frac{E(T)S_3(T)}{V} \qquad (20)$$

where β_1, β_2, and β_3 are the reaction rate constants, V is the volume of the reactor, $E(T)$ is the time-dependent volumetric efflux, and $H(T)$ is the fraction of the reaction chamber which is homogeneous at time T.

The conditions under which the oscillator will oscillate in a CSTR have been examined previously [5]. To simplify things, this examination assumed that the concentration of substrate molecules in the chamber was constant, because, although these concentrations do oscillate, they are always much higher than the oscillating concentrations of the products. Using this assumption, linear approximations can be made to explicitly solve the differential equations for the oscillating product concentrations. These approximations give us a way to specify the center and period of the oscillations by setting an appropriate influx of substrate molecules and an appropriate concentration of gates. Our circumstances differ from the CSTR in that the efflux alternates between off and on, and the system is almost never completely homogeneous. We recognize that the system is never less than 76% homogeneous at any given time, however, and so it is reasonable to assume constant, complete homogeneity, and constant efflux, in order to use the approximation from our previous work as a starting point for specifying the period and center of the oscillator.

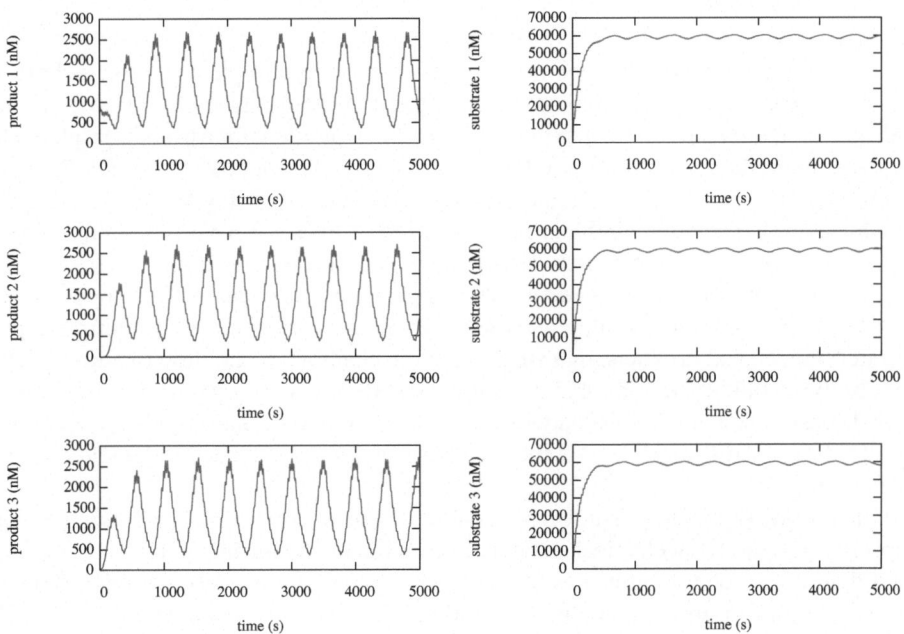

Fig. 6. The oscillator system operating with a period of 480 s and a center of 1.5 μM

We set the efflux rate for the charging cycle equal to the rate we used for the flip-flop, $0.12\,\text{nL s}^{-1}$. We use the same period (15 seconds in the charging phase and 15 seconds in the mixing phase) which worked well for the flip-flop. Based on the efflux rate, we use the linear approximations derived from the CSTR simulation research to calculate an estimate for the gate concentration and substrate influx needed for oscillations of period 250 seconds, centered at $1\,\mu\text{M}$. We find we should keep each of the gate concentrations steady at $1500\,\text{nM}$, while the molecular influx for each substrate should be set to $7.29 \times 10^{-6}\,\text{nM s}^{-1}$. Figure 6 shows the results of integration over a 5000 second period with these initial values. We can see that the actual period is 480 seconds, and the actual center is close to $1.5\,\mu\text{M}$. The linear approximations were off by about 20% in the CSTR simulation; in our simulation, the period estimation is just over half the actual period, and the center estimation is off by about 50%. There are two reasons for this. One is the fact that we assumed our efflux rate and reactor homogeneity to be constant in order to use the same approximations that worked in the CSTR setting. Another, more instrumental reason stems from the fact that reactions happen much more quickly in our microfluidic system, since we have a much higher concentration of reagents. This causes the nonlinear terms that are not taken into account in the linear approximations to become much more prominent. More analysis is required to find a more accurate way to specify the period and center of our oscillations.

7   Related Work

Microfluidics has previously been proposed as a laboratory implementation technique for automating DNA-based combinatorial computation algorithms [9,10,11]. McCaskill and van Noort have solved the maximum clique graph problem for a 6-node graph in the lab using microfluidic networks and DNA [12,13,14]. Their approach uses DNA not as an enzyme but as an easily selectable carrier of information (using Watson-Crick base pair matching). The computational network which solves the maximum clique problem requires a large number of micro-channels, proportional to the number of gates in the system, which grows as the number of graph nodes squared. Our approach, in contrast, may allow one to implement complex logic, performed with multiple types of gates, inputs, and products, in a single reaction chamber, in addition to allowing the possibility of linking several chambers together. Recently, van Noort and McCaskill have discussed systematic flow pattern solutions in support of microfluidic network design [15]; it remains to be seen if these techniques can be extended to handle designs such as ours.

Other work shows that it is even possible to use microfluidics for computational purposes as a purely mechanical substrate, i.e., without chemical reactions [16,17,18]. That fluidics can be used thus has been known for a long time [19], but microfluidics for the first time offers the potential for building relatively complex devices [20,21,22].

Microfluidic mixing is a difficult problem. While we have opted for the rotary mixing chamber design as one for which modeling the kinetics of mixing is within reach, other designs have been proposed; droplet-based mixing [23,24,25] is especially attractive [26]. Analysis of mixing remains a challenging problem [27,28]. Related to mixing, or achieving uniform concentration, is the problem of achieving particular spatiotemporally nonuniform concentrations [29, 30, 31].

Numerous oscillatory chemical and biochemical processes have been reported in the past decades, starting with the famous Belousov-Zhabotinsky reaction [32, 33, 34, 35], via studies of hypothetical systems of coupled chemical reactions (some even intended as computational devices) [36, 37, 38, 39, 40, 41, 42, 43, 44], to the recent remarkable demonstration by Elowitz and Leibler of a gene transcription oscillatory network [45].

8 Conclusions

Networks of deoxyribozyme-based logic gates can function correctly in a microfluidic environment. This is the first feasible setting in which open-reactor experiments using these gates may be conducted in the laboratory. The immediate and obvious advantage of this approach, compared to using a larger open reactor, is a massive savings of cost and time. Our simulations of a flip-flop and an oscillator in such a setting show that useful microfluidic experiments could be conducted in mere hours, rather than the days or weeks it would take to see results in a large, continuous-flow stirred tank reactor. Perhaps most significantly, the extremely small volume of a microfluidic reactor means that a three-hour experiment could cost less than $50 in reagents, even though deoxyribozyme-based gates and the oligonucleotide substrates and inputs which they react with can cost as much as $40 per nanomole. The materials cost for the flip-flop experiment can thus be around $1,000; the cost of microfluidic chip fabrication is estimated at $20,000 [S. Han, personal communication], assuming an existing facility.

Our microfluidic reaction chambers are also very conducive to being networked together, with control logic outside the system operating valves on the channels connecting them. We will investigate the possibility of attaching gate molecules to beads, and keeping them within a chamber by placing semi-permeable membranes at the chamber entrances and exits. With such a system, we could keep discrete sections of logic separate from each other when desired, and redirect outputs and inputs selectively. This may be especially useful if certain types of gates whose logic we wish to connect actually conflict undesirably with each other if they are placed in the same chamber (by partially binding to each others' input or substrate molecules, for example). We believe that using microfluidic rotary mixing chambers to implement complex logic with deoxyribozyme-based gates in actual laboratory experiments is the first step toward completely understanding their potential, and eventually even deploying them in situations as complex as living cells.

Acknowledgments

We are grateful to Plamen Atanassov, Sang Han, Elebeoba May, Sergei Rudchenko, and Milan Stojanovic for helpful advice, to Clint Morgan for his simulation code, and to the anonymous reviewers for their detailed comments, and especially for alerting us to the work of van Noort. This material is based upon work supported by the National Science Foundation (grants CCR-0219587, CCR-0085792, EIA-0218262, EIA-0238027, and EIA-0324845), Sandia National Laboratories, Microsoft Research, and Hewlett-Packard (gift 88425.1). Any opinions, findings, and conclusions or recommendations expressed in this material are those of the author(s) and do not necessarily reflect the views of the sponsors.

References

1. Stojanovic, M.N., Mitchell, T.E., Stefanovic, D.: Deoxyribozyme-based logic gates. Journal of the American Chemical Society **124** (2002) 3555–3561
2. Stojanovic, M.N., Stefanovic, D.: Deoxyribozyme-based half adder. Journal of the American Chemical Society **125** (2003) 6673–6676
3. Stojanovic, M.N., Stefanovic, D.: A deoxyribozyme-based molecular automaton. Nature Biotechnology **21** (2003) 1069–1074
4. Stojanovic, M.N., Semova, S., Kolpashchikov, D., int Morgan, C., Stefanovic, D.: Deoxyribozyme-based ligase logic gates and their initial circuits. Journal of the American Chemical Society **127** (2005) 6914–6915
5. Morgan, C., Stefanovic, D., Moore, C., Stojanovic, M.N.: Building the components for a biomolecular computer. In: DNA Computing: 10th International Meeting on DNA-Based Computers. (2004)
6. Chou, H.P., Unger, M.A., Quake, S.R.: A microfabricated rotary pump. Biomedical Microdevices **3** (2001) 323–330
7. Unger, M.A., Chou, H.P., Thorsen, T., Scherer, A., Quake, S.R.: Monolithic microfabricated valves and pumps by multilayer soft lithography. Science **298** (2000) 580–584
8. Crank, J.: The Mathematics of Diffusion. Second edn. Oxford University Press (1975)
9. Livstone, M.S., Landweber, L.F.: Mathematical considerations in the design of microreactor-based DNA computers. In: DNA Computing: 9th International Meeting on DNA-Based Computers. Volume 2943 of Lecture Notes in Computer Science., Madison, Wisconsin, Springer-Verlag (2003) 180–189
10. Gehani, A., Reif, J.: Micro-flow bio-molecular computation. BioSystems **52** (1999)
11. Ledesma, L., Manrique, D., Rodríguez-Patón, A., Silva, A.: A tissue P system and a DNA microfluidic device for solving the shortest common superstring problem. (2004)
12. van Noort, D., Gast, F.U., McCaskill, J.S.: DNA computing in microreactors. In: DNA Computing: 7th International Meeting on DNA-Based Computers. (2001)
13. Wagler, P., van Noort, D., McCaskill, J.S.: Dna computing in microreactors. Proceedings of SPIE **4590** (2001) 6–13
14. van Noort, D., Wagler, P., McCaskill, J.S.: Hybrid poly(dimethylsoloxane)-silicon microreactors used for molecular computing. Smart Materials and Structures **11** (2004) 756–760
15. van Noort, D., McCaskill, J.S.: Flows in micro fluidic networks: From theory to experiment. Natural Computing **3** (2004) 395–410
16. Chiu, D.T., Pezzoli, E., Wu, H., Stroock, A.D., Whitesides, G.M.: Using three-dimensional microfluidic networks for solving computationally hard problems. Proceedings of the National Academy of Sciences of the USA **98** (2001) 2961–2966
17. Fuerstman, M.J., Deschatelets, P., Kane, R., Schwartz, A., Kenis, P.J.A., Deutsch, J.M., Whitesides, G.M.: Solving mazes using microfluidic networks. Langmuir **19** (2003) 4714–4722
18. Vestad, T., Marr, D.W.M., Munakata, T.: Flow resistance for microfluidic logic operations. Applied Physics Letters **84** (2004) 5074–5075
19. Foster, K., Parker, G.A., eds.: Fluidics: components and circuits. Wiley-Interscience, London and New York (1970)

20. Thorsen, T., Maerkl, S.J., Quake, S.R.: Microfluidic large-scale integration. Science **298** (2002) 580–584
21. Groisman, A., Enzelberger, M., Quake, S.R.: Microfluidic memory and control devices. Science **300** (2003) 955–958
22. Hong, J.W., Quake, S.R.: Integrated nanoliter systems. Nature Biotechnology **21** (2003) 1179–1183
23. Paik, P., Pamula, V.K., Fair, R.B.: Rapid droplet mixers for digital microfluidic systems. Lab on a Chip **3** (2003) 253–259
24. Tice, J.D., Song, H., Lyon, A.D., Ismagilov, R.F.: Formation of droplets and mixing in multiphase microfluidics at low values of the Reynolds and the capillary numbers. Langmuir **19** (2003) 9127–9133
25. Song, H., Tice, J.D., Ismagilov, R.F.: A microfluidic system for controlling reaction networks in time. Angewandte Chemie International Edition **42** (2003) 768–772
26. Gerdts, C.J., Sharoyan, D.E., Ismagilov, R.F.: A synthetic reaction network: Chemical amplification using nonequilibrium autocatalytic reactions coupled in time. Journal of the American Chemical Society **126** (2004) 6327–6331
27. Wiggins, S.: Integrated nanoliter systems. Nature Biotechnology **21** (2003) 1179–1183
28. Solomon, T.H., Mezić, I.: Uniform resonant chaotic mixing in fluid flows. Nature **425** (2003) 376–380
29. Jeon, N.L., Dertinger, S.K.W., Chiu, D.T., Choi, I.S., Stroock, A.D., Whitesides, G.M.: Generation of solution and surface gradients using microfluidic systems. Langmuir **16** (2000) 8311–8316
30. Dertinger, S.K.W., Chiu, D.T., Jeon, N.L., Whitesides, G.M.: Generation of gradients having complex shapes using microfluidic networks. Analytical Chemistry **73** (2001) 1240–1246
31. Jeon, N.L., Bakaran, H., Dertinger, S.K.W., Whitesides, G.M., Van De Water, L., Toner, M.: Neutrophil chemotaxis in linear and complex gradients of interleukin-8 formed in a microfabricated device. Nature Biotechnology **20** (2002) 826–830
32. Field, R.J., Körös, E., Noyes, R.: Oscillations in chemical systems. II. Thorough analysis of temporal oscillation in the bromate-cerium-malonic acid system. Journal of the American Chemical Society **94** (1972) 8649–8664
33. Noyes, R., Field, R.J., Körös, E.: Oscillations in chemical systems. I. Detailed mechanism in a system showing temporal oscillations. Journal of the American Chemical Society **94** (1972) 1394–1395
34. Tyson, J.J.: The Belousov-Zhabotinskii Reaction. Volume 10 of Lecture Notes in Biomathematics. Springer-Verlag, Berlin (1976)
35. Epstein, I.R., Pojman, J.A.: An Introduction to Nonlinear Chemical Dynamics. Oxford University Press, New York (1998)
36. Hjelmfelt, A., Ross, J.: Chemical implementation and thermodynamics of collective neural networks. Proceedings of the National Academy of Sciences of the USA **89** (1992) 388–391
37. Hjelmfelt, A., Ross, J.: Pattern recognition, chaos, and multiplicity in neural networks of excitable systems. Proceedings of the National Academy of Sciences of the USA **91** (1994) 63–67
38. Hjelmfelt, A., Schneider, F.W., Ross, J.: Pattern recognition in coupled chemical kinetic systems. Science **260** (1993) 335–337
39. Hjelmfelt, A., Weinberger, E.D., Ross, J.: Chemical implementation of neural networks and Turing machines. Proceedings of the National Academy of Sciences of the USA **88** (1991) 10983–10987
40. Hjelmfelt, A., Weinberger, E.D., Ross, J.: Chemical implementation of finite-state machines. Proceedings of the National Academy of Sciences of the USA **89** (1992) 383–387
41. Laplante, J.P., Pemberton, M., Hjelmfelt, A., Ross, J.: Experiments on pattern recognition by chemical kinetics. The Journal of Physical Chemistry **99** (1995) 10063–10065

42. Rössler, O.E.: A principle for chemical multivibration. Journal of Theoretical Biology **36** (1972) 413–417

43. Rössler, O.E., Seelig, F.F.: A Rashevsky-Turing system as a two-cellular flip-flop. Zeitschrift für Naturforschung **27 b** (1972) 1444–1448

44. Seelig, F.F., Rössler, O.E.: Model of a chemical reaction flip-flop with one unique switching input. Zeitschrift für Naturforschung **27 b** (1972) 1441–1444

45. Elowitz, M.B., Leibler, S.: A synthetic oscillatory network of transcriptional regulators. Nature **403** (2000) 335–338

DNA Recombination by XPCR

Giuditta Franco[1], Vincenzo Manca[1], Cinzia Giagulli[2], and Carlo Laudanna[2]

[1] Department of Computer Science, University of Verona, Italy
franco@sci.univr.it, vincenzo.manca@univr.it
[2] Section of General Pathology, Department of Pathology, University of Verona, Italy
cinzia.giagulli@univr.it, carlo.laudanna@univr.it

Abstract. The first step of the Adleman-Lipton extract model in DNA computing is the combinatorial generation of libraries. In this paper a new method is proposed for generating a initial pool, it is a quaternary recombination of strings via application of null context splicing rules. Its implementation, based on a kind of PCR called XPCR, results to be convenient with respect to the standard methods, in terms of efficiency, speed and feasibility. The generation algorithm we propose was tested by a lab experiment here described, since the presence of few sequences is enough for checking the completeness of the library. The simple technology of this approach is interesting in and of itself, and it can have many useful applications in biological contexts.

Keywords: SAT, PCR, XPCR, generation of DNA initial pool.

1 Introduction

DNA Computing is an emerging area where information is stored in biopolymers, and enzymes manipulate them in a massively parallel way according to strategies that can produce computationally universal operations [7, 1, 9]. The research in this field has already taken different pathways of theoretical and experimental interest, also developing into new sub-disciplines of science and engineering, for example nanotechnology and material design.

One of the ambition of DNA computation was solving NP complete problems, and since the exponential amount of DNA, typical of the Adleman-Lipton generate-and-search model, became a limit to the scale up of the procedures for problems of realistic size, then alternative approaches have been explored recently. For example, one of them is focused on building directly the solutions of the problem by means of 3D graph self-assembly [10], while another one evolves approximate solutions of instances of the NP-complete problems by means of the evolution of a population of finite state machines [13].

Nevertheless, the production of combinatorial libraries as the solution space of instances of SAT problem [9, 3, 12] is a typical recombination which is fundamental, not only to DNA computation, but to various types of in vitro selection experiments, for selecting new DNA or RNA enzymes (such as ribozymes), or for performing crossover of homologous genes [14] and mutagenesis.

A. Carbone and N.A. Pierce (Eds.): DNA11, LNCS 3892, pp. 55–66, 2006.

In [4] a special type of PCR called XPCR was introduced to implement a procedure for extracting, from an heterogeneous pool, all the strands containing a given substrand. The XPCR technique was tested in different situations and was shown working as expected [4]. Here we focus on the combinatorial generation of libraries, and propose an economic and efficient XPCR-based procedure for generating DNA solution spaces; a lab experiment is reported that confirm the correctness and the completeness of this method.

The classic two methods for initial pool generation were introduced in 1994. The *hybridization-ligation* method was introduced by Adleman in [1] for solving a Hamiltonian path problem; it links oligonucleotides hybridized by complementarity and ligase reaction. The other method, called *parallel overlap assembly (POA)* was introduced by Stemmer [14] to perform crossover between homologous sequences of genes. Its implementation was based on hybridization/polymerase extension, and it was applied successfully by Kaplan et al. in DNA computing for solving maximal clique problem [11]. The second method demands less time and is a better choice in terms of generation speed and material consumption than the first one [8].

More recently, *mix-and-split* method was introduced [6] to generate an initial RNA pool for the knight problem, that is a combinatorial library of binary numbers. It was used by Braich et al. in [2] to generate an initial pool for a 20-variable 3-SAT problem, the biggest instance solved in lab. It combines a chemical synthesis and an enzymatic extension method, in fact, two half libraries are combined by primer extension while each of them was synthesized chemically on two columns by repetitive mixing and splitting steps. Therefore this method takes the advantages of an extension method but performs a big part of the work by chemical synthesis, which can become quite expensive.

Finally, a modified version of PNA-mediated Whiplash PCR (PWPCR) was presented in [13], with an in vitro algorithm to evolve approximate solutions of Hamiltonian Path problem. The PWPCR procedure is basically implemented by the recursive polymerase extension of a mixture of DNA hairpins, and here a similar approach is pursued with an easier and cheaper technology.

Let us consider a solution space of dimension n, that is, given by all possible sequences of type $\alpha_1\alpha_2\cdots\alpha_n$, where α_i can be, for each i, one of two different strings X_i or Y_i.

Our method starts from four types of strands and, by using only polymerase extension, generates the whole DNA solution space, where all types of possible sequences of the pool are present. Therefore, with respect to the other methods it is very easy and cheap: it does not use PNA molecules and no chemical synthesis is required apart that one for the initial strands.

In next sections we give, with examples, the intuition behind our XPCR generation algorithm, and recall the XPRC procedure (for more details see [4]) which XPCR generation is based on. Then, the algorithm is described, and its implementation in laboratory is presented with an experiment which confirms the validity of the generation procedure. In a further paper [5] the mathematical correctness and completeness of the algorithm is proved in a formal setting. In particular, in the case of solution spaces of dimension n where each element of the sequence can be of k different types, the following general proposition holds.

Proposition 1. *Starting from four sequences and for any value of n, by XPCR generation algorithm all 2^n combinations are generated in $2(n-2)$ steps.*

Moreover in [5] we prove the existence of some special types of sequences, such that if they are present in the pool, then the pool contains all the solutions of the n-dimensional DNA solution space.

Our approach takes all the advantages and the efficiency of an enzymatic elongation method. Moreover, with respect to the POA, in any step of recombination, the unwanted side products are eliminated by the electrophoresis during the XPCR procedure, and the completeness of the library can be tested very easily by checking the presence of two special strands. The existence of such strands [5] is a consequence of the null context splicing theory [7], and it holds because the procedure is completely based on splicing rules.

Finally, the recombination of strings modeled by a null context splicing rule is performed in nature by few enzymes, thus it can be done for a very limited number of restriction sites, while the XPCR scales up this kind of recombination to any site.

2   Quaternary Recombination

The main intuition underlying the quaternary recombination was inspired by [2], where the following sequences were studied to avoid mismatch phenomena. We start from the sequences I_1, I_2, I_3, I_4, each constituted by n different strings that can occur in two distinct forms.

For the sake of simplicity we describe this method for a solution space of dimension 6. From a mathematical point of view, the method could be generalized to any dimension, and to cases where the variables can assume k values with $k > 2$. Consider the sequences:

1. *Positive:* $I_1 = X_1 X_2 X_3 X_4 X_5 X_6$
2. *Negative:* $I_2 = Y_1 Y_2 Y_3 Y_4 Y_5 Y_6$
3. *Positive-Negative:* $I_3 = X_1 Y_2 X_3 Y_4 X_5 Y_6$
4. *Negative-Positive:* $I_4 = Y_1 X_2 Y_3 X_4 Y_5 X_6$

Starting from these four initial sequences, every *solution* $\alpha_1 \alpha_2 \alpha_3 \alpha_4 \alpha_5 \alpha_6$ of the solution space is generated by means of *null context splicing rules* [7], that are of type

$$r_\gamma: \quad \phi \gamma \psi, \ \delta \gamma \eta \quad \longrightarrow \quad \phi \gamma \eta, \ \delta \gamma \psi$$

where ϕ, γ, ψ, δ, η are strings on the considered alphabet.

In fact, any string $\alpha_1 \alpha_2 \alpha_3 \alpha_4 \alpha_5 \alpha_6$ can be seen as the concatenation of substrings of I_1, I_2, I_3, I_4 that suitable splicing rules cut and recombine along common substrings. For example, the sequence $X_1 X_2 Y_3 X_4 Y_5 Y_6$, can be obtained starting from the initial ones in the following way:

1. $r_{X_2} : I_1, I_4 \longrightarrow X_1 X_2 Y_3 X_4 Y_5 X_6, Y_1 X_2 X_3 X_4 X_5 X_6$
2. $r_{Y_5} : I_2, X_1 X_2 Y_3 X_4 Y_5 X_6 \longrightarrow Y_1 Y_2 Y_3 Y_4 Y_5 X_6, \mathbf{X_1 X_2 Y_3 X_4 Y_5 Y_6}$

The generation of any solution can be associated to a *generation sequence* of rules, that is, a sequence on the alphabet $R = \{r_{X_2}, r_{X_3}, r_{X_4}, r_{X_5}, r_{Y_2}, r_{Y_3}, r_{Y_4}, r_{Y_5}\}$. For example, the generation sequence of the string $X_1 X_2 Y_3 X_4 Y_5 Y_6$ considered above is $r_{X_2} r_{Y_5}$. We observe that, in this context, the order of application of the rules is not relevant. For example, the same string $X_1 X_2 Y_3 X_4 Y_5 Y_6$ can be also obtained by permuting the application order of the rules (from the same initial strings):

1. $r_{Y_5} : I_4, I_2 \longrightarrow Y_1 X_2 Y_3 X_4 Y_5 Y_6, Y_1 Y_2 Y_3 Y_4 Y_5 X_6$
2. $r_{X_2} : I_1, Y_1 X_2 Y_3 X_4 Y_5 Y_6 \longrightarrow \mathbf{X_1 X_2 Y_3 X_4 Y_5 Y_6}, Y_1 X_2 X_3 X_4 X_5 X_6$

The following *Recombination Canonic Procedure* gives a general insight of the validity of the method outlined above.

- **input:** a sequence $\alpha_1 \alpha_2 \alpha_3 \ldots \alpha_n$
- **put** $i = 1$ and H_1 equal to the initial sequence (among the four ones) where the subsequence $\alpha_1 \alpha_2$ occurs
- **for** $i = 1, \ldots, n-2$, **do**
  begin
    • **put** L_i equal to the initial sequence where $\alpha_{i+1} \alpha_{i+2}$ occurs
    • **if** $H_i = L_i$ **then** put $H_{i+1} = H_i$
    • **else** put H_{i+1} equal to the first product of the null context splicing $r_{\alpha_{i+1}}$ applied to strings H_i, L_i.
  end
- **output:** the sequence of null context splicing rules that were applied during *for* instruction.

The sequence of rules given in output by the previous algorithm is a generation sequence for the input string. In fact, if we apply these rules to the initial pool, then we get a set of strings which surely contains the input string. In the general case of n variables, a generation sequence is at most $n-2$ long, since there can be only one occurrence of rules r_{α_i} for each i belonging to $\{2, 3, \ldots, n-1\}$.

To implement null context splicing rules r_γ we use a kind of PCR called $XPCR_\gamma$, which was introduced as tool of extraction in [4], and which we recall briefly in the next section.

2.1   XPCR Procedure

We suppose to have an heterogeneous pool of DNA double strands having the same length and sharing a common prefix α and a common suffix β. Given a specified string γ, by means of $XPCR_\gamma$, we can recombine all the strings of the pool that contain γ as substring.

The $XPCR_\gamma$ procedure is described by the following steps. We indicate by $PCR(\xi, \overline{\eta})$ a standard PCR performed by forward primer ξ and backward primer $\overline{\eta}$, where $\overline{\eta}$ is the reversed and complemented sequence of η.

- **input** a pool P of strings having α as prefix and β as suffix
- **split** P into P_1 and P_2 (with the same approximate size)
- **apply** $PCR(\alpha, \overline{\gamma})$ to P_1 and $PCR(\gamma, \overline{\beta})$ to P_2 (*cutting step*, see Figure 1)
- perform **electrophoresis** on P_1 and on P_2 to select γ-prefixed or γ-suffixed strings, that corresponds to eliminate the sequences of the initial length.
- **mix** the two pools resulting from the previous step in a new pool P
- **apply** $PCR(\alpha, \overline{\beta})$ to P (*recombination step*, see Figure 2)
- **output** the pool resulting from the previous step.

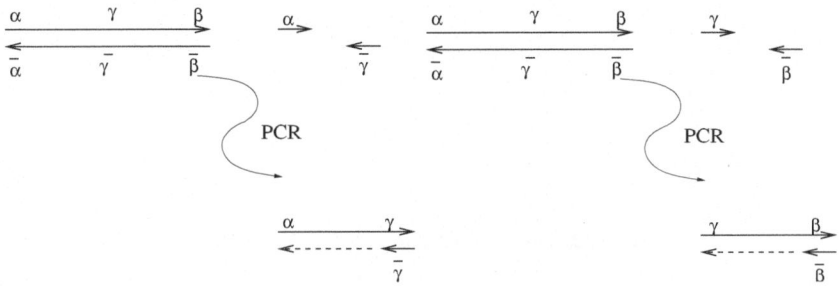

Fig. 1. Cutting step of $XPCR_\gamma$

In the recombination step, left parts $\alpha \cdots \gamma$ and right parts $\gamma \cdots \beta$ of the sequences of the pool having γ as subsequence are recombined in all possible ways, regardless to the specificity of the sequences between α and γ, or γ and β. Therefore, not only the whole sequences containing γ are restored but also new sequences are generated by recombination (see Figure 2).

Note that, if the hybridization between the two strands $\alpha \cdots \gamma$ and $\gamma \cdots \beta$ happens in a different location (other than at point γ), the 'wrong' product has not the length of the sequences in the initial pool, so it is eliminated by final electrophoresis [4].

However, in the XPCRs performed by the XPCR Recombination Algorithm described in the next section, the interactions of hybridization/polymerase extension between the strands $\alpha \cdots \gamma$ and $\gamma \cdots \beta$ can happen only along the codeword γ, because by construction they do not have more complementary regions (in fact, if γ represents the $i-th$ variable, than the strand $\alpha \cdots \gamma$ contains the codewords of the first i variables, and $\gamma \cdots \beta$ contains the complementary codewords of the last $n - i + 1$ variables).

Procedure $XPCR_\gamma$ implements a version of null context splicing rule with common substring γ, where strings are assumed to share a common prefix and a common suffix (both ϕ and δ start with a certain prefix, and both ψ and η end with a certain suffix) and strings obtained after their recombination are added to the strings available before it

$$r_\gamma: \quad \phi\gamma\psi, \ \delta\gamma\eta \quad \longrightarrow \quad \phi\gamma\eta, \ \delta\gamma\psi, \ \phi\gamma\psi, \ \delta\gamma\eta.$$

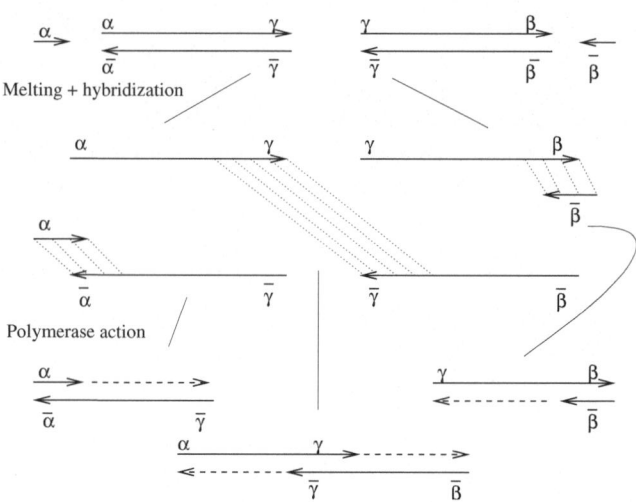

Fig. 2. Recombination step of $XPCR_\gamma$

2.2   XPCR Recombination Algorithm

We start with a pool having only the four initial sequences I_1, I_2, I_3, I_4 that are extended by a common prefix α and a common suffix β (for performing the XPCR procedure).

- **Let** P_1 and P_2 be two copies of the pool

$$\{\alpha \cdots I_1 \cdots \beta, \ \alpha \cdots I_2 \cdots \beta, \ \alpha \cdots I_3 \cdots \beta, \ \alpha \cdots I_4 \cdots \beta\}$$

- **for** $i = 2, 3, 4, 5$ **do**
  begin
  - **perform** $XPCR_{X_i}$ on P_1 and $XPCR_{Y_i}$ on P_2
  - **mix** the two pools obtained in the previous step in a unique pool P
  - **split** P randomly in two new pools P_1 and P_2 (with the same approximate size)
  end

If all steps are performed correctly we obtain a pool where all 2^6 combinations $\alpha_1\alpha_2 \cdots \alpha_6$ with $\alpha_i \in \{X_i, Y_i\}$ are present.

It can be proved that, for any dimension n of the solution space, two sequences exist such that their simultaneous presence in the pool guarantees that all the possible recombinations happened. The proof of this fact is based essentially on the following lemma (see [5] for the formal details).

Let us call *i-trio-factor* any substring $\alpha_{i-1}\alpha_i\alpha_{i+1}$ where exactly two consecutive variables are Xs or Ys, and consider the corresponding null context splicing rule r_{α_i}, then it holds that

Lemma 1. *For any* $i = 2, \ldots, n - 1$ *the rule* r_{α_i} *has been applied in the pool iff in the pool there is a string including a corresponding i-trio-factor.*

A set W of strings is a recombination witness if the strings of W include all the possible i-trio-factors. In our case, we have chosen the following strings that, according to the lemma, are clearly a set of recombination witnesses.

1. *XXYY alternating:* $W_1 = X_1 X_2 \, Y_3 Y_4 \, X_5 X_6$
2. *YYXX alternating:* $W_2 = Y_1 Y_2 \, X_3 X_4 \, Y_5 Y_6$

In order to verify that all the expected recombinations happened, in the experiment described in the next section, it was enough to check that in the final pool the sequences W_1 and W_2 were present.

3 Experiment

The following experiment showed that, given any SAT problem with 6 variables, the solution pool with sequences encoding all 64 assignments can be generated by starting from 4 specific sequences and by using only XPCR and electrophoresis. The success of this experiment is meaningful because it can be scaled up easily to any number n of variables: starting from 4 specific sequences, all 2^n combinations can be generated by XPCR and electrophoresis in a linear number of biosteps.

The experiment started on an initial pool containing the sequences 150bp long[1] (in the following all the sequences are written with respect to the usual $5' - 3'$ orientation):

$$S_1 = \alpha \, W_a \, X_1 \, X_2 \, X_3 \, X_4 \, X_5 \, X_6 \, Z_a \, \beta$$

$$S_2 = \alpha \, W_b \, Y_1 \, Y_2 \, Y_3 \, Y_4 \, Y_5 \, Y_6 \, Z_b \, \beta$$

$$S_3 = \alpha \, W_a \, X_1 \, Y_2 \, X_3 \, Y_4 \, X_5 \, Y_6 \, Z_b \, \beta$$

$$S_4 = \alpha \, W_b \, Y_1 \, X_2 \, Y_3 \, X_4 \, Y_5 \, X_6 \, Z_a \, \beta$$

where α and β, necessary to perform XPCR procedure, were 20b long[2], for $i = 1, 2, \ldots, 6$, X_i and Y_i, representing the boolean values of the variables of the problem, were 15b long[3], and the elongating sequences W_j and Z_j for $j = a, b$ were 10b long[4]. The steps of the experiment follow.

---

[1] S_1 = GCAGTCGAAGCTGTTGATGC CAAGAGATGG TCGTCTGCTAGCATG TCACGC-
CACGGAACG GTGAGCGCGAGTGTG ATATGCAATGATCTG ATCCGTCCCGATAAG
CAAGTCAGATTGACC GCACGTAACT AGACGCTGCCGTAGTCGACG
S_2 = GCAGTCGAAGCTGTTGATGC CAAGATATGG CCCGATTAGTACAGC TACT-
GATAAGTTCCG TCGCTCCGACACCTA TCAGCCGGCTTGCAC AACTGATACGACTCG
TATTGTCACGCATCG GTACGTAACT AGACGCTGCCGTAGTCGACG
S_3 = GCAGTCGAAGCTGTTGATGC CAAGAGATGG TCGTCTGCTAGCATG TACT-
GATAAGTTCCG GTGAGCGCGAGTGTG TCAGCCGGCTTGCAC ATCCGTCCCGATAAG
TATTGTCACGCATCG GTACGTAACT AGACGCTGCCGTAGTCGACG
S_4 = GCAGTCGAAGCTGTTGATGC CAAGATATGG CCCGATTAGTACAGC TCACGC-
CACGGAACG TCGCTCCGACACCTA ATATGCAATGATCTG AACTGATACGACTCG
CAAGTCAGATTGACC GCACGTAACT AGACGCTGCCGTAGTCGACG

[2] $\alpha = GCAGTCGAAGCTGTTGATGC$, $\beta = AGACGCTGCCGTAGTCGACG$

[3] X_1 = TCGTCTGCTAGCATG X_2 = TCACGCCACGGAACG X_3 = GTGAGCGCGAGTGTG
X_4 = ATATGCAATGATCTG, X_5 = ATCCGTCCCGATAAG, X_6 = CAAGTCAGATTGACC.
Y_1 = CCCGATTAGTACAGC, Y_2 = TACTGATAAGTTCCG, Y_3 = TCGCTCCGACACCTA, Y_4
= TCAGCCGGCTTGCAC, Y_5 = AACTGATACGACTCG, Y_6 = TATTGTCACGCATCG

[4] W_a = CAAGAGATGG, W_b = CAAGATATGG, Z_a = GCACGTAACT, Z_b =GTACGTAACT

1. The initial pool was split randomly in four test tubes (a) (b) (c) (d), and the following PCRs were performed respectively
   (a) $PCR(\alpha, \overline{X_2})$
   (b) $PCR(X_2, \overline{\beta})$
   (c) $PCR(\alpha, \overline{Y_2})$
   (d) $PCR(Y_2, \overline{\beta})$
   so obtaining amplification of four types of sequences $\alpha \ldots X_2$, $X_2 \ldots \beta$, $\alpha \ldots Y_2$, and $Y_2 \ldots \beta$, that are 60, 105, 60, 105 long respectively (see lanes 2, 3, 4, 5 of Figure 3).
2. Two $PCR(\alpha, \overline{\beta})$ were performed in parallel, one after having put together the product of (a) and (b), and the other one after having put together the product of (c) and (d) (sequences 150 long were amplified, see lanes 6 and 7 of Figure 3), then the two pools were mixed.
3. For $i = 3, 4, 5$ the analogues of the previous steps (1) and (2) were performed by replacing index 2 with $i = 3, 4, 5$, and by referring to Figure 4, Top left for $i = 3$, Top right for $i = 4$, and Bottom for $i = 5$ respectively.
4. An electrophoresis was performed to select the sequences 150 long among longer ones (generated by unspecific amplification).

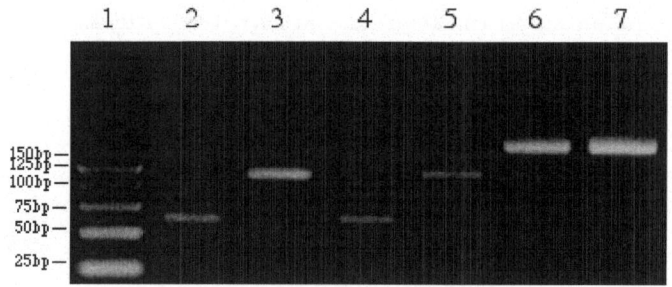

Fig. 3. Electrophoresis results. *Lane 1*: molecular size marker ladder (25bp). *Lane 2*: amplification of $\alpha \cdots X_2$ strands (60bp) and *lane 3*: amplification of $X_2 \cdots \beta$ strands (105bp), both PCRs performed at 52°C. *Lane 4*: amplification of $\alpha \cdots Y_2$ strands (60bp) and *lane 5*: amplification of $Y_2 \cdots \beta$ strands (105bp), both PCRs performed at 45°C. *Lane 6*: cross pairing amplification of $\alpha \cdots X_2$ and $X_2 \cdots \beta$ (150bp) and *lane 7*: cross pairing amplification of $\alpha \cdots Y_2$ and $Y_2 \cdots \beta$ (150bp), both XPCRs performed at 63°C.

Remark. By the electrophoresis results one can note that the sequences 150 long present in the pool were amplified (in linear or exponential manner) during each step. Although this phenomenon did not disturb the correctness of the computation, it caused noise and useless occupation of space in a test tube by increasing the unspecific matter, as one can see clearly in the figure 4. An improved experiment could be performed very easily by inserting intermediate electrophoresis steps to clean the signal from the noise caused by these amplifications.

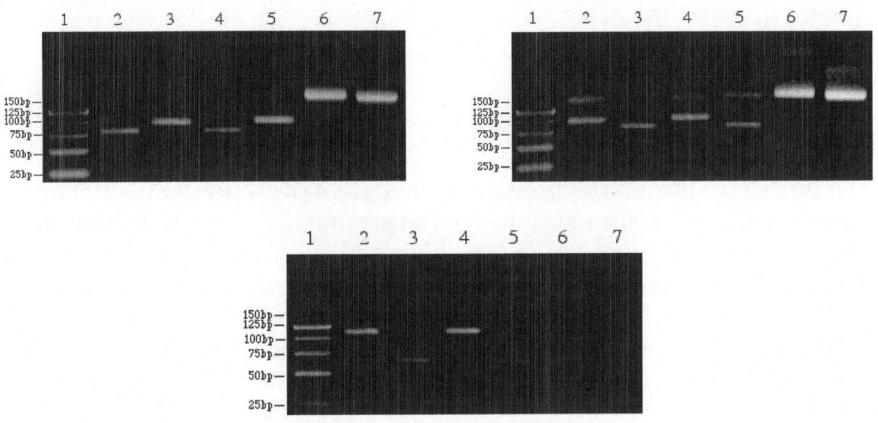

Fig. 4. Electrophoresis results. *Lane 1*: molecular size marker ladder (25bp). **Top left.** *Lane 2*: amplification of $\alpha \cdots X_3$ strands (75bp), *lane 3*: amplification of $X_3 \cdots \beta$ strands (90bp), *lane 4*: amplification of $\alpha \cdots Y_3$ strands (75bp), *lane 5*: amplification of $Y_3 \cdots \beta$ strands (90bp), all PCRs performed at 52°C. *Lane 6*: cross pairing amplification of $\alpha \cdots X_3$ and $X_3 \cdots \beta$ (150bp) and *lane 7*: cross pairing amplification of $\alpha \cdots Y_3$ and $Y_3 \cdots \beta$ (150bp), both XPCRs performed at 63°C. **Top right.** *Lane 2*: amplification of $\alpha \cdots X_4$ strands (90bp), *lane 3*: amplification of $X_4 \cdots \beta$ strands (75bp), *lane 4*: amplification of $\alpha \cdots Y_4$ strands (90bp), *lane 5*: amplification of $Y_4 \cdots \beta$ strands (75bp), all PCRs performed at 42°C. *Lane 6*: cross pairing amplification of $\alpha \cdots X_4$ and $X_4 \cdots \beta$ (150bp) and *lane 7*: cross pairing amplification of $\alpha \cdots Y_4$ and $Y_4 \cdots \beta$ (150bp), both XPCRs performed at 63°C. **Bottom.** *Lane 2*: amplification of $\alpha \cdots X_5$ strands (105bp), *lane 3*: amplification of $X_5 \cdots \beta$ strands (60bp), *lane 4*: amplification of $\alpha \cdots Y_5$ strands (105bp), *lane 5*: amplification of $Y_5 \cdots \beta$ strands (60bp), all PCRs performed at 45°C. *Lane 6*: cross pairing amplification of $\alpha \cdots X_5$ and $X_5 \cdots \beta$ (150bp) and *lane 7*: cross pairing amplification of $\alpha \cdots Y_5$ and $Y_5 \cdots \beta$ (150bp), both XPCRs performed at 63°C.

The success of the experiment was tested by the presence of the recombination witnesses $X_1 X_2 \, Y_3 Y_4 \, X_5 X_6$ and $Y_1 Y_2 \, X_3 X_4 \, Y_5 Y_6$ in the final pool, guaranteed by the amplification of the following PCRs[5].

We indicated with $Z_1 Z_2 Z_3 Z_4 Z_5 Z_6$ a generic recombination witness, and for each of them the following procedure was executed in lab on a distinct copy of the pool P resulting from the experiment. That is, firstly the pool was split in two pools P_1 and P_2 with the same approximate size, and then on each of them the presence of a recombination witness $Z_1 Z_2 Z_3 Z_4 Z_5 Z_6$ was checked by means the following steps:

1. **perform** $PCR(Z_1, \overline{Z_6})$
2. perform **electrophoresis** and select strands 90 long

---

[5] The authors thank an anonymous referee for his/her interesting comments and for suggesting us to find a better procedure checking the complete recombination of the final pool.

3. **perform** $PCR(Z_2, \overline{Z_5})$
4. perform **electrophoresis** and select strands 60 long
5. **perform** $PCR(Z_3, \overline{Z_4})$

   output: YES if the last PCR amplifies (sequences 30 long), NO otherwise.

The procedure proved the presence of $Z_1 Z_2 Z_3 Z_4 Z_5 Z_6$ in the pool P because
i) after the first two steps all and only the strands $Z_1 \cdots Z_6$ of P were present
in the resulting pool (lines 2 and 5 of Figure 5), ii) after the second PCR and
electrophoresis all and only the strands $Z_2 \cdots Z_5$ from strands $Z_1 Z_2 \cdots Z_5 Z_6$
of P were present in the resulting pool (lines 3 and 6 of Figure 5), iii) and
the last PCR amplified the portions $Z_3 Z_4$ of such strands, that are found if
and only if the sequence $Z_1 Z_2 X_3 Z_4 Z_5 Z_6$ was present in P (lines 4 and 7 of
Figure 5).

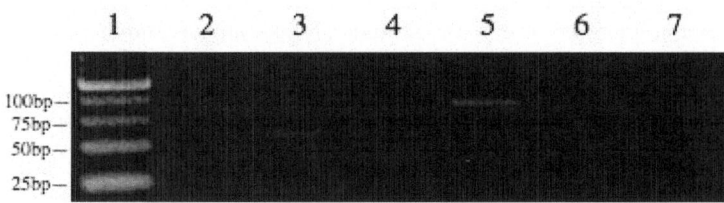

Fig. 5. Electrophoresis results. *Lane 1*: molecular size marker ladder (25bp). *Lane 2*: positive control by PCR $(X_1, \overline{X_6})$ (90bp) performed at 44°C. *Lane 3*: positive control by $PCR(X_2, \overline{X_5})$ (60bp) performed at 46°C. *Lane 4*: positive control by $PCR(Y_3, \overline{Y_4})$ (30bp) performed at 42°C. *Lane 5*: positive control by $PCR(Y_1, \overline{Y_6})$ (90bp) performed at 44°C. *Lane 6*: positive control by $PCR(Y_2, \overline{Y_5})$ (60pb) performed at 42°C. *Lane 7*: positive control by $PCR(X_3, \overline{X_4})$ (30pb) performed at 42°C.

Appendix (Experimental Protocols)

Reagents. 25 bp marker DNA ladder and agarose (Promega); PCR buffer, $MgCl_2$ and dNTP (Roche); Taq DNA Polymerase (produced in laboratory); all the synthetic DNA oligonucleotides 150 bp long and all the primers were from Primm s.r.l.(Milano, Italy).

Annealing of synthetic DNA oligonucleotides. Two complementary synthetic 150 bp long DNA oligonucleotides $(5' - 3'$ and $3' - 5')$ were incubated at 1:1 molar ratio at 90°C for 4 min in presence of 2.5 mM of $MgCl_2$ and then at 70°C for 10 min. The annealed oligos were slowly cooled to 37°C, then further cooled to 4°C until needed.

PCR amplification. PCR amplification was performed on a PE Applied Biosystems GeneAmp PCR System 9700 (Perkin Elmer, Foster City, CA) in a 50 μl final reaction volume containing 1.25U of Taq DNA Polymerase, 1.5 mM $MgCl_2$, 200 μM each dNTP, PCR buffer, 80 ng DNA template, 0.5-1 μM of forward and reverse primers. The reaction mixture was preheated to 95∘C for 5 min. (initial denaturing), termocycled 30 times: 95°C for 30 sec (denaturing), different temperatures (see captures of the figures) for 30 sec. (annealing), 72°C for 15 sec. (elongation); final extension was performed at 72°C for 5 min.

Preparation and running of gels. Gels were prepared in 7×7 cm plastic gel cassettes with appropriate combs for well formation. Approximately 20 ml of 4% agarose solutions were poured into the cassettes and allowed to polymerize for 10 min. Agarose gels were put in the electrophoresis chamber and electrophoresis was carried out at 10 volt/cm^2, then the bands of the gels are detected by a gel scanner. The DNA bands (final PCR products) of interest were excised from the gel and the DNA was purified from the gel slices by Promega Kit (Wizard SV Gel and PCR Clean-Up System).

References

1. Adleman, L. M.: Molecular Computation of solutions to combinatorial problems. Science **266** (1994) 1021–1024
2. Braich, R. S., Chelyapov, N., Johnson, C., Rothemund, P. W. K., Adleman, L. M.: Solution of a 20-variable 3-SAT problem on a DNA computer. Science **296** (2002) 499–502
3. Braich, R. S., Johnson, C., Rothemund, P. W. K., Hwang, D., Chelyapov, N., Adleman, L. M: Solution of a Satisfiability Problem on a Gel-Based DNA Computers. A. Condon, G. Rozenberg eds, Proceedings of 6th International Workshop On DNA Based Computers, Leiden Netherlands, (2000) 31–42
4. Franco, G., Giagulli, C., Laudanna, C., Manca, V.: DNA Extraction by XPCR. C. Ferretti G. Mauri C. Zandron et al. eds, DNA 10, LNCS 3384, Springer-Verlag Berlin Eidelberg, (2005) 106–114
5. G. Franco, *Combinatorial Aspects of DNA Solution Spaces generated by XPCR Recombination*, in preparation.
6. Faulhammer, D., Cukras, A. R., Lipton, R. J., Landweber, L. F.: Molecular computation: RNA solution to chess problems. Proc. Natl. Acad. Sci. USA **98** (2000) 1385–1389
7. Head, T.: Formal language theory and DNA: An analysis of the generative capacity of specific recombinant behaviors. Bulletin of Mathematical Biology **49** (1987) 737–759
8. Lee, J. Y., Lim, H-W., Yoo, S-I., Zhang, B-T., Park, T. H.: Efficient Initial Pool Generation for Weighted Graph Problems Using Parallel Overlap Assembly. G. Mauri G. Rozenberg C. Zandron eds, Preliminary Proceedings of the 10th International Meeting on DNA Based Computers, Milan, Italy, (2004) 357–364
9. Lipton, R. J.: DNA solutions of hard computational problems. Science **268** (1995) 542–544
10. Jonoska, N., Sa-Ardyen, P., Seeman, N.C.: Computation by self-assembly of DNA graphs. Journal of Genetic Programming and Evolvable Machines **4** (2003) 123–137
11. Kaplan, P. D., Ouyang, Q. O., Thaler, D. S., Libchaber, A.: Parallel overlap assembly for the construction of computational DNA libraries. J. Theor. Biol. **188** (1997) 333–341

12. Manca, V., Zandron, C.: A Clause String DNA Algorithm for SAT. N. Jonoska N.C. Seeman eds, Proceedings of the 7th International Workshop on DNA-Based Computers: DNA 7, LNCS 2340, Springer, (2002) 172–181
13. Rose, J. A., Hagiya, M., Deaton, R. J., Suyama, A.: A DNA-based in vitro Genetic program. Journal of Biological Physics **28** 3 (2002) 493–498
14. Stemmer, W.: DNA shuffling by random fragmentation and reassembly: in vitro recombination for molecular evolution. Proc. Natl. Acad. Sci. USA **91** (1994) 10747–10751

An Algorithm for SAT
Without an Extraction Phase

Pierluigi Frisco[1], Christiaan Henkel[2], and Szabolcs Tengely[3]

[1] Dept. of Comp. Sci., School of Eng., C. S. and Math., University of Exeter,
Harrison Building, North Park Road, Exeter, EX4 4QF, UK
P.Frisco@exeter.ac.uk
[2] Institute of Biology, Leiden University, Wassenaarseweg 64,
2333AL Leiden, The Netherlands
henkel@rulbim.leidenuniv.nl
[3] Mathematical Institute, Leiden University, Niels Bohrweg 1,
2333CA Leiden, The Netherlands
tengely@math.leidenuniv.nl

Abstract. An algorithm that could be implemented at a molecular level
for solving the satisfiability of Boolean expressions is presented.

This algorithm, based on properties of specific sets of natural numbers, does not require an extraction phase for the read out of the solution.

1 Introduction

Adleman's solution of an instance of the direct Hamiltonian path problem with
the implementation of an algorithm at a molecular level [1] has been of inspiration for many to pursue other algorithms that can be implemented in the same
way to solve instances of hard computational problems.

A problem is said to be *hard* if it cannot be solved by a deterministic Turing machine with a polynomial time algorithm in function of its input [8, 14].
For many of this kind of problems the number of possible solutions increases
exponentially in function to the size of the input.

The algorithm described in [1] is related to the research of all Hamiltonian
paths in a graph. The algorithm proposed by Adleman can be simplified in
a two-phase process: first a library of DNA molecules encoding the input of
the problem is *created* and is put in a test tube such that the DNA molecules
can, under appropriate conditions, anneal and ligate, then the DNA molecules
encoding solutions to the problem are *extracted* from the test tube.

During annealing and ligation other, 'new', DNA molecules, different from
the ones present in the input library, can be created. Because of the massive
parallelism and the nondeterminism of the annealing process the creation of
the 'new' DNA molecules is quite fast and can lead to DNA molecules encoding
solutions for the considered instance of the problem. As the name suggests during
the extraction phase the solutions are extracted from the pool.

It should be clear that this kind of algorithms does not guarantee that a
solution will be created even if it could. This because the annealing between

A. Carbone and N.A. Pierce (Eds.): DNA11, LNCS 3892, pp. 67–80, 2006.

complementary single stranded DNA molecules is a genuinely nondeterministic operation. Anyhow even if present in the pool a solution could not be detected during the extraction phase. More than error prone this last phase can be quite laborious and expensive.

Algorithms based on this two-phase process are common in Molecular Computing [2, 4, 6, 11, 12, 13, 16, 17, 20].

In Section 3 we describe how a specific creation of the input library of DNA molecules can be used to implement an algorithm without an extraction phase for satisfiability of Boolean expression (SAT), a hard computational problem, stated as decision problem (a problem remains hard if it is stated as decision, enumeration or research problem [14]). The presented algorithm is based on specific sets of natural numbers defined in Section 2.

The first algorithm for DNA computing without an extraction phase has been introduced in [10]. Here the authors define LOD (Length Only Discrimination), that is the concept of not having an extraction phase, and give an experimental result on a small instance of Hamiltonian path problem (HPP). In [19] another algorithm for HPP based on LOD is presented. In Section 5.2 we indicate the elements of novelty of our algorithms compared to the ones already present in the literature.

We did not implement in a biological laboratory the algorithm presented in Section 3, anyhow the biochemical specifications related to the creation of the input library of DNA molecules and to the implementation of the presented algorithms are sketched in Section 4.

2   Unique-Sum Sets

In this section we define unique-sum sets used in the algorithm presented by us in Section 3. Moreover we give some examples, we indicate some properties and results related to these sets, and we define a family of unique-sum sets.

Let $\mathbb{N}$ be the set of natural numbers.

Definition 1 (unique-sum set, ordered unique-sum set). *Let $G = \{n_1, ..., n_p\}$ be a set of different positive integers, and $s = \sum_{i=1}^{p} n_i$ the sum of the elements of G. G is said to be a unique-sum set if the equation $\sum_{i=1}^{p} c_i n_i = s, c_i \in \mathbb{N} \cup \{0\}$, has only the solution $c_i = 1, i \in \{1, \ldots, n\}$.*

A unique-sum set $G = \{n_1, \ldots, n_p\}$ is an ordered unique-sum set if $n_i < n_{i+1}$ for $1 \leq i \leq p - 1$.

In what follows we will only consider ordered unique-sum sets.

An example of a unique-sum set is $G = \{4, 6, 7\}$, $4 + 6 + 7 = 17$ and 17 cannot be written in a different way as a non-negative integer linear combination of the elements in G. An example of a set that is not a unique-sum set is $G' = \{3, 4, 5\}$, $3 + 4 + 5 = 12 = 4 + 4 + 4 = 3 + 3 + 3 + 3$.

The concept of unique-sum set resembles that of subset-sum-distinct set (see e.g. [3]), but there one requires that for any two distinct finite subsets $G_1, G_2 \subseteq G$ the sum of all elements of G_1 is distinct from the sum of all elements of G_2.

Lemma 1. *Given a unique-sum set* $G = \{n_1, \ldots, n_p\}$, *any proper subset of* G *is a unique-sum set.*

Lemma 2. *Let* k *be a positive integer, and* $kG = \{k \cdot n_1, \ldots, k \cdot n_p\}$. *If* G *is a unique-sum set, then* kG *is also a unique-sum set.*

Definition 2 (maximal unique-sum set). *Given a unique-sum set* $G = \{n_1, \ldots, n_p\}$, *it is maximal if there exists no positive integer* $n_{p+1} \notin G$, *such that* $G \cup \{n_{p+1}\}$ *is a unique-sum set.*

It is easy to check that $G = \{2, 3\}$ is a maximal unique-sum set, but $G = \{4, 6\}$ is not, since $\{4, 6, 7\}$ is a unique-sum set too.

Now we describe a method to verify if a set is a unique-sum set. It is based on generating functions (see [15]). We consider the function

$$F_G(x) = \prod_{i=1}^{p} (1 - x^{n_i})^{-1}.$$

Using the identity $(1-x)^{-1} = 1 + x + x^2 + x^3 + \ldots, x \in \mathbb{R}, |x| < 1$, we can rewrite $F_G(x)$ as a power series, having rational integers as coefficients, in the following form: $F_G(x) = P_0 + P_1 x + P_2 x^2 + \ldots + P_k x^k + \ldots$, and, by construction, the coefficient of x^k is the number of solutions of the equation

$$\sum_{i=1}^{p} c_i n_i = k, \quad c_i \in \mathbb{N} \cup \{0\}.$$

Therefore G is unique-sum set, if and only if $P_s = 1$, where $s = \sum_{i=1}^{p} n_i$. We do not have to use infinite expansions, since we are interested in the value of P_s. The coefficient of x^s in

$$\prod_{i=1}^{p} (1 + x^{n_i} + x^{2n_i} + \ldots + x^{\lfloor \frac{s}{n_i} \rfloor n_i})$$

is exactly P_s, where $[\cdot]$ denotes the integer part of the rational number $\frac{s}{n_i}$. Let us see two examples. Let $G = \{8, 12, 14, 15\}$, thus $s = 49$ and we have to compute the coefficient of x^{49} in $F_G(x) = (1 - x^8)^{-1}(1 - x^{12})^{-1}(1 - x^{14})^{-1}(1 - x^{15})^{-1} = f_1(x)f_2(x)f_3(x)f_4(x)$, where

$$f_1(x) = 1 + x^8 + x^{16} + x^{24} + x^{32} + x^{40} + x^{48},$$
$$f_2(x) = 1 + x^{12} + x^{24} + x^{36} + x^{48},$$
$$f_3(x) = 1 + x^{14} + x^{28} + x^{42},$$
$$f_4(x) = 1 + x^{15} + x^{30} + x^{45}.$$

It turns out to be 1, thus G is a unique-sum set. Let $G = \{8, 12, 14, 15, 19\}$, thus $s = 68$ and we have to compute the coefficient of x^{68} in $F_G(x) = (1 - x^8)^{-1}(1 - x^{12})^{-1}(1 - x^{14})^{-1}(1 - x^{15})^{-1}(1 - x^{19})^{-1} = f_1(x)f_2(x)f_3(x)f_4(x)f_5(x)$, where

$$f_1(x) = 1 + x^8 + x^{16} + x^{24} + x^{32} + x^{40} + x^{48} + x^{56} + x^{64},$$
$$f_2(x) = 1 + x^{12} + x^{24} + x^{36} + x^{48} + x^{60},$$
$$f_3(x) = 1 + x^{14} + x^{28} + x^{42} + x^{56},$$
$$f_4(x) = 1 + x^{15} + x^{30} + x^{45} + x^{60},$$
$$f_5(x) = 1 + x^{19} + x^{38} + x^{57}.$$

It turns out to be 12, thus G is not a unique-sum set.

Now we will deal with the construction of unique-sum sets. Given a set of different positive integers $G = \{n_1, \ldots, n_p\}$, such that $\gcd(n_1, \ldots, n_p) = 1$, it is known (see e.g. [5]) that for suitable large integer M, the equation

$$\sum_{i=1}^{p} c_i n_i = M, c_i \in \mathbb{N} \cup \{0\}, \tag{1}$$

has at least one solution. Let us denote by Φ_G the greatest positive integer for which (1) is not solvable. Wilf [18] gave an algorithm to compute Φ_G efficiently. We can use this constant to find possible extensions of a given unique-sum set (in the case when $\gcd(n_1, \ldots, n_p) = 1$), or to prove that it is maximal. First suppose that $\gcd(n_1, \ldots, n_p) = 1$, then we can compute Φ_G using the algorithm described in [18]. By the definition of Φ_G we know that if there exists an integer n_{p+1} such that $G \cup \{n_{p+1}\}$ is a unique-sum set, then $n_{p+1} \leq \Phi_G$. Thus we have to check only finitely many sets using the method mentioned previously. We have checked that the set $G = \{8, 12, 14, 15\}$ is a unique-sum set. In this case $\Phi_G = 33$, but there is no positive integer $k \leq 33$ such that $G \cup \{k\}$ is a unique-sum set, therefore G is maximal. If $\gcd(n_1, \ldots, n_p) = d > 1$ and the new element n_{p+1} is such that $\gcd(n_1, \ldots, n_p, n_{p+1}) = d' > 1$, then we still can succeed, since $\frac{1}{d'}G$ has to be a unique-sum set. In the remaining case, when $\gcd(n_1, \ldots, n_p) = d > 1$ and $\gcd(n_1, \ldots, n_p, n_{p+1}) = 1$, we show an example. Let $G = \{4, 6\}$ and n_3 is odd, then $s = n_3 + 10$ is also odd, thus if we have a solution of $4x_1 + 6x_2 + n_3 x_3 = s$, then $x_3 > 0$. We obtain that $4x_1 + 6x_2 + n_3(x_3 - 1) = 10$, that is $x_1 = x_2 = x_3 = 1$ if $n_3 > 6$. In this way we obtained infinitely many unique-sum sets in the form $\{4, 6, 2k + 1\}, k > 2$.

Now we give a family of sets. Let $G_k = \cup_{m=1}^{k}\{2^k - 2^{k-m}\}$, the sum of the elements of G_k is $s_k = (k - 1)2^k + 1$. The first sets in this family are:

$$G_1 = \{1\},$$
$$G_2 = \{2, 3\},$$
$$G_3 = \{4, 6, 7\},$$
$$G_4 = \{8, 12, 14, 15\},$$
$$G_5 = \{16, 24, 28, 30, 31\},$$
$$G_6 = \{32, 48, 56, 60, 62, 63\},$$

Theorem 1. *For all $k \in \mathbb{N}$ the set G_k is a unique-sum set.*

The proofs of Lemma 1, Lemma 2 and Theorem 1, the proof that each element in the family of sets previously given is the unique-sum set having the smallest sum in function of the number of elements and other properties and results related to unique-sum sets can be found in [7].

3 An Algorithm for the Satisfiability of Boolean Expressions

The *satisfiability of Boolean expressions* (SAT) problem can be formulated as: given a Boolean expression ϕ with variables $X = \{x_1, \ldots, x_n\}$, is there an assignment $A : X \to \{T, F\}$ such that A satisfies ϕ?

If the Boolean expression ϕ is given by a conjunction of clauses $C_1 \wedge C_2 \wedge \ldots \wedge C_p$ (where '$\wedge$' is the logical AND operator) each being a disjunction of at most k literals (a literal is a variable x_i or its negation $\neg x_i$, for $1 \leq i \leq n$), then the problem is called k-SAT.

In [11] the author demonstrates that 3-SAT is well suited to take advantage of the massive parallelism present in molecular computation. At the present time SAT is probably the problem with the most number of algorithms implemented [12, 20, 16, 4] or implementable [11, 9, 13] at a molecular level.

Let ϕ be an instance for k-SAT having n variables and p clauses, let $L = \{l_1, l_2, \ldots, l_q\}$ ($q \leq 2n$), an ordered set of literals satisfying at least one clause of ϕ such that if $l_i, \neg l_i \in L$ for $1 \leq i \leq q$, then $l_i = l_j, \neg l_i = l_{j+1}$ for a $1 \leq j \leq q - 1$. Moreover let $C = \{C_1, \ldots, C_p\}$ the set of clauses present in ϕ, and let $G = \{n_1, \ldots, n_{p+2}\}$ be a unique-sum set having sum s_G.

The input library of molecules is composed by:

edges: Each pair (l_i, l_j), $i \leq j, l_i \neq \neg l_j, 1 \leq i, j \leq q$, of literals in L is encoded by an ordered (from 5' to 3') single stranded DNA molecule composed by the 8-mer s_{l_i} (encoding l_i) followed by the 8-mer s_{l_j} (encoding l_j). It is important to notice now that these pairs define a partial order in L. The order is partial and not total as there is no pair for a literal and its negation if both literals are present in L.

Moreover there are going to be two additional 8-mer single stranded DNA molecules: s_b and s_e.

For each literal $l \in L$ there will be ordered (from 5' to 3') single stranded DNA molecules composed by the 8-mer s_b followed by the 8-mer s_l and single stranded DNA molecules composed by the 8-mer s_l followed by the 8-mer s_e.

All the $s_l, l \in L$, s_b and s_e are different sequences of nucleotides.

vertices: We associate to each clause $C_j \in C$ a unique number $n_k \in G$. We will consider C_j associated to n_{j+1} for $1 \leq j \leq p$. For each literal l in L there will be a set of ordered (from 5' to 3') partially double DNA molecules composed by: a single stranded 8-mer $\bar{s}_l$ complementary to s_l; a double stranded $(n_{j+1} - 16)$-mer for each clause $C_j, 1 \leq j \leq p$ satisfied by l; a single stranded 8-mer $\bar{s}_l$ complementary to s_l.

begin: Ordered (from 5' to 3') partially double DNA molecules composed by: a single stranded 8-mer $\bar{s}_b$ complementary to s_b followed by a double stranded $(n_1 - 8)$-mer.

end: Ordered (from 5' to 3') partially double DNA molecules composed by: a double stranded $(n_{p+2} - 8)$-mer followed by a single stranded 8-mer $\bar{s}_e$ complementary to s_e.

The following example is meant to clarify the above. Let the Boolean expression $\phi = C_1 \wedge C_2 \wedge C_3 = (x_1 \vee x_2 \vee x_3) \wedge (\neg x_1 \vee \neg x_2 \vee \neg x_3) \wedge (\neg x_1 \vee \neg x_2 \vee x_3)$ be an instance of 3-SAT. An ordered set of literals of ϕ satisfying at least one clause is $L = \{x_1, \neg x_1, x_2, \neg x_2, x_3, \neg x_3\} = \{l_1, l_2, l_3, l_4, l_5, l_6\}$, while the set of clauses of ϕ is $C = \{C_1, C_2, C_3\}$.

The set of single stranded DNA molecules encoding **edges** is depicted in Figure 1.

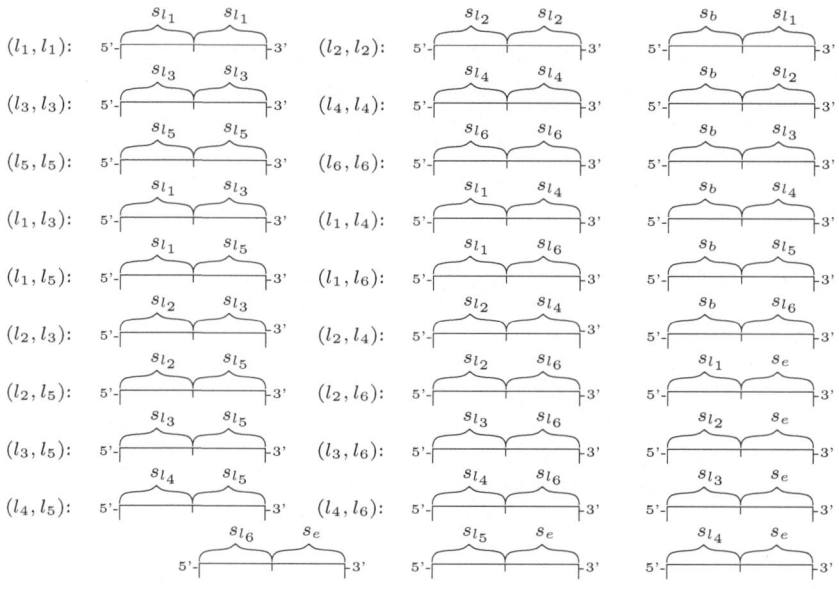

Fig. 1. Encoding of **edges** for the example of 3-SAT

Let us consider now the unique-sum set $G = \{16, 24, 28, 30, 31\}$, having sum $s_G = 129$. We associate 24 to C_1, 28 to C_2 and 30 to C_3. The literals l_1 and l_3 both satisfy only C_1; the literals l_2 and l_4 both satisfy C_2 and C_3; the literal l_5 satisfies C_1 and C_3; the literal l_6 satisfies C_2. Considering this we give now the lengths of the double stranded DNA molecules present in the encodings of **vertices**. As C_1 is associated to 24, then the double stranded DNA molecule is 8-bp (result of 24-16); as C_2 is associated to 28, then the double stranded DNA molecule is 12-bp (result of 28-16); as C_3 is associated to 30, then the double stranded DNA molecule is 14-bp (result of 30-16). The double stranded DNA molecule present in the encoding of **begin** is 8-bp (result of 16-8), while the one present in the encoding of **end** is 23-bp (result of 31-8). These molecules are depicted in Figure 2.

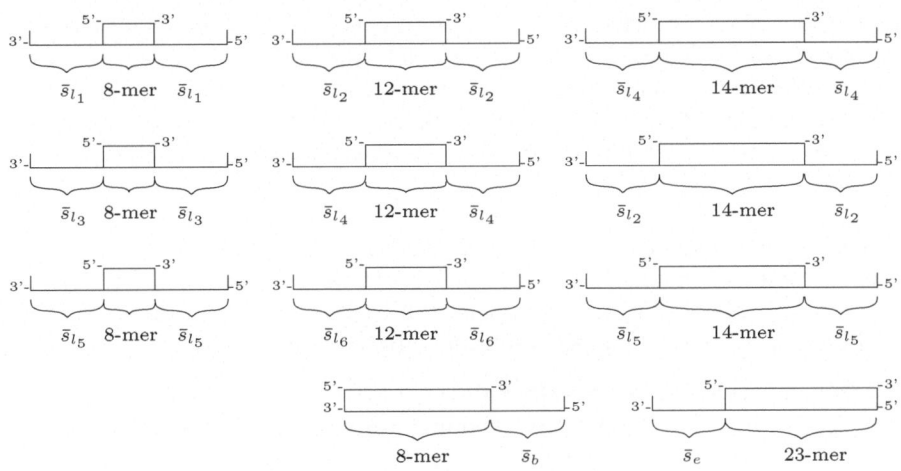

Fig. 2. Encoding of **vertices** for the example of 3-SAT

The described encoding for this example can be visualised as the graph depicted in Figure 3, where $C_i(l_j)$ indicates that the clause C_i is satisfied by the literal l_j (for $1 \leq i \leq 3, 1 \leq j \leq 6$). In this graph black dots indicate hubs: they have been introduced to decrease the number of arrows present in the graph and make it more readable, hubs have no relation with the encoding described by us.

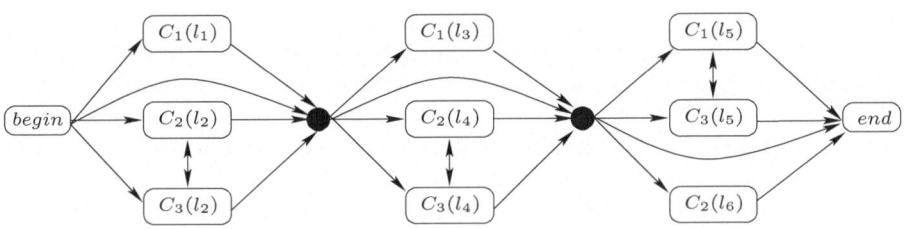

Fig. 3. Graph related to the example of 3-SAT

The annealing and ligation of the library of molecules is likely to form DNA molecules s_G-bp long only if there is an assignment A satisfying ϕ. Considering the graph depicted in Figure 3 such molecules can be visualised as paths starting at *begin* and ending at *end* and passing by nodes encoding clauses satisfied by literals where the encoding of each clause is present only once. Examples of such paths are: $begin - C_1(l_1) - C_2(l_4) - C_3(l_5) - end, begin - C_2(l_4) - C_3(l_4) - C_1(l_5) - end$.

If a resulting molecule is s_G-bp long, then it will start with a sequence encoding **begin** and it will end with a sequence encoding the **end**, the intermediate part will be composed by encodings of **vertices** (clauses satisfied by literals) annealed and ligated to **edges**. This intermediate part cannot contain both the encoding of a clause satisfied by a literal l and the encoding of a clause satisfied by a literal $\neg l$, for $l \in L$. Moreover in the intermediate part the encoding of a clause can be present only once.

Any assignment A satisfying ϕ can be encoded (by the annealing and ligation of the molecules in the input library) in a double stranded DNA molecule s_G-bp long.

The presence of such a molecule can be detected by one run of gel electrophoresis independent of the size of the instance of the problem.

For a Boolean formula ϕ, instance of k-SAT, with p clauses and n variables (so at most $2n$ literals), in the worst case (all literals are present in a clause and each literal satisfies each clause) the input library of molecules is composed by:

$2n$ DNA molecules encoding **edges** of the form (l_i, l_i);

$2n \sum_{i=1}^{n-1}(2n - 2i)$ DNA molecules encoding **edges** of the form $(l_i, l_j), i > j, l_i \neq \neg l_j, 1 \leq i, j \leq 2n$;

$4n$ DNA molecules encoding **edges** of the form (b, l) and (l, e) for $l \in L$;

$2np$ (each of the $2n$ literals can satisfy each of the p clauses) DNA molecules encoding **vertices**;

1 DNA molecule encoding **begin**;

1 DNA molecule encoding **end**.

In the following section we describe how the initial library of DNA molecules can be created.

4   Biochemical Specifications

As presented in the previous section unique-sum sets allow the creation of algorithms where part of the instance of the problem is encoded in the length of partially double DNA molecules. The actual sequence of the double part of these molecules is then of only minor importance. This fact can be exploited in the efficient production of these molecules.

Each element of the family of unique-sum sets presented in Section 2 can be written as $G_k = \{2^{k-1}, 2^{k-1}+2^{k-2}, 2^{k-1}+2^{k-2}+2^{k-3}, \ldots, 2^{k-1}+2^{k-2}+\cdots+2^0\}$. If moreover we consider that $2^h = 2^{h-1} + 2^{h-1}$, then it is possible to devise an efficient algorithm for the creation of long double stranded DNA molecules by controlled concatenation of two shorter ones. Only the short ($\leq$ 8-bp) DNA molecules need to be chemically synthesised.

The concatenation of two molecules requires tight control of the reaction as a simple ligation of molecules in solution will also produce many longer multimers. One way to perform controlled reactions is making the ends of the double stranded DNA molecules unavailable for ligation.

The following steps will create a specific concatenation of two generic double stranded DNA molecules A and B:

1. attach one end of A to a solid support. For example, use a 5' biotin label and streptavidin coated beads;
2. ensure the free 5' end is phosphorylated;

3. remove phosphates from B by alkaline phosphatase treatment;
4. mix and ligate;
5. remove all unbound molecules;
6. remove the molecules from the beads. This can be accomplished by simple endonuclease digestion if a DNA linker is used between the biotin label and molecule A;
7. if necessary, PCR (with or without biotynilated primers) can be used as an amplification procedure.

This procedure ensures that only one copy of molecule B can be attached to the immobilized A. However, some small chances of error still exist. For example, two molecules A can be ligated, creating a tether between two beads. Another possibility is incomplete ligation, i.e. some molecules A may not be ligated to B. Such errors are inevitable, but the chances can be minimized by optimization of laboratory protocols. If measurable quantities of erroneous molecules are formed, the correct molecules can be purified by preparative gel electrophoresis.

Very small molecules ($\leq$ 8-bp) can be added in an alternative way, using an extra sequence which is recognized by a type IIs restriction endonuclease. The sequence recognized by the restriction enzyme should be concatenated only at the two ends of the double stranded DNA molecule. The rest of the DNA molecule could be easily constructed so not to contain the restriction site. For example, one base pair can be added by ligation to 5' NNNNNNGACTC, and subsequent digestion with *MlyI* (New England Biolabs). This enzyme recognises the sequence 5' GAGTC and produces a blunt cut five bp to the 3' end. The result is 5' N, or any one base pair added. A similar technique can be used to produce different single stranded extensions necessary for programmable ligation. The enzyme used should then produce a staggered cut outside its recognition sequence. Using this method, the only molecules that need to be synthesized chemically are the 2 original 8 nucleotide strands and in total 6 oligonucleotides for adding 1, 2, or 4-bp.

The following example should clarify the strategy outlined above. Let us imagine that we want to create DNA molecules long as the elements in the unique-sum set $G_6 = \{32, 48, 56, 60, 62, 63\}$. Let us also consider that the two ends of each molecule have to be single stranded (each 8 bases long) while the rest of the molecule has to be double stranded. So, considering the elements in G_6, the part of the molecules that is double stranded has to be as long as the elements of the set $G'_6 = \{16, 32, 40, 44, 46, 47\} = \{8 + 8, 16 + 16, 32 + 8, 40 + 4, 44 + 2, 46 + 1\}$.

1. synthesize a molecule 8-bp long (such a molecule is stable enough and long enough to be ligated);
2. generate a molecule 16-bp long (element of G'_6) concatenating two molecules 8-bp long;
3. generate a molecule 32-bp long (element of G'_6) concatenating two molecules 16-bp long;
4. generate a molecule 40-bp long (element of G'_6) concatenating a molecule 32-bp long with one 8-bp long;
5. generate a molecule 44-bp long (element of G'_6) concatenating a molecule 40-bp long with one 4-bp long;

6. generate a molecule 46-bp long (element of G'_6) concatenating a molecule 44-bp long with one 2-bp long;
7. generate a molecule 47-bp long (element of G'_6) concatenating a molecule 46-bp long with one 1-bp long.

The single stranded molecules used in the algorithm presented in Section 3 need to be chemically synthesized and concatenated to the two sides of the double stranded DNA molecules.

5   Discussions

5.1   Biological

Experimental implementation of the algorithm presented in Section 3 is subject to some constraints. Thermodynamics dictates a certain minimum length for the DNA molecules present in the input library. DNA molecules of only a few bp do not anneal at room temperature: if, for example the unique-sum set $G_3 = \{4, 6, 7\}$ is considered for the encoding, then all members of the set should be multiplied by a constant to yield to DNA molecules long enough to be stable. The set obtained by the multiplication is ensured to be a unique-sum set by Lemma 2.

Length separation by electrophoresis imposes an upper limit on the size of the DNA molecules associated to the elements of a unique-sum set considered for encoding an instance of a problem. DNA electrophoresis has a maximum resolution of about 0.1%: discriminating between DNA fragments that have a difference in length of 1-bp per 1000 is realistic using large polyacrylamide gels or capillary electrophoresis. This limitation is due to current technology and not on DNA itself. Let us consider the set G_7, having sum $s_{G_7} = 769$, indicated in Section 2. The number $768 = 12 \cdot 64$ can be obtained as sum of elements in G_7. The difference between s_{G_7} and 768 represents the 0.13% of s_{G_7}. Similar computation for G_8 gives a value of 0.05% of its sum, already below the maximal resolution of the just described DNA electrophoresis.

We can envisage three possibilities to overcome this limit in the implementation of algorithms based on unique-sum sets:

1. other families of unique-sum sets may be found having a bigger difference between the sum of the set and the smaller or bigger number that can be obtained summing elements in the set;
2. different algorithms based on unique-sum sets can be devised;
3. the technology of DNA analysis can be improved so to increase the resolution.

The algorithm devised for the decision problem presented in Section 3 can be easily modified for research problems. If the presence of a solution is detected by gel electrophoresis, the precise sequence of it (telling in the case of SAT the sequence of clauses satisfied by a literal) can be found by DNA sequencing, multiplex PCR or restriction analysis. The analysis techniques themselves also entail some sequence design considerations.

5.2   Algorithmic

The creation of algorithms in DNA computing without an extraction phase is not new. Length-only discrimination (LOD) was introduced in [10] where the authors present experimental confirmations of this technique.

In [10] the algorithm giving the length of the molecules encoding the vertices is: "... if we need to find n different lengths, then starting with an arbitrary number for the lengths of the first vertex, we can produce the sequence of length with desired properties by making a gap between the lengths of the ith and the $(i+1)^{th}$ vertices be $(n+i)$.". So, if for instance we want to find the lengths of the molecules for a graph with 9 vertices we have:

1: k, $k \in \mathbb{N}$
2: $(k) + 9 + 1 = k + 10$
3: $(k + 10) + 9 + 2 = k + 2 \cdot 9 + 3 = k + 21$
4: $(k + 21) + 9 + 3 = k + 3 \cdot 9 + 6 = k + 33$
5: $(k + 33) + 9 + 4 = k + 4 \cdot 9 + 10 = k + 46$
6: $(k + 46) + 9 + 5 = k + 5 \cdot 9 + 15 = k + 60$
7: $(k + 60) + 9 + 6 = k + 6 \cdot 9 + 21 = k + 75$
8: $(k + 75) + 9 + 7 = k + 7 \cdot 9 + 28 = k + 91$
9: $(k + 91) + 9 + 8 = k + 8 \cdot 9 + 36 = k + 108$

So we obtain the set $K_9 = \{k, k + 10, k + 21, k + 33, k + 46, k + 60, k + 75, k + 91, k + 108\}$ having sum $s_{K_9} = 9k + 444$ (so we are considering the coefficients $f_1 = <1, 1, 1, 1, 1, 1, 1, 1>$, notice that the sum of these coefficients is 9). But this sum can also be written as $k + 3(k+10) + (k+33) + 3(k+91) + (k+108)$ which means that it can be obtained also by the coefficients $f_2 = <1, 3, 0, 1, 0, 0, 0, 3, 1>$ (notice that also the sum of these coefficients is 9). So, if in this example we consider that the initial vertex (having no incoming edges and only one outgoing edge) is associated to **1**, that the final vertex (having only one incoming edge and no outgoing edges) is associated to **9**, and that the rest of the graph is totally connected, then **1-2-8-2-8-4-2-8-9** would be interpreted as an Hamiltonian path (while it is not). This implies that the just presented algorithm to generate sets of numbers for algorithms based on LOD is not always valid.

The fact that the two sets of coefficients have both sum 9 is essential as also molecules encoding edges are present. In [10] edges are encoded such that the relative molecules are: "...longer than any vertex encoding.". This implies that any two sets of coefficients (as the ones indicated in the above) having the same sum would bring to accepted solutions (this would not be the case if the sets of coefficients had different sums as the associated DNA molecules would have different lengths). This affirmation is wrong if we consider f_2.

The other sets of coefficients for K_9 having the same properties of f_2 are: $f_3 = <1, 2, 0, 0, 1, 2, 2, 0, 1>$, $f_4 = <1, 0, 2, 2, 1, 0, 0, 2, 1>$, $f_5 = <1, 0, 0, 0, 6, 1, 0, 0, 1>$. These sets of coefficients can be used to find other sets of coefficients for $K_n, n \geq 10$, that is for sets obtained by the algorithm described in [10].

Let us list the elements found by the algorithm described in [10] from the second to the eighth for a set with $n \geq 9$ elements:

2: $k + n + 1$
3: $k + 2n + 3$
4: $k + 3n + 6$
5: $k + 4n + 10$
6: $k + 5n + 15$
7: $k + 6n + 21$
8: $k + 7n + 28$

The sum of these elements is $7k + 28n + 84$ but this sum can also be obtained by $2(k + 2n + 3) + 2(k + 3n + 6) + k + 4n + 10 + 2(k + 7n + 28)$ (we just used the set of coefficients f_2 but we could have used also f_3, f_4 or f_5).

This means that for $n = 10$ the set of coefficients $< 1, 0, 2, 2, 1, 0, 0, 2, 1, 1 >$ (having sum 10) gives the sum $s_{K_{10}}$; for $n = 11$ the set of coefficients $< 1, 0, 2, 2, 1, 0, 0, 2, 1, 1, 1 >$ (having sum 11) gives the sum $s_{K_{11}}$, etc..

The just given description does not render all the sets of coefficients for sets with $n \geq 10$ elements. For instance other sets of coefficients giving the sum $s_{K_{10}}$ are $< 1, 3, 0, 1, 0, 0, 1, 1, 2, 1 >, < 1, 2, 0, 1, 0, 0, 4, 1, 0, 1 >$, etc..

In [19] the authors describe algorithms based on LOD. Also in this paper sets with a unique sum are considered. The elements of such sets $G = \{n_1, \ldots, n_p\}$ are defined as follows:

$$\begin{cases} n_1 = 1 \\ n_k = kn_{k-1} + 1 - \sum_{i=1}^{k-1} n_i \end{cases}$$

The numbers in these sets grow (from n_1 to n_p) as $p!$. It is possible to see this if we express n_k as a function of n_{k-1}. We have that $n_{k-1} = (k-1)n_{k-2} + 1 - \sum_{i=1}^{k-2} n_i$, so $n_k = kn_{k-1} + 1 - \sum_{i=1}^{k-1} n_i = k(k-1)n_{k-2} + k - k\sum_{i=1}^{k-2} n_i + 1 - \sum_{i=1}^{k-1} n_i$. So $n_p = p(p-1)(p-2)\ldots 1 - x$ where x is a polynomial in n_i $(1 \leq i \leq p-1)$. This implies that the sum of a set with p elements grows as $p!$, while the sum of a set with p elements in the family of sets given in Section 2 grows as an exponential (power of 2).

As proved in [7] the family of unique-sum sets given in Section 2 is the one giving unique-sum sets with the smallest sum in relation to the number of elements in the set. So given a unique-sum set G' with n elements its sum $s_{G'}$ cannot be smaller than s_G the sum of the smallest set with n elements in the family presented in Section 2. A consequence of this is that the algorithm for SAT we presented is not of practical use because of the exponential increase in length of the DNA molecules needed to encode large instances of the considered problem.

The presented research is a starting point in creating algorithms that can be implemented at a molecular level based on properties of specific sets of numbers. Some natural continuations of this research are identified by the following questions:

Is it possible to relax the definition of unique-sum set (to, for instance, sets whose sum can be obtained with only a constant number of non-negative linear combinations of the elements in the set) and create algorithms implementable at a molecular level that can take advantage of this relaxed definition?

Are there other kind of sets that can be considered when we take in account the specific problem we want to solve and the way the algorithm is devised?

Acknowledgements

We thank J. Khodor for the interesting discussions about sets with a unique sum. The work of P. Frisco has been supported by the research grant NAL/01143/G of The Nuffield Foundation.

References

1. L. M. Adleman. Molecular computation of solutions to combinatorial problems. *Science*, 266:1021–1024, November 11, 1994.
2. L. M. Adleman. On constructing a molecular computer. volume 27 of *DIMACS: Series in Discrete Mathematics and Theoretical Computer Science*, pages 1–22. American Mathematical Society, 1995.
3. J. Bae. On generalized subset-sum-distinct sequences. *Int. J. Pure Appl. Math.*, 1(3):343–352, 2002.
4. R. S. Braich, N. Chelyapov, C. Johnson, P. W. K. Rothemund, and L. Adleman. Solution to a 20-variable 3-SAT problem on a DNA computer. *Science*, 296(5567):499–502, 2002.
5. A. Brauer. On a problem of partitions. *Amer. J. Math.*, 64:299–312, 1942.
6. D. Faulhammer, A. R. Cukras, , R. J. Lipton, and L. F. Landweber. Molecular computation: RNA solutions to chess problems. In *Proc. Nat. Acad. Sci. USA*, volume 97, pages 13690–13695, 2000.
7. P. Frisco and Sz. Tengely. On unique-sum sets. *Manuscript in preparation*, 2005.
8. M. R. Garey and D. S. Johnson. *Computers and intractability*. W. H. Freeman and Co., San Francisco, 1979.
9. N. Jonoska, S. A. Karl, and M. Saito. Three dimensional DNA structures in computing. *BioSystems*, 52:243–253, 1999.
10. Yevgenia Khodor, Julia Khodor, and T. F. Knight Jr. Experimental conformation of the basic principles of length-only discrimination. Poster at 7th International Workshop on DNA-Based Computers, DNA 2001, Tampa, U.S.A, 10-13 June 2001.
11. R. J. Lipton. Using DNA to solve NP-complete problems. *Science*, 268:542–545, April 28, 1995.
12. Q. Liu, L. Wang, A. G. Frutos, A. E. Condon, R. M. Corn, and L. M. Smith. DNA computing on surfaces. *Nature*, 403, 2000.
13. V. Manca and C. Zandron. *A DNA algorithm for 3-SAT(11,20)*, volume 2340 of *Lecture Notes in Computer Science*. Springer Verlag, Berlin, Heidelberg, New York, 2001.
14. C. H. Papadimitriou. *Computational complexity*. Addison-Wesley Pub. Co., 1994.
15. G. Pólya. On picture-writing. *Amer. Math. Monthly*, 63:689–697, 1956.

16. K. Sakamoto, H. Gounzu, K. Komiya, D. Kiga, S. Yokoyama, T. Yokomori, and M. Hagiya. Molecular computation by DNA hairpin formation. *Science*, 288: 1223–1226, May 19, 2000.
17. K. A. Schmidt, C. V. Henkel, G. Rozenberg, and H. P. Spaink. DNA computing using single-molecule hybridization detection. *Nucleic acid research*, 32:4962–4968, 2004.
18. H. S. Wilf. A circle-of-lights algorithm for the "money-changing problem". *Amer. Math. Monthly*, 85(7):562–565, 1978.
19. T. Yokomori, Y. Sakakibara, and S. Kobayashi. A magic pot : Self-assembly computation revisited. In W. Brauer, H. Ehrig, J. Karhumki, and A. Salomaa, editors, *Formal and Natural Computing: Essays Dedicated to Grzegorz Rozenberg*, volume 2300 of *Lecture Notes in Computer Science*, pages 418–429, 2002.
20. H. Yoshida and A. Suyama. Solution to 3-SAT by breadth first search. volume 54 of *DIMACS Series in Discrete Mathematics and Theoretical Computer Science*, pages 9–22. American Mathematical Society, 1999.

Sensitivity and Capacity of Microarray Encodings

Max H. Garzon, Vinhthuy Phan,
Kiran C. Bobba, and Raghuver Kontham

Computer Science, The University of Memphis,
Memphis, TN 38152-3240, USA
{mgarzon, vphan, kbobba, rkontham}@memphis.edu

Abstract. Encoding and processing information in DNA-, RNA- and other biomolecule-based devices is an important topic in DNA-based computing with potentially important applications to fields such as bioinformatics, and, conceivably, microbiology and genetics. New methods to encode large data sets compactly on DNA chips has been recently proposed in (Garzon & Deaton, 2004) [18]. The method consists of shredding the data into short oligonucleotides and pouring it over a DNA chip with spots populated by copies of a basis set of noncrosshybridizing strands. In this paper, we provide an analysis of the sensitivity, robustness, and capacity of the encodings. First, we provide preliminary experimental evidence of the degree of variability of the representation and show that it can be made robust despite reaction conditions and the uncertainty of the hybridization chemistry *in vitro*. Based on these simulations, we provide an empirical estimate of the capacity of the representation to store information. Second, we present a new theoretical model to analyze and estimate the sensitivity and capacity of a given DNA chip for information discrimination. Finally, we briefly discuss some potential applications, such as genomic analysis, classification problems, and data mining of massive amounts of data in abiotic form *without* the onerous cost of massive synthesis of DNA strands.

Keywords: Data representation, Gibbs energy, h-distance, fault-tolerant computing, DNA chips, microarrays, genomic analysis, data mining, classification and discrimination.

1 Introduction

Biomolecular computing (BMC) was originally motivated by computational and engineering purposes. This endeavour would not be possible without some type of representation of data and information, directly or indirectly, onto biomolecules, both as input and as output in a computation. Virtually every application of DNA computing maps data to appropriate sequences to achieve intended reactions, reaction products, and yields. DNA molecules usually process information by intramolecular and (more often) intermolecular reactions, usually hybridization in DNA-based computing. The problem of data and information encoding

A. Carbone and N.A. Pierce (Eds.): DNA11, LNCS 3892, pp. 81–95, 2006.
© Springer-Verlag Berlin Heidelberg 2006

on DNA bears an increasing interest for both biological and non-biological applications.

Most of prior work in this area has been restricted to the so-called *word design problem*, or even the encoding problem (Garzon et al., 1997) [10]. In this paper, however, we address a fairly distinct issue, herein called the *representation problem*. The problem is to find a systematic (i.e., application independent) procedure to map both symbolic (abiotic) and nonsymbolic (e.g., biological) information onto biomolecules for massively parallel processing in wet test tubes for real world problems. Mapping of non-biological information for processing *in vitro* is an enormous challenge. Even the easier direct readout problem, i.e., converting genomic data into electronic form for conventional analysis, is an expensive and time-consuming process in bioinformatics (Mount, 2001) [19]. Moreover, the results of these analyses are usually only available in manual form that cannot be directly applied to feedback on the carriers of genomic information.

Three properties are deemed critical for eventual success of a mapping algorithm/protocol. It must be (Blain and Garzon, 2004)[3]:

- **Universal**
  Any kind of symbolic data/pattern can be mapped, in principle, to DNA. Otherwise the mapping will restrict the kind of information mapped, and the processing capabilities in DNA form may be too peculiar or too constrained to be useful in arbitrary applications.
- **Scalable**
  Mapping can only be justified in massive quantities that cannot be processed by conventional means. Therefore it must be scalable to the tera-bytes and higher orders it will eventually encounter. Currently, no cost-effective techniques exist for transferring these volumes by manual addition and extraction of patterns one by one. Ordinary symbolwise transductions require manually manufacturing the corresponding DNA strands, an impossible task with current technology.
- **Automatic and high-speed**
  Manual mapping (e.g., by synthesis of individual strands) is also very costly timewise. An effective strategy must be automatable (from and back to the user) and eventually orders of magnitude faster than processing of the data *in silico*.

The purpose of this paper is to provide an analysis of a new approach recently proposed to represent data (Garzon & Deaton, 2004)[18, 8] that is readily implementable in practice on the well developed technology of DNA chips (Steckel, 2003) [21]. The method has the potential to represent in appropriately chosen DNA olignucleotides massive amounts of arbitrary data in the order of tera- and peta-byte scales for efficient and biotechnologically feasible processing. Direct encoding into DNA strands (Garzon et al., 2003d) [18], (Baum, 1995) [1] is not a very efficient method for storage or processing of such massive amounts of data not already given in DNA form because of the enormous implicit cost of DNA synthesis to produce the encoding sequences, even if their composition were available. The more indirect, but more efficient, approach is reviewed in Section 2, assuming the existence of a large basis of noncrosshybridizing DNA molecules, as provided by good codeword sets recently obtained through several sources (Deaton et al., 2002a; Garzon et Al, 2003) [7, 2]. The method appears at first sight to be plagued by the uncertainty and fuzzyness inherent in the reactions among biomolecular ensembles. In Section 2.2, we establish that these concerns are not justified by establishing, somewhat surprisingly, that it is possible to factor out noise and map symbolic data in a very "linear" fashion with respect to the properties of concatenation and set multiplicity on the symbolic side, and hybridization and amplification on the biochemical side. We further provide a preliminary experimental assessment of the sensitivity of the representation for problems such as recognition, discrimination, and classification. In Section 3, we also provide a theoretical analysis of the sensitivity and potential capacity of the method. Finally, in Section 4 we briefly discuss some advantages and potential applications, such as genomic analysis, classification problems, and data mining of massive amounts of data in abiotic form, as well as some problematic issues that require further study for wide implementation and application of the method.

2   Encoding Data and Information in DNA Spaces

The obvious method to encode data on DNA, namely a one-one mapping of alphabet symbols (e.g., bits) or words (e.g., bytes or English words in a dictionary) to DNA fragments could possibly be used to encode symbolic data (strings) in DNA single strands. Longer texts can be mapped homomorphically by ligation of these segments to represent larger concatenations of symbolic text. A fundamental problem with this approach is that abiotic data would appear to require massive synthesis of DNA strands of the order of the amount of data to be encoded. Current lab methods may produce massive amounts of DNA copies of the same species, but not of too many diverse species selected and assembled in very specific structures such as English sentences in a corpus of data (e.g., a textbook), or records in a large data warehouse. Even if the requisite number of species were available, the mapping between the data and the DNA strands is hard to establish and maintain, as the species get transformed by the reactions they must get involved in and they must be translated back to humanly usable expression.

An alternative more effective representation using recently available large sets of noncrohybridzying oligonucleotides obtainable *in vitro* (Chen et. al., 2005; Bi et. al, 2003) [4, 2] has been suggested in (Garzon and Deaton, 2004) [18]. We repeat next the basic definitions to make this paper self-contained. This method can be regarded as a new implementation of the idea in (Head et al., 1999; 2001) [16, 15] of aqueous computing for writing on DNA molecules, although through a simpler set of operations (only hybridization.) Since binary strings can be easily mapped to a four letter alphabet, we will simply assume that the data are given in DNA form over $\{a, c, g, t\}$. Representations using sets with crosshybridization present are usually ambiguous and cannot be reliably used. More details on this point can be can be found in (Garzon and Deaton, 2004) [17, 18].

2.1   Representation Using a Non-crosshybridizing Basis

Let B be a set of DNA molecules (the encoding basis, or "stations" in Head's terminology (Head et al., 1999) [15], here not necessarily bi-stable), which is assumed to be finite and noncrosshybridizying according to some model of hybridization, denoted $|*, *|$ (for example, the Gibbs energy, or the h-distance in (Garzon et al, 1997) [10, 9]). We will also assume that we are provided some parameter coding for the stringency of reaction conditions τ (for example, a threshold on the Gibbs energy or the h-distance) under which hybridization will take place. For simplicity, it is further assumed that the length of the strands in B is a fixed integer n, and that B contains no hairpins. For example, if the h-distance is the hybridization criterion and $\tau = 0$, two strands x, y can only hybridize if they are perfectly complementary (i.e., $h(x, y) \leq 0$), so a maximal such set B can be obtained by selecting one strand from every (non-palindromic) pair of Watson-Crick complementary strands; but if, on the othr hand, $\tau = n$, the mildest hybridization condition, any two strands can hybridize, so a maximal set B consists of only one strand of length n, to which every other strand may hybridize without further restrictions. Let $m = |B|$ be the cardinality of B. The basis strands will also be referred as *probes*. For easy visualization, we will assume in the illustrating examples below that m is a perfect square $m = 36$ and that the base set of probess has been affixed onto a DNA chip.

Given a string x (ordinarily much longer than the probe length n and even perhaps the number of probes m), x is said to be *h-dependent on B* is there is some concatenation c of elements of B that will hybridize to x under stringency τ, i.e., such that $|x, c| \leq \tau$. Shredding x to the corresponding fragments according to the components of c in B leads to the following slightly weaker but more manageable definition. The *signature* of x with respect to B is a vector X of dimension m that is obtained as follows. Shredding x to $|x|/n$ fragments of size n or less, X_i is the number f of fragments of x that are within threshold τ from a strand i in B, i.e., such that $|f, i| < \tau$. The value X_i will thus be referred to as a *pixel* at probe spot i. The input strands x will also be referred as *targets*.

The only difference between a DNA-memory device and a DNA microarray is that the spots on the microarray consist of carefully chosen non-crosshybridizing DNA *basis* oligonucleotides rather than entire genes. Signatures can, however, be just as easily easily implemented in practice using currently available microarray technology.

For practical applications, a number of questions arise about this representation. First, the vector X may appear not to be well-defined, since it is clear that its expression depends on the various ways to find matching segments c in the input target x, the basis strands, and their concentrations. To start with, the number r of strands per spot, here called the *resolution*, can be varied at will and so change the intensity of each pixel and the resolution ability of the representation to distiguish various inputs. To avoid some of these techincal difficulties, we will assume a relatively low resolution ($r = 6$ in the experiments below and $r = 1$ in the theoretical analysis of capacity.) On DNA chips, this resolution can be as high as the concentration (number of strands) of the basis strands (in solution), or as large as the number of strands per spot (on a chip.) More seriously, however, is the inherent uncertainty in hybridization reactions that make a signature dependent on the specific reaction conditions used in an experiment to "compute" it. From previous results in (Garzon and Deaton, 2004) [18], it is known that this problem disappears if a noncrosshybridizing set of high quality is used for the basis set. Experimentally, the signal to noise-ratio (precisely defined below) in the signature (given by the pixelwise ratio of signature signal to standard deviation of the same variable over all runs of the experiment) appears to be maximum. The hybridization likelihood between any pair of strands in a noncrosshybridizing set is minimized or even eliminated (by setting an appropriate stringency condition τ), regardless of the strands involved, the essential reason being that *a given fragment will can then only hybridize to at most one probe.* By assuming that either the test tube is small or that the reaction time is long enough that all possible hybridizations are exhausted within the experiment's time regardless of "kinetic bottlenecks", the basic problem thus becomes that of determining the set of possible signatures one may obtain by shredding the input in different ways, or even by using on different basis set.

In order to shed light on these questions, we performed a series of experiments with six target plasmids (described below) and three basis sets of different noncrosshybridizing qualities. The first set, $H40$, was obtained by randomly generating $40-$mers and filtering out strands that whose h-distance is less than a given threshold ($\tau = 19\%$ of the shorter strands.) The second set, Ark, was obtained bottom up, by concatenating pairs of $20-$mers randomly chosen from a set of $20-$mers obtained by similar filtering and adding the resultant strand to the current membes of the set if its h-distance is greater than or equal to τ. The original set of $20-$mers was obtained by using a more sophisticated genetic algorithm search using a Gibbs energy model (Deaton et al., 2002) [6] as fitness function. The third set, Hyb, was obtained by concatenating $40-$mers from $H40$ and $20-$mers from Ark and again adding the resultant strand to the set if its h-distance is greater than or equal to threshold h-distance ($\tau = 29$.) The noncrosshybridizing quality of these sets is high, as measured by the pairwise Gibbs energy of strands in the sets shown in Fig. 1.

Once the basis set and the reactions conditions have been optimized, the most important question remains, i.e., how *unique* is the signature for a given target x? To gain some insights into this question, six(6) large plasmids of lengths varying between $2.9K$ and $3.2K$ bps were chosen for targets and shredded into fragments of size 35 bps or less. Regarding the protocols as a stochastic process, experiments were conducted *in simulation* to obtain their signatures on a basis B as described above. Each experiment was run 10 times in a tested simulation environment, *Edna* (Garzon and Blain, 2004) [17] and (Garzon and Rose, 2004) [13]. As expected, we obtained a range of different signatures on different runs. Therefore, to make this concept precise, it is necessary to re-define a signature

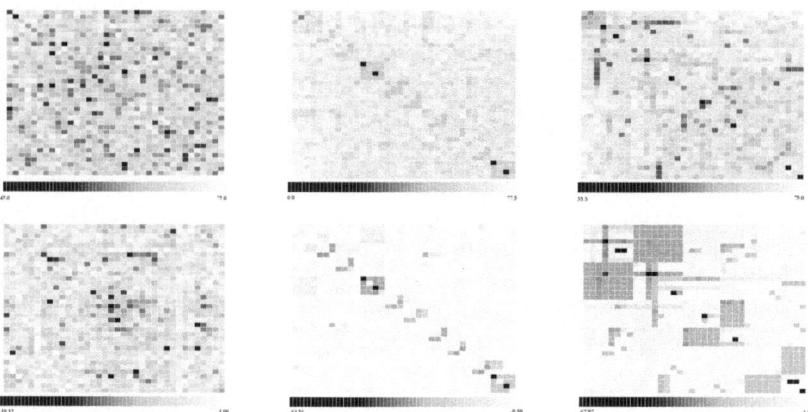

Fig. 1. Noncrosshybridization quality of a selection of three basis sets H40 (left column), Ark (middle column), and Hyb (right column) measured by the combinatorial h-distance (Garzon et al, 1997) [10] (the top row), and, the Gibbs energy model of (Deaton et al., 2002) [6] (bottom row). Their quality is high since lighter colors represent pairs far apart in hybridization distance or Gibbs energy (which is shown normalized to a comparable scale), i.e. lower hybridization affinity.

as a *sphere* in a high-dimensional euclidean space of dimension m (the number of spots on the microarray, i..e, number of noncrosshybridizng strands in the basis set.) The center of this sphere (below called the *ideal point signature*) is the componentwise average in mD-euclidean space of the outcomes of all possible point signatures obtained in running an experiment to find the signature. The radius of the sphere will be some measure of the variability of the all possible point signatures obtained in a given set of conditions. Here we use the *average euclidean distance* (i.e., the L_2-average) of all possible point signatures to the ideal signature.

With this definition of a signature as a sphere in mD-Euclidean space of radius given by the average distance from the ideal point signature, the problem of translating arbitrary data is resolved. We will refer to this sphere as the *volume* signature to distinguish it from the point signatures in the original definition. Examples can be seen in Fig. 5 (left). Fixing a basis set B, every target x determines a unique (volume) signature.

2.2   Sensitivity and Robustness of the DNA-Chip Representation

The critical question now about the signature of a given target x is the amount of information it contains, particularly to what extent it determines the target x uniquely, or, at least, whether it can distinguish it from other input targets. In this section we address these questions.

How much information about the target x does its volume signature provide? A comparison can be made using the so-called chipwise SNR (Signal-to-Noise Ratio) defined as follows. For each pixel X_i, SNR_i defined as the ratio of the

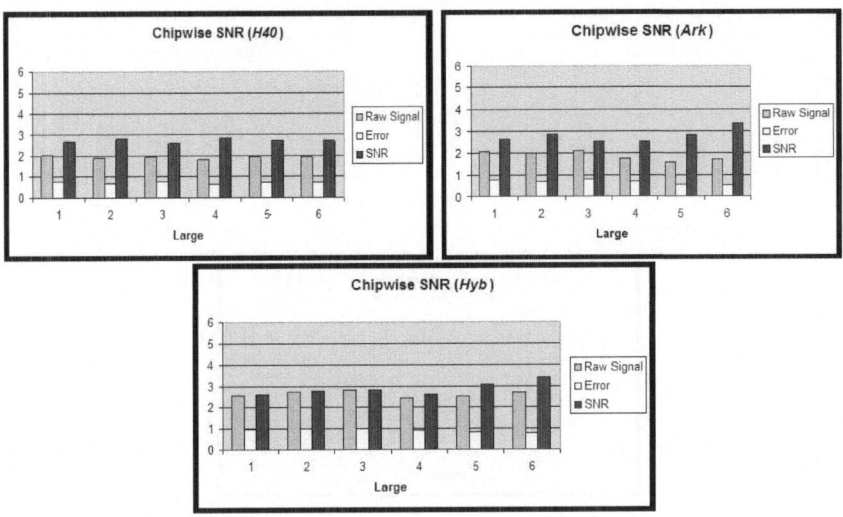

Fig. 2. Signal to Noise ratios (SNR) in six experiments with plasmids genomes over three sets H40 (top left), Ark (top right), Hyb (bottom) of various noncrosshybridization qualities

pixel's average value divided by the standard deviation of the same random variable X_i. The SNR of the target x (with respect to a given basis) is given by ratio of the L_2-average of the pixelwise signals divided by the L_2-average of pixelwise standard deviations. Fig. 2 shows the chipwise SNR comparison for all plasmids used in the experiments. The SNRs for the chosen plasmids are shown in Fig. 2. Some of them can be clearly distinguishable even if we just look at their SNRs alone, although it is too raw an average to expect full distinction among all plasmids. Nonetheless, the SNR gives a sense of the sensitivity of this representation.

There are other factors determining the radius of a volume signature that impact the variability of the representation. It is clear that slicing the input x into different fragments might change its volume radically, and that, conversely, re-assembling the fragments in a different order may yields the same representation for a different input x'. How much does the representation depend on the lengths of the shredding x into pieces? The results described next provide an intuition on how Euclidean spheres radii change in representation signatures across a range of plasmid sizes ($2.9K$ to $3.2K$). Again, all experiments for sensitivity were performed ten times. Only results on the $H40$ probe set are shown below.

Fig. 3 shows the variations in the signature's radius obtained by varying the lengths of the the fragments shredding the target x. The radius increased for smaller fragments (15-25bp) compared to the original fragment size (25-35bp.) This increase is to be expected because more fragments are availability for hybridization, which results in higher signal and proportionately higher variability. The higher the standard deviation the bigger the radius of euclidean sphere.

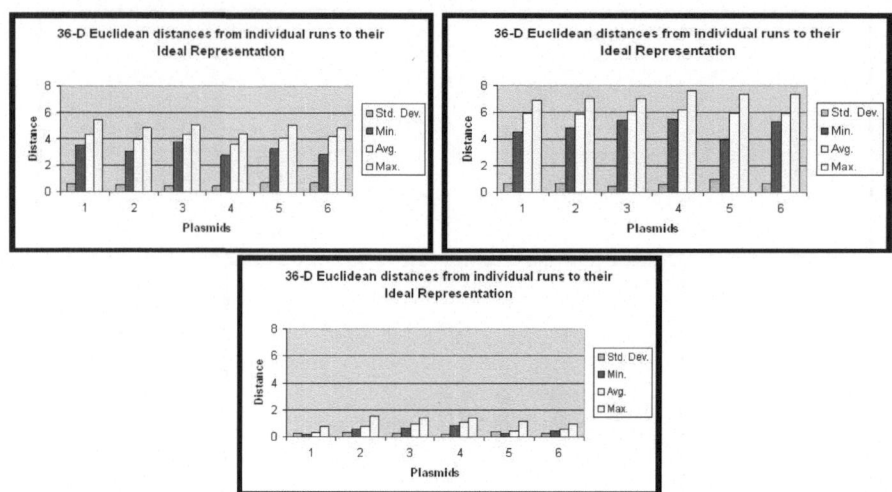

Fig. 3. Volume signature variability for original fragments of $25 - 35$bps (top left); small fragments of $15 - 25$bps (top right); and large fragments of $35 - 45$bp (bottom). Larger fragments yield a crisper signature (smaller radius) while shorter fragments yield fuzzier signatures (larger radius).

The converse argument can be given to explain the decrease in radius with large fragments (35-45bp).

Further experiments were performed to determine the sensitivity of the signature through contamination of targets in several ways. The contamination will be referred to as "noise." The noise introduced into original plasmids was of three types. The results described next provide a quantitative idea of the change expected in sensitivity of the signatures for plasmid 1. The target plasmid 1 was varied by introducing three types of noise:

- *Substitution*: Plasmids fragments are replaced by other random fragments of equal length;
- *Addition*: Random fragments were inserted in the plasmid;
- *Reduction*: Random fragments were removed from the plasmid.

Fig. 4 shows the variations in signatures of the resulting plasmid targets. The signatures' radii do not change much with substitution noise regardless of the amount substituted. However, the radii increased with increase in noise in the case of added noise and radii decreased with increased reduction noise. This behavior is similar for small and large fragments. This is additional evidence of sensitivity of the volume signature to changes in the length of and number of target fragments.

In order to determine the robustness of the representation, i.e., how much change must be made to a target for it to produce a different volume signature, we used the so-called *overlap* of volume signatures. This measure attempts to

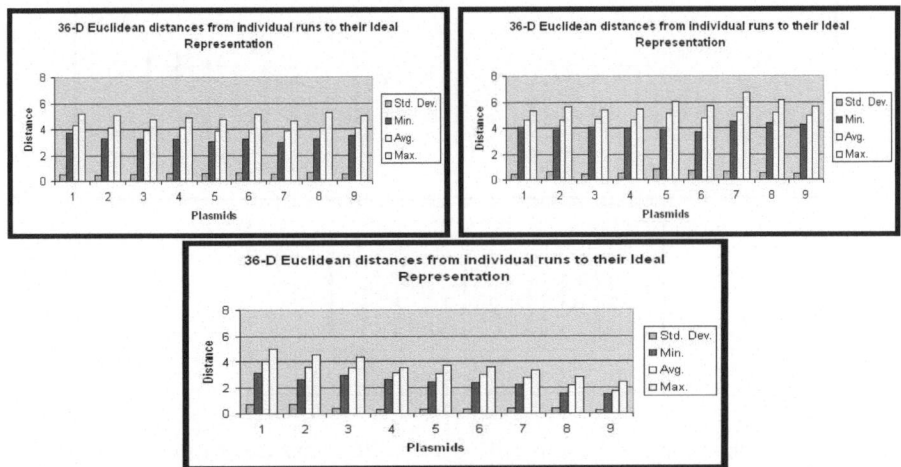

Fig. 4. Volume signature variability for target basis H40 for noise that has been substituted (top left), added (top right) or reduced (bottom). Volume signatures are also sensitive to changes in the length of and number of residues in the probe. However, the radii vary in proportion to noise. This behavior is similar for small and large fragments.

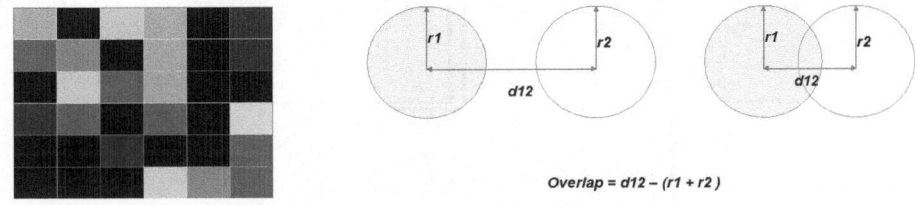

Fig. 5. The ideal representation of plasmid 1 (left). The overlap between two representations (spheres) is the excess (or defect) of the distance between ideal representations and the sum of the radii of the individual signatures. If the overlap is positive, the volume signatures do not intersect (middle), while they will if the overlap is negative (right).

capture the displacement in the ideal representation from its original parent with various types of noise, as shown in Fig. 5. *Overlap* is the difference between the distance between ideal representation and the sum of the average radii of their volume signatures. Fig. 6 shows the euclidean distances traveled from the ideal signature by variation of plasmid 1. Increasing substitution noise smoothly shifts the ideal signature but maintains overlap up to 60%. Only at 70% does the volume signature become nearly disjoint. With added noise, the threshold for the same phenomenon is about 90% noise, and with reduced noise it is about 60% noise. An overlap distance of −1 can be considered enough for two spheres to separate. So, it can be concluded that representations are sensitive to noise from Fig. 6 as the distance to original ones increase with increase in noise.

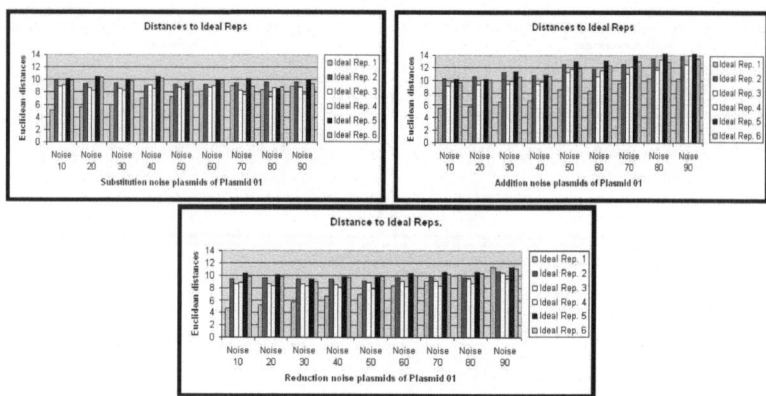

Fig. 6. Volume signatures are robust. It requires 70% for substitution noise (top left) and reduction noise (bottom) for a probe to become closer to others away from itself, while it remains closest to the original even with 90% added (top right) straneous fragments.

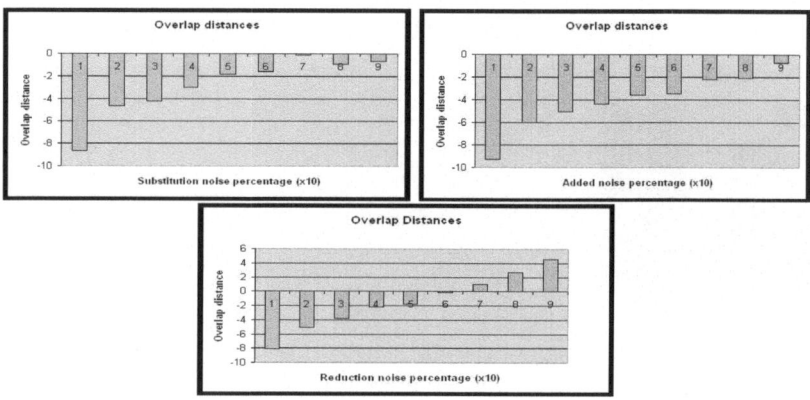

Fig. 7. Overlaps of volume signatures of noisy variations of plasmid 1 to its original volume signature. Substitution noise of 70% (top left) is required for the volume signature to become nearly disjoint. With added noise (top right), the threshold for the same phenomenon is about 90% noise. With reduced noise (bottom), the threshold it is about 60% noise. An overlap distance of −1 can be considered enough for two spheres to separate. Thus, representations are fairly insensitive to a small amounts of noise, while remaining sensitive to larger changes.

Fig. 7 shows a further analysis of the same experiment by considering the distances of the noisy plasmid 1 to the ideal signature of all six plasmids. The most interesting threshold is the amount of noise required for varying plasmid 1 to become closer to another plasmid than to its original. That number is 70% for substitution noise and reduction noise, but the noisy plasmids remain closest to the original even under 90% added straneous fragments. This is is a remarkable robustness.

3   Theoretical Analysis of Sensitivity and Capacity of DNA-Based Chips

We now provide an abstraction of the concept of a signature in order to provide a theoretical model to estimate the capacity of DNA chips under optimal conditions. First, due to the fact that, under realistic conditions, it is infeasible to expose very long uncut copies of an input sequence to the chip, we assumed in the definition of signature that the targets are shredded by restriction enzymes into manageable fragments before they are exposed to the chip. To simplify the analysis in the theoretical model, however, we will assume that no shredding of targets will be carried out.

To justify this assumption, we observe that we can disregard all basis strands that are Watson-Crick complementary to the cleaving restriction sites used for shredding since hybridizations of targets to basis strands in the vicinity of the restriction sites will not be happen. Therefore, we can eliminate shredding if we guarantee that the basis set contains no restriction site used by shredding enzymes and still get an identical signature for the same target. Second, we will assume that basis strands float freely in solution instead of being affixed to a chip, which is justified given the nonhybridization property of the basis set. Third, we will also assume that a fixed concentration of basis strand and targets is placed in the tube. Thus, target strands are exposed, in principle, to hybridization of all basis strands at many places, and, consequently, many copies of the same basis strand may hybridize to several parts of the input sequence. Thus, even though target sequences can be arbitrarily long, there can only be a bounded number of point signatures, and so different targets may yield the same point signature. Under these assumptions, the volume signature produced by an uncut input target is essentially the same as the one produced by the the original definition above.

In this model, the chip capacity (i.e., the number of distinguishable target signatures) becomes a function of σ, *i.e.*, the total number of copies of all basis strands that hybridize to a target.. Realistically, when σ varies slightly, so does its capacity. Given a set $B = \{i_1, i_2, \cdots, i_k\}$ of *basis strands* and a target sequence X, its *signature* is $x_B = (x_1, x_2, \cdots, x_m)$, where x_i is the number of times basis oligo i hybridizes to (different parts) of X. Under these conditions, input targets X and Y are indistinguishable if and only if $x_B = y_B$, i.e. $x_i = y_i$, for all $1 \leq i \leq m$.

The basis B used to create a DNA chip relates to the capacity of the chip in interesting ways. We observed that the arguments in (Phan & Garzon, 2004) [20] show that the memory capacity of the noncrosshybridizing basis B is large if (a) its oligo distribution as substrings of the input sequences is as far from uniform as possible; and, (b) they *cover* the input targets as much as possible. Specifically, we found that

Proposition 1. *The probability of two different input sequences being indistinguishable from each other is*

$$P(X_B = Y_B | X \neq Y) = \frac{\binom{\sigma}{x_1, x_2, \cdots, x_k}}{k^\sigma} \leq \frac{2^{\sigma H(P)}}{k^\sigma} = \frac{1}{2^{\sigma(\log_2 k - H(P))}} \qquad (1)$$

where $\sum_{i=1}^{k} x_i = \sum_{i=1}^{k} y_i = \sigma$, and $H(P) = -\frac{x_i}{\sigma} \sum_{i=1}^{k} \frac{x_i}{\sigma}$, the Shannon entropy of the distribution of B in X (and Y).

In other words, the capacity of the chip based on B is small if one of two conditions are true:
(1) σ is small, or (2) the distribution of the bases as substrings of the inputs sequences approaches random (i.e. $H(P)$ approaches $\log_2 k$). When B covers the input sequences completely, every substring of an input of the same length as the $|s_i|'s$ hybridizes to one of the bases, and consequently $\sigma \approx \frac{|X|}{|i|}$, where $|i|$ is the length of basis oligo i. Conversely, when B covers the input sequences sparsely, $\sigma \ll \frac{|X|}{|i|}$ and the probability of two different input sequences being indistinguishable increases.

Using these arguments, we can also provide a theoretical estimate of the capacity of the DNA chip for volume signatures as defined above. The limit of a DNA chip's capacity is the number of distinguishable signatures that the chip can possibly produce. Since the total number of occurrences of each basis strand (x_i's) in X adds up to σ, we have the following conditions:

$$\forall i, (x_i \geq 0), \qquad \text{and} \qquad x_1 + x_2 + \cdots + x_k = \sigma \qquad (2)$$

Using an elementary combinatorial argument, we can show that

Proposition 2. *The optimal capacity a DNA chip is $\binom{\sigma+m-1}{m-1}$, if defined as the maximum number of distinguishable point signatures.*

As mentioned above, it is not the case that exposing an input target a number of times will get an identical signature each time. In the current mode, where the chip is not affixed but in solution, this sensitivity to distinguish input is decreased. because the signatures of different but similar input sequences are likely indistinguishable. The sensivity of the chip can be collectively captured by two parameters r and r_σ, regardless of the sources of noise. The capacity of the chip is estimated indirectly via the size of a maximal set, called C, in signature space. This set C can be thought of as a *maximal* collection of centers of non-intersecting spheres with a fixed radius r. Hence, the sphere of radius r specifically captures the uncertainty of telling signatures of similar sequences apart; sequences whose signatures are within a radius r are not distinguishable. The other parameter, r_σ captures the fact that due to noise or other factors, even when the bases *cover* well input sequences, the number of basis strands hybridized to these inputs may not always be exactly σ. Hence, we assume that the total number of basis strands hybridized to the input sequences vary from $\sigma - r_\sigma$ to $\sigma + r_\sigma$. On these considerations, we have established the following estimate, where the set V consists of the signatures of all input targets that the chip could distinguish under the sensitivity parameter r.

Theorem 1. *The maximal set C of signatures that are distinguishable on a DNA-based (m, σ, r)-chip is of size $|C|$ bounded by*

$$\frac{|V|}{v(2r+1)} \leq |C| \leq \frac{|V|}{v(r)} \tag{3}$$

A full proof is omitted. Briefly, these bounds are obtained by determining the upper and lower bounds of a maximal code, in a similar fashion as the Hamming and Gilbert-Varshamov bounds, respectively. Intuitively, the input sequences in V include those whose signatures fall inside the hyperplane in equation 2 and those input sequences whose signatures fall within a distance r of the hyperplane.

Lemma 1

$$|V| = \binom{\sigma + m - 1}{m - 1} + \sum_{i=1}^{r_\sigma} 2\binom{\sigma + i + m - 1}{m - 1}$$

$|V|$ is, however, not the same as $|C|$; i.e. it is not the capacity of the chip because two signatures within a distance of r from each other are not disintiguishable. To estimate $|C|$, we need to know, $v(r)$, the number of signatures inside a sphere of radius r.

Lemma 2

$$v(r) = 1 + \sum_{e=1}^{r} \sum_{i=1}^{min\{e,k\}} 2^i \binom{k}{i} \binom{e - i - 1}{i - 1}$$

Proof. A full proof is omitted for space reasons. Briefly, the sum accounts for all points at distance exactly e from a center, for $0 \leq e \leq r$. $\square$

4 Conclusions and Future Work

This paper gives experimental (in simulation) and theoretical analyzes of a recenly proposed method (Garzon and Deaton, 2004) [18] to represent abiotic information onto DNA molecules in order to make processing data at massive scales *efficient* and *scalable*. The mapping is readily implementable with current microarray technology (Stekel, 2003) [21], bypasses synthesis of all but a few strands, and it's promising for the tera- and peta-byte scopes volumes required for a meaningful applications (more below.) Furthermore, we quantify the sensitivity of the representations and show that it can be made robust despite the uncertainty of the hybridization chemistry. Third, we show a theoretical analysis of the capacity of this type of representation to code information, as well as an information-theoretic estimate of the number of distinguishable targets that can be represented on a given chip under reaction conditions characterized by hybridization stringency parameters.

A direct application of this method in bioinformatics is a new approach to genomic analysis that increases the signal-to-noise ratio in microarrays commonly used in bioinformatics. The method yields higher resolution and accuracy in the analysis of genomica data, and only requires some processing in what can be termed an "orthogonalization" procedure to the given set of targets/genes before placing them on the microarrays. These advantages may be critical for problems such as classification problems (disease/healthy data). More details can be found in (Garzon et al., 2005) [12].

Further applications can be expected in the analysis and data mining of abiotic data, whose representation is automatically defined with respect to a given basis set B. Given a noncrosshybridizing basis and adequate thresholds on the stringency of reaction condition and acceptable levels of variability of the representation (i.e., the capacity to distinguish inputs through their representations), the signatures of arbitrary inputs are completely determined and require no precomputation or synthesis of any DNA strands, other than the basis strands. In other words, this method provides a *universal* and *scalable* method to represent data of any type. For example, because of the superposition (linearity) property (module the variability implicit in the representation), a corpus of English text can be automatically encoded just by finding representations for the words in the basic vocabulary (words) in the corpus. Thereafter, the representation of a previously unknown piece of text can be inferred by superposition of the component words. There is evidence that these representations can be used for semantic processing of text corpora in lieu of the original text [11]. Given the newly available large basis sets [4, 5, 6] in the order of megasets, device with the ability to process data for information extraction appear now within reach in a relatively short time.

Acknowledgements

Much of the work presented here has been done in collaboration with a molecular computing consortium that includes Russell Deaton, Jin Wu (U. of Arkansas), Junghuei Chen, and David Wood (U. Delaware). Support from the National Science Foundation grant QuBiC/EIA-0130385 is gratefully acknowledged.

References

1. E. Baum. Building an associative memory vastly larger than the brain. *Science*, 268:583–585, 1995.
2. H. Bi, J. Chen, R. Deaton, M. Garzon, H. Rubin, and D. Wood. A pcr-based protocol for in vitro selection of non-crosshybridizing oligonucleotides. *J. of Natural Computing*, 2003.
3. Derrel Blain and M. Garzon. Simulation tools for biomolecular computing. *In: (Garzon and Rose, 2004)*, 3:4:117–129, 2004.
4. J. Chen, R. Deaton, M. Garzon, J.W. Kim, D.H. Wood, H. Bi, D. Carpenter, J.S. Le, and Y.Z. Wang. Sequence complexity of large libraries of dna oligonucleotides. *In these proceedings*, 2005.
5. J. Chen, R. Deaton, Max Garzon, D.H. Wood, H. Bi, D. Carpenter, and Y.Z. Wang. Characterization of non-crosshybridizing dna oligonucleotides manufactured in vitro. Proc. 8th Int Conf on DNA Computing DNA8.

6. R. Deaton, J. Chen, H. Bi, and J. Rose. A software tool for generating non-crosshybridizing libraries of dna oligonucleotides. pages 252–261, 2002. In: [14].

7. R.J. Deaton, J. Chen, H. Bi, M. Garzon, H.Rubin, and D.H. Wood. A pcr-based protocol for in vitro selection of non-crosshybridizing oligonucleotides. *In: (Hagiya & Ohuchi, 2002)*, pages 105–114, 2002a.

8. M. Garzon, K. Bobba, and B. Hyde. Digital information encoding on dna. *Springer-Verlag Lecture Notes in Computer Science 2590(2003)*, pages 151–166, 2003b.

9. M. Garzon, R. Deaton, P. Neathery, R.C. Murphy, D.R. Franceschetti, and E. Stevens Jr. On the encoding problem for dna computing. pages 230–237, 1997. Poster at The Third DIMACS Workshop on DNA-based Computing, U of Pennsylvania. Preliminary Proceedings.

10. M. Garzon, P.I. Neathery, R. Deaton, R.C. Murphy, D.R. Franceschetti, and S.E. Stevens Jr. A new metric for dna computing. *In: (Koza et al., 1997)*, pages 472–478, (1997a).

11. M. Garzon, A. Neel, and K. Bobba. Efficiency and reliability of semantic retrieval in dna-based memories. pages 157–169, 2003.

12. M. Garzon, V. Phan, K. Bobba, and R. Kontham. Sensitivity analysis of microaary data: A new approach. In *Proc. IBE Conference, Athens GA.*, 2005. Biotechnology Press.

13. M. Garzon and John Rose. Simulation tools for biomolecular computing. *Special Issue of the Journal of Natural Computing*, 4:3, 2004.

14. M. Hagiya and A. Ohuchi. In *Proc. 8th Int. Meeting on DNA-Based Computers.*, 2002. Springer-Verlag Lecture Notes in Computer Science LNCS 2568. Springer-Verlag.

15. T. Head, M. Yamamura, and S. Gal. Aqueous computing: Writing on molecules. 1999. Proceedings of the Congress on Evolutionary Computing (CEC'99).

16. T. Head, M. Yamamura, and S. Gal. Relativized code concepts and multi-tube dna dictionaries. In *Finite vs Infinite: COntrobutions to an eternal dilemma (Discrete math and Theoretical Computer SCience)*, pages 175–186, 2001.

17. Garzon M, D. Blain, and A. Neel. Virtual test tubes for biomolecular computing. *In: (Garzon and Rose, 2004)*, 3:4:460–477, 2004.

18. Garzon M and R. Deaton. Codeword design and information encoding in dna ensembles. *J. of Natural Computing*, 3:4:253–292, 2004.

19. D. Mount. Bioinformatics: sequence and genome analysis. *Spring Harbor Lab Press, MD*, 2001.

20. V. Phan and M. Garzon. Information encoding using dna. *Proc. 10th Int Conf on DNA Computing DNA10*, 2004.

21. D. Stekel. *Microarray Bioinformatics*. Cambridge University Press, 2003.

Simple Operations for Gene Assembly

Tero Harju[1,4], Ion Petre[2,3,4],
Vladimir Rogojin[3,4], and Grzegorz Rozenberg[5,6]

[1] Department of Mathematics, University of Turku,
Turku 20014, Finland
harju@utu.fi
[2] Academy of Finland
[3] Department of Computer Science, Åbo Akademi University,
Turku 20520, Finland
ipetre@abo.fi, vrogojin@abo.fi
[4] Turku Centre for Computer Science,
Turku 20520, Finland
[5] Leiden Institute for Advanced Computer Science, Leiden University,
Niels Bohrweg 1, 2333 CA Leiden, The Netherlands
[6] Department of Computer Science, University of Colorado at Boulder,
Boulder, Co 80309-0347, USA
rozenber@liacs.nl

Abstract. The intramolecular model for gene assembly in ciliates considers three operations, ld, hi, and dlad that can assemble any gene pattern through folding and recombination: the molecule is folded so that two occurrences of a pointer (short nucleotide sequence) get aligned and then the sequence is rearranged through recombination of pointers. In general, the sequence rearranged by one operation can be arbitrarily long and consist of many coding and non-coding blocks. We consider in this paper some simpler variants of the three operations, where only one coding block is rearranged at a time. We characterize in this paper the gene patterns that can be assembled through these variants. Our characterization is in terms of signed permutations and dependency graphs. Interestingly, we show that simple assemblies possess rather involved properties: a gene pattern may have both successful and unsuccessful assemblies and also more than one successful assembling strategy.

1 Introduction

The ciliates have a very unusual way of organizing their genomic sequences. In the macronucleus, the somatic nucleus of the cell, each gene is a contiguous DNA sequence. Genes are generally placed on their own very short DNA molecules. In the micronucleus, the germline nucleus of the cell, the same gene is broken into pieces called MDSs (macronuclear destined sequences) that are separated by noncoding blocks called IESs (internally eliminated sequences). Moreover, the order of MDSs is shuffled, with some of the MDSs being inverted. The structure is particularly complex in a family of ciliates called *Stichotrichs* – we concentrate in this paper on this family. During the process of sexual reproduction,

A. Carbone and N.A. Pierce (Eds.): DNA11, LNCS 3892, pp. 96–111, 2006.

ciliates destroy the old macronuclei and transform a micronucleus into a new macronucleus. In this process, ciliates must assemble all genes by placing in the orthodox order all MDSs. To this aim they are using *pointers*, short nucleotide sequences that identify each MDS. Thus, each MDS M begins with a pointer that is exactly repeated in the end of the MDS preceding M in the orthodox order. The ciliates use the pointers to splice together all MDSs in the correct order.

The intramolecular model for gene assembly, introduced in [9] and [27] consists of three operations: ld, hi, and dlad. In each of these operations, the molecule folds on itself so that two or more pointers get aligned and through recombination two or more MDSs get combined into a bigger composite MDS. The process continues until all MDSs have been assembled. For details related to ciliates and gene assembly we refer to [15], [20], [21], [22], [23], [24], [25], [26] and for details related to the intramolecular model and its mathematical formalizations we refer to [3], [4], [7], [8], [11], [12], [13], [28], [29], as well as to the recent monograph [5]. For a different intermolecular model we refer to [17], [18], [19].

In general there are no restrictions on the number of nucleotides between the two pointers that should be aligned in a certain fold. However, all available experimental data is consistent with restricted versions of our operations, in which between two aligned pointers there is never more than one MDS, see [5] and [6]. We propose in this paper a mathematical model for simple variants of ld, hi, and dlad. The model, in terms of signed permutations, is used to answer the following question: which gene patterns can be assembled by the simple operations? As it turns out, the question is difficult: the simple assembly is a non-deterministic process, with more than one strategy possible for certain patterns and in some cases, with both successful and unsuccessful assemblies. We completely answer the question in terms of sorting signed permutations. Here, a signed permutation represents the sequence of MDSs in a gene pattern, including their orientation.

There is rich literature on sorting (signed and unsigned) permutations, both in connection to their applications to computational biology in topics such as genomic rearrangements or genomic distances, but also as a classical topic in discrete mathematics, see, e.g., [1], [2], [10], [16].

2 Mathematical Preliminaries

For an alphabet Σ we denote by Σ^* the set of all finite strings over Σ. For a string u we denote $\text{dom}(u)$ the set of letters occurring in u. We denote by Λ the empty string. For strings u, v over Σ, we say that u is a *substring* of v, denoted $u \leq v$, if $v = xuy$, for some strings x, y. We say that u is a *subsequence* of v, denoted $u \leq_s v$, if $u = a_1 a_2 \ldots a_m$, $a_i \in \Sigma$ and $v = v_0 a_1 v_1 a_2 \ldots a_m v_m$, for some strings v_i, $0 \leq i \leq m$, over Σ. For some $A \subseteq \Sigma$ we define the morphism $\phi_A : \Sigma^* \to A^*$ as follows: $\phi_A(a_i) = a_i$, if $a_i \in A$ and $\phi_A(a_i) = \Lambda$ if $a_i \in \Sigma \setminus A$. For any $u \in \Sigma^*$, we denote $u|_A = \phi_A(u)$. We say that the *relative positions* of letters from set $A \subseteq \Sigma$ are the same in strings $u, v \in \Sigma^*$ if and only if $u|_A = v|_A$.

Let $\Sigma_n = \{1, 2, \ldots, n\}$ and let $\overline{\Sigma}_n = \{\overline{1}, \overline{2}, \ldots, \overline{n}\}$ be a *signed copy* of Σ_n. For any $i \in \Sigma_n$ we say that i is a *unsigned letter*, while $\overline{i}$ is a *signed* letter. Let $\|.\|$ be the morphism from $(\Sigma_n \cup \overline{\Sigma}_n)^*$ to Σ_n^* that unsigns the letters: for all $a \in \Sigma_n$, $\|\overline{a}\| = \|a\| = a$. For a string u over $\Sigma_n \cup \overline{\Sigma}_n$, $u = a_1 a_2 \ldots a_m$, $a_i \in \Sigma_n \cup \overline{\Sigma}_n$, for all $1 \leq i \leq m$, we denote its *inversion* by $\overline{u} = \overline{a}_m \ldots \overline{a}_2 \overline{a}_1$, where $\overline{\overline{a}} = a$, for all $a \in \Sigma_n$.

Consider a *bijective mapping* (called *permutation*) $\pi : \Delta \rightarrow \Delta$ over an alphabet $\Delta = \{a_1, a_2, \ldots, a_l\}$ with the order relation $a_i \leq a_j$ for all $i \leq j$. We often identify π with the string $\pi(a_1)\pi(a_2)\ldots\pi(a_l)$. The domain of π, denoted $\mathsf{dom}(\pi)$, is Δ. We say that π is *(cyclically) sorted* if $\pi = a_k\, a_{k+1} \ldots a_l\, a_1\, a_2 \ldots a_{k-1}$, for some $1 \leq k \leq l$.

A *signed permutation* over Δ is a string ψ over $\Delta \cup \overline{\Delta}$ such that $\|\psi\|$ is a permutation over Δ. We say that ψ is *(cyclically) sorted* if $\psi = a_k\, a_{k+1} \ldots a_l\, a_1\, a_2 \ldots a_{k-1}$ or $\psi = \overline{a}_{k-1} \ldots \overline{a}_2\, \overline{a}_1\, \overline{a}_l \ldots \overline{a}_{k+1}\, \overline{a}_k$, for some $1 \leq k \leq l$. Equivalently, ψ is sorted if either ψ, or $\overline{\psi}$ is a sorted unsigned permutation. In the former case we say that ψ is sorted in the *orthodox order*, while in the latter case we say that ψ is sorted in the *inverted order*.

3   The Intramolecular Model

Three molecular operations, ld, hi, dlad were conjectured in [9] and [27] for gene assembly. We only show here the folding and the recombinations taking place in each case, referring for more details to [5]. It is important to note that all foldings are aligned by pointers, some relatively short nucleotide sequences at the intersection of MDSs and IESs. The pointer at the end of an MDS M coincides (as a nucleotide sequence) with the pointer in the beginning of the MDS following M in the assembled gene.

3.1   Simple Operations

Note that all three operations ld, hi, dlad are *intramolecular*, that is, a single molecule folds on itself to rearrange its coding blocks. Thus, since ld excises one circular molecule, that circular molecule can only contain noncoding blocks (or, in a special case, contain the entire gene, see [5] for details on boundary ld): we say that ld must always be *simple* in a successful assembly. As such, the effect of ld is that it combines two consecutive MDSs into a bigger composite MDS. E.g., consider that $M_i M_{i+1}$ is part of the molecule, i.e., MDS M_{i+1} succeeds M_i being separated by one IES I. Thus, pointer $i + 1$ has two occurrences that flank I. Then ld makes a fold as in Fig. 1 aligned by pointer $i + 1$, excises IES I as a circular molecule and combines M_i and M_{i+1} into a longer coding block.

In the case of hi and dlad, the rearranged sequences may be arbitrarily large. E.g., the actin I gene in S.nova has the following sequence of MDSs: $M_3 M_4 M_6 M_5 M_7 M_9 \overline{M}_2 M_1 M_8$, where MDS M_2 is inverted. Here, pointer 3 has two occurrences: one in the beginning of M_3 and one, inverted, in the end of M_2. Thus, hi is applicable to this sequence with the hairpin aligned on pointer 3, even though five MDSs separate the two occurrences of pointer 3. Similarly, dlad

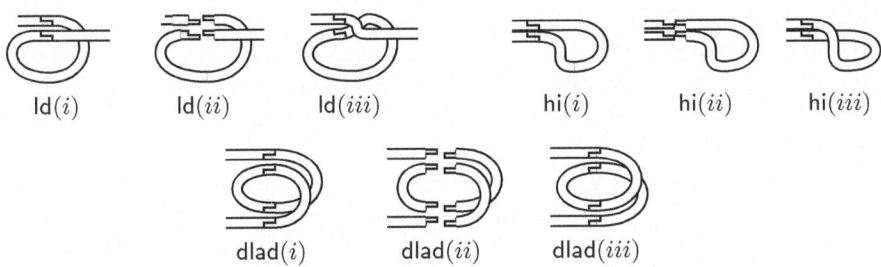

Fig. 1. Illustration of the ld, hi, dlad molecular operation showing in each case: (i) the folding, (ii) the recombination, and (iii) the result

is applicable to the MDS sequence $M_2 M_8 M_6 M_5 M_1 M_7 M_3 M_{10} M_9 M_4$, with the double loops aligned on pointers 3 and 5. Here the first two occurrences of pointers 3, 5 are separated by two MDSs (M_8 and M_6) and their second occurrences are separated by four MDSs (M_3, M_{10}, M_9, M_4).

As it turns out, all available experimental data is consistent with applications of so-called "simple" hi and dlad: particular instances of hi and dlad where the folds and thus, the rearranged sequences contain only one MDS. We define the simple operations in the following.

An application of the hi-operation on pointer p is *simple* if the part of the molecule that separates the two copies of p in an inverted repeat contains only one MDS (and one IES). We have here two cases, depending on whether the first occurrence of p is incoming or outgoing. The two possibilities are illustrated in Fig. 2, where the MDSs are indicated by rectangles and their flanking pointers are shown.

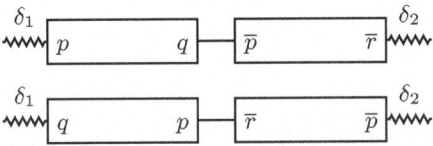

Fig. 2. The MDS/IES structures where the *simple hi*-rule is applicable. Between the two MDSs there is only one IES.

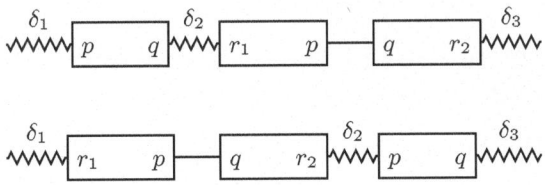

Fig. 3. The MDS/IES structures where the *simple dlad*-rule is applicable. Straight line denotes one IES.

An application of dlad on pointers p, q is *simple* if the sequence between the first occurrences of p, q and the sequence between the second occurrences of p, q consist of either one MDS or one IES. We have again two cases, depending on whether the first occurrence of p is incoming or outgoing. The two possibilities are illustrated in Fig. 3.

One immediate property of simple operations is that they are not universal, i.e., there are sequences of MDSs that cannot be assembled by simple operations. One such example is the sequence $(\overline{2}, \overline{b})(4, e)(3, 4)(2, 3)$. Indeed, neither ld, nor simple hi, nor simple dlad is applicable to this sequence.

4   Gene Assembly as a Sorting of Signed Permutations

The gene structure of a ciliate can be represented as a signed permutation, denoting the sequence and orientation of each MDS, while omitting all IESs. E.g., the signed permutation associated to gene actin I in S.nova is $3\,4\,6\,5\,7\,9\,\overline{2}\,1\,8$. The rearrangements made by ld, hi, dlad at the molecular level leading to bigger composite MDSs have a correspondent on permutations in combining two already sorted blocks into a longer sorted block. Assembling a gene is equivalent in terms of permutations to sorting the permutation associated to the micronuclear gene as detailed below.

When formalizing the gene assembly as a sorting of permutations we effectively ignore the operation ld observing that once such an operation becomes applicable to a gene pattern, it can be applied at any later step of the assembly, see [3] and [7] for a formal proof. In particular, we can assume that all ld operations are applied in the last stage of the assembly, once all MDSs are sorted in the correct order. In this way, the process of gene assembly can indeed be described as a process of sorting the associated signed permutation, i.e., arranging the MDSs in the proper order, be that orthodox or inverted.

The simple hi is formalized on permutations through operation sh. For each $p \geq 1$, sh_p is defined as follows:

$$\mathsf{sh}_p(x\,(p+1)\,\overline{p}\,y) = x\,\overline{(p+1)\,\overline{p}}\,y, \qquad \mathsf{sh}_p(x\,\overline{p}\,(p-1)\,y) = x\,\overline{p}\,\overline{(p-1)}\,y,$$
$$\mathsf{sh}_p(x\,(p-1)\,\overline{p}\,y) = x\,(p-1)\,p\,y, \qquad \mathsf{sh}_p(x\,\overline{p}\,(p+1)\,y) = x\,p\,(p+1)\,y,$$

where x, y are signed strings over Σ_n. We denote $\mathsf{Sh} = \{\mathsf{sh}_i \mid 1 \leq i \leq n\}$.

The simple dlad is formalized on permutations through operation sd. For each p, $2 \leq p \leq n-1$, sd_p is defined as follows:

$$\mathsf{sd}_p(x\,p\,y\,(p-1)\,(p+1)\,z) = x\,y\,(p-1)\,p\,(p+1)\,z,$$
$$\mathsf{sd}_p(x\,(p-1)\,(p+1)\,y\,p\,z) = x\,(p-1)\,p\,(p+1)\,y\,z,$$

where x, y, z are signed strings over Σ_n. We also define $\mathsf{sd}_{\overline{p}}$ as follows:

$$\mathsf{sd}_{\overline{p}}(x\,\overline{(p+1)}\,\overline{(p-1)}\,y\,\overline{p}\,z) = x\,\overline{(p+1)}\,\overline{p}\,\overline{(p-1)}\,y\,z,$$
$$\mathsf{sd}_{\overline{p}}(x\,\overline{p}\,y\,\overline{(p+1)}\,\overline{(p-1)}\,z) = x\,y\,\overline{(p+1)}\,\overline{p}\,\overline{(p-1)}\,z,$$

where x, y, z are signed strings over Σ_n. We denote $\mathsf{Sd} = \{\mathsf{sd}_i, \mathsf{sd}_{\overline{i}} \mid 1 \leq i \leq n\}$.

We say that a signed permutation π over the set of integers Σ_n is *sortable* if there are operations $\phi_1, \ldots, \phi_k \in \mathsf{Sh} \cup \mathsf{Sd}$ such that $(\phi_1 \circ \ldots \circ \phi_k)(\pi)$ is a sorted permutation. In this case $\Phi = \phi_1 \circ \ldots \circ \phi_k$ is a *sorting strategy* for π. Permutation π is Sh-*sortable* if $\phi_1, \ldots, \phi_k \in \mathsf{Sh}$ and π is Sd-*sortable* if $\phi_1, \ldots, \phi_k \in \mathsf{Sd}$. We say that ϕ_i is *part* of Φ and also that ϕ_i is used in Φ before ϕ_j for all $1 \leq j < i \leq k$.

Example 1. (i) Permutation $\pi_1 = 3\,\overline{4}\,\overline{5}\,6\,\overline{1}\,2$ is sortable and a sorting strategy is $\mathsf{sh}_1(\mathsf{sh}_5(\mathsf{sh}_4(\pi_1))) = 3\,4\,5\,6\,1\,2$. Permutation $\pi'_1 = 3\,4\,5\,6\,\overline{1}\,\overline{2}$ is unsortable. Indeed, no sh operations and no sd operation is applicable to π'_1.

(ii) Permutation $\pi_2 = 1\,3\,4\,2\,\overline{5}$ is sortable and has only one sorting strategy: $\mathsf{sh}_5(\mathsf{sd}_2(\pi_2)) = 1\,2\,3\,4\,5$.

(iii) There exist permutations with several successful strategies, even leading to different sorted permutations. One such permutation is $\pi_3 = 3\,5\,1\,2\,4$. Indeed, $\mathsf{sd}_3(\pi_3) = 5\,1\,2\,3\,4$. At the same time, $\mathsf{sd}_4(\pi_3) = 3\,4\,5\,1\,2$.

(iv) The simple operations yield a nondeterministic process: there are permutations having both successful and unsuccessful sorting strategies. One such permutation is $\pi_4 = 1\,3\,5\,7\,9\,2\,4\,6\,8$. Note that $\mathsf{sd}_3(\mathsf{sd}_5(\mathsf{sd}_7(\pi_4))) = 1\,9\,2\,3\,4\,5\,6\,7\,8$ is a unsortable permutation. However, π_4 can be sorted, e.g., by the following strategy: $\mathsf{sd}_2(\mathsf{sd}_4(\mathsf{sd}_6(\mathsf{sd}_8(\pi_4)))) = 1\,2\,3\,4\,5\,6\,7\,8\,9$.

(v) Permutation $\pi_5 = 1\,3\,5\,2\,4$ has both successful and unsuccessful sorting strategies. Indeed, $\mathsf{sd}_3(\pi_5) = 1\,5\,2\,3\,4$, a unsortable permutation. However, $\mathsf{sd}_2(\mathsf{sd}_4(\pi_5)) = 1\,2\,3\,4\,5$ is sorted.

(vi) Applying a cyclic shift to a permutation may render it unsortable. Indeed, permutation $2\,1\,4\,3\,5$ is sortable, while $5\,2\,1\,4\,3$ is not.

(vii) Consider the signed permutation $\pi_7 = 1\,11\,3\,9\,5\,7\,2\,4\,13\,6\,15\,8\,10\,12\,14\,16$. Operation sd may be applied to π_7 on integers 3, 6, 9, 11, 13, and 15 . Doing that however leads to a unsortable permutation:

$$\mathsf{sd}_3(\mathsf{sd}_6(\mathsf{sd}_9(\mathsf{sd}_{11}(\mathsf{sd}_{13}(\mathsf{sd}_{15}(\pi_7)))))) = 1\,5\,6\,7\,2\,3\,4\,8\,9\,10\,11\,12\,13\,14\,15\,16.$$

However, omitting sd_3 from the above composition leads to a sorting strategy for π_7: let

$$\pi'_7 = \mathsf{sd}_6(\mathsf{sd}_9(\mathsf{sd}_{11}(\mathsf{sd}_{13}(\mathsf{sd}_{15}(\pi_7))))) = 1\,3\,5\,6\,7\,2\,4\,8\,9\,10\,11\,12\,13\,14\,15\,16.$$

Then $\mathsf{sd}_2(\mathsf{sd}_4(\pi'_7))$ is a sorted permutation.

Lemma 1. *Let π be a signed permutation over Σ_n and $i \in \Sigma_n \cup \overline{\Sigma_n}$. Then we have the following properties:*

(i) If sd_i is applicable to π, then $\mathsf{sd}_{\overline{i}}$ is applicable to $\overline{\pi}$ and $\overline{\mathsf{sd}_i(\pi)} = \mathsf{sd}_{\overline{i}}(\overline{\pi})$.

(ii) If sh_i, where i is unsigned, is applicable to π, then sh_{i-1} or sh_{i+1} is applicable to $\overline{\pi}$ and $\overline{\mathsf{sh}_i(\pi)} = \mathsf{sh}_{i-1}(\overline{\pi})$ or $\overline{\mathsf{sh}_i(\pi)} = \mathsf{sh}_{i+1}(\overline{\pi})$.

5   Sh-Sortable Permutations

We characterize in this section all signed permutations that can be sorted using only the Sh operations. As it turns out, their form is easy to describe since the Sh operations do not change the relative positions of the letters in the permutation.

The following result characterizes all Sh-sortable signed permutations.

Theorem 1. *A signed permutation $\pi = p_1 \ldots p_n$, $p_i \in \Sigma_n \cup \overline{\Sigma_n}$, is sh-sortable if and only if*

(i) $\|\pi\| = k\,(k+1)\ldots n\,1\ldots(k-1)$, *for some $1 \leq k \leq n$ and there are i, j, $1 \leq i \leq k-1$, $k \leq j \leq n$ such that p_i and p_j are unsigned letters, or*

(ii) $\|\pi\| = (k-1)\ldots 1\,n\ldots(k+1)\,k$, *for some $1 \leq k \leq n$ and there are i, j, $1 \leq i \leq k-1$, $k \leq j \leq n$ such that p_i and p_j are signed letters.*

In Case (i), π sorts to $k\,(k+1)\ldots n\,1\ldots(k-1)$, while in Case (ii), π sorts to $(\overline{k-1})\ldots\overline{1}\,\overline{n}\ldots\overline{(k+1)}\,\overline{k}$.

Example 2. (i) Permutation $\pi_1 = \overline{5}\,\overline{6}\,\overline{7}\overline{8}\,\overline{1}\,2\overline{3}\overline{4}$ is Sh sortable and an Sh-sorting for π_1 is $\mathsf{sh}_4(\mathsf{sh}_3(\mathsf{sh}_1(\mathsf{sh}_8(\mathsf{sh}_5(\mathsf{sh}_6(\pi_1)))))) = 5\,6\,7\,8\,1\,2\,3\,4$. Note that sh_5 can be applied only after sh_6 and also, sh_4 can be applied only after sh_3.

(ii) Permutation $\pi_2 = \overline{5}\,\overline{6}\,\overline{7}\overline{8}\,\overline{1}\,\overline{2}\,\overline{3}\,\overline{4}$ is unsortable, since we cannot unsign 1, 2, 3 and 4.

6   Sd-Sortable Permutations

We characterize in this section the Sd-sortable permutations. A crucial role in our result is played by the dependency graph of a signed permutation.

6.1   The Dependency Graph

This is in general a directed graph with self-loops: there may be edges from a node to itself. The dependency graph describes for a permutation π the order in which Sd-operations can be applied to π.

For a permutation π over Σ_n we define its dependency graph as the directed graph $G_\pi = (\Sigma_n, E)$, where $(i, j) \in E$, $1 \leq i \leq n$, $2 \leq j \leq n-1$, if and only if $(j-1)i(j+1) \leq_s \pi$. Also, if $(j+1)(j-1) \leq_s \pi$, then $(j, j) \in E$. Intuitively, the edge (i, j) represents that the rule sd_j may be applied in a sorting strategy for π only after rule sd_i has been applied. A loop (i, i) represents that sd_i can never be used in a sorting strategy for π. Note that G_π may also have a loop on node i if $(i-1)i(i+1) \leq_s \pi$.

Example 3. (i) The graph associated to permutation $\pi_1 = 1\,4\,3\,6\,5\,7\,2$ is shown in Fig. 4(a). It can be seen, e.g., that sd_3 can never be applied in a sorting strategy for π and because of edge $(3, 5)$, neither can sd_5. Also, the graph suggests that sd_6 should be applied before sd_4 and this one before sd_2. Indeed, $\mathsf{sd}_2(\mathsf{sd}_4(\mathsf{sd}_6(\pi))) = 1\,2\,3\,4\,5\,6\,7$.

(ii) The graph associated to permutation $\pi_2 = 1\,4\,3\,2\,5$ is shown in Fig. 4(b). Thus, the graph has a cycle with nodes 2 and 4. Indeed, to apply sd_2 in a strategy for π_2, sd_4 should be applied first and the other way around.

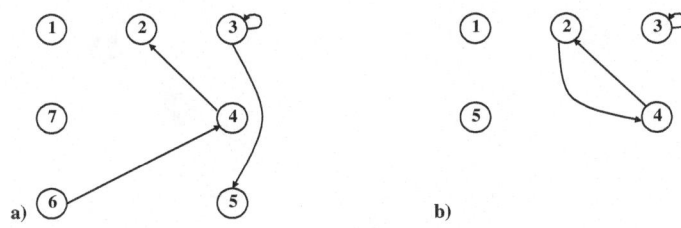

Fig. 4. Dependency graphs (a) associated to $\pi_1 = 1\,4\,3\,6\,5\,7\,2$ and (b) associated to $\pi_2 = 1\,4\,3\,2\,5$

Lemma 2. *Let π be a unsigned permutation over Σ_n and $G_\pi = (\Sigma_n, E)$ its dependency graph.*

(i) There exists no sorting strategy Φ for π such that sd_i and sd_{i+1} are both used in Φ, for some $1 \leq i \leq n-1$.

(ii) If sd_j is used in a sorting strategy for π and $(i,j) \in E$, for some $i, j \in \Sigma_n$, then sd_i is also used, before sd_j, in the same sorting strategy.

(iii) If there is a path from i to j in G_π, then in any strategy where sd_j is used, sd_i is also used, before sd_j.

(iv) If G_π has a cycle containing $i \in \Sigma_n$, then sd_i cannot be applied in any sorting strategy of π.

(v) There is no strategy where sd_1 and sd_n can be applied.

6.2   The Characterization

We characterize in this subsection the Sd-sortable permutations. We first give an example.

Example 4. Consider the dependency graph G_π for $\pi = 1\,11\,3\,9\,5\,7\,2\,4\,13\,6\,15\,8\,10$ $12\,14\,16$, shown in Fig. 5. Based on Lemma 2 and G_π we build a sorting strategy Φ for π. We label all nodes i for which sd_i is used in Φ by M and the other nodes by U. Nodes labelled by M are shown with a white background in Fig. 5, while nodes labelled by U are shown with black one.

By Lemma 2(iv)(v) operations sd_1, sd_8, sd_{10} and sd_{16} cannot be applied in any strategy of π. Thus, $1, 8, 10, 16 \in U$. Now, to apply operation sd_2, since we have edge $(11, 2)$ in the dependency graph G_π, it follows by Lemma 2(ii) that sd_{11} must be applied in the same strategy as sd_2. Thus, $2, 11 \in M$. According to Lemma 2(i) we cannot apply sd_2 and sd_3 in the same strategy, thus we label 3 by U. To use sd_4, since edge $(9, 4)$ is present in the dependency graph, we need to label both 4 and 9 by M. It follows then by Lemma 2(i) that $5 \in U$. Then 6 can be labelled by M and then, necessarily, $7 \in U$. Note now, that if $12 \in M$, since $(3, 12)$ is an edge in G_π, then by Lemma 2(ii), $3 \in M$, which contradicts our labelling of 3. Thus, $12 \in U$. Then 13 can be labelled by M and necessarily, $14 \in U$. Also, 15 can now be labelled by M.

In this way we obtain $M = \{2, 4, 6, 9, 11, 13, 15\}$ and $U = \{1, 3, 5, 7, 8, 10, 12, 14, 16\}$. Note that, since elements in U do not change their relative positions in the strategy Φ we are building, $\pi|_U$ has to be sorted: $\pi|_U = 1\,3\,5\,7\,8\,10\,12\,14\,16$.

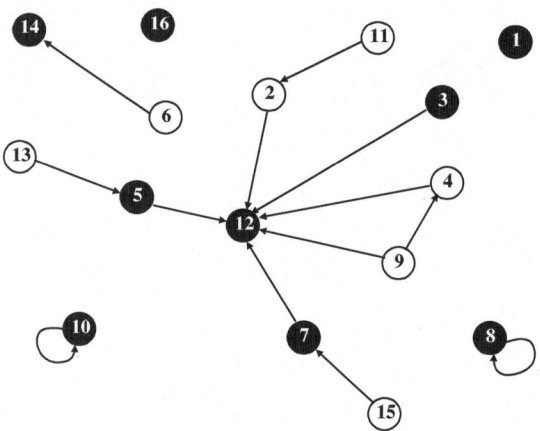

Fig. 5. The dependency graph associated to $\pi = 1\,11\,3\,9\,5\,7\,2\,4\,13\,6\,15\,8\,10\,12\,14\,16$

Our strategy Φ is now a composition of operations sd_i, with $i \in M$. The dependency graph shows the order in which these operations must be applied, i.e., sd_2 can be applied only after sd_{11} and sd_4 can be applied only after sd_9. In this way, we can sort π by applying the following sorting strategy:

$$(\mathsf{sd}_2 \circ \mathsf{sd}_4 \circ \mathsf{sd}_7 \circ \mathsf{sd}_{15} \circ \mathsf{sd}_{13} \circ \mathsf{sd}_{11} \circ \mathsf{sd}_9)(\pi) = 1\,2\,3\,4\,5\,6\,7\,8\,9\,10\,11\,12\,13\,14\,15\,16.$$

Clearly, our choice of M and U is not unique. For instance, we may have $M = \{2, 4, 7, 9, 11, 13, 15\}$ and $U = \{1, 3, 5, 6, 8, 10, 12, 14, 16\}$ as shown in Fig. 6. The strategy will be in this case $\mathsf{sd}_2 \circ \mathsf{sd}_4 \circ \mathsf{sd}_6 \circ \mathsf{sd}_{15} \circ \mathsf{sd}_{13} \circ \mathsf{sd}_{11} \circ \mathsf{sd}_9$.

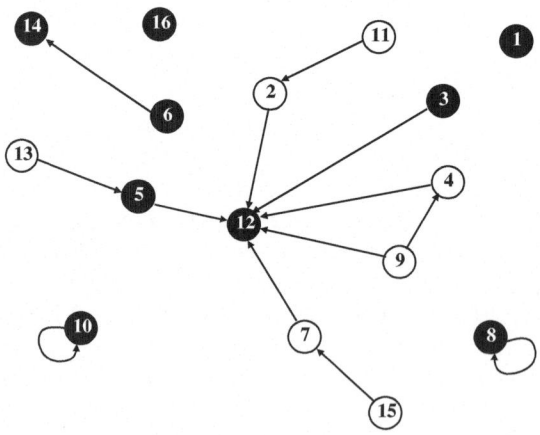

Fig. 6. The dependency graph associated to $\pi = 1\,11\,3\,9\,5\,7\,2\,4\,13\,6\,15\,8\,10\,12\,14\,16$

- Case 1: $S \cap T = \emptyset$

We have $m^3(m-1)^3$ choices for this kind of sets, and in this case

$$
\begin{aligned}
E(X_S X_T) &= P(X_S = 1, X_T = 1) \\
&= P(X_S = 1 | X_T = 1) P(X_T = 1) \\
&= \frac{r}{(m-1)^2} \frac{r}{m^2} = \frac{r^2}{m^2(m-1)^2}, \text{ and}
\end{aligned}
$$

$$
E(X_S)E(X_T) = \frac{r}{m^2} \frac{r}{m^2} = \frac{r^2}{m^4}, \text{ hence}
$$

$$
\mathrm{Cov}(X_S, X_T) = \frac{r^2}{m^2(m-1)^2} - \frac{r^2}{m^4} \sim \frac{2r^2}{m^3(m-1)^2}, \text{ and hence}
$$

$$
\sum_{S \cap T = \emptyset} \mathrm{Cov}(X_S, X_T) = m^3(m-1)^3 \frac{2r^2}{m^3(m-1)^2}
$$

$$
= 2(m-1)r^2 \sim 2mr^2.
$$

- Case 2: $|S \cap T| = 1$

We have $m^3 \binom{3}{1}(m-1)^2 = 3m^3(m-1)^2$ choices for those kind of set, and we get:

$$
\begin{aligned}
E(X_S X_T) &= P(X_S = 1, X_T = 1) \\
&= P(X_S = 1 | X_T = 1) P(X_T = 1) = 0, \text{ as}
\end{aligned}
$$

$$
P(X_S = 1 | X_T = 1) = 0, \text{ and hence}
$$

$$
E(X_S)E(X_T) = \frac{r}{m^2} \frac{r}{m^2} = \frac{r^2}{m^4}, \text{ and hence}
$$

$$
\sum_{|S \cap T| = 1} \mathrm{Cov}(X_S, X_T) = -3m^3(m-1)^2 \frac{r^2}{m^4} \sim -3mr^2.
$$

- Case 3: $|S \cap T| = 2$

We have $m^3 \binom{3}{2}(m-1) = 3m^3(m-1)$ choices for those kind of sets, so

$$
\begin{aligned}
E(X_S X_T) &= P(X_S = 1, X_T = 1) \\
&= P(X_S = 1 | X_T = 1) P(X_T = 1) = 0, \text{ and}
\end{aligned}
$$

$$
E(X_S)E(X_T) = \frac{r}{m^2} \frac{r}{m^2} = \frac{r^2}{m^4}, \text{ and hence}
$$

$$
\sum_{|S \cap T| = 2} \mathrm{Cov}(X_S, X_T) = -3m^3(m-1) \frac{r^2}{m^4}
$$

$$
= -3 \frac{(m-1)r^2}{m} \sim -3r^2.
$$

that three connected junctions by two sticky ends will close in a complete K_3. Then the expected number of K_3 complete complexes in the pot is

$$E(X) = mr,$$

moreover

$$\lim_{r \to 1} P(X = m) = 1$$

Proof. For the proof we use the following notation:

S: a set of 3 junctions from J_3, one of each type,
A_S: the event that the junctions from S form a complete K_3,
X_S: the associated indicator random variable for A_S,
B_i: the event that $\mathbf{h}_i$ and $\mathbf{h}_i$ will connect, for $i = 1, 2, 3$ and
ξ_i: the associated indicator random variable for B_i.

For the set of sticky end types $H = \{\mathbf{h}_1, \mathbf{h}_2, \mathbf{h}_3, \mathbf{h}_1, \mathbf{h}_2, \mathbf{h}_3\}$ in S, the probability that the junctions from S will form a complete K_3 is equal to the probability that all junctions of S would connect. That is,

$$P(\xi_1 = 1, \xi_2 = 1, \xi_3 = 1) = P(\xi_1 = 1)P(\xi_2 = 1 | \xi_1 = 1)P(\xi_3 = 1 | \xi_1, \xi_2 = 1)$$
$$= ppr = p^2 r.$$

(Notice we assume the conditional probability of the second connection is the same as the probability of the first)

Since X_S is the indicator random variable for the event A_S, $X = \sum X_S$ will denote the number of complete K_3's in the pot, and

$$E(X_S) = P(A_S) = p^2 r.$$

We have m^3 sets S and by the linearity of expectation the expected number of complete K_3's in the pot is

$$E(X) = m^3 p^2 r.$$

Ignoring the thermodynamic properties of the solution, the probability of one sticky end connecting with its complementary is $p = \frac{1}{m}$, from which it follows that $E(X) = mr$.

To calculate the variance for the number of complete K_3's in the pot

$$Var(X) = \sum_{S,T} Cov(X_S, X_T)$$

we need to calculate the covariances first:

$$Cov(X_S, X_T) = E(X_S X_T) - E(X_S)E(X_T).$$

In order to do that we need to look at two sets S and T, each one consisting of the three different junctions from the pot, one from each type. Again, for the analysis of the covariance we consider the case when $p = \frac{1}{m}$.

when Π and C_{start} and C_{stop} are restricted. An interesting case is when Π is a timed symport/antiport system which operates in such a way that *no* symbol is exported into the environment (thus there are no rules of the form (u, out) and $(u, out; v, in)$ in the skin membrane). Call this system a restricted symport/antiport system. This type of (un-timed) system was studied in [17]. We can show the following (due to the space restrictions we leave the proof to the reader).

Theorem 4. *Let Π be a restricted timed symport/antiport system and C_{start} and C_{stop} be Presburger relations. Then $N(\Pi)$ can be accepted by a deterministic polynomial-time multicounter machine. (This means that the multicounter machine, when given n, can decide whether or not n is in $N(\Pi)$ in time polynomial in n).*

3   A P System Model of Timed Rules and Combinatorial Gene Expression

The goal of this section is to define a P system model that is close to the biology of the cell and, at the same time, keep some of the features of the P systems so that it can be studied with the now widely used mathematical tools for P systems. With a similar goal in mind we have recently defined a successful model in [15] which has been adopted as one of the natural/biological models of P systems by the research community; the model has become one of the major paradigms in the field.

We now extend the model proposed in [15] with several new ideas: different reactions can take different amounts of time; objects in the system can bind/dissociate according to their physical properties (3D shape, polarities); the cell contains its genetic material, enabling it to produce new objects according to the blueprints provided in genes.

Another interesting modeling effort in the direction of defining more realistic, i.e. more bio-compatible P systems was reported recently in the membrane computing annual conference, where two papers [5], [14] were suggesting approaches to make the P systems "time independent". Both the authors take in consideration rules in the P system that can take various amounts of time, in contrast with the other definitions of P systems that are modeling the rules as taking each 1 clock-cycle.

Our proposed model is different from the model in [5] or [14] by the fact that we will have in the system the idea of binding two molecules together. We are also interested in models that "behave" as close as possible to reality, in contrast with the aforementioned papers that were focussing on finding systems "time independent"; i.e. systems that have the same output even though the time associated with a rule is changing.

Another novelty of our proposed model is the introduction of the genetic material: one of the regions of the P system will be labeled *nucleus*, and will contain, among other things, the genes of that cell. As far as we know, there are no other

P system models that are describing the interactions between various molecules in the system and genes; especially gene activation process, gene activators, gene repressors, etc.

The genes will be denoted by G_1, G_2, G_3, ..., G_n, and they will be either activated or de-activated. Since it is still an ongoing debate in the biology community about how exactly is the mRNA built from an activated gene, we chose to model the process in the following way: genes are by default deactivated, they become activated only when some specific activator molecules bind to a gene. In that moment the gene is activated, and the mRNA is produced and sent to cytoplasm to be translated into several copies of the protein. After all this process takes place, we consider that the gene becomes deactivated and some of its activator molecules (or all) have left the nucleus. If more activator molecules are/were present in the nucleus, then at the next clock-cycle they can start binding to the gene, and the gene is activated once more.

The model will contain the activator rules as well as repressor rules for each gene plus, when activated, the gene will be able to produce new objects in the system.

Definition 1. *A genetic P system is a construct* $\Pi = (V, \mu, G, w_{cyt}, w_{ER}, w_{nucleus}, R_{cyt}, R_{ER}, R_{nucleus})$, *where*

- *V is the alphabet of Π representing the set of all possible molecules that can appear in the system.*
- *μ gives the membrane structure of the system; the plasma membrane (labeled* cyt*) contains two different sub-regions labeled with* ER *(for endoplasmic reticulum) and* nucleus*. In standard membrane systems notation μ is written as* $\mu = [_{cyt}[_{ER}]_{ER} [_{nucleus}]_{nucleus}]_{cyt}$.
- *G is the set of genes for the cell.*
- *w_{cyt}, w_{ER}, $w_{nucleus}$ are words over V that represent the initial multiplicities of the objects in their corresponding regions in the system. Please note that in the initial configuration we assume all the objects in the system to be "unbound" with any other object. We will call from now on as objects/molecules elements from V and also complexes of bound together elements from V. They will be written in the form $\langle XYZ \rangle$ when X, Y, $Z \in V$; thus $\langle XYZ \rangle$ is a single object in the system.*
- *R_{cyt}, R_{ER}, $R_{nucleus}$ are finite sets of rules associated with each of the three regions defined by the P system. We will describe in the following the types of rules that can be found in each of the three sets of rules.*

The rules are of four different categories: general rules (g1, g2, g3), cytoplasm rules (c1, c2), endoplasmic reticulum rules (e1, e2, e3, e4) and nucleus rules (n1, n2, n3).

The general rules are specifying the types of rules that can appear in any region of the cell; they model the binding/unbinding of molecules (g1, g2) and the catalytic reactions from the cell (g3).

The cytoplasm rules can only be applied in cytoplasm; they are modeling the creation of new proteins from mRNA by the ribosomes (c1) and the destruction of proteins by the proteases (c2).

In the endoplasmic reticulum (ER) we have rules that model the movement of objects between CYT and ER. We model the work of ion channels (e1), uniport (e2), symport/antiport (e3/e4).

The last type of rules are the ones that can only appear in nucleus, they model the work of activators/repressors binding to genes (n1, n2) and the transcription of a gene followed by the expel of the mRNA into the cytoplasm (n3) so that the protein-building mechanism (c1) can start.

The general rules in the cytoplasm, endoplasmic reticulum and nucleus will have the forms:

g1. association (binding) rules: $\langle X \rangle + \langle Y \rangle \rightarrow_t \langle XY \rangle$ where $X, Y \in V^+$ and t specifies the number of clock-cycles it takes for the binding to take place for the specified molecules. It must be stressed the fact that the product $\langle XY \rangle$ is the same with the product $\langle YX \rangle$; we will write the products in the lexicographic order.

g2. dissociation (unbinding) rules: $\langle XY \rangle \rightarrow_t \langle X \rangle + \langle Y \rangle$ where $X, Y \in V^+$ and t specifies the number of clock-cycles required for the unbinding operation.

g3. catalysis rules: $\langle X \rangle + \langle Y \rangle \rightarrow_t \langle X \rangle + \langle Z \rangle$ where $X, Y, Z \in V^+$ and t specifies the number of clock-cycles required for the enzyme $\langle X \rangle$ to perform the catalysis.

We will apply the previous rules in a nondeterministically parallel manner with the only remark that the binding rules have higher priority than other rules such as the catalatic rules, ion channel rules, etc.; in this way the model accounts (among other things) for the allosteric changes of enzymes.

The following types of rules will be associated only with the cytoplasm:

c1. creation of proteins by the ribosome: $\langle A^n \rangle \rightarrow_t \langle A^{n-1} \rangle + \langle A \rangle_l$, where $l \in \{here, in\text{-}ER, in\text{-}nucleus\}$ for all $1 \leq n$, and $A^1 = A$.

c2. destruction of objects: $\langle PY_u \rangle \rightarrow_t \langle P \rangle$, where $P, Y_u \in V^+$ and P is a protease and Y_u is a protein marked for destruction by ubiquitin.

The following types of rules will be associated only with the endoplasmic reticulum:

e1. ion channels: $\langle Ion \rangle \rightarrow_{t_1} \langle \overline{Ion} \rangle$, $\langle \overline{Ion} \rangle \rightarrow_{t_2} \langle Ion \rangle$, $\langle Ion \rangle + \langle X \rangle \rightarrow_{t_3} \langle IonX \rangle$, $\langle IonX \rangle \rightarrow_{t_4} \langle Ion \rangle + \langle X \rangle_{in/out}$ where $Ion, \overline{Ion}, X \in V$. The rules defined for the ion channels take in consideration the fact that the channels have a periodical transition between the on and off configurations.

e2. uniport: $\langle Uni \rangle + \langle X \rangle \rightarrow_t \langle Uni \rangle + \langle X \rangle_{out}$, or $\langle Uni \rangle + \langle X \rangle_{cyt} \rightarrow_t \langle Uni \rangle + \langle X \rangle_{in}$, where $Uni, X \in V$ and $\langle X \rangle_{cyt}$ means that the object X is in that moment in the cytoplasm.

e3. symport: $\langle Sim \rangle + \langle XY \rangle \rightarrow_t \langle Sim \rangle + \langle X \rangle_{out} + \langle Y \rangle_{out}$, or $\langle Sim \rangle + \langle XY \rangle_{cyt} \rightarrow_t \langle Sim \rangle + \langle X \rangle_{in} + \langle Y \rangle_{in}$, where $Sim, X, Y \in V$ and $\langle XY \rangle_{cyt}$ means that the objects are in the cytoplasm.

e4. antiport: $\langle Anti \rangle + \langle X \rangle + \langle Y \rangle_{cyt} \rightarrow_t \langle Anti \rangle + \langle X \rangle_{out} + \langle Y \rangle_{in}$ where we have that $Anti, X, Y \in V$ and the subscript cyt specifies that the respective molecule is in the cytoplasm.

The following types of rules will be associated only with the nucleus:

n1. activator/repressor binding to a gene: $\langle A \rangle + \langle G_i \rangle \rightarrow_t \langle AG_i \rangle$ for $A \in V^+$ and $G_i \in G$.

n2. more activators binding to the gene G_i: $\langle A \rangle + \langle BG_i \rangle \rightarrow_t \langle ABG_i \rangle$ for $A, B \in V^+$ and $G_i \in G$.

n3. gene activation and mRNA move to cytoplasm: $\langle XG_i \rangle \rightarrow_t \langle YG_i \rangle + \langle A^k \rangle_{out}$, where $X, Y \in V^*$, $A \in V$, $Y \subseteq X$ and k is the number of copies of the protein A (codified in the mRNA) that will be produced by the ribosomes in the cytoplasm.

One can note that the system is defined flexibly enough so that much more biological processes can be expressed using the given rules. For example, the phosphorylation reaction can be expressed with several objects in V using the catalytic reaction of the type g3: $\langle X \rangle + \langle Y \rangle \rightarrow_t \langle X \rangle + \langle Y_p \rangle$ and can be continued for several steps: $\langle X \rangle + \langle Y_p \rangle \rightarrow_{t'} \langle X \rangle + \langle Y_{pp} \rangle$ by using other type g3 rules.

Remark. We assume that a complex of objects (several objects are bound together after a repeated use of rules of type g1) is working as a whole; thus if we have the complex object $\langle ABC \rangle$, then a rule defined only for $\langle AB \rangle$ cannot be applied to $\langle ABC \rangle$. This is due to the fact that the 3D shape of the complex $\langle AB \rangle$ can be changed dramatically by the binding with $\langle C \rangle$.

We note that the previous remark helps the rules n1, n2 simulate also the work of the gene regulation repressors, since the gene cannot be activated if the repressor is bound to it.

The rule allows for modeling the enzymatic regulation that takes place in cells: if the enzyme $\langle A \rangle$ is catalyzing the reaction from $\langle X \rangle$ into $\langle Y \rangle$ at some rate of 3 clock-cycles, we can write it as a rule of type g3: $\langle A \rangle + \langle X \rangle \rightarrow_3 \langle A \rangle + \langle Y \rangle$. Now, let us assume that the cell decides to down-regulate the enzyme to a ten times slower speed by using the molecule $\langle B \rangle$. In this case we would model the reactions using a binding rule $\langle A \rangle + \langle B \rangle \rightarrow_2 \langle AB \rangle$, and a catalytic rule $\langle AB \rangle + \langle X \rangle \rightarrow_{30} \langle AB \rangle + \langle Y \rangle$.

We now briefly discuss the different types of output for the new model. The first type of output of the system could be associated with the number of times each rule of the type n3 is applied for each of the genes contained in a predefined string $w \in G^*$ in a given amount of time. In other words, we are interested in a specific number of genes, and we are looking at how many times these genes have been activated. The system in this case can compute a vector of integer values; the number of components and the order of the components being given by the word w.

Another type of "termination" for the computation for such a system could be viewed as a combination of activated genes at a given moment as well as minimal multiplicities for several molecules in each component of the system. Such a configuration could be viewed as a "final state" of the machinery. In that moment one could use the first idea of the output of the system; i.e. counting the number of activations for particular genes.

Yet another idea is to consider as we have done for timed symport/antiport systems in Section 2 the time that passed between two distinguished configurations as the the result of the computation. This method of considering the computation of the system seems quite flexible and elegant. It does not require a cell to "accumulate" a large quantity of a specific molecule that would encode the output.

4 Final Remarks

In this paper, we considered (as in the previous papers that have been mentioned) time as an "active" participant in a computation. We also reviewed two known models – one using spiking neurons and the other, the time independent P systems. We then defined and obtained several results concerning some new models – the timed P system, P system with timed rules, and gene expression models. For the newly introduced timed P systems we improved or matched two of the best known results for "regular" symport/antiport P systems. We are currently working on proving the remaining two results ($NTP_3(sim_2, anti_0)$ and $NTP_2(sim_3, anti_0)$). It is worth noting that the new feature of outputting the result using time is more flexible than the previously considered methods, thus the previous results could be even improved by using completely different techniques that take advantage of the flexibility of the time as a framework of outputting the result of a computation.

Acknowledgements

We would like to acknowledge the fruitful discussions with B. Tănasă (MIT) and the insightful suggestions received from the anonymous referees. O. H. Ibarra gratefully acknowledges the support in part by NSF Grants CCR-0208595, CCF-0430945 and CCF-0524136; A. Păun gratefully acknowledges the support in part by LA BoR RSC grant LEQSF (2004-07)-RD-A-23 and NSF Grants IMR-0414903 and CCF-0523572.

References

1. L.M. Adleman, Molecular Computation of Solutions to Combinatorial Problems, Science 266 (1994), 1021–1024.
2. B. Alberts, *Essential Cell Biology. An Introduction to the Molecular Biology of the Cell*, Garland Publ. Inc., New York, London, 1998.
3. F. Bernardini, M. Gheorghe, On the Power of Minimal Symport/Antiport, Workshop on Membrane Computing, A. Alhazov et al. (eds.), WMC-2003, Tarragona, July 17-22, 2003, Technical Report N. 28/03, Research Group on Mathematical Linguistics, Universitat Rovira i Virgili, Tarragona (2003), 72–83.
4. F. Bernardini, A. Păun, Universality of Minimal Symport/Antiport: Five Membranes Suffice, WMC03 revised papers in *Lecture Notes in Computer Science* 2933, Springer (2004), 43–54.

5. M. Cavaliere, Towards Asynchronous P Systems, *Pre-proceedings of the Fifth Workshop on Membrane Computing (WMC5)*, Milano (Italy), June 14-16, 2004, 161–173.
6. M. Cavaliere, R. Freund, Gh. Păun, Event–Related Outputs of Computations in P Systems, personal communication, (manuscript).
7. R. Freund, A. Păun, Membrane Systems with Symport/Antiport: Universality Results, in *Membrane Computing. Intern. Workshop WMC-CdeA 2002, Revised Papers* (Gh. Păun, G. Rozenberg, A. Salomaa, C. Zandron, eds.), *Lecture Notes in Computer Science*, 2597, Springer-Verlag, Berlin, 2003, 270–287.
8. P. Frisco, J.H. Hogeboom, Simulating Counter Automata by P Systems with Symport/Antiport, in *Membrane Computing. Intern. Workshop WMC-CdeA 2002*, revised papers (Gh. Păun, G. Rozenberg, A. Salomaa, C. Zandron, eds.), *Lecture Notes in Computer Science*, 2597, Springer-Verlag, Berlin, 2003, 288–301.
9. P. Friso, About P Systems with Symport/Antiport, Second Brainstorming Week in Membrane Computing, Sevilla, February 2004, Technical Report 01/2004 of the Research Group in Natural Computing, University of Sevilla, Spain, 2004, 224–236.
10. J. Hopcroft, J. Ulmann, *Introduction to Automata Theory, Languages, and Computation*, Addison-Wesley, 1979.
11. L. Kari, C. Martin-Vide, A. Păun, On the Universality of P Systems with Minimal Symport/Antiport Rules, *Lecture Notes in Computer Science* 2950, Berlin, (2004), 254–265.
12. W. Maas: Computing with Spikes. *Spec. Iss. on Found. of Inf. Processing of TELEMATIK*, 8, 1 (2002), 32–36.
13. M.L. Minsky, Recursive Unsolvability of Post's Problem of "Tag" and Other Topics in Theory of Turing Machines, *Annals of Mathematics*, 74 (1961), 437–455.
14. D. Sburlan, Clock-free P Systems, *Pre-proceedings of the Fifth Workshop on Membrane Computing (WMC5)*, Milano (Italy), June 14-16, 2004, 372–383.
15. A. Păun, Gh. Păun, The Power of Communication: P Systems with Symport/Antiport, *New Generation Computing*, 20, 3 (2002) 295–306.
16. Gh. Păun: *Membrane Computing – An Introduction*. Springer-Verlag, Berlin, 2002.
17. Gh. Păun, M. Perez-Jimenez, F. Sancho-Caparrini, On the Reachability Problem for P Systems with Symport/Antiport, *Proc. Automata and Formal Languages Conf.*, Debrecen, Hungary, 2002.
18. Gh. Păun, Further Twenty-six Open Problems in Membrane Computing, the Third Brainstorming Meeting on Membrane Computing, Sevilla, Spain, February 2005.
19. G. Rozenberg, A. Salomaa, eds., *Handbook of Formal Languages*, 3 volumes, Springer-Verlag, Berlin, 1997.
20. G. Vazil, On the Size of P Systems with Minimal Symport/Antiport, Preproceedings of International Workshop WMC04, Milan, June 2004, 422–431.

On Bounded Symport/Antiport P Systems[*]

Oscar H. Ibarra and Sara Woodworth

Department of Computer Science,
University of California, Santa Barbara, CA 93106, USA
ibarra@cs.ucsb.edu

Abstract. We introduce a restricted model of a one-membrane symport/antiport system, called *bounded S/A system*. We show the following:

1. A language $L \subseteq a_1^* \ldots a_k^*$ is accepted by a bounded S/A system if and only if it is accepted by a $\log n$ space-bounded Turing machine. This holds for both deterministic and nondeterministic versions.
2. For every positive integer r, there is an $s > r$ and a unary language L that is accepted by a bounded S/A system with s objects that cannot be accepted by any bounded S/A system with only r objects. This holds for both deterministic and nondeterministic versions.
3. Deterministic and nondeterministic bounded S/A systems over a unary input alphabet are equivalent if and only if deterministic and nondeterministic linear-bounded automata (over an arbitrary input alphabet) are equivalent.

We also introduce a restricted model of a multi-membrane S/A system, called *special S/A system*. The restriction guarantees that the number of objects in the system at any time during the computation remains constant. We show that for every nonnegative integer t, special S/A systems with environment alphabet E of t symbols (note that other symbols are allowed in the system if they are not transported into the environment) has an infinite hierarchy in terms of the number of membranes. Again, this holds for both deterministic and nondeterministic versions. Finally, we introduce a model of a one-membrane bounded S/A system, called *bounded SA acceptor*, that accepts string languages. We show that the deterministic version is strictly weaker than the nondeterministic version.

Clearly, investigations into complexity issues (hierarchies, determinism versus nondeterminism, etc.) in membrane computing are natural and interesting from the points of view of foundations and applications, e.g., in modeling and simulating of cells. Some of the results above have been shown for other types of restricted P systems (that are not symport/antiport). However, these previous results do not easily translate for the models of S/A systems we consider here. In fact, in a recent article, "Further Twenty Six Open Problems in Membrane Computing" (January 26, 2005; see P Systems Web Page at http://psystems.disco.unimib.it), Gheorghe Paun poses the question of whether the earlier results, e.g., concerning determinism versus nondeterminism can be proved for restricted S/A systems.

Keywords: Symport/antiport system, communicating P system, Turing machine, multihead two-way finite automaton, multicounter machine, hierarchy, deterministic, nondeterministic.

[*] This work was supported in part by NSF Grants CCR-0208595, CCF-0430945, and CCF-0524136.

1 Introduction

Membrane computing is a relatively new computing paradigm which abstracts the activities of biological cells to find a new model for computing. While humanity has learned to compute mechanically within the relatively recent past, other living processes have computed naturally for millions of years. One such natural computing process can be found within biological cells. Cells consist of membranes which are used to contain, transfer, and transform various enzymes and proteins in a naturally decentralized and parallel manner. By modeling these natural processes of cells, we can create a new model of computing which is decentralized, nondeterministic, and maximally parallel.

Using biological membranes as an inspiration for computing was first introduced by Gheorghe Paun in a seminal paper [10] (see also [11, 12]). He studied the first membrane computing model, called P system, which consists of a hierarchical set of *membranes* where each membrane contains both a multiset of *objects* and a set of *rules* which determine how these objects interact within the system. The rules are applied in a *nondeterministic* and *maximally parallel* fashion. At each step of the computation, a maximal multiset of rules is chosen nondeterministically (note that several instances of a rule may be selected) and the rules applied simultaneously (i.e., in parallel). Maximal here means that in each step, no additional rule instance not already in the multiset of rules is applicable and could be added to the multiset of rules. The system is nondeterministic because the maximal multiset may not be unique. The outermost membrane is often referred to as the *skin* membrane and the area surrounding the system is referred to as the *environment*. A membrane not containing any membrane is referred to as an *elementary* membrane. As a branch of Natural Computing which explores new models, ideas, and paradigms from the way nature computes, membrane computing has been quite successful: many models have been introduced, most of them Turing complete. (See http://psystems.disco.unimb/it for a large collection of papers in the area, and in particular the monograph [12].) Due to the maximal parallelism inherent in the model, P systems have a great potential for implementing massively concurrent systems in an efficient way that would allow us to solve currently intractable problems (in much the same way as the promise of quantum and DNA computing) once future biotechnology gives way to a practical bio-realization. Given this potential, the Institute for Scientific Information (ISI) has selected membrane computing as a fast "Emerging Research Front" in Computer Science (http://esi-topics.com/erf/october2003.html).

One very popular model of a P system is called a symport/antiport system (introduced in [9]). It is a simple system whose rules closely resemble the way membranes transport objects between themselves in a purely communicating manner. Symport/antiport systems (S/A systems) have rules of the form (u, out), (u, in), and $(u, out; v, in)$ where $u, v \in \Sigma^*$. Note that u, v are multisets that are represented as strings (the order in which the symbols are written is not important, since we are only interested in the multiplicities of each symbol). A rule of the form (u, out) in membrane i sends the elements of u from membrane i out to the membrane (directly) containing i. A rule of the form (u, in) in membrane i transports the elements of u into membrane i from the membrane enclosing i. Hence this rule can only be used when the elements of u exist in the outer membrane. A rule of the form $(u, out; v, in)$ simultaneously sends u out of the membrane i while transporting v into membrane i. Hence this rule cannot be applied unless membrane i contains the elements in u and the membrane surrounding i contains the elements in v. Formally an S/A system is defined as

$$M = (V, H, \mu, w_1, \cdots, w_{|H|}, E, R_1, \cdots, R_{|H|}, i_o)$$

where V is the set of objects (symbols) the system uses, H is the set of membrane labels, μ is the membrane structure of the system, w_i is the initial multiset of objects within membrane i, and the rules are given in the set R_i. E is the set of objects which can be found within the environment, and i_o is the designated elementary output membrane. (When the system is used as a recognizer or acceptor, there is no need to specify i_o.) A large number of papers have been written concerning symport/antiport systems. It has been shown that "minimal" such systems (with respect to the number of membranes, the number of objects, the maximum "size" of the rules) are universal.

Initially, membrane systems were designed to be nondeterministic systems. When multiple, maximal sets of rules are applicable, nondeterminism decides which maximal set to apply. Recently, deterministic versions of some membrane models have been studied to determine whether they are as computationally powerful as the nondeterministic versions [5, 7]. Deterministic models guarantee that each step of the computation consists of only one maximal multiset of applicable rules. In some cases, both the nondeterministic and deterministic versions are equivalent in power to Turing Machines (see, e.g., [5]). In some non-universal P systems, the deterministic versus the nondeterministic question has been shown to be equivalent to the long-standing open problem of whether deterministic and nondeterministic linear-bounded automata are equivalent [7]; for another very simple class of systems, deterministic is strictly weaker than nondeterministic [7]. However, these two latter results do not easily translate for S/A systems.

In this paper, we look at restricted models of symport/antiport systems. Two models, called *bounded S/A system* and *special S/A system*, are acceptors of multisets with the restriction that the multiplicity of each object in the system does not change during the computation. These models differ in whether they also bound the number of membranes within the system or bound the number of distinct objects that can occur abundantly in the environment. Another model, called *bounded S/A acceptor*, is an acceptor of string languages. Again, this model has the property that at any time during the computation, the number of objects in the system is equal to the number of input symbols that have been read so far (in addition to a fixed number of objects given to the system at the start of the computation). We study the computing power of these models. In particular, we investigate questions concerning hierarchies (with respect to the number of distinct objects used in the system or number of membranes in the system) and whether determinism is strictly weaker than nondeterminism.

2 One-Membrane Bounded S/A System

Let M be a one-membrane symport/antiport system over an alphabet V, and let $\Sigma = \{a_1, ..., a_k\} \subseteq V$ be the *input* alphabet. M is restricted in that all rules are of the form $(u, out; v, in)$, where $u, v \in V^*$ with $|u| = |v| \geq 1$. Thus, the number of objects in the system at any time during the computation remains the same. Note that all the rules are antiport rules.

There is a fixed string (multiset) w in $(V - \Sigma)^*$ such that initially, the system is given a string $wa_1^{n_1}...a_k^{n_k}$ for some nonnegative integers $n_1, ..., n_k$ (thus, the input multiset is $a_1^{n_1}...a_k^{n_k}$). If the system halts, then we say that the string $a_1^{n_1}...a_k^{n_k}$ is accepted. The set of all such strings is the language $L(M)$ accepted by M. We call this system a *bounded S/A system*. M is *deterministic* if the maximally parallel multiset of rules applicable at each step in the computation is unique. We will show the following:

1. A language $L \subseteq a_1^*...a_k^*$ is accepted by a deterministic (nondeterministic) bounded S/A system if and only if it is accepted by a deterministic (nondeterministic) *log n* space-bounded Turing machine (with a two-way read-only input with left and right end markers).
2. For every r, there is an $s > r$ and a unary language L (i.e., $L \subseteq o^*$) accepted by a bounded S/A system with an alphabet of s symbols that cannot be accepted by any bounded S/A system with an alphabet of r symbols. This result holds for both deterministic and nondeterministic versions.
3. Deterministic and nondeterministic bounded S/A systems over a unary input alphabet are equivalent if and only if deterministic and nondeterministic linear-bounded automata (over an arbitrary alphabet) are equivalent. This later problem is a long-standing open problem in complexity theory [16].

The restriction $|u| = |v| \geq 1$ in the rule $(u, out; v, in)$ can be relaxed to $|u| \geq |v| \geq 1$, but the latter is equivalent in that we can always introduce a dummy symbol d and add $d^{|u|-|v|}$ to v to make the lengths the same and not use symbol d in any rule. We note that a similar system, called bounded P system (BPS) with cooperative rules of the form $u \rightarrow v$ where $|u| \geq |v| \geq 1$, was also recently studied in [3] for their model-checking properties.

For ease in exposition, we first consider the case when the input alphabet is unary, i.e., $\Sigma = \{o\}$. Thus, the bounded S/A system M has initial configuration wo^n (for some n). The idea is to relate the computation of M to a restricted type of multicounter machine, called *linear-bounded multicounter machine*.

A *deterministic* multicounter machine Z is linear-bounded if, when given an input n in one of its counters (called the input counter) and zeros in the other counters, it computes in such a way that the sum of the values of the counters at any time during the computation is at most n. One can easily normalize the computation so that every increment is preceded by a decrement (i.e., if Z wants to increment a counter C_j, it first decrements some counter C_i and then increments C_j) and every decrement is followed by an increment. Thus we can assume that every instruction of Z, which is not 'Halt', is of the form:

$$p : \text{If } C_i \neq 0, \text{ decrement } C_i \text{ by } 1, \text{ increment } C_j \text{ by } 1, \text{ and goto } k \text{ else goto state } l.$$

where p, k, l are labels (states). We do not require that the contents of the counters are zero when the machine halts.

If in the above instruction, there is a "choice" for states k and/or l, the machine is *nondeterministic*. We will show that we can construct a deterministic (nondeterministic) bounded S/A system M which uses a fixed multiset w such that, when M is started

2. $d_1, d_2, d_3, d_4, d_5, d_6$ — These objects are used to delay various objects from being used for a number of steps. The objects d_1 and d_2 are used to delay an action for 1 step. The remaining objects are used to delay an action for 3 steps.
3. c_1, c_2, c_3 — These objects are called check objects and are used to guarantee a subtract instruction expels at most one o object from the appropriate counter membrane.
4. l_{i1}, l_{i2} for each instruction $l_i : (ADD(r), l_j)$.
   The object l_{i1} signifies that the next instruction we will execute is l_j. The object l_{i2} is used in executing instruction l_i.
5. $l_{i1}, l_{i2}, l_{i3}, l_{i4}$ for each instruction $l_i : (SUB(r), l_j, l_k)$.
   The object l_{i1} signifies that the next instruction we will execute is l_j. The objects l_{i2}, l_{i3}, and l_{i4} are used in executing instruction l_i and are used to signify which branch of l_i will determine the next instruction.
6. l_{i1} for each instruction $l_i : (HALT)$.

The sets of rules for the R_i's are created as follows:

1. The set $R_1 = \emptyset$.
2. The set R_2 contains the following delay rules: $(d_1, out; d_2, in)$; $(d_3, out; d_4, in)$; $(d_4, out; d_5, in)$; $(d_5, out; d_6, in)$.
3. For each instruction $l_i : (ADD(r), l_j)$ in Z:
   The set R_2 contains the following rules: $(l_{i1}, out; l_{i2}d_1o, in)$; $(l_{i2}d_2, out; l_{j1}, in)$.
   The set R_{r+2} contains the following rules: $(l_{i2}o, in)$; (l_{i2}, out).
4. For each instruction $l_i : (SUB(r), l_j, l_k)$ in Z:
   The set R_2 contains the following rules: $(l_{i1}, out; l_{i2}c_1d_3, in)$; $(l_{i2}o, out; c_2l_{i3}, in)$; $(c_1c_2d_6l_{i3}, out; l_{j1}, in)$; $(l_{i2}d_6, out; c_3l_{i4}, in)$; $(c_1c_3l_{i4}, out; l_{k1}, in)$.
   The set R_{r+2} contains the following rules: $(l_{i2}c_1, in)$; $(l_{i2}o, out)$; $(c_1, out; c_2, in)$; (c_2, out); $(l_{i2}, out; d_6, in)$; (d_6, out); $(c_1, out; c_3, in)$; (c_3, out).
5. For each instruction $l_i : (HALT)$ no rules are added.

Informally, these special S/A system rules work using the following ideas. Initially the system is started with the first instruction label object l_{01} in the program membrane and the input o^n within membrane 3 (corresponding to counter 1). To execute an add instruction, the initial instruction object is replaced with the objects needed to execute the instruction - l_{i2}, d_1, and o. If the instruction is a subtract instruction the instruction l_{i1} is replaced with l_{i2} along with a delay object d_3 and a check object c_1. Once the appropriate objects are in the program membrane, a o object is appropriately moved into or out of the counter membrane corresponding to the current instruction. In the case where the current instruction tries to decrement a zero counter, the check objects cooperate with the delay objects to detect the situation and bring the appropriate objects into and out of the active membranes. Finally, the instruction executing objects are expelled from the program membrane and the correct next instruction object is brought into the program membrane.

Note that when a counter is decremented, an o object is removed from the corresponding membrane and moved into the skin membrane. When a counter is incremented, an o

object is brought into the corresponding membrane from the skin membrane. Since the multicounter machine being simulated is, by definition, guaranteed to always decrement before incrementing, we are guaranteed to have thrown an o object into membrane 1 before we ever try bringing an o object from membrane 1 to membrane 2. This guarantees that the special S/A system will operate through the multicounter machine's program instructions correctly.   □

Corollary 2. *Let t be any positive integer. Then multi-membrane special S/A systems with an environment alphabet of t symbols are equivalent to multi-membrane special S/A systems with no symbols in the environment. This holds for deterministic and non-deterministic versions.*

Proof. This follows from the above theorem and the observation that a system with an environment of t symbols can be simulated by a two-way FA with $2m(t+1)$ heads.   □

The proof of the next result is similar to that of Theorem 3.

Theorem 7. *For every r, there exist an $s > r$ and a unary language L (i.e., subset of o^*) accepted by an s-membrane special S/A system that cannot be accepted by any r-membrane special S/A system. This result holds for both deterministic and nondeterministic versions.*

4   One-Membrane Bounded S/A Systems Accepting String Languages

Let M be a (one-membrane) S/A system with alphabet V and input alphabet $\Sigma \subseteq V$. We assume that Σ contains a distinguished symbol \$, called the (right) end marker. The rules are restricted to be of the form:

1. $(u, out; v, in)$,
2. $(u, out; vc, in)$

where both u and v are in V^+, $|u| = |v| \geq 1$, and c is in Σ. Note that because of the requirement that $|u| = |v|$, the only way that the number of symbols in the membrane can grow is when a rule of type 2 is used. The second type of of rule is called a *read-rule*. We call M a *bounded S/A acceptor*. There is an abundance of symbols from V in the environment. The symbol c in a rule of type 2 can only come from the input string $z = a_1...a_n$ (where a_i is in $\Sigma - \{\$\}$ for $1 \leq i < n$, and $a_n = \$$), which is provided online externally; none of the symbols in v in the rules come from z.

There is a fixed string w in $(V - \Sigma)^*$, which is the initial configuration of M. Maximal parallelism in the application of the rules is assumed as usual. Hence, in general, the size of the multiset of rules applicable at each step is unbounded. In particular, the number of instances of read-rules (i.e., rules of the form $(u, out; vc, in)$) applicable in a step is unbounded. However, if a step calls for reading k input symbols (for some k), these symbols must be consistent with the next k symbols of the input string z that have not yet been processed. Note that rules of type 1 do not consume any input symbol from z.

The input string $z = a_1...a_n$ (with $a_n = \$$) is accepted if, after reading all the input symbols, M eventually halts. The language accepted is $L(M) = \{a_1...a_{n-1} \mid a_1...a_n$ is accepted by $M\}$ (we do not include the end marker).

We have two versions of the system described above: deterministic and nondeterministic bounded S/A acceptors. Again, in the deterministic case, the maximally parallel multiset of rules applicable at each step of the computation is unique. We will show that the deterministic version is strictly weaker than the nondeterministic version. The proof uses some recent results in [7] concerning a simple model of a CPS, called SCPA.

An SCPA M has multiple membranes, with the skin membrane labeled 1. The symbols in the initial configuration (distributed in the membranes) are *not* from Σ (the input alphabet). The rules (similar to those of a CPS) are of the form:

1. $a \rightarrow a_x$
2. $ab \rightarrow a_x b_y$
3. $ab \rightarrow a_x b_y c_{come}$

The input to M is a string $z = a_1...a_n$ (with $a_n = \$$, the end marker), which is provided externally online. The restrictions on the operation of M are the following:

1. There are no rules in membrane 1 with a_{out} or b_{out} on the right-hand side of the rule (i.e., no symbol can be expelled from membrane 1 into the environment).
2. A rule of type 3 (called a read-rule) can only appear in membrane 1. This brings in c if the next symbol in the input string $z = a_1...a_n$ that has not yet been processed (read) is c ; otherwise, the rule is not applicable.
3. Again, in general, the size of the maximally parallel multiset of rules applicable at each step is unbounded. In particular, the number of instances of read-rules (i.e., rules of the form $ab \rightarrow a_x b_x c_{come}$) applicable in a step is unbounded. However, if a step calls for reading k input symbols (for some k), these symbols must be consistent with the next k symbols of the input string z that have not yet been processed (by the semantics of the read-rule described in the previous item).

The system starts with an initial configuration which consists of some symbols from $V - \Sigma$ distributed in the membranes. The input string $z = a_1...a_n$ is accepted if, after reading all the input symbols, the SCPA eventually halts. The language accepted by M is $L(M) = \{a_1...a_{n-1} \mid a_1...a_n$ is accepted by $M\}$ (we do not include the end marker).

A *restricted 1-way linear-space DCM (NCM)* M is a deterministic (nondeterministic) finite automaton with a one-way read-only input tape with right delimiter (end marker) $\$$ and a number of counters. As usual, each counter can be tested for zero and can be incremented/decremented by 1 or unchanged. The counters are restricted in that there is a positive integer c such that at any time during the computation, the amount of space used in any counter (i.e., the count) is at most ck, where k is the number of symbols of the input that have been read so far. Note that the machine need not read an input symbol at every step. An input $w = a_1...a_n$ (where a_n is the end marker, $\$$, which only occurs at the end) is accepted if, when M is started in its initial state with all counters zero, it eventually enters an accepting state while on $\$$.

We note that although the machines are restricted, they can accept fairly complex languages. For example, $\{a^n b^n c^n \mid n \geq 1\}$ and $\{a^{2^n} \mid n \geq 0\}$ can both be accepted by restricted 1-way linear-space DCMs. (We usually do not include the end marker, which is part of the input, when we talk about strings/languages accepted.) It can be shown that a restricted 1-way linear-space DCM (NCM) is equivalent to a restricted 1-way $\log n$-space deterministic (nondeterministic) Turing machine that was studied in [2].

We will need the following result that was recently shown in [7]:

Theorem 8. *A language L is accepted by a restricted 1-way linear-space DCM (NCM) if and only if it is accepted by a deterministic SCPA (nondeterministic SCPA).*

Theorem 9. *Deterministic (nondeterministic) bounded S/A acceptors are equivalent to deterministic (nondeterministic) SCPAs.*

Proof. First we show that a deterministic (nondeterministic) SCPA M can be simulated by a deterministic (nondeterministic) bounded S/A acceptor M', which has only one membrane. Suppose M has membranes $1, ..., m$, with index 1 representing the skin membrane. For every symbol a in the system and membrane i, create a new symbol a_i. We construct M' by converting the rules to one-membrane rules as described in the proof of Lemma 2, except that now we have to handle rules of the form $ab \to a_x b_y c_{come}$ in membrane 1. We transform such a rule to $(a_1 b_1, out; a_j b_k c_1, in)$, where j and k are the indices of the membranes into which a and b are transported to, as specified by x and y. After we have constructed M', modify it slightly by deleting the subscripts of all symbols with subscript 1 (in the rules and initial configuration). Thus unsubscripted symbols are associated with symbols in membrane 1 of the SCPA M.

For the converse, we need only show (by Theorem 8) that a deterministic (nondeterministic) bounded S/A acceptor M can be simulated by a restricted 1-way linear-space DCM (NCM) Z. The construction of Z is like in Lemma 3, except that now, Z uses counters (instead of heads), and in the maximally parallel step, the read-rules are the first ones to be processed. Define an atomic read-rule process as follows: Z systematically cycles through the read-rules and finds (if it exists) the first one that is applicable (note that for a read-rule $(u, out; vc, in)$ to be applicable, the next input symbol that has yet to be processed must be c). Z applies a sequence of these read-rules until no more read-rule is applicable. Then all the other rules are processed. We omit the details. If M is a nondeterministic SCPA, the construction of a nondeterministic Z is similar, in fact, easier. □

From Theorem 8 and the fact that deterministic SCPAs are strictly weaker than nondeterministic SCPAs [7], we have:

Theorem 10. *Deterministic bounded S/A acceptors are strictly weaker than nondeterministic bounded S/A acceptors.*

Let $L = \{x\#^p \mid x$ is a binary number with leading bit 1 and $p \neq 2val(x)\}$, where $val(x)$ is the value of x. It was shown in [7] that L can be accepted by a nondeterministic SCPA but not by any deterministic SCPA. Hence, L is an example of a language that can be accepted by a nondeterministic bounded S/A acceptor that cannot be accepted by any deterministic bounded S/A acceptor.

The following follows from Theorem 9 and the fact that similar results hold for SCPAs [7].

Expectation and Variance of Self-assembled Graph Structures

Nataša Jonoska, Gregory L. McColm, and Ana Staninska

Department of Mathematics,
University of South Florida
{jonoska, mccolm, staninsk}@math.usf.edu

Abstract. Understanding how nanostructures are self-assembled into more complex forms is a crucial component of nanotechnology that shall lead towards understanding other processes and structures in nature. In this paper we use a model of self-assembly using flexible junction molecules and describe how it can in some static conditions be used to predict the outcome of a graph self-assembly. Using probabilistic methods, we show the expectation and the variance of the number of self-assembled cycles, K_3, and discuss generalization of these results for C_n. We tie this analysis to previously observed experimental results.

1 Introduction

The study of molecular self-assembly is rapidly developing in one of the most important aspects of nanotechnology. Self-assembly is a physico-chemical process by which simpler structures reorganize and combine into more complicated forms without any external intervention. Because of its nature, DNA molecule uses the natural Watson-Crick mechanism to change, transform and self-assemble into different structures. Although naturally occurring DNA molecule has a double helix structure, it can be configured in many other forms, like: hairpin, branched 3 and 4 junction molecules, stick cube, truncated octahedron, etc. [3, 11, 18, 22]. These newly formed molecules have been proposed for computational purposes [8] as well as for scaffolding for other structures [3, 18, 22].

Several models for DNA self-assembly have appeared, mostly using rigid square tiles [1, 2, 10, 14, 15]. In this paper, we discuss another model, that uses flexible tiles (each tile composed of a single junction molecule) of various sizes initially proposed in [8]. Flexible junction molecules have been used to obtain experimentally regular graph structure, such as the cube [3] and truncated octahedron [18, 22] and non-regular graph structures [7, 8]. Junction molecules with 5 or 6 arms have been reported [20] and 4 armed junction molecules have been used in a lattice [17]. Recently an octahedron has been configured whose edges (junction arms) are made of DX and PX molecules [18].

This model is based on DNA branched junction molecules with flexible arms extending to free sticky ends. By imposing restrictions on the number of types of tiles, one can get DNA computability classes that correspond to extant complexity classes. A "polynomial" restriction produces precisely the NPTIME queries;

A. Carbone and N.A. Pierce (Eds.): DNA11, LNCS 3892, pp. 144–157, 2006.

Theorem 11. *Let NBSA (DBSA) be the class of languages accepted by nondeterministic (deterministic) bounded S/A acceptors. Then:*

1. *NBSA is closed under union and intersection but not under complementation.*
2. *DBSA is closed under union, intersection, and complementation.*

References

1. C. S. Calude and Gh. Paun. *Computing with Cells and Atoms: After Five Years (new text added to Russian edition of the book with the same title first published by Taylor and Francis Publishers, London, 2001).* To be published by Pushchino Publishing House, 2004.
2. E. Csuhaj-Varju, O. H. Ibarra, and G. Vaszil. On the computational complexity of P automata. In *Proc. DNA 10* (C. Ferretti, G. Mauri, C. Zandron, eds.), Univ. Milano-Bicocca, 97–106,2004.
3. Z. Dang, O. H. Ibarra, C. Li, and G. Xie. On model-checking of P systems. *Proc. 4th International Conference on Unconventional Computation*, to appear, 2005.
4. R. Freund, L. Kari, M. Oswald, and P. Sosik. Computationally universal P systems without priorities: two catalysts are sufficient. *Theoretical Computer Science*, 330(2): 251–266, 2005.
5. R. Freund and Gh. Paun. On deterministic P systems. See P Systems Web Page at http://psystems.disco.unimib.it, 2003.
6. O. H. Ibarra. The number of membranes matters. In *Proc. 4th Workshop on Membrane Computing*, Lecture Notes in Computer Science 2933, Springer-Verlag, 218-231, 2004. Journal version to appear in *Theoretical Computer Science*, 2005.
7. O. H. Ibarra. On determinism versus nondeterminism in P systems. Theoretical Computer Science, to appear, 2005.
8. B. Monien, Two-way multihead automata over a one-letter alphabet, *RAIRO Informatique theorique*, 14(1):67–82, 1980.
9. A. Paun and Gh. Paun. The power of communication: P systems with symport/antiport. *New Generation Computing* 20(3): 295–306, 2002.
10. Gh. Paun. Computing with membranes. *Turku University Computer Science Research Report No. 208*, 1998.
11. Gh. Paun. Computing with membranes. *Journal of Computer and System Sciences*, 61(1):108–143, 2000.
12. Gh. Paun. *Membrane Computing: An Introduction.* Springer-Verlag, 2002.
13. Gh. Paun. Further twenty six open problems in membrane computing. See P Systems Web Page at http://psystems.disco.unimib.it, 1-12, January 20, 2005.
14. Gh. Paun, J. Pazos, M. J. Perez-Jimenez, and A. Rodriguez-Paton. Symport/antiport P systems with three objects are universal. *Fundamenta Informaticae*, 64(1-4): 353–367, 2005.
15. Gh. Paun and G. Rozenberg. A guide to membrane computing. *Theoretical Computer Science*, 287(1):73–100, 2002.
16. W. Savitch. Relationships between nondeterministic and deterministic tape complexities. *J. Comput. Syst. Sci.*, 4(2): 177–192, 1970.
17. W. Savitch. A note on multihead automata and context-sensitive languages. *Acta Informatica*, 2:249–252, 1973.
18. P. Sosik. P systems versus register machines: two universality proofs. In *Pre-Proceedings of Workshop on Membrane Computing (WMC-CdeA2002), Curtea de Arges, Romania*, pages 371–382, 2002.

2   Junction Graph Model

The main building blocks for the junction graph model, are stable branched junction DNA molecules, which are molecules that have flexible arms with sticky ends (see Fig. 1 to the left). Each arm has two parts: a body and a sticky end extended from the body. The body part is a double stranded DNA molecule, while the sticky end part is a linear DNA strand that hasn't connected to a Watson-Crick complement. When two arms from two different junction molecules connect, they glue their sticky ends together, forming a more complex structure. A 1-junction molecule is a hairpin structure with only one sticky end, a 2-junction molecule is a double helix with two sticky ends, one at each end. In order to permit a junction to connect to a nearby junction, regardless of spatial orientation, we will follow [8, 9] and [12] by supposing that bulged T's can be added in the junction sequences to add flexibility of the branches.

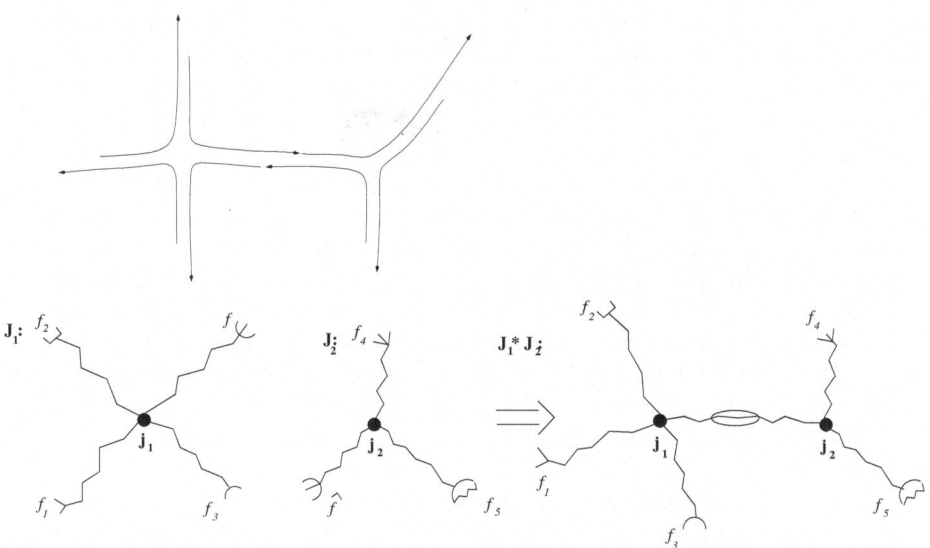

Fig. 1. Above: Watson-Crick bonding of two DNA junction molecules. Below: Junction graph that represents bonding of the two DNA junction molecules depicted on the left.

For the simplicity we will suppress some of the technicalities and represent complexes as labelled graphs.

Definition 1. *A* junction graph *is a graph $G = (V, E)$ with finite set of vertices V and edges E. The set of vertices V is partitioned into two disjoined sets J and F such that all vertices in F have degree 1, and no two vertices of F are adjacent.*

The elements of J are vertices that represent centers of junction molecules, while the elements of F are vertices that represent the free sticky ends on the

no restriction at all produces the classes of all computable queries [9, 10]. Using this model, for a given problem, a solution is obtained if and only if a graph-like complex of appropriate size can appear.

A graph-like complex is a closed DNA molecule that does not contain any single stranded sticky ends. Whereas in [9] the main concern is about the kind of complexes to be expected when an experiment is running, this paper is concerned with the amount of each type of complete complexes and what portion of these complexes correspond to the encoded structure. Answers to these questions shall provide a better understanding of the self-assembly process as well as possibly predict what we should expect to obtain within a pot with DNA molecules.

This paper seeks to raise these questions and uses probabilistic analysis to provide answers in some special cases. In addition it relates the answers to previously observed experimental results. We begin with a pot containing uniformly distributed DNA molecules compatible of forming a cyclic graph structure. After the annealing process it is assumed that no free sticky ends remain. We prove that under certain probability conditions almost all structures represent the originally encoded graph i.e., the appearance of dimer (double cover) or trimer (triple cover) molecules is with small probability. (It is assumed that the reader is familiar with basic probability theory, but for those who are not [13] gives a good introduction to the field).

Experimentally this has been observed by several assembly processes [6, 7, 16, 12]. Our theoretical analysis is the first attempt to understand the stochastic self-assembly of these junction molecules into complete structures. We tie our results with the detailed experimental analysis of a similar set-up in [4] and generating triangles in [12]. These observations provide a good estimate of the probability values that we use in the proof of the main result.

The model used for theoretical analysis is static model and does not consider any thermodynamic properties of the solution. It deals only with the input and the output of an experiment. We hope to extend this static model to a dynamic version.

The description of the model is presented in Section 2. It contains the main definitions of complexes and structures that are built up by junction molecules. The probabilistic model is presented in Section 3. The main assumption of the model is that the probability r that the last of the possible connections within a complex appears after the other connections have been established is very high. Under these conditions we show that probability of appearance of dimers and trimers in a pot designed to form monomer cycles approaches 0. In Section 4 we provide comparison with two experimental results already observed. The first experiment is from [4] and deals with linear duplex DNA that can close into a cyclic molecule. From the molecular concentration used in the experiment we can deduce the probability value for r. The second comparison is with the experiment in [12] that deals with construction of triangles i.e. precisely with the special case considered in the main result. Under the conditions of this experiment no dimers or trimers were observed which coincides with our findings. We end the paper with some concluding remarks.

5. S. Janson, T. Luczak, A. Rucinski. *Random Graphs*, New York : John Wiley, 2002.
6. N. Jonoska, S. Liao, N.C.Seeman, *Transducers with Programmanle Input by DNA Self-Assembli*, in: Aspects of Molecular Computing (N. Jonoska, Gh. Paun, G.Rozenberg eds.), Springer LNCS **2950** (2004) 219-240.
7. N. Jonoska, S. Karl, M. Saito. *Three dimensional DNA structures in computing*, BioSystems **52** (1999) 143-153.
8. N. Jonoska, P. Sa-Ardyen, N.C. Seeman. *Computation by self-assembly of DNA graphs*, Genetic Programming and Evolvable Machines **4** (2003) 123-137.
9. N. Jonoska, G.L. McColm. *Self-assembly by DNA Junction Molecules: The Theoretical Model*, Foundations of Nanoscience. J.Reif (edi) (2004).
10. M-Y. Kao, V. Ramachandran. *DNA self-assembly for constructing 3D boxes*. Algorithms and Computations, ISAC 2001 Preceedings, Springer LNCS **2223** (2001) 429-440.
11. C. Mao, W. Sun, N.C. Seeman. *Designed Two-Dimensional DNA Holliday Junction Arrays Visualized by Atomic Force Microscopy*, Journal of American Chemical Society 121(23) (1999) 5437-5443.
12. J. Qi, X. Li, X. Yang, N.C. Seeman. *Ligation of Triangles Built from bulged 3-arm DNA Branched Junctions.*, Journal of American Chemical Society 120 (1996) 6121-6130.
13. S. M. Ross. *A First Course in Probability*, Prentice Hall, (2001).
14. P.W.K. Rothemund, P. Papadakis, E. Winfree. *Algorithmic Self-Assembly of DNA Sierpinski Triangles*, Preproceedings of 9_{th} DNA Based Computers, Madison Wisconsin June 1-4 (2003).
15. P.W.K. Rothemund, E. Winfree. *The Program-Size Complexity of Self-Assembled Squares*, Proceedings of 33rd ACM meeting STOC 2001, Portland, Oregon, May 21-23 (2001) 459-468.
16. P. Sa-Ardyen, N. Jonoska, N.C. Seeman. *Self-Assembly of graphs represented by DNA helix axis topology*, Journal of American Chemical Society 126(21) (2004) 6648-6657.
17. Seeman, N.C., *DNA junctions and lattices*, Journal of theoretical biology 99 (1982) 237-247.
18. W.M. Shih, J.D. Quispe, G.F. Joyce., *A 1.7-kilobase single stranded DNA folds into a nanoscale octahedron*, Nature **427** (2004) 618-621.
19. J.H. Spencer. *Ten lectures on the probabilistic method*, SIAM, Philadelphia, PA (1987).
20. Y. Wang, J.E. Mueller, B. Kemper, N.C. Seeman. *The assembly and characterization of 5 arm and 6 arm DNA junctions*, Biochemistry 30 (1991) 5667-5674.
21. E. Winfree, F. Liu, L.A. Wenzler, N.C. Seeman. *Design and self-assembly of two-dimentsional DNA crystals*, Nature 394 (1998) 539-544.
22. Y. Zhang, N.C. Seeman. *The construction of a DNA truncated octahedron*, Journal of American Chemical Society 116(5) (1994) 1661-1669.

Appendix A

Proposition 1. *Let P be a pot type which contains $3m$ uniformly distributed 2-branched junctions of three different types capable of admitting K_3 complex. Let X denote the number of complete K_3 complexes in P and r the probability*

were obtained, at least not detectable by the non-denaturing gel. Moreover, no linear and non-complete structures were obtained as well. This corresponds to the conditional probability r, as in the second half of Proposition 3, being close to 1. At the same time it shows that the analysis of considering only complete complexes as a result of the annealing can provide a valid model to study.

It is interesting to note that the experiment showed that at least two different topoisomers can be obtained in the process of "gluing". Our model considers complexes that represent isomorphic graphs as having the same complete structure type, hence distinct topoisomers cannot be detected by the model. This shows real limitations of the model in studying the topological properties of the products.

Notice that in the example where $r < 1$, we have a variety of complete cycles, while in the example where $r \to 1$, we only have minimal cycles.

5   Concluding Remarks

The paper provides a first step in detailed analysis of self-assembly of flexible junction molecules. The case that is covered by Proposition 3 and Theorem 4 although basic, it still shows to be close to the results of experiments that are already known in the literature. The assumption for the conditional probability r can be extended to cases of regular and non-regular graphs. The computation of the expectation and the variance can be done in a similar way. However, we believe that more complicated structures have additional geometric and other intrinsic constraints that would make such simplification of our assumptions non realistic and superfluous. Good conditions for studying such complicated structures remain to be discovered.

Acknowledgment

Authors thank N. Seeman for providing valuable information and references. The work is supported in part by NSF Grants CCF #0432009 and EIA#0086015.

References

1. L.M. Adleman, Q. Cheng, A. Goel, M-D. Huang, D. Kempe, P. Moisset de Espanes, P.W.K. Rothemund. *Combinatorial optimization problems in self-assembly*, STOC'02 Proceedings, Montreal Quebec, Canada, 2002.
2. L.M. Adleamn, J. Kari, L. Kari, D. Reishus. *On the decidability of self-assembly of infinite ribons* Proceedings of FOCS 2002, IEEE Symposium on Foundations of Computer Science, Washington (2002) 530-537.
3. J.H. Chen, N.C. Seeman. *Synthesis from DNA of a molecule with the connectivity of a cube*, Nature **350** (1991) 631-633.
4. A. Dugaiczyk, H.W. Boyer, H.M. Goodman. *Ligation of EcoRI endonuclease-generated DNA fragments into linear and circular structures.* Journal of Molecular Biology **96**(1) (1975) 171-178.

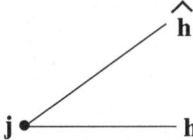

Fig. 3. The junction structure used in the pot of the experiment described in [4]

appearing in the process increases with the time of the reaction. As an example, for the same ratio as in the previous case, $j/i = 1.56$, after 15 minutes of ligation no circular structures were obtained, only linear ones.

For theoretical analysis, we can consider $\phi 80(8)$ DNA molecules as 2-branched junction molecules with two sticky end types (Fig. 3). The probability that one junction will form a circular monomer will be r. For our case that is the probability that both sticky ends on a junction to connect

$$P(\mathbf{h} \text{ and } \hat{\mathbf{h}} \text{ to connect }) = r$$

Since we have m junctions, the expected number of circular monomers is $EX = mr$. Apparently r is positively correlated to the ratio j/i. As we saw in the example, when $j/i = 12.4$, then $r = .64$. If the molecule length is the same, we can increase the number of circular monomers by decreasing the molar concentration. So the probability of obtaining circular complexes would depend mainly on the molar concentration.

Major difference between this model and the experimental results is the presence of linear fragments, which could be regarded as potentially large cycles. Our model reflects the situation when the molar concentration is very low, that means mainly small cycles are to be formed. We expect in reality for a not diluted solution to get lots of linear fragments, which will not close. This case is not considered in our static model.

4.2   Annealing Triangles

In the experimental process described in [12], three 2-branched junctions capable of forming a triangle (K_3) were designed (precisely the case considered in Proposition 3). Although the original design of the molecules had a purpose to be a substructure of a more complex motif, the experimental analysis of the triangles was performed as well. The triangle structures used in the experiment are presented in Fig. 2 (a). The central portion of the 2-armed junctions are designed to have buldged T sequence that provided extra flexibility at the arms. The main purpose of the experiment was to test the rigidity of the triangular shape, which proved to be lacking.

However, in the process of building a more complex structure, the characterization of the triangular shape was performed. The molar concentration used in the experiment was one picomole per strand in $10\mu L$ solution and in the ligation process of the experiment no dimers, trimers or other higher cover structure

$$E(X) = mr,$$

where m denotes the amount of junctions in **P** *of each type.*

Remark. For the pot $\mathbf{P}= \{\mathbf{j_1},\dots,\mathbf{j_n}\}$ described above the junction types are chosen in such a way that the cycle C_n of length n is the smallest complete complex that can be formed from the junctions in the pot.

4   Comparison with Results from Prior Experiments

4.1   Annealing of Linear and Circular Molecules

Experimental results concerning cyclization of DNA molecules have appeared in literature and we choose to compare our results with the ones by A. Dugaiczyk, H. W. Boyer and H. M. Goodman [4]. They analyzed the relation between the length as well as concentration of DNA fragments from one side, and the distribution of different molecular structure types obtained by ligation of those fragments from another side. The experiment measures for the rate of ligation of *Eco*RI-cleaved simian virus (SV 40), ligating the obtained DNA fragments of the virus. For different DNA concentration, the starting DNA fragments produced either linear multimers or circular structures made of those fragments. The amount of linear or circular products were measured under controlled conditions, varying on two parameters j and i. The parameter j corresponds to the local concentration of one sticky end in the neighborhood or the volume of the other sticky end of the same DNA molecule. The parameter i is the total concentration of the sticky ends of a DNA molecule. The local concentration j depends on the contour length of the molecule l, while the total concentration i depends on the molar concentration M of the molecule. Thus j/i depends on the contour length of the molecule l and on the molar concentration M.

It was shown experimentally that circular structures were favored when the ratio j/i was greater than 1, while more linear structures were favored when the ratio j/i was less then 1. When $j = i$, equal amounts of linear and circular types would be expected, but in the experiments of A. Dugaiczyk, H. W. Boyer and H. M. Goodman most of the obtained molecules were linear. The number of circular complexes was shown to be proportional with the ratio j/i and with significant increase of the ratio more of the obtained complexes were circular.

For example, consider their results for the $\phi 80(8)$ DNA molecule, which has molecular weight $M_r = 0.46 \times 10^6$. (In comparing their results to our model, they had one type of two-armed tile, whose sticky ends were complementary.) When $j/i = 12.4$, after 4 hours of ligation, they obtained 64% circular monomers, 12% circular dimers, 8% linear dimers, 4% circular trimers, 4% linear trimers, 8% linear tetramers and no other structures. When $j/i = 1.56$ after 4 hours of ligation, they obtained 35% circular monomers, 20% circular dimers, 5% linear dimers, 5% circular trimers, 5% linear trimers, and 30% molecules were linear tetramers, pentamers or hexamers and there were no other structures. Also, the time of ligation often plays a significant role. The amount of circular structures

– Case 4: $|S \cap T| = 3$, i.e. $S = T$

We have m^3 choices for those kind of sets, so

$$E(X_S, X_T) = P(X_S = 1, X_T = 1)$$
$$= P(X_S = 1 | X_T = 1)P(X_T = 1) = \frac{r}{m^2}, \text{ and}$$

$$EX_S EX_T = \frac{r}{m^2}\frac{r}{m^2} = \frac{r^2}{m^4}, \text{ and hence}$$

$$\text{Cov}(X_S, X_T) = \frac{r}{m^2} - \frac{r^2}{m^4} = \frac{r}{m^2}(1 - \frac{r}{m^2}), \text{ so}$$

$$\sum_{|S\cap T|=3} \text{Cov}(X_S, X_T) = m^3 \frac{r(m^2 - r)}{m^4}$$

$$= \frac{r(m^2 - r)}{m} \sim mr.$$

From the obtained information above,

$$Var(X) = \sum_{S,T} Cov(X_S, X_T) \sim 2mr^2 - 3mr^2 - 3r^2 + mr \sim mr(1 - r).$$

(More precisely, $Var(X) = \frac{r}{m}(m^2 - m^2 r + mr + r) \geq 0$ if $m \geq 0$ and $r \leq 1$.)

When $r \to 1$, $VarX = E(X - EX)^2 \to 0$, and since X is a nonegative random variable it follows that almost surely $X = E(X) = m$. That means almost surely only K_3 complexes are obtained in $\mathbf{P}$.

Hairpin Structures in DNA Words

Lila Kari[1], Stavros Konstantinidis[2], Elena Losseva[1],
Petr Sosík[3,4,*], and Gabriel Thierrin[5]

[1] Department of Computer Science,
The University of Western Ontario, London, ON, N6A 5B7 Canada
{lila, elena}@csd.uwo.ca
[2] Dept. of Mathematics and Computing Science,
Saint Mary's University, Halifax, Nova Scotia, B3H 3C3 Canada
s.konstantinidis@stmarys.ca
[3] Facultad de Informática, Universidad Politécnica de Madrid,
Campus de Montegancedo s/n, Boadilla del Monte 28660,
Madrid, Spain
[4] Institute of Computer Science, Silesian University, Opava, Czech Republic
petr.sosik@fpf.slu.cz
[5] Department of Mathematics,
The University of Western Ontario, London, ON, N6A 5B7 Canada
thierrin@uwo.ca

Abstract. We formalize the notion of a DNA hairpin secondary structure, examining its mathematical properties. Two related secondary structures are also investigated, taking into the account imperfect bonds (bulges, mismatches) and multiple hairpins. We characterize maximal sets of hairpin-forming DNA sequences, as well as hairpin-free ones. We study their algebraic properties and their computational complexity. Related polynomial-time algorithms deciding hairpin-freedom of regular sets are presented. Finally, effective methods for design of long hairpin-free DNA words are given.

1 Introduction

A single strand of deoxyribonucleic acid (DNA) consists of a sugar-phosphate backbone and a sequence of nucleotides attached to it. There are four types of nucleotides denoted by A, C, T, and G. Two single strands can bind to each other if they have opposite polarity (strand's orientation in space) and are pairwise Watson-Crick complementary: A is complementary to T, and C to G. The binding of two strands is also called annealing. The ability of DNA strands to anneal to each other allows for creation of various secondary structures. A DNA hairpin is a particular type of secondary structure investigated in this paper. An example of a DNA hairpin structure is shown in Figure 1.

The reader is referred to [1, 16] for an overview of the DNA computing paradigm. The study of DNA secondary structures such as hairpin loops is motivated by finding reliable encodings for DNA computing techniques. These

---

* Corresponding author.

A. Carbone and N.A. Pierce (Eds.): DNA11, LNCS 3892, pp. 158–170, 2006.
© Springer-Verlag Berlin Heidelberg 2006

$$\text{GTCAGCGATAG}^{\text{A}}{}^{\text{C}}{}^{\text{C}}{}_{\text{A}}$$
$$\text{CAGTCGCTATC}_{\text{A}}{}_{\text{C}}{}_{\text{C}}{}_{\text{T}}$$

Fig. 1. An example of a DNA hairpin loop

techniques usually rely on a certain set of DNA bonds and secondary structures, while other types of bonds and structures are undesirable. Various approaches to the design of DNA encodings without undesirable bonds and secondary structures are summarized in [14] and [11]. For more details we refer the reader e.g. to [5, 12, 13]. Here we apply the formal language approach which has been used in [2, 7, 8, 10, 11] and others.

Hairpin-like secondary structures are of special importance for DNA computing. For instance, they play an important role in insertion/deletion operations with DNA. Hairpins are the main tool used in the Whiplash PCR computing techniques [18]. In [20] hairpins serve as a binary information medium for DNA RAM. Last, but not least, hairpins are basic components of "smart drugs" [3].

The paper is organized as follows. Section 2 introduces basic definitions, Section 3 presents results on hairpins, and in Section 4 we study two important variants of the hairpin definition. The first one takes into the account imperfect DNA bonds (mismatches, bulges), the second one is related to hairpin-based nanomachines. We study algebraic properties of (maximal) hairpin-free languages. The hairpin-freedom problem and the problem of maximal hairpin-free sets are both shown to be decidable in polynomial time for both regular and context-free languages. The last section provides methods of constructing long hairpin-free words.

2   Preliminary Definitions

We denote by X a finite alphabet and by X^* its corresponding free monoid. The cardinality of the alphabet X is denoted by $|X|$. The empty word is denoted by 1, and $X^+ = X^* - \{1\}$. A *language* is an arbitrary subset of X^*. For a word $w \in X^*$ and $k \geq 0$, we denote by w^k the word obtained as catenation of k copies of w. Similarly, X^k is the set of all words from X^* of length k. By convention, $w^0 = 1$ and $X^0 = \{1\}$. We also denote $X^{\leq k} = X^0 \cup X^1 \cup \ldots \cup X^k$. A *uniform*, or *block*, *code* is a language all the words of which are of the same length k, for some $k \geq 0$, and is therefore contained in X^k.

A mapping $\psi : X^* \to X^*$ is called a *morphism* (*anti-morphism*) of X^* if $\psi(uv) = \psi(u)\psi(v)$ (respectively $\psi(uv) = \psi(v)\psi(u)$) for all $u, v \in X^*$, and $\psi(1) = 1$. See Chapter 7 in [19] for a general overview of morphisms. An involution $\theta : X \longrightarrow X$ is defined as a map such that θ^2 is the identity function. An involution θ can be extended to a morphism or an antimorphism over X^*. In both cases θ^2 is the identity over X^* and $\theta^{-1} = \theta$. The simplest involution is the identity function ϵ. A *mirror involution* μ is an antimorphic involution which maps each letter of the alphabet to itself.

We shall refer to the DNA alphabet $\Delta = \{A, C, T, G\}$, over which two involutions of interest are defined. The DNA *complementarity involution* γ is a morphism given by $\gamma(A) = T$, $\gamma(T) = A$, $\gamma(C) = G$, $\gamma(G) = C$. For example, $ACGCTG = \mu(GTCGCA) = \gamma(TGCGAC)$.

The antimorphic involution $\tau = \mu\gamma$ (the composite function of μ and γ, which is also equal to $\gamma\mu$), called the *Watson-Crick involution*, corresponds to the DNA bond formation of two single strands. If for two strings $u, v \in \Delta^*$ it is the case that $\tau(u) = v$, then the two DNA strands represented by u, v anneal as Watson-Crick complementary sequences.

A nondeterministic finite automaton (*NFA*) is a quintuple $A = (S, X, s_0, F, P)$, where S is the finite and nonempty set of states, s_0 is the start state, F is the set of final states, and P is the set of productions of the form $sx \to t$, for $s, t \in S$, $x \in X$. If for every two productions $sx_1 \to t_1$ and $sx_2 \to t_2$ of an NFA we have that $x_1 \neq x_2$ then the automaton is called a *DFA* (deterministic finite automaton). The language accepted by the automaton A is denoted by $L(A)$. The *size* $|A|$ of the automaton A is the number $|S| + |P|$.

Analogously we define a *pushdown automaton (PDA)* and a *deterministic pushdown automaton (DPDA)*. We refer the reader to [6, 19] for detailed definitions and basics of formal language theory.

3   Hairpins

Definition 1. *If θ is a morphic or antimorphic involution of X^* and $k > 0$, then a word $u \in X^*$ is said to be θ-k-hairpin-free or simply hp(θ,k)-free if $u = xvy\theta(v)z$ for some $x, v, y, z \in X^*$ implies $|v| < k$.*

Notice that words of length less than $2k$ are hp(θ,k)-free. If we interpret this definition for the DNA alphabet Δ and the Watson-Crick involution τ, then a hairpin structure with the length of bond at least k is a word that is not hp(θ,k)-free.

Definition 2. *Denote by hpf(θ, k) the set of all hp(θ,k)-free words in X^*. The complement of hpf(θ, k) is hp$(\theta, k) = X^* - hpf(\theta, k)$.*

Notice that $hp(\theta, k + 1) \subseteq hp(\theta, k)$ for all $k > 0$.

Definition 3. *A language L is called θ-k-hairpin-free or simply hp(θ, k)-free if $L \subseteq hpf(\theta, k)$.*

It follows by definition that a language L is hp(θ, k)-free iff $X^*vX^*\theta(v)X^* \cap L = \emptyset$ for all $|v| \geq k$. An analogous definition was given in [7], where a θ-k-hairpin-free language is called θ-subword-k-code. The authors focused on their coding properties and relations to other types of codes. They consider also the restriction on the length of the hairpin, namely that $1 \leq |y| \leq m$ for some $m \geq 1$. The reader can verify that many of the results given in this paper remain valid if we apply this additional condition.

Example. Recall that γ is the DNA complementary involution over Δ^*, then:

$$hpf(\gamma, 1) = \{A, C\}^* \cup \{A, G\}^* \cup \{T, C\}^* \cup \{T, G\}^*$$

We give the necessary and sufficient conditions for finiteness of the languages $hpf(\theta, k)$, $k \geq 1$. Proofs of the following results can be found in [9]. Recall that $hpf(\mu, k)$ is the set of all words which do not contain any two non-overlapping mirror parts of length at least k.

Proposition 4. *Let X be a binary alphabet. For every word $w \in X^*$ in $hpf(\mu, 4)$ we have that $|w| \leq 31$. Moreover the following word of length 31 is in $hpf(\mu, 4)$:*

$$a^7 b a^3 b a b a b a b^2 a b^2 a^2 b^7.$$

Proposition 5. *Consider a binary alphabet X. Then $hpf(\mu, k)$ is finite if and only if $k \leq 4$.*

Proposition 6. *Let θ be a morphic or antimorphic involution. The language $hpf(\theta, k)$ over a non-singleton alphabet X is finite if and only if one of the following holds:*

(a) $\theta = \epsilon$, the identity involution;
(b) $\theta = \mu$, the mirror involution, and either $k = 1$ or $|X| = 2$ and $k \leq 4$.

3.1   Properties of hp(θ, 1)-Free Languages

Recall the definition of an embedding order: $u \leq_e v$ if and only if $u = u_1 u_2 \cdots u_n$, $v = v_1 u_1 v_2 u_2 \cdots \cdots v_n u_n v_{n+1}$ for some integer n with $u_i, v_j \in X^*$.

A language L is called *right $\leq_e$-convex* [21] if $u \leq_e w$, $u \in L$ implies $w \in L$. The following result is well known: *All languages (over a finite alphabet) that are right $\leq_e$-convex are regular.*

Proposition 7. *The language $hp(\theta, 1)$ is right $\leq_e$-convex (and hence regular).*

Proof. Observe that if $u = u_1 u_2 \in hp(\theta, 1)$ and $w \in X^*$ then $u_1 w u_2 \in hp(\theta, 1)$. Hence, for $u \in hp(\theta, 1)$, $u \leq_e v$ implies $v \in hp(\theta, 1)$.

Let $L \subseteq X^*$ be a nonempty language and let $S(L) = \{w \in X^* | u \leq_e w, u \in L\}$. Recall further that a set H with $\emptyset \neq H \subseteq X^+$ is called a *hypercode* over X^* *iff* $x \leq_e y$ and $x, y \in H$ imply $x = y$. That is, a hypercode is an independent set with respect to the embedding order.

Proposition 8. *Let θ be a morphic or antimorphic involution. Then there exists a unique hypercode H such that $hp(\theta, 1) = S(H)$.*

Proof. Let $H = \bigcup_{a \in X} a\theta(a)$, then $S(H) = \bigcup_{a \in X} X^* a X^* \theta(a) X^* = hp(\theta, 1)$. The uniqueness of H is immediate.

3.2   Properties of hp(θ, k)-Free Languages

Proposition 7, true for the case $k = 1$, cannot in general be extended to the case $k > 1$. Consider, for example, $X = \{a, b\}$ and a morphism $\theta(a) = b$, $\theta(b) = a$. If $u = a^2 b^2$, then $u = a^2 \theta(a^2)$ and hence $u \in hp(\theta, 2)$. But $u \leq_e w$ for $w = abab^2$, and $w \notin hp(\theta, 2)$. Therefore, the language $hp(\theta, 2)$ is not $\leq_e$-convex. However, the following weaker result is proven in [9].

Proposition 9. *The languages $hp(\theta, k)$ and $hpf(\theta, k)$, $k \geq 1$, are regular.*

Proposition 9 suggests an existence of fast algorithms solving some problems important from the practical point of view. We investigate two such problems now. Let θ be a fixed morphic or antimorphic involution and let $k \geq 1$ be an arbitrary but fixed integer.

Hairpin-Freedom Problem.

Input: A nondeterministic automaton M.
Output: Yes/No depending on whether $L(M)$ is hp(θ, k)-free.

Maximal Hairpin-Freedom Problem.

Input: A deterministic automaton M_1 accepting a hairpin-free language, and a NFA M_2.
Output: Yes/No depending on whether there is a word $w \in L(M_2) - L(M_1)$ such that $L(M_1) \cup \{w\}$ is hp(θ, k)-free.

We assume that M and M_1 are finite automata in the case of regular languages, and pushdown automata in the case of context-free languages.

Proposition 10. *The hairpin-freedom problem for regular languages is decidable in linear time (w.r.t. $|M|$).*

Proof. By definition, $L(M)$ is hp(θ, k)-free iff $L(M) \subseteq hpf(\theta, k)$ iff $L(M) \cap hp(\theta, k) = \emptyset$. This problem is solvable in time $\mathcal{O}(|M_k| \cdot |M|)$ for regular languages, where M_k is a NFA accepting $hp(\theta, k)$. The automaton M_k is fixed for a chosen k.

Proposition 11. *The maximal hairpin-freedom problem for regular languages is decidable in time $\mathcal{O}(|M_1| \cdot |M_2|)$.*

Proof. We want to determine whether there exists a word $w \in hpf(\theta, k)$ such that $w \notin L(M_1)$, but $w \in L(M_2)$. It is decidable in time $\mathcal{O}(|M_1| \cdot |M_2| \cdot |M_k'|)$ whether $(hpf(\theta, k) \cap L(M_2)) - L(M_1) = \emptyset$. The size of an NFA accepting $hpf(\theta, k)$ is denoted by $|M_k'|$. The automaton M_k' is fixed for a chosen k.

As an immediate consequence, for a given block code K of length l it is decidable in linear time with respect to $|K| \cdot l$, whether there is a word $w \in X^l - K$ such that $K \cup \{w\}$ is hp(θ, k)-free. This is of particular interest since the lab sets of DNA molecules form often a block code.

Notice also that for a finite set S of DNA sequences (which is the case of practical interest) the size of the automaton M (or M_1) is in the worst case proportional to the total length of all sequences in S.

Proposition 12. *The hairpin-freedom problem for context-free languages is decidable in cubic time (w.r.t. $|M|$).*

Proposition 13. *The maximal hairpin-freedom problem for deterministic context-free languages is decidable in time $\mathcal{O}((|M_1| \cdot |M_2|)^3)$.*

Proof. We want to determine if $\exists w \in hpf(\theta, k)$ such that $w \notin L(M_1)$, but $w \in L(M_2)$. Denote $M_1 = (Q_1, X, \Gamma, q_1, Z_0, F_1, P_1)$, and let $M_2' = (Q_2, X, q_2, F_2, P_2)$ be a NFA accepting the language $hpf(\theta, k) \cap L(M_2)$. Consider the PDA $M = (Q, X, \Gamma, q_0, Z_0, F, P)$, where $Q = Q_1 \times Q_2$, $q_0 = (q_1, q_2)$. For $p \in Q_1, q \in Q_2$, and $Z \in \Gamma$ we define:

(1) $(p, q)aZ \underset{P}{\rightarrow} (p', q')\alpha$ iif $paZ \underset{P_1}{\rightarrow} p'\alpha$ and $qa \underset{P_2}{\rightarrow} q'$,

(2) $(p, q)1Z \underset{P}{\rightarrow} (p', q)\alpha$ iif $p1Z \underset{P_1}{\rightarrow} p'\alpha$

Let $F = \{(p, q) | p \notin F_1 \text{ and } q \in F_2\}$. Then $L(M) = (hpf(\theta, k) \cap L(M_2)) - L(M_1)$, and the size of M is $\mathcal{O}(|M_1| \cdot |M_2|)$. Let G be a CFG such that $L(G) = L(M)$. Note that the construction of G takes cubic time w.r.t. $|M|$, see Theorem 7.31 of [6]. Finally, it is possible to decide in linear time w.r.t. $|G|$ (see Section 7.4.3 of [6]) whether $L(G) = \emptyset$ or not.

The time complexity of the above mentioned algorithms is furthermore proportional to the (constant) size of a NFA accepting the language $hp(\theta, k)$ or $hpf(\theta, k)$, respectively. Therefore we recall results from [9] characterizing the minimal size of these automata.

Proposition 14. *The number of states of a minimal NFA accepting the language $hp(\theta, k)$, $k \geq 1$, over an alphabet X with the cardinality ℓ, is between ℓ^k and $3\ell^k$. Its size is at most $3(\ell^k + \ell^{k+1})$.*

Proposition 15. *Let there be distinct letters $a, b \in X$ such that $a = \theta(b)$. Then the size of a minimal NFA accepting $hpf(\theta, k)$, $k \geq 1$, over an alphabet X with the cardinality ℓ, is at least $2^{(\ell-2)^k/2}$.*

Corollary 16. *Consider the DNA alphabet $\Delta = \{A, C, T, G\}$ and the Watson-Crick involution τ.*

(i) *The size of a minimal NFA accepting $hp(\tau, k)$ is at most $15 \cdot 4^k$. The number of its states is between 4^k and $3 \cdot 4^k$.*

(ii) *The number of states of either a minimal DFA or an NFA accepting $hpf(\tau, k)$ is between $2^{2^{k-1}}$ and $2^{3 \cdot 2^{2k}}$.*

The above results show that the size of a minimal NFA for $hp(\tau, k)$ grows exponentially w.r.t. k. However, one should recall that k is the *minimal* length of bond allowing for a stable hairpin. Therefore k is rather low in practical applications and the construction of the mentioned automaton remains computationally tractable.

4   Variants of Hairpins

4.1   Scattered Hairpins

It is a known fact that parts of two DNA molecules could form a stable bond even if they are not exact mutual Watson-Crick complements. They may contain some mismatches and even may have different lengths. Hybridizations of this type are addressed e.g. in [2] and [11]. Motivated by this observation, we consider now a generalization of hairpins.

Definition 17. *Let θ be an involution of X^* and let k be a positive integer. A word $u = wy \in X^*$ is θ-k-scattered-hairpin-free or simply $shp(\theta, k)$-free if for all $t \in X^*$, $t \leq_e w$, $\theta(t) \leq_e y$ implies $|t| < k$.*

$$
\begin{array}{l}
\text{GTCAG}^{\text{T}}{}^{\text{C}}{}_{\text{A}}\text{CGATAG}^{\text{A}}{}^{\text{C}}{}^{\text{C}}{}^{\text{A}} \\
\text{CAGTC}_{\text{C}}{}_{\text{A}}{}_{\text{A}}\text{GCTATC}_{\text{A}}{}_{\text{C}}{}_{\text{C}}{}^{\text{T}}
\end{array}
$$

Fig. 2. An example of a scattered hairpin – a word in $shp(\tau, 11)$

Definition 18. *We denote by $shpf(\theta, k)$ the set of all $shp(\theta, k)$-free words in X^*, and by $shp(\theta, k)$ its complement $X^* - shpf(\theta, k)$.*

Definition 19. *A language L is called θ-k-scattered-hairpin-free or simply $shp(\theta,k)$-free if $L \subseteq shpf(\theta, k)$.*

Lemma 20. $shp(\theta, k) = S\left(\bigcup_{w \in X^k} w\theta(w) \right).$

Based on the above immediate result, analogous statements as in Section 3 hold also for scattered hairpins. Proofs are straightforward and left to the reader.

Proposition 21. *(i) The language $shp(\theta, k)$ is right $\leq_e$ -convex.*
(ii) The languages $shp(\theta, k)$ and $shpf(\theta, k)$ are regular.
(iii) There exists a unique hypercode H such that $shp(\theta, k) = S(H)$.

Analogously as in Section 3 we can also define the *scattered-hairpin-freedom problem* and *maximal scattered-hairpin-freedom problem*. Then we easily obtain the following results whose proofs are analogous to those in Section 3.

Corollary 22. *(i) The scattered-hairpin-freedom problem is decidable in linear time for regular languages and in cubic time for context-free languages.*
(ii) The maximal scattered-hairpin-freedom problem is decidable in time $\mathcal{O}(|M_1| \cdot |M_2|)$ for regular languages and in time $\mathcal{O}((|M_1| \cdot |M_2|)^3)$ for deterministic context-free languages.

Also the size of the minimal automaton accepting the language $shp(\theta, k)$ is similar to the case of $hp(\theta, k)$ in Section 3.2.

For the proof of the next proposition we recall the following technical tools from [4].

Definition 23. *A set of pairs of strings* $\{(x_i, y_i) \,|\, i = 1, 2, \ldots, n\}$ *is called a fooling set for a language* L *if for any* i, j *in* $\{1, 2, \ldots, n\}$,

(1) $x_i y_i \in L$, *and*
(2) if $i \neq j$ *then* $x_i y_j \notin L$ *or* $x_j y_i \notin L$.

Lemma 24. *Let* F *be a fooling set of a cardinality* n *for a regular language* L. *Then any NFA accepting* L *needs at least* n *states.*

Proposition 25. *The number of states of a minimal NFA accepting the language* $shp(\theta, k)$, $k \geq 1$, *over an alphabet* X *with the cardinality* ℓ, *is between* ℓ^k *and* $3\ell^k$, *its size is at most* $7\ell^k + 3\ell^{k+1}$.

Proof. Let $M_k = (S, X, s_1, F, P)$ be an NFA accepting $shp(\theta, k)$. The statement is trivial for the cases $\ell = 1$ or $k = 1$. Assume for the rest of the proof that $k \geq 2$ and $\ell \geq 2$.

(i) The reader can easily verify that the set $F = \{(w, \theta(w)) | w \in X^k\}$ is a fooling set for $hp(\theta, k)$. Therefore $|S| \geq \ell^k$.
(ii) Let

$$S = \{s_w, p_w \,|\, w \in X^{\leq k-1}\} \cup \{q_w \,|\, w \in X^k\}.$$

Let further $F = \{p_1\}$. The set of productions P is defined as follows:

$$
\begin{aligned}
s_v a &\to s_w \ \text{iif } va = w, &&\text{for each } v \in X^{\leq k-2},\ a \in X;\\
s_v a &\to q_w \ \text{iif } va = w, &&\text{for each } v \in X^{k-1},\ a \in X;\\
q_w a &\to p_v \ \text{iif } \theta(av) = w, &&\text{for each } v \in X^{k-1},\ a \in X;\\
p_w a &\to p_v \ \text{iif } av = w, &&\text{for each } v \in X^{\leq k-2},\ a \in X.\\
r a &\to r &&\text{for all } r \in S,\ a \in X.
\end{aligned}
$$

The reader can verify that $L(M_k) = shp(\theta, k)$, and that $|S| \leq 3\ell^k$, $|P| \leq 4\ell^k + 3\ell^{k+1}$, therefore $|M_k| \leq 7\ell^k + 3\ell^{k+1}$.

Note: An example of a similar automaton accepting the language $hp(\theta, k)$ can be found in [9].

4.2   Hairpin Frames

In this section we point out the following two facts. First, long DNA and RNA molecules can form complicated secondary structures as that shown in Figure 3. Second, simple hairpins can be useful in various DNA computing techniques and nanotechnologies, as in [3, 18, 20] and others. Hence it may be desirable to design DNA strands forming simple hairpins but avoiding more complex structures. This motivates another extension of the results from Section 3.

Definition 26. *The pair $(v, \theta(v))$ of a word u in the form $u = xvy\theta(v)z$, for $x, v, y, z \in X^*$, is called an* hp-pair *of u. The sequence of hp-pairs $(v_1, \theta(v_1))$, $(v_2, \theta(v_2)), \cdots, (v_j, \theta(v_j))$ of the word u in the form:*

$$u = x_1 v_1 y_1 \theta(v_1) z_1 x_2 v_2 y_2 \theta(v_2) z_2 \cdots x_j v_j y_j \theta(v_j) z_j$$

is called an hp-frame of degree j *of u or simply an* hp(j)-frame *of u.*

An hp-pair is an hp-frame of degree 1. The definition of hairpin frames characterizes secondary structures containing several complementary sequences such as that in Fig. 3.

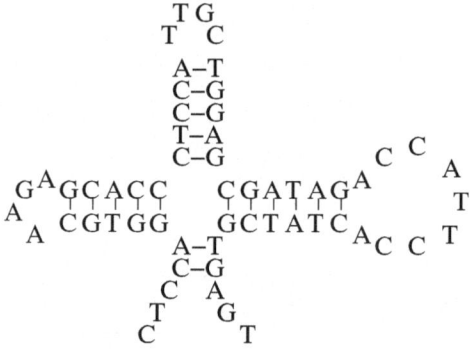

Fig. 3. An example of a hairpin frame – a word in $hp(\tau, fr, 3)$

A word u is said to be an *hp(fr,j)-word* if it contains at least one hp-frame of degree j. Observe that there may be more ways of finding hp-pairs in u, resulting in hp-frames of various degrees. Obviously, any hp(fr,j)-word is also hp(fr,i) for all $1 \le i \le j$.

Definition 27. *For an involution θ we denote by $hp(\theta, fr, j)$ the set of all hp(fr,j)-words $u \in X^*$, and by $hpf(\theta, fr, j)$ its complement in X^*.*

The results in Section 3, concerning the languages $hp(\theta, 1)$ and $hpf(\theta, 1)$, can easily be extended for the case of hairpin frames. Proofs are left to the reader.

Lemma 28. $hp(\theta, fr, j) = hp(\theta, 1)^j = \left(\bigcup_{a \in X} X^* a X^* \theta(a) X^* \right)^j.$

Proposition 29. *(i) The language $hp(\theta, fr, j)$ is right $\le_e$ -convex.*
(ii) The languages $hp(\theta, fr, j)$ and $hpf(\theta, fr, j)$ are regular.
(iii) There exists a unique hypercode H such that $hp(\theta, fr, j) = S(H)$.

Corollary 30. *(i) The* hp(fr,j)-freedom *problem is decidable in linear time for regular languages and in cubic time for context-free languages.*

(ii) *The* maximal hp(fr,j)-freedom *problem is decidable in time* $\mathcal{O}(|M_1| \cdot |M_2|)$ *for regular languages and in time* $\mathcal{O}((|M_1| \cdot |M_2|)^3)$ *for deterministic context-free languages.*

Proposition 31. *The size of a minimal NFA accepting the language* $hp(\theta, fr, j)$, $j \geq 1$, *over an alphabet* X *with the cardinality* ℓ, *is at most* $4\ell j + 2j + 1$.

Proof. The statement follows by the construction of an NFA $M = (S, X, s_1, F, P)$ accepting the language $hp(\theta, fr, j)$. Let

$$S = \{s_0, s_1, \ldots, s_j\} \cup \{p_i^k \mid 1 \leq i \leq j, \ 1 \leq k \leq \ell\}.$$

Let further $F = \{s_j\}$, and denote $X = \{a_1, \ldots, a_\ell\}$. The set of productions P is defined as follows:

$$s_{i-1} a_k \rightarrow p_i^k, \ p_i^k \theta(a_k) \rightarrow s_i \text{ for all } 1 \leq i \leq j, \ 1 \leq k \leq \ell;$$
$$sa \rightarrow s \qquad \qquad \text{for all } s \in S, \ a \in X.$$

The reader can verify that $L(M_k) = hp(\theta, fr, j)$, and that $|M| = 4\ell j + 2j + 1$.

Unlike the cases of hairpins or scattered hairpins, the size of the minimal NFA accepting $hp(\theta, fr, j)$ is $\mathcal{O}(j\ell)$. However, if we considered also a minimal length k of the hairpin bonds, we would obtain the same exponential size of the automaton as in Section 3.2, but multiplied by j.

5   Construction of Long Hairpin-Free Words

In this section we discuss the problem of constructing long hp(θ, k)-free words for the cases where θ is the Watson-Crick involution and $\theta = \epsilon$. This question is relevant to various encoding problems of DNA computing. For example, in [20] the authors consider n-bit memory elements that are represented by DNA words of the form

$$u_1 v_1 w_1 \theta(v_1) \cdots u_n v_n w_n \theta(v_n) u_{n+1},$$

such that (i) all the u's and v's have length 20 and the w's have length 7, and (ii) the only bonds permitted in a word of this form are the bonds between v_i and $\theta(v_i)$ for all $i = 1, \ldots, n$. This encoding problem can be solved if we first construct a long hp(θ, k)-free word w of length $(20 + 20 + 7)n + 20 = 47n + 20$. Then w can be written in the form

$$u_1 v_1 w_1 \cdots u_n v_n w_n u_{n+1}$$

and is such that no bonds can occur between any two subwords of length k of w. Here k is the parameter that represents the smallest length of a block of nucleotides that can form a stable bond with a corresponding block of complementary nucleotides – see also the relevant discussion in [11].

For the case where θ is the Watson-Crick involution we consider the method of [11] for constructing ($\theta, H_{0,k}$)-bond-free languages L. Such a language L has the property that, for any two subwords u and v of L of length k, one has that $u \neq \theta(v)$. Note that each word of L is a hp(θ, k)-free word. Moreover, if L is infinite then it contains arbitrarily long words, hence, also words of length $47n + 20$, for any n, as required in the encoding problem discussed in the beginning of this section. We also note that if L is (θ, k)-bond-free then it is (θ, k')-bond-free for any $k' \geq k$. The method of [11] is based on the *subword closure* language operation $\otimes$: Let S be a set of words of length k. Then $S^{\otimes}$ is the set of all words w of length at least k such that any subword of w of length k belongs to S. We note that given the set S one can construct a deterministic finite automaton accepting $S^{\otimes}$ in linear time [11]. The method is as follows. Let S be any set of words of length k such that $S \cap \theta(S) = \emptyset$. Then $S^{\otimes}$ is a $(\theta, H_{0,k})$-bond-free language. In our case, we wish to choose S such that $S^{\otimes}$ is infinite. For example, let S_2 be the set $\{AA, AC, CA, CC, AG, GA\}$. In [11] the authors show an automaton accepting $S_2^{\otimes}$. As $S_2^{\otimes}$ contains the set $(ACCAGAC)^+$ it follows that $S_2^{\otimes}$ is infinite as well.

For the case of $\theta = \epsilon$, we consider a totally different approach. Let $H(K)$ denote the minimum Hamming distance between any two different codewords of a code K. A language K is said to be a *solid code* if (i) no word of K is a subword of another word of K, and (ii) a proper and nonempty prefix of K cannot be a suffix of K. See [17] or Chapter 8 in [19] for background information on codes.

Proposition 32. *Let $k \geq 2$ and let K be a uniform solid code of length k. If $H(K) > \lfloor k/2 \rfloor$, or $H(K) = \lfloor k/2 \rfloor$ and there are no different codewords with a common prefix of length $\lfloor k/2 \rfloor$, then the word $w_1...w_n$ is hp(θ, k)-free for all $n \leq \mathrm{card}(K)$ and for all pairwise different codewords $w_1, ..., w_n$.*

Proof. Assume there is $v \in X^k$ such that $w_1...w_n = xvyvz$ for some words x, y, z. If $|x|$ is a multiple of k then $v = w_j$ for some $j \geq 1$. As the w_i's are different, $|y|$ cannot be a multiple of k. Hence, $v = s_t p_{t+1}$, where $t > j$ and s_t is a proper and nonempty suffix of w_t and p_{t+1} is a proper and nonempty prefix of w_{t+1}; a contradiction. Now suppose $|x|$ is not a multiple of k. Then, $v = s_j p_{j+1}$ for some nonempty suffix s_j and prefix p_{j+1}. Again, the second occurrence of v cannot be in K. Hence, $v = s_t p_{t+1}$ for some $t \geq j$. Hence, $s_j p_{j+1} = s_t p_{t+1}$. If $|s_j| \neq |s_t|$, say $|s_j| > |s_t|$, then a prefix of p_{t+1} is also a suffix of s_j; which is impossible. Hence, $s_j = s_t$ and $p_{j+1} = p_{t+1}$.

Note that $H(K) \geq \lfloor k/2 \rfloor$ and, therefore, $\lfloor k/2 \rfloor \leq H(p_{j+1}s_{j+1}, p_{t+1}s_{t+1}) = H(s_{j+1}, s_{t+1}) \leq |s_{j+1}| = k - |p_{j+1}|$. Hence, $|p_{j+1}| \leq \lceil k/2 \rceil$. Similarly, $|s_j| \leq \lceil k/2 \rceil$. Also, as $k = |s_j| + |p_{j+1}|$, one has that $|s_j|, |p_{j+1}| \in \{\lfloor k/2 \rfloor, \lceil k/2 \rceil\}$. If $H(K) = \lfloor k/2 \rfloor$ then $p_{j+1} = p_{t+1}$ implies that w_{j+1} and w_{t+1} have a common prefix of length $\lfloor k/2 \rfloor$; a contradiction. If $H(K) > \lfloor k/2 \rfloor$ then both p_{j+1} and s_j are shorter than $\lceil k/2 \rceil$ which contradicts with $k = |s_j| + |p_{j+1}|$.

Suppose the alphabet size $|X|$ is $l > 2$. We can choose any symbol $a \in X$ and consider the alphabet $X_1 = X - \{a\}$. Then for any uniform code $F \subseteq X_1^{k-1}$ it follows that the code Fa is a uniform solid code of length $k : Fa \subseteq X^k$. We are interested in cases where the code F is a linear code of type $[k-1, m, d]$. That is, F is of length $k-1$, cardinality $(l-1)^m$, and $H(F) = d$, and there is an m

by $k-1-m$ matrix G over X_1 such that $F = \{w * [I_m|G] : w \in X_1^m\}$, where I_m is the identity m by m matrix and $*$ is the multiplication operation between a 1 by m vector and an m by m matrix. Thus, $u \in F$ iff $u = wx$ for some $w \in X_1^m$ and $x \in X_1^{k-1-m}$ and $x = wG$.

Proposition 33. *Let F be a linear code over X_1 of type $[k-1, m, \lfloor k/2 \rfloor]$. If $m \leq \lfloor k/2 \rfloor$ or k is even then the word $w_1..w_n$ is hp(θ, k)-free for all $n \leq \mathrm{card}(F)$ and for all pairwise different codewords $w_1, ..., w_n$ in Fa.*

Proof. It is sufficient to show that $H(Fa) = \lfloor k/2 \rfloor$ and there are no different words in Fa with a common prefix of length $\lfloor k/2 \rfloor$. Obviously $H(Fa) = H(F) = \lfloor k/2 \rfloor$. As F is generated by a matrix $[I_m|G]$, where G is a matrix in $X_1^{m \times (k-1-m)}$, it follows that there can be no different words in F with a common prefix of length m. If $m \leq \lfloor k/2 \rfloor$ then there can be no different words in Fa with a common prefix of length $\lfloor k/2 \rfloor$. If k is even, consider the well known bound on $|F|$: $|F| \leq |X_1|^{k-1-\lfloor k/2 \rfloor + 1}$. Hence, $|X_1|^m \leq |X_1|^{\lfloor k/2 \rfloor}$ which gives $m \leq \lfloor k/2 \rfloor$. Hence, again, we are done.

By the above one can construct an hp(θ, k)-free word of length nk, for some $n \leq \mathrm{card}(F)$, for every choice of n different words in Fa. It is interesting that, for $k = 13$ and $|X| = 4$, the famous Golay code G_{12} of type $[12, 6, 6]$ satisfies the premises of the above Proposition.

Acknowledgements

Research was partially supported by the Canada Research Chair Grant to L.K., NSERC Discovery Grants R2824A01 to L.K. and R220259 to S.K., and by the Grant Agency of Czech Republic, Grant 201/06/0567 to P.S.

References

1. M. Amos, *Theoretical and Experimental DNA Computations.* Springer-Verlag, Berlin, 2005.
2. M. Andronescu, D. Dees, L. Slaybaugh, Y. Zhao, A. Condon, B. Cohen, S. Skiena, Algorithms for testing that sets of DNA words concatenate without secondary structure. In *Proc. 8th Workshop on DNA-Based Computers*, M. Hagiya, A. Ohuchi, Eds., *LNCS* 2568 (2002), 182–195.
3. Y. Benenson, B. Gil, U. Ben-Dor, R. Adar, E. Shapiro, An autonomous molecular computer for logical control of gene expression. *Nature* 429 (2004), 423–429.
4. J.C. Birget, Intersection and union of regular languages and state complexity. *Information Processing Letters* 43 (1992), 185–190.
5. J. Chen, R. Deaton, M. Garzon, J.W. Kim, D. Wood, H. Bi, D. Carpenter, Y.-Z. Wang, Characterization of non-crosshybridizing DNA oligonucleotides manufactured *in vitro*. In [15], 132–141.
6. J. Hopcroft, J. Ullman, R. Motwani, Introduction to Automata Theory, Languages, and Computation, 2nd ed., Addison-Wesley, 2001.

7. N. Jonoska, D. Kephart, K. Mahalingam, Generating DNA code words. *Congressus Numerantium* 156 (2002), 99–110.
8. N. Jonoska, K. Mahalingam, Languages of DNA based code words. In *DNA Computing, 9th International Workshop on DNA Based Computers*, J. Chen and J.H. Reif, Eds., *LNCS* 2943 (2004), 61–73.
9. L. Kari, S. Konstantinidis, P. Sosík, G. Thierrin, On hairpin-free words and languages. In *Developments in Language Theory, 9th Int. Conf.*, C. de Felice and A. Restivo, Eds., *LNCS* 3572 (2005), 296–307.
10. L. Kari, S. Konstantinidis, E. Losseva, G. Wozniak, Sticky-free and overhang-free DNA languages. *Acta Informatica* 40, 2003, 119–157.
11. L. Kari, S. Konstantinidis, P. Sosík, Bond-free languages: formalizations, maximality and construction methods. In [15], 16–25.
12. S. Kobayashi, Testing Structure Freeness of Regular Sets of Biomolecular Sequences. In [15], 395–404.
13. A. Marathe, A. Condon, R. Corn, On combinatorial DNA word design. *DNA based Computers V*, DIMACS Series, E.Winfree, D.Gifford Eds., AMS Press, 2000, 75–89.
14. G. Mauri, C. Ferretti, Word Design for Molecular Computing: A Survey. In *DNA Computing, 9th International Workshop on DNA Based Computers*, J. Chen and J.H. Reif, Eds., LNCS 2943 (2004), 37–46.
15. G. Mauri, C. Ferretti, Eds., *DNA 10, Tenth International Meeting on DNA Computing*. Preliminary proceedings, University of Milano-Bicocca, 2004.
16. G. Paun, G. Rozenberg, A. Salomaa, *DNA Computing: New Computing Paradigms*, Springer Verlag, Berlin, 1998.
17. S. Roman, *Coding and Information Theory,* Springer-Verlag, New York, 1992.
18. J. A. Rose, R. J. Deaton, M. Hagiya, A. Suyama, PNA-mediated Whiplash PCR. In *DNA Computing, 7th International Workshop on DNA Based Computers*, N. Jonoska and N. C. Seeman, Eds., LNCS 2340 (2002), 104–116.
19. G. Rozenberg, A. Salomaa, Eds., *Handbook of Formal Languages*, Vol. 1, Springer Verlag, Berlin, 1997.
20. N. Takahashi, A. Kameda, M. Yamamoto, A. Ohuchi, Aqueous computing with DNA hairpin-based RAM. In [15], 50–59.
21. G. Thierrin, Convex languages. *Proc. IRIA Symp. North Holland* 1972, 481–492.

Efficient Algorithm for
Testing Structure Freeness of
Finite Set of Biomolecular Sequences

Atsushi Kijima and Satoshi Kobayashi

Graduate School of University of Electro-Communications,
1-5-1, Chofugaoka, Chofu, Tokyo 182-8585, Japan
kijiman@comp.cs.uec.ac.jp, satoshi@cs.uec.ac.jp

Abstract. In this paper we will focus on the structure freeness test problem of finite sets of sequences. The result is an extension of Andronescu's algorithm which can be applied to the sequence design of various DNA computing experiments. We will first give a general algorithm for this problem which runs in $O(n^5)$ time. Then, we will give an evaluation method for sequence design system, which requires $O(n^5)$ time for precomputation, and $O(n^4)$ time and $O(n^5)$ space for each evaluation of sequence sets. The authors believe that this result will give an important progress of efficient sequence design systems.

1 Introduction

Since Adleman's novel biological experiment for solving directed Hamiltonian path problem by DNA molecules was reported([1]), DNA computing paradigm has emerged and progressed while communicating with related fields, such as DNA nanotechnology([19], [15], [7]), biotechnology([5]), etc. One of the most important problems in DNA computing experiments include the design of *structure free* biomolecular sequences which can avoid unwanted secondary structure([6], [8]). In order to develop a sequence design system, we need to devise an efficient algorithm to test the structure freeness of a given set of biomolecular sequences.

Concerning sequence design for DNA computing, there have been many interesting and important works which propose some variants of Hamming distance over biomolecular sequences. And these metrics are used for the evaluation of the sequences([3], [12], [10], etc.). Comparing those Hamming distance approaches, Condon, et al. mathematically formulated a structure freeness test problem of biomolecular sequences at the secondary structure level([8], [2]). This problem is closely related to the prediction problem of RNA secondary structures([9], [11], [16], [20]), and is important in that its efficient algorithms can be applied to the evaluation of sequence sets in sequence design systems.

Andronescu, et al., proposed an $O(m^2 n^3)$ time algorithm for testing the structure freeness of a sequence set $S_1 \cdots S_k$, where each S_i is a finite set of sequences of length l_i, $n = \sum_{i=1}^{k} l_i$, and $m = \max\{|S_i| \mid i = 1, ..., k\}$([2]). Kobayashi, et al., gave an $O(m^6 n^6)$ time algorithm for testing the structure freeness of a sequence set S^+, where S is a finite set of sequences of length n and $m = |S|$

A. Carbone and N.A. Pierce (Eds.): DNA11, LNCS 3892, pp. 171–180, 2006.
© Springer-Verlag Berlin Heidelberg 2006

([14]). Furthermore, Kobayashi devised an $O(n^8)$ time algorithm for testing the structure freeness of a regular set of sequences, where n is the number of vertices of graphs for representing the set([13]). (Note that Condon proposed to use a graph for representing a regular set of sequences.)

In spite of this progress in evaluation methods, we still need more efficient algorithms in order to develop an efficient sequence design system. In this paper, we will focus on the structure freeness test problem of finite sets of sequences. The obtained result is an extension of Andronescu's algorithm and can be applied to the sequence design of various DNA computing experiments. We will first give a general algorithm for this problem which runs in $O(n^5)$ time. Then, we will give an evaluation method for sequence design system, which requires $O(n^5)$ time for precomputation, and $O(n^4)$ time and $O(n^5)$ space for each evaluation of sequence sets. The authors believe that this result will give an important progress of efficient sequence design systems.

2 Preliminaries

Σ is an alphabet $\{A, C, G, T\}$ or $\{A, C, G, U\}$. A symbol in Σ is called a *base*. A string over Σ represents a DNA or RNA strand with $5' \to 3'$ direction. Consider a string α over Σ. By $|\alpha|$ we denote the length of α. For a finite set X, by $|X|$ we denote the number of elements of X. For an integer i such that $1 \le i \le |\alpha|$, by $\alpha[i]$ we denote the ith base of α.

2.1 Secondary Structure

We will partly follow the terminologies and notations used in ([17]). We introduce a relation $\theta \subseteq \Sigma \times \Sigma$ defined by $\theta = \{(A, T), \ldots, (T, G)\}$ for representing Watson-Crick and non-Watson-Crick base pairs of a DNA strand. For the case of an RNA strand, the symbol T is replaced by U. By (i, j) we denote a hydrogen bond between the ith base and the jth base of a string α. A hydrogen bond is also called a *base pair*. A base pair (i, j) of a string α can be formed only if $(\alpha[i], \alpha[j]) \in \theta$ holds. Without loss of generality, we may assume that $i < j$ for a base pair (i, j). A finite set of base pairs of string α is called a *secondary structure* of α. A string α with its secondary structure T is called a *structured string* and denoted by $\alpha(T)$. For representing the ith base in $\alpha(T)$, we often use the integer i.

In this paper, we consider secondary structures T such that there exist no base pairs $(i, j), (k, l) \in T$ satisfying $i < k < j < l$. In the sequel, we assume that every secondary structure is *pseudo-knot free*.

For a base pair $(i, j) \in \alpha(T)$ and a base r in $\alpha(T)$, we say that (i, j) *surrounds* r if $i < r < j$ holds. For a base pair $(p, q) \in T$, we say that (i, j) surrounds (p, q) if $i < p < q < j$ holds. A base pair (p, q) or an unpaired base r is said to be *accessible* from (i, j), if it is surrounded by (i, j) and is not surrounded by any base pair (k, l) such that (k, l) is surrounded by (i, j). If (p, q) is accessible from (i, j), we write $(p, q) < (i, j)$.

For each base pair $bp = (i, j) \in T$, we define a *cycle* $c(bp)$ as a substructure consisting of the base pair (i, j) together with any base pairs $(p_1, q_1), (p_2, q_2), \ldots,$ $(p_{k-1}, q_{k-1}) \in T$ accessible from (i, j) and any unpaired bases accessible from (i, j). If a cycle $c(bp)$ contains k base pairs including the base pair (i, j), it is said to be k-cycle. In case $k = 1$, we often call it a *hairpin*. In case $k = 2$, it is called *internal loop*. In case $k > 2$, it is called *multiple loop*. In these definitions, the base pair (i, j) is called a *closing base pair* of the cycle. (See Fig. 1).

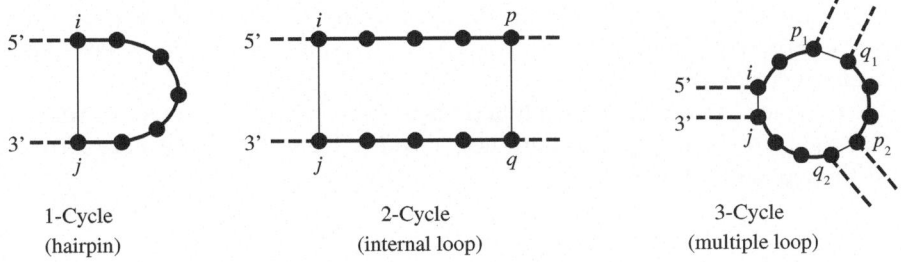

1-Cycle	2-Cycle	3-Cycle
(hairpin)	(internal loop)	(multiple loop)

Fig. 1. Secondary Structure

In case of $(1, |\alpha|) \notin T$, the substructure of $\alpha(T)$ consisting of the base pair (i, j) such that (i, j) is not surrounded by any $(p, q) \in T$ $((i, j) \neq (p, q))$ and the unpaired bases such that they are not surrounded by (i, j) is called a *free end structure* of $\alpha(T)$. We do not consider a free end structure because of space constraint.

The *loop length* of a 1-cycle c with a base pair (i, j) is defined as $j - i + 1$. For a 2-cycle c with base pairs $(i, j), (p, q)$ $((p, q) < (i, j))$, we define *loop length* of c as $p - i + j - q + 2$ and define *loop length mismatch* of c as $|(p - i) - (j - q)|$.

By $\uparrow \alpha \downarrow$ we denote a 1-cycle consisting of a string α with a base pair between $\alpha[1]$ and $\alpha[|\alpha|]$. By $\overline{\uparrow \alpha \ \beta \downarrow}$ we denote 2-cycle consisting of strings α and β with two base pairs between $\alpha[1]$ and $\beta[|\beta|]$ and between $\alpha[|\alpha|]$ and $\beta[1]$.

3   Free Energy of Secondary Structure

In this paper, we use following simplified functions to assign free energy values to each substructures. We use these simplifications only for the clarity of the algorithm. Experimental evidence is used to determine such free energy values.

1. The free energy $E(c)$ of a 1-cycle c with a base pair (i, j) is dependent on the base pair (i, j) and its loop length l:

$$E(c) = f_1(\alpha[i], \alpha[j]) + g_1(l) . \tag{1}$$

2. The free energy $E(c)$ of a 2-cycle c with two base pair $(i, j), (p, q)((p, q) < (i, j))$ is dependent on the base pairs $(i, j), (p, q)$, its loop length l and its loop length mismatch d:

$$E(c) = f_2(\alpha[i], \alpha[j], \alpha[p], \alpha[q]) + g_2(l) + g_3(d) . \tag{2}$$

3. The free energy $E(c)$ of a k-cycle c $(k > 2)$ with a closing base pair (i, j) and the base pairs $(p_1, q_1), (p_2, q_2), \ldots, (p_{k-1}, q_{k-1})$ accessible from (i, j) is dependent on the base pairs $(i, j), (p_l, q_l)$ $(l = 1, \ldots, k - 1)$, the number n_b $(= k)$ of base pairs in c and the number n_u of unpaired bases in c:

$$E(c) = m_1(\alpha[i], \alpha[j]) + \sum_{l=1}^{k}(m_1(\alpha[q_l], \alpha[p_l])) + M_b * n_b + M_u * n_u + C_M \ . \ (3)$$

In these definitions, the functions $f_1, g_1, f_2, g_2, g_3, m_1$ and the constants M_b, M_u, C_M are experimentally obtained. We assume that M_b, M_u, C_M are non-negative. For each function g_i $(i = 1, 2, 3)$, we assume that g_i is weakly mono-tonically increasing[1].

We assume that all the above functions are computable in constant time.

Let $c_1, \ldots, c_k$ be the cycles contained in $\alpha(T)$. Then, the free energy $E(\alpha(T))$ of $\alpha(T)$ is given by following:

$$E(\alpha(T)) = \sum_{i=1}^{k} E(c_i) \ . \tag{4}$$

4   Structure Freeness of Finite Regular Set

We will consider the problem of testing whether a given finite regular set of strings is structure free or not. The problem is formally defined in the following way:

Let R be a regular language over Σ. Then, we say that R is *structure free with threshold* D if for any structured string $\alpha(T)$ such that $\alpha \in R$ and T is pseudo-knot free, it holds that $E(\alpha(T)) \geq D$. We have interests in deciding for given R, whether or not R is structure free with threshold D. In Sect. 6, we will give a polynomial time algorithm for solving this problem in the case that R is finite.

For specifying a regular language R, we use a labeled directed graph with initial and final vertices. Let $M = (V, E, \sigma, I, F)$, where V is a finite set of vertices, E is a subset of $V \times V$, σ is a label function from V to Σ, and $I, F \in V$. For $p, q \in V$ and $x \in \Sigma^*$, we write $p \xrightarrow{x} q$ if there is a path with labels x from p to q in M. Note that x contains the labels $\sigma(p)$ and $\sigma(q)$. We write $p \to q$ if $p \xrightarrow{x} q$ for some $x \in \Sigma^*$. A string α is *accepted* by M if $p \xrightarrow{\alpha} q$ for some $p \in I$ and $q \in F$. This graph representation could be regarded as a Moore type machine with no edge labels. Thus, a set of strings is regular iff it is accepted by a graph M. A graph M is said to be *trimmed* if every vertex is reachable from an initial vertex and has a path to a final vertex.

In this paper, we have interests in testing structure freeness of a finite regular set. Note that a set of strings is finite iff it is accepted by a trimmed and acyclic graph.

---

[1] This assumption can be extended so that $g_i(l)$ is weakly monotonically increasing within the range $l > L_i$ for some constant L_i. Because of space constraint, we use the simplified assumption.

5   Minimum Free Energy of Substructure

Let R be a finite regular language over Σ and $M = (V, E, \sigma, I, F)$ be a trimmed and acyclic graph accepting R.

We can topologically sort vertices in V in $O(|V| + |E|)$ time. By an integer i, we denote the ith vertex in the topological order. Let $\alpha(T)$ be a structured string such that $\alpha \in R$ and T is pseudo-knot free.

Definition 1. *For $i, j, p, q \in V$, we define:*

(1) $minH(i, j) = \min \left\{ E(\overline{\uparrow x \downarrow}) \mid i \xrightarrow{x} j \right\}$,

(2) $minI(i, j, p, q) = \min \left\{ E(\overline{\uparrow x\ y \downarrow}) \mid i \xrightarrow{x} p,\ q \xrightarrow{y} j,\ p \to q \right\}$.

For each $i, j, p, q \in V$ such that there is no $\overline{\uparrow x \downarrow}$ or $\overline{\uparrow x\ y \downarrow}$, the value of $minH(i, j)$ or $minI(i, j, p, q)$ is defined as $+\infty$.

For each pair of vertices i, j, we define $Len(i)(j)$ as a set of the length $|x|$ such that $i \xrightarrow{x} j$. For a given graph M, we compute the array Len by the algorithm shown in Fig. 2, where every vertices are sorted in the topological order.

```
Make-Len(M)
begin
    for i, j ∈ V do Len(i)(j) := φ; end
    for (i, j) ∈ E do Len(i)(j) := {2}; end
    for d = 2 to |V| - 1 do
        for i = 1 to |V| - d do
            j := i + d;
            Len(i)(j) := Len(i)(j) ∪   ⋃      {x + 1 | x ∈ Len(i)(k)};
                                     i<k<j
                                    (k,j)∈E
        end
    end
end
```

Fig. 2. The algorithm `Make-Len`

Since a given graph M is acyclic, for any $i, j \in V$, $|Len(i)(j)| \leq |V|$ holds. So, we can compute an array Len in $O(|V|^2|E|)$ time.

By Definition 1 and by using the array Len, we get the following Proposition 1.

Proposition 1. *For $i, j, p, q \in V$, we define:*

(1) $minH(i, j) = \min \left\{ f_1(\sigma(i), \sigma(j)) + g_1(l) \mid l \in Len(i)(j) \right\}$

(2) $minI(i, j, p, q) = \min \{\ f_2(\sigma(i), \sigma(j), \sigma(p), \sigma(q)) +$
$$g_2(l_1 + l_2) + g_3(|l_1 - l_2|) \mid l_1 \in Len(i)(p),\ l_2 \in Len(q)(j) \} \ .$$

In case of $Len(i, j) = \phi$ for some $i, j \in V$, we define $minH(i, j) = +\infty$. In case of $Len(i, j) = \phi$ or $Len(p, q) = \phi$ for some $i, j, p, q \in V$, we also define $minI(i, j, p, q) = +\infty$.

Note that the number of elements of a set $\{(x+y, |x-y|) \mid x \in Len(i)(p),\ y \in Len(q)(j)\}$ is $O(|V|^2)$ for $i, j, p, q \in V$.

5.1   Minimum Free Energy of Internal Loop

By Proposition 1, it takes $O(|V|^6)$ time to compute $minI(i,j,p,q)$ for all i,j,p,q $\in V$. We can compute $minI(i,j,p,q)$ more efficiently by computing an array $S_{X,Y}$, $\overline{S}_{X,Y}$ defined as follows:

Definition 2. *Let X and Y be finite sets of positive integers. We define $S_{X,Y}$ as follows:*

$$S_{X,Y} = \big\{ (x, \min\{y \in Y \mid x \leq y\}) \mid x \in X \big\} \ .$$

Note that we can compute $S_{X,Y}$ in $O(|X| + |Y|)$ time by using the algorithm shown in Fig. 3, and the number of elements of $S_{X,Y}$ is $O(|X| + |Y|)$.

In order to apply $S_{X,Y}$ to computing minimum free energy of strings, we define $\overline{S}_{X,Y}$ as follows:

$$\overline{S}_{X,Y} = \big\{ (x,y) \mid (x,y) \in S_{X,Y} \big\} \cup \big\{ (x,y) \mid (y,x) \in S_{Y,X} \big\} \ . \tag{5}$$

Theorem 1. *For $i,j,p,q \in V$, we can compute $minI(i,j,p,q)$ in the following way:*
$$minI(i,j,p,q) = \min \big\{ \ f_2(\sigma(i),\sigma(j),\sigma(p),\sigma(q)) + g_2(x+y) + g_3(|x-y|) \ | \\ (x,y) \in \overline{S}_{Len(i)(p),Len(q)(j)}, Len(p)(q) \neq \phi \big\} \ .$$

Proof. Let $X = Len(i)(q)$ and $Y = Len(q)(j)$. It suffices to show that for any $(x,y) \in X \times Y$, there exists $(x',y') \in \overline{S}_{X,Y}$ such that $g_2(x+y) + g_3(|x-y|) \geq g_2(x'+y') + g_3(|x'-y'|)$.

```
Make-S(X, Y)
begin
    S_{X,Y} := φ
    i := |X|;
    j := |Y|;
    y_0 = -∞;
    while i ≥ 1 and j ≥ 1 do
        if x_i ≤ y_j then
            while x_i ≤ y_j and y_{j-1} < x_i and i ≥ 1 do
                S_{X,Y} := S_{X,Y} ∪ (x_i, y_j);
                i := i - 1;
            end
            j := j - 1;
        else
            i := i - 1;
        end
    end
    return S_{X,Y};
end
```

Fig. 3. The algorithm Make-S

We consider two cases:

(1) In case of $x \leq y$, let $x' = x$ and $y' = \min\{y'' \in Y \mid x \leq y''\}$. Note that $(x', y') \in S_{X,Y}$. We have $x = x' \leq y' \leq y$. Then, we have $x' + y' \leq x + y$ and $0 \leq y' - x' \leq y - x$. Since functions g_2 and g_3 are weakly monotonically increasing, we have $g_2(x' + y') \leq g_2(x + y)$ and $g_3(y' - x') \leq g_3(y - x)$. Therefore, we can compute minimum free energy $minI(i, j, p, q)$ by using $S_{X,Y}$.

(2) In case of $x > y$, let $x' = \min\{x'' \in X \mid y \leq x''\}$ and $y' = y$. Note that $(y', x') \in S_{Y,X}$. In the same way above, we can also say that we can compute $minI(i, j, p, q)$ by using $S_{Y,X}$.

We can compute $minI$ by using $S_{X,Y}$ or $S_{Y,X}$ in both cases (1) and (2). Therefore, we can compute $minI$ by using $\overline{S}_{X,Y}$. $\qquad\square$

Theorem 2. *For each $i, j, p, q \in V$, $minI(i, j, p, q)$ can be computed in $O(|V|)$ time.*

Proof. Since for any $i, j \in |V|$, the number of elements of $Len(i)(j)$ is less than or equal to $|V|$, the number of elements in $S_{Len(i)(p),Len(q)(j)}$ and $S_{Len(q)(j),Len(i)(p)}$ are $O(|V|)$. Therefore, the number of elements of $\overline{S}_{Len(i)(p),Len(q)(j)}$ is $O(|V|)$. $\qquad\square$

In real applications of RNA secondary structure prediction([11], [20]), the loop length of internal loops is assumed to be bounded by some constant in order to make the prediction algorithms more efficient. This assumption also enables us to compute $minI(i, j, p, q)$ in constant time for each $i, j, p, q \in V$.

6   Algorithm for Testing Structure Freeness

We will give the algorithm SFT-FS for testing the structure freeness of a given finite set of strings represented by a graph $M = (V, E, \sigma, I, F)$. Let an integer i represent the ith element of V in topological order. The algorithm is shown in Fig. 4, where $a(i, j) = m_1(\sigma(i), \sigma(j))$ is the energy contribution of a base pair in a multiple loop.

Our algorithm is based on the dynamic programming approach used in various RNA secondary structure prediction algorithms([11], [20], [17], etc.). While a base adjacent to another base can be determined uniquely for a strand, it does not hold for a set of strands. We consider all possible bases adjacent to a base. Correctness of the algorithm is informally understood as follows:

Let R be a finite set of strings and M be a graph accepting R. Let $\alpha \in R$ be a structured strand $\alpha(T)$ with the minimum free energy $E(\alpha(T))$ such that $i \xrightarrow{\alpha} j$ for some $i \in I, j \in F$. For some $p, q \in V$ such that $p \xrightarrow{\beta} q$, if β is a substring of α, β has the minimum free energy $E(\beta(\hat{T}))$ among all substrands in R such that $p \xrightarrow{\beta} q$, where $\hat{T} \subseteq T$. Otherwise there exist $p \xrightarrow{\beta'} q$ such that $E(\beta'(T')) < E(\beta(\hat{T}))$ and $T' \subseteq T$. Then, we can replace β in α to β' and have $E(\alpha'(T'')) < E(\alpha(T))$, which contradicts the minimality of the free energy $E(\alpha(T))$. The algorithm computes such minimum free energy from smaller to

```
Init(M)
begin
    Topological-Sort(M);
    compute Len(i)(j) for all i, j ∈ V by calling Make-Len(M);
    compute S̄_{Len(i)(p),Len(q)(j)}
      by calling Make-S(Len(i)(p), Len(q)(j)) for all i, j, p, q ∈ V;
    compute minH(i,j), minI(i,j,p,q) for all i, j, p, q ∈ V;
end
```

```
SFT-FS(M)
begin
    Init(M);
    for d = 1 to |V| - 1 do
        for i = 1 to |V| - d do
            j = i + d;
```

$$(I) \quad C[i,j] = \min \begin{cases} minH(i,j), \\ \min_{i<p<q<j} \{minI(i,j,p,q) + C[p,q]\}, \\ \min_{\substack{i<i'<j'<j \\ (i,i'),(j',j)\in E}} \{FM[i',j'] + a(i,j)\}. \end{cases}$$

$$(II) \quad F[i,j] = \min \begin{cases} C[i,j], \\ \min_{\substack{i<k<k'<j \\ (k,k')\in E}} \{FM[i,k] + FM[k',j]\}. \end{cases}$$

$$(III) \quad FM[i,j] = \min \begin{cases} M_b + C[i,j], \\ \min_{\substack{i<i'<j \\ (i,i')\in E}} \{M_c + FM[i',j]\}, \\ \min_{\substack{i<j'<j \\ (j',j)\in E}} \{M_c + FM[i,j']\}, \\ \min_{\substack{i<k<k'<j \\ (k,k')\in E}} \{FM[i,k] + FM[k',j]\}. \end{cases}$$

```
        end
    end
    if there exist F[i,j] < threshold D for some i ∈ I, j ∈ F return 'No';
    else return 'Yes';
end
```

Fig. 4. The algorithm SFT-FS

larger substructures with the recurrences (I)–(III) applied to topologically sorted vertices.

We can run $\texttt{Init}(M)$ in $O(|V|^5)$ time, $\texttt{SFT-FS}(M)$ in $O(|V|^4)$ time, and it costs $O(|V|^5)$ time in total. By using the constant upper bound assumption on loop length in Sect. 5.1, we can run $\texttt{Init}(M)$ in $O(|V|^4)$ time.

7 Application to Strand Design

Our algorithm requires more time in $\texttt{Init}(M)$ than $\texttt{SFT-FS}(M)$. Once the initialization $\texttt{Init}(M)$ is done, we can evaluate strands more efficiently. Even if we change a label function σ for a vertex, it is not necessary to compute $\overline{S}_{Len(i)(p),Len(q)(j)}$ again. Furthermore, we can compute $minI(i,j,p,q)$ for all possibilities of label function σ which has four possibilities $\sigma(i) = \texttt{A}$, $\sigma(i) = \texttt{C}$,

$\sigma(i) = $ G or $\sigma(i) = $ T for a vertex i. Then, time to compute $minI$ is $O(|V|^5)$. These observations lead us to a method for strand design shown in Fig. 5. In this search algorithm, we can evaluate a set of strings R.

```
Strand-Design(M)
begin
    Init(M) with all possibilities of label function;

    while SFT-FS(M) returns 'No' do
        change a label of a randomly selected vertex;
    end
    return M;
end
```

Fig. 5. Random strand design algorithm

In this Strand-Design(M), a random search is used for finding a structure free set of sequences. In real applications to sequence design, we should use more sophisticated search strategies, such as stochastic local search([18]), genetic algorithm([4]), etc.

8   Conclusion

We give an efficient algorithm for testing the structure freeness of a finite set of strands. We also give a method for strand design generating a finite set of structure free strands. Our future works will include the improvement of the algorithm for computing $minI$ and the implementation of the strand design system based on the results presented in this paper.

References

1. L. Adleman, Molecular Computation of Solutions to Combinatorial Problems. *Science*, vol.266, pp.1021-1024, 1994.
2. M. Andronescu, D. Dees, L. Slaybaugh, Y. Zhao, A. Condon, B. Cohen, and S. Skiena, Algorithms for Testing That Sets of DNA Words Concatenate without Secondary Structure. *Proc. of The 9th International Meeting on DNA Based Computers*, LNCS, vol.2568, pp.182-195, 2003.
3. M. Arita and S. Kobayashi, DNA sequence design using templates. *New Generation Computing*, vol.20, pp.263-277, 2002.
4. M. Arita, A. Nishikawa, M. Hagiya, K. Komiya, H. Gouzu, and K. Sakamoto, Improving sequence design for DNA computing. *Proc. of Genetic and Evolutionary Computation Conference 2000*, pp.875-882, 2000.
5. Y. Benenson, B. Gil, U. Ben-Dor, R. Adar, E. Shapiro, An autonomous molecular computer for logical control of gene expression. *Nature*, vol.429, pp.423-429, 2004.
6. A. Brenneman and A. E. Condon, Strand Design for Bio-Molecular Computation. *Theoretical Computer Science*, vol.287, pp.39-58, 2002.

7. A. Carbone and N. C. Seeman, Circuits and programmable self-assembling DNA structures. *Proc. Natl. Acad. Sci. USA*, vol.99, pp.12577-12582, 2002.

8. A. E. Condon, Problems on RNA Secondary Structure Prediction and Design. *Proc. of ICALP'2003*, Lecture Notes in Computer Science, vol.2719, pp.22-32, 2003.

9. R. M. Dirks, N. A. Pierce, An algorithm for computing nucleic acid base-pairing probabilities including pseudoknots. *Journal of Computational Chemistry*, vol.25, pp.1295-1304, 2004.

10. A. G. D'yachkov, A. J. Macula, W. K. Pogozelski, T. E. Renz, V. V. Rykov, D. C. Torney, A weighted insertion-deletion stacked pair thermodynamic metric for DNA codes. *Preliminary Proc. of Tenth International Meeting on DNA Computing*, pp.142-151, 2004.

11. I. L. Hofacker, W. Fontana, P. F. Stadler, L. S. Bonhoeffer, M. Tacker, and P. Schuster, Fast Folding and Comparison of RNA Secondary Structures (The Vienna RNA Package). *Monatshefte für Chemie*, vol.125, pp.167-188, 1994.

12. L. Kari, S. Konstantinidis, and P. Sosík, Bond-free languages: formalizations, maximality and construction methods. *Preliminary Proc. of Tenth International Meeting on DNA Computing*, pp.16-25, 2004.

13. S. Kobayashi, Testing structure freeness of regular sets of biomolecular sequence. *Preliminary Proc. of Tenth International Meeting on DNA Computing*, pp.395-404, 2004.

14. S. Kobayashi, T. Yokomori, and Y. Sakakibara, An Algorithm for Testing Structure Freeness of Biomolecular Sequences. *Aspects of Molecular Computing — Essays dedicated to Tom Head on the occasion of his 70th birthday*, Springer-Verlag, LNCS, vol.2950, pp.266-277, 2004.

15. C. Mao, T. H. LaBean, J. H. Reif, and N. C. Seeman, Logical computation using algorithmic self-assembly of DNA triple-crossover molecules. *Nature*, vol.407, pp.493-496, 2000.

16. J. S. McCaskill, The equilibrium partition function and base pair binding probabilities for RNA secondary structure. *Biopolymers*, vol.29, pp.1105-1119, 1990.

17. D. Sankoff, J. B. Kruskal, S. Mainville, and R. J. Cedergen, Fast Algorithms to Determine RNA Secondary Structures Containing Multiple Loops. *Time Warps, String Edits, and Macromolecules : The Theory and Practice of Sequence Comparison*, D. Sankoff and J. Kruskal, Editors, Chapter 3, pp.93-120, 1983.

18. D. C. Tulpan, H. H. Hoos, and A. E. Condon, Stochastic local search algorithms for DNA word design. *Proc. 8th International Workshop on DNA-Based Computers*, LNCS 2568, pp.229-241, 2002.

19. E. Winfree, F. Liu, L. A. Wenzler, and N. C. Seeman, Design self-assembly of two-dimensional DNA crystals. *Nature*, vol.394, pp.539-544, 1998.

20. M. Zuker, On finding all suboptimal foldings of an RNA molecule. *Science*, vol.244, pp.48-52, 1989.

Communicating Distributed H Systems: Optimal Results with Efficient Ways of Communication

Shankara Narayanan Krishna

Department of Computer Science and Engineering,
Indian Institute of Technology, Bombay,
Powai, Mumbai 400 076, India
krishnas@cse.iitb.ac.in

Abstract. Distributed H systems and several variants of distributed H systems have been studied extensively [1, 2, 3, 4]. This paper is an effort in the direction of obtaining efficient distributed systems. To this end, a universality result using 2 components is obtained using two-level distributed H systems. This is an improvement over the existing universality result with 3 components. Further, we propose *lazy* communicating distributed H systems (LCDH systems), a variant of communicating distributed H systems, with lesser communication. A universality result is obtained with this variant, using only 2 components. This improves the universality result $RE = CDH_3$ by reducing the number of components as well as the communication between components.

1 Introduction

Communicating distributed H (CDH) systems were introduced in [1] as efficient extensions of splicing systems. In CDH systems, parts of the model which are able to work independently can be separated, and the result can be obtained by synthesizing the partial results produced by the individual parts. However, the communication in CDH systems is rather inefficient since they allow transport of possibly the entire contents of each component in every step. Distributed H systems [2, 3, 4] have been studied extensively, with different means of communication, one of them being two-level distributed H systems. These systems do not allow communication between components in the sense of CDH systems, and so are more efficient.

In this paper, we concentrate on two-level distributed H systems and CDH systems. We introduce *lazy* CDH systems as a variant of CDH systems, wherein, some components are classified as *lazy*, depending on the way they communicate. The idea of having *lazy* components is to reduce the number of strings that can be considered for communication in every step. We also obtain an unexpected improved universality result for two-level distributed H systems (without any laziness conditions), as well as a universality result for *lazy* CDH systems, both in 2 components, which show that with better means of communication, the number of components can be reduced. In the following subsection, we give some basic definitions and notions of formal language theory used in this paper; more details can be found in [4].

A. Carbone and N.A. Pierce (Eds.): DNA11, LNCS 3892, pp. 181–192, 2006.

1.1   Prerequisites and Basic Definitions

An alphabet is a finite nonempty set of symbols. For an alphabet V, we denote by V^* the set of strings of symbols over V. The empty string is denoted by λ. V^* is the free monoid generated by V under the operation of concatenation. (The unit element of this monoid is λ). Each subset of V^* is called a *language* over V.

Let $x \in V^*$. If $x = x_1 x_2$, for some $x_1, x_2 \in V^*$, then x_1 is called a *prefix* of x and x_2 is called a *suffix* of x. If $x = x_1 x_2 x_3$, for some $x_1, x_2, x_3 \in V^*$, then x_2 is called a *substring* of x. The length of a string x is denoted by $|x|$. The number of occurrences of a symbol a in x denoted $|x|_a$.

Consider an alphabet V and two special symbols $\#, \$ \notin V$. A splicing rule over V is a string $u_1 \# u_2 \$ u_3 \# u_4$ where $u_1, u_2, u_3, u_4 \in V^*$. For a splicing rule $r = u_1 \# u_2 \$ u_3 \# u_4$, the result of splicing two strings $x = x_1 u_1 u_2 x_2, y = y_1 u_3 u_4 y_2$ is defined as $(x, y) \models_r (z, w)$ where $z = x_1 u_1 u_4 y_2, w = y_1 u_3 u_2 x_2$.

An H scheme is a pair $\sigma = (V, R)$ where V is an alphabet and $R \subseteq V^* \# V^* \$ V^* \# V^*$ is a set of splicing rules. For an H scheme $\sigma = (V, R)$, and a language L, the set obtained by using the splicing operation on L is denoted by $\sigma_2(L) = \{z \in V^* \mid (x, y) \models_r (z, w) \text{ or } (x, y) \models_r (w, z)\}$, for some $x, y \in L$, and $r \in R$. $\sigma_2^i(L)$ is defined inductively: $\sigma_2^0(L) = L$, $\sigma_2^{i+1}(L) = \sigma_2^i(L) \cup \sigma_2(\sigma_2^i(L)), i \geq 0$. Hence, $\sigma_2^*(L) = \bigcup_{i \geq 0} \sigma_2^i(L)$.

An extended H system is a quadruple $\gamma = (V, T, A, R)$ where $T \subseteq V$ is the terminal alphabet, R is the set of splicing rules and A is the set of axioms. Thus, γ has an underlying H scheme $\sigma = (V, R)$, augmented with a subset of V and a set of axioms. The language generated by γ is defined as $L(\gamma) = \sigma_2^*(A) \cap T^*$.

The power of extended H systems as well as some extensions of H systems have been studied extensively in the literature. In this paper, we are interested in two such extensions viz., communicating distributed H systems and two level H systems. We give the definitions of these systems in sections 2 and 3.

We denote by RE the family of recursively enumerable languages. A recursively enumerable language can be generated by a type-0 grammar $G = (N, T, S, P)$ where N is a set of non-terminals, $T \subseteq N$ is the set of terminal symbols, S is the start symbol, and P consists of productions of the form $u \to v, u, v \in (N \cup T)^*, |u|_N > 0$.

Notation: In the following sections, a splicing rule is represented by $x \# y \$ a \# b$. However, while explaining the functionality of such a rule in the proofs, we represent them by $(x|y, a|b) \models (xb, ay)$.

2   Two-Level Distributed H Systems

Two-level distributed H systems were introduced in [2], [3]. In [2], two-level distributed systems were considered in the non-separated form, whereas in [3], separated systems were considered.

A two-level (non-separated) communicating distributed H system of degree $n, n \geq 1$ is a construct $\Gamma = (V, T, (w_1, A_1, I_1, E_1), \ldots (w_n, A_n, I_n, E_n))$, where V is the alphabet, $T \subseteq V$ is the terminal alphabet, $w_i \in V^*$, $A_i \subseteq V^*$, and

$I_i, E_i \subseteq V^* \# V^* \$ V^* \# V^*$, for symbols $\#, \$$ not in V. All sets $A_i, I_i, E_i, 1 \leq i \leq n$ are finite; (w_i, A_i, I_i, E_i) is the ith component of the system; w_i is the active axiom, A_i is the set of passive axioms, I_i and E_i are the sets of internal and external splicing rules respectively.

The contents of a component i is described by a pair (x_i, M_i), where $x_i \in V^*$ is the active string and $M_i \subseteq V^*$ is the set of passive strings. An n-tuple $\pi = [(x_1, M_1), \ldots, (x_n, M_n)]$ is called a configuration of the system. For $1 \leq i \leq n$ and a given configuration π as above, we define $\mu(x_i, \pi) =$ external if there are $r \in E_i$ and $x_j, j \neq i$ such that $(x_i, x_j) \models_r (u, v)$ for some $u, v \in V^*$. Otherwise, $\mu(x_i, \pi)$ is internal.

For two configurations π, π' as above, we write $\pi \Rightarrow_{int} \pi'$ if the following conditions hold: (i) for all $i, 1 \leq i \leq n$, we have $\mu(x_i, \pi) =$ internal, (ii) for each $i, 1 \leq i \leq n$, either $(x_i, z) \models_r (x_i', z')$ for some $z \in M_i, z' \in V^*, r \in I_i$, and $M_i' = M_i \cup \{z'\}$, or (iii) no rule $r \in I_i$ can be applied to (x_i, z), for any $z \in M_i$, and then $(x_i', M_i') = (x_i, M_i)$.

The relation $\Rightarrow_{ext}$ defines an external splicing, and $\Rightarrow_{int}$ defines an internal splicing. In both cases, splicing is performed in parallel and all components not able to use a splicing rule do not change their contents. External splicing has priority over internal splicing and all operations have as their first term an active string; the first string obtained by splicing becomes the new active string of the component and the second string becomes an element of the set of passive strings of that component.

The language generated by a two-level distributed H system Γ is defined by $L(\Gamma) = \{w \in T^* \mid [(w_1, A_1), \ldots, (w_n, A_n)] \Rightarrow^* [(x_1, M_1), \ldots, (x_n, M_n)]\}$, for $w = x_1, x_i \in V^*, 2 \leq i \leq n$, and $M_i \subseteq V^*, 1 \leq i \leq n$. LDH_n denotes the family of languages generated by two level distributed H systems with utmost n components. If in the above, we consider all the sets E_i to be the same, i.e, if $E_i = E$, for all $1 \leq i \leq n$, then we get a separated two-level distributed H system. The family of languages generated by separated two-level distributed H systems with n components is denoted by $SLDH_n$. When no restriction is imposed on the number of components, n is replaced by $*$. In the following, we improve the universality result in [3, 4].

Theorem 1. $RE = SLDH_n = LDH_n$ for all $n \geq 2$.

Proof. The idea behind the proof is very close to the one used in [3] and the proof is much simpler. Consider a type-0 grammar $G = (N, T, S, P)$. We construct the $SLDH$ system $\Gamma = (V, T, (w_1, A_1, I_1), (w_2, A_2, I_2), E)$ with

$V = N \cup T \cup \{X, Z, Z_s, Z_l, Z_r, C_1, C_2\},$

$w_1 = SXXC_1,$

$A_1 = \{ZvXZ_s \mid u \to v \in P\}$

$\quad \cup \{ZXX\alpha Z_l, Z\alpha XXZ_r \mid \alpha \in N \cup T\}$

$I_1 = \{\#uXZ\$Z\#vXZ_s \mid u \to v \in P\}$

(Replacing u by v simulating $u \to v$. Z_s is introduced after replacement)

$\cup \{\#\alpha XZ\$Z\#XX\alpha Z_l \mid \alpha \in N \cup T\}$

(shifting α to the right of X. Z_l is introduced after the right shift)

$\cup \{\#XZ\$Z\#\alpha XXZ_r \mid \alpha \in N \cup T\},$

(shifting α to the left of X. Z_r is introduced after the left shift)

$w_2 = C_2 Z,$

$A_2 = \{Z_s Z, Z_l Z, Z Z_r\},$

$I_2 = \{C_2\#Z_s\$Z_s\#Z, C_2\#Z_l\$Z_l\#Z, C_2\#Z_r\$Z_r\#Z\},$

(Changing Z_s, Z_l, Z_r back to Z to start a new simulation)

E consists of the rules

E1. $X\#X\$C_2\#Z, C_2\#Z\$X\#X,$

(First step while simulating a rule $u \to v$, or while shifting)

E2. $X\#Z_s\$C_2\#X, C_2\#X\$X\#Z_s,$

(Replace Z_s by a string ending in C_1 in w_1; replace the suffix of w_2 by Z_s)

E3. $XX\alpha\#Z_l\$C_2 X\#, C_2\#X\$XX\alpha\#Z_l, \alpha \in N \cup T,$

(Replace Z_l by a string ending in C_1 in w_1; replace the suffix of w_2 by Z_l)

E4. $\alpha XX\#Z_r\$C_2 X\alpha\#, C_2\#X\alpha\$\alpha XX\#Z_r, \alpha \in N \cup T,$

(Replace Z_r by a string ending in C_1 in w_1; replace the suffix of w_2 by Z_r)

E5. $\#XXC_1\$C_2 Z\#.$

(To terminate, cut off the symbols XXC_1 from the right)

Component 1 simulates rules of P and also shifts symbols to the right and left of the marker XX. Component 2 saves the suffix of the active string w_1 that is cut while simulation and shifting, and also checks that the shifting done in component 1 is correct.

To start with, we have $w_1 = SXXC_1, w_2 = C_2 Z$. In general, assume that $w_1 = z_1 u XXz_2 C_1, w_2 = C_2 Z$, where $u = u'a$, where $a \in V, u \in V^*$. (initially, $z_1 u' = \lambda, a = S, z_2 = \lambda$).

<u>Case 1:</u> Simulation of a rule $u \to v \in P$. To begin, $E1$ is the only applicable rule. $E1$ is applied in parallel to both components.

1. $E1 \Rightarrow w_1 = z_1 u XZ, C_2 Xz_2 C_1 \in M_1, w_2 = C_2 Xz_2 C_1, z_1 u XZ \in M_2$. In the next step, no external rules are applicable, since w_1 does not contain XX or XZ_s or XXC_1. Note that $E1$ cuts the suffix $Xz_2 C_1$ of w_1 and appends it to w_2; it also cuts the suffix Z of w_2 and appends it to w_1.
2. Use the internal rule $(z_1|uXZ, Z|vXZ_s) \models (z_1 vXZ_s, ZuXZ)$ in component 1 obtaining $w_1 = z_1 vXZ_s, w_2 = C_2 Xz_2 C_1$, to simulate $u \to v$. Component 2 is idle. In the next step, only $E2$ is applicable, and it acts in parallel on both components.
3. Now, $E2 \Rightarrow w_1 = z_1 vXXz_2 C_1, w_2 = C_2 Z_s$, re adjoining the suffix $z_2 C_1$ to w_1. No external rules are applicable in the next step since $w_2 \neq C_2 X, C_2 Z$.
4. To get back to $C_2 Z$, use the internal rule $(C_2|Z_s, Z_s|Z) \models (C_2 Z, Z_s Z_s)$ in component 2 (component 1 is idle) giving $w_1 = z_1 vXXz_2 C_1, w_2 = C_2 Z$.

Case 1 handles $w_1 = z_1 u X X z_2 C_1$, when there is a rule $u \rightarrow v \in P$. Assume now that for $u = u'a$, there exists no rule in P for a, but there exists $u' \rightarrow v' \in P$. To simulate $u' \rightarrow v'$ as above, we need u' to be adjacent to XX in w_1. To obtain this, we need to shift a to the right of XX obtaining $z_1 u' X X a z_2 C_1$. Case 2 handles this situation.

<u>Case 2:</u> Transforming $w_1 = z_1 u' a X X z_2 C_1$ into $z_1 u' X X a z_2 C_1$, given $w_2 = C_2 Z$.

1. To begin, only $E1$ is applicable in both components. $E1 \Rightarrow w_1 = zu'aXZ, C_2 X z' C_1 \in M_1, w_2 = C_2 X z' C_1, zu'aXZ \in M_2$. In the next step, no external rules are applicable since w_1 does not contain $XX, X Z_s, X C_1$.

2. By assumption (since there is no rule in P for a), we choose any of the two internal rules (different from the one chosen in case 1, step 2). Component 2 will remain idle in this step. Using $(zu'|aXZ, Z|XXaZ_l) \models (zu'XXaZ_l, ZaXZ)$ in component 1, we obtain $w_1 = zu'XXaZ_l$, and $w_2 = C_2 X z' C_1$. $E3$ is only applicable in the next step, and it acts in parallel on both components.

3. Now, $E3 \Rightarrow w_1 = zu'XXaz'C_1, C_2 X Z_l \in M_1, w_2 = C_2 Z_l, zu'XXaxz'C_1 \in M_2$, shifting a to the right of XX in w_1. In the next step, no external rules are applicable, since $w_2 \neq C_2 X, C_2 Z$.

4. To get back to $C_2 Z$, use the internal rule $(C_2|Z_l, Z_l|Z) \models (C_2 Z, Z_l Z_l)$ in component 2 (component 1 remains idle) giving $w_1 = z_1 u' X X a z_2 C_1, w_2 = C_2 Z$.

After cases 1 and 2, one more situation needs to be handled. Assume that we have $w_1 = z_1 X X a z_2 C_1$, with rules $z_1 a \rightarrow z \in P$, and no rules in P for any substring of z_1. Clearly, case 2 is not useful, and to simulate a rule as in case1, we need $z_1 a$ to the left of XX. To do this, the a should be shifted to the left of XX obtaining $w_1 = z_1 a X X z_2 C_1$.

<u>Case 3:</u> Transforming $w_1 = z_1 X X a z_2 C_1$ into $z_1 a X X z_2 C_1$, given $w_2 = C_2 Z$.

1. As in the above cases, we start with $E1$. $E1 \Rightarrow w_1 = z_1 X Z, C_2 X a z_1' C_1 \in M_1, w_2 = C_2 X a z_1' C_1, z_1 X Z \in M_2$. No external rules are applicable in the next step.

2. We can choose an internal rule in component 1 involving Z_l or Z_r, lets choose the one with Z_r. Using $(z_1|XZ, Z|\alpha XXZ_r) \models (z_1 \alpha XXZ_r, ZXZ), \alpha \in N \cup T$ in component 1, we obtain $w_1 = z_1 \alpha X X Z_r$. Component 2 is idle, and hence $w_2 = C_2 X a z_1' C_1$. $E4$ is only applicable in the next step to both components.

3. $E4 \Rightarrow w_1 = z_1 a X X z_1' C_1, C_2 X a Z_r \in M_1, w_2 = C_2 Z_r, z_1 a X X X a z_1' C_1 \in M_2$, shifting a to the left of XX. Note that $E4$ can be applied only if $\alpha = a$ in the previous step. No external rules are applicable in the next step since $w_2 = C_2 Z_r$.

4. Using the internal rule $(C_2|Z_r, Z_r|Z) \models (C_2 Z, Z_r Z_r)$ in component 2 (component 1 being idle), we obtain $w_1 = z_1 a X X z_1' C_1, w_2 = C_2 Z$.

Now, any of the three cases can be iterated. To terminate, we have only one choice: to remove the substring XX of w_1 which facilitates simulation of rules or shifting. This is done by using $E5$, when all symbols are to the left of XX. This will cut off from the active string $wXXC_1$ in component 1, the tail XXC_1, leaving w as the active string. Now, no more rules can be applied to w. If $w \in T^*$, it gets listed in the language, otherwise, nothing is computed. $\square$

Note the almost symmetric nature of the external rules in the above theorem. It helps in applying the external rules simultaneously in both components, with no waiting. Even when applying internal rules, there is a minimal wait of exactly one step for the other component.

3   Communicating Distributed H (CDH) Systems

CDH systems have been explored extensively in [1, 3, 5, 6], obtaining universality results with arbitrarily many components, six components, three components and nine components respectively. In the following section, we briefly recall the basics of CDH systems [4] and introduce the concept of laziness into CDH systems. We then consider an example and prove that universality can be obtained with 2 components.

A CDH system is a construct $\Gamma = (V, T, (A_1, R_1, V_1), \ldots, (A_n, R_n, V_n))$, where V is an alphabet, $T \subseteq V$, A_i are finite languages over V, R_i are finite sets of splicing rules over V, and $V_i \subseteq V, 1 \leq i \leq n$. Each triple $(A_i, R_i, V_i), 1 \leq i \leq n$, is called a component of Γ; A_i, R_i, V_i are the sets of axioms, the sets of splicing rules, and the selector of the component i, respectively. Let $\mathcal{B} = V^* - \bigcup_{i=1}^n V_i^*$. The pair $\sigma^{(i)} = (V, R_i)$ is the underlying H scheme associated to the component i of the system.

An $n-$tuple $(L_1, L_2, \ldots, L_n), L_i \subseteq V^*$, is called a configuration of the system. The initial configuration of the system is $(A_1, \ldots, A_n)$. For two configurations $(L_1, \ldots, L_n), (L'_1, \ldots, L'_n)$, we define $(L_1, \ldots, L_n) \Rightarrow (L'_1, \ldots, L'_n)$ iff $L'_i = \bigcup_{j=1}^n (\sigma_2^{(j)^*}(L_j) \cap V_i^*) \cup (\sigma_2^{(i)^*}(L_i) \cap \mathcal{B})$, for each i, $1 \leq i \leq n$.

In words, the contents of each component are spliced according to the set of rules (we pass from L_i to $\sigma_2^{(i)^*}(L_i)$) and the result is redistributed among the n components according to the selectors $V_1, \ldots, V_n$; the part which cannot be redistributed remains in the component. As no conditions are imposed on the alphabets V_i, when a string in $\sigma_2^{(j)^*}(L_j)$ belongs to several languages V_i^*, then copies of the string will be distributed to all components i with this property.

The language generated by Γ is defined as $L(\Gamma) = \{w \in T^* \mid w \in L_1 \text{ for } L_1, \ldots, L_n \subseteq V^* \text{ such that } (A_1, \ldots, A_n) \Rightarrow^* (L_1, \ldots, L_n)\}$. The family of languages generated by communicating distributed H systems of degree utmost $n, n \geq 1$ is denoted by CDH_n. When n is not specified, then we replace n by $*$.

3.1   Introducing Laziness

Let Γ be a CDH_n system. We now define three kinds of strings viz., *active*, *passive* and *inactive* based on Γ as follows:

1. A string $w \in \sigma_2^{(i)^*}(L_i)$ is said to be *active* if there exists (i) a splicing rule $(w_2'\#w_2''\$a\#b)$ in R_i, (ii) a string $xaby$ in $\sigma_2^{(i)^*}(L_i)$, and (iii) a substring $w_2'w_2''$ of w. Note that we can also describe w having ab as a substring such that there exists a string $xw_2'w_2''y$ in $\sigma_2^{(i)^*}(L_i)$. Clearly, if w is an active string, it can be spliced using rules of R_i to obtain further strings.

2. A string $w \in \sigma_2^{(i)^*}(L_i)$ is said to be *passive* if for all splicing rules $(w_2'\#w_2''\$a\#b)$ in R_i, such that $w_2'w_2''$ (or ab) is a substring of w, there does not exist any string $xaby$ (or $x_1 w_2' w_2'' x_2$) in $\sigma_2^{(i)^*}(L_i)$.

3. A string $w \in \sigma_2^{(i)^*}(L_i)$ is said to be *inactive* if for all splicing rules $(w_2'\#w_2''\$a\#b)$ in R_i, w does not contain $w_2'w_2''$ or ab as a substring.

A *lazy* communicating distributed H system is a construct

$$\Gamma = (V, T, (A_1, R_1, V_1, \gamma_1), \ldots, (A_n, R_n, V_n, \gamma_n)),$$

where V is an alphabet, $T \subseteq V$, A_i are finite languages over V, R_i are finite sets of splicing rules over V, and $V_i \subseteq V, 1 \leq i \leq n$. Each tuple $(A_i, R_i, V_i, \gamma_i), 1 \leq i \leq n$, is called a component of Γ; A_i, R_i, V_i are the sets of axioms, the sets of splicing rules, and the selector of the component i, respectively; γ_i is a parameter taking values l or e, depending on whether the component is *lazy* or *eager*; T is the terminal alphabet of the system. Let $\mathcal{B} = V^* - \bigcup_{i=1}^n V_i^*$.

There are two kinds of components : *lazy* components and *eager* components. The two kinds of components differ in the way they communicate : *eager* components behave the same way as the components in a CDH system, whereas *lazy* components communicate only their *inactive* strings, provided they pass the necessary filters.

The pair $\sigma^{(i)} = (V, R_i)$ is the underlying H scheme associated to the component i of the system.

An n-tuple $(L_1, L_2, \ldots, L_n), L_i \subseteq V^*$, is called a configuration of the system. L_i is also called the contents of component i. The initial configuration of the system is $(A_1, \ldots, A_n)$. For two configurations $(L_1, \ldots, L_n), (L_1', \ldots, L_n')$, we define $(L_1, \ldots, L_n) \Rightarrow (L_1', \ldots, L_n')$ iff

1. $L_i' = \bigcup_{j=1}^n (S_j \cap V_i^*) \cup (\sigma_2^{(i)^*}(L_i) \cap \mathcal{B})$, for each eager $i, 1 \leq i \leq n$, and $S_j = \sigma_2^{(j)^*}(L_j)$ if j is eager, and $S_j \subseteq \sigma_2^{(j)^*}(L_j)$ is the set consisting of all *inactive* strings of $\sigma_2^{(j)^*}(L_j)$, if j is lazy.

2. $L_j' = \bigcup_{i=1}^n (S_i \cap V_j^*) \cup (S_j \cap \mathcal{B}) \cup (L_j \backslash S_j)$ for each lazy $j, 1 \leq j \leq n$, and $S_i = \sigma_2^{(i)^*}(L_i)$ if i is eager, and $S_j \subseteq \sigma_2^{(j)^*}(L_j)$ is the set of *inactive* strings of $\sigma_2^{(j)^*}(L_j)$, if j is lazy.

In words, the contents of a component i are spliced according to the associated set of rules, and,

- If i is eager, the result is redistributed among the n components according to the selectors $V_1, \ldots, V_n$; the part which cannot be redistributed (which does not belong to some $V_i^*, 1 \leq i \leq n$) remains in the component.

– If i is lazy, the subset of ${\sigma_2^{(i)}}^*(L_i)$ consisting of the *inactive* strings of ${\sigma_2^{(i)}}^*(L_i)$ is redistributed among the n components according to the selectors $V_1, \ldots, V_n$, and the part of the subset which cannot be redistributed remains in the component.

The language generated by Γ is defined by $L(\Gamma) = \{w \in T^* \mid w \in L_1 \text{ for some } L_1, \ldots, L_n \subseteq V^* \text{ such that } (A_1, \ldots, A_n) \Rightarrow^* (L_1, \ldots, L_n)\}$.

We denote by LCDH_n the family of languages generated by lazy communicating distributed H systems of degree utmost $n, n \geq 1$. When n is not specified, we replace n by $*$.

Note that an LCDH system with all components *eager* is the same as a CDH system. Let us consider an example.

Example 1. Consider the system Γ

$$(\{a, b, c, X, Y, Z, Z', F_1, F_2, F\}, \{a, b, c\}, (A_1, R_1, V_1, e), (A_2, R_2, V_2, l), (A_3, R_3, V_3, l)),$$

$$A_1 = \{XY, aX\}, R_1 = \{a\#X\$X\#Y, c\#F\$\#aY, c\#F\$aY\#\}, V_1 = T \cup \{F\},$$
$$A_2 = \{bZ, Z'Z'\}, R_2 = \{a\#Y\$\#bZ, ab\#Z\$Z'\#Z'\}, V_2 = T \cup \{Y\},$$
$$A_3 = \{F_1cF_1, F_2, FF\}, R_3 = \{b\#Z'\$F_1\#cF_1, c\#F_1\$\#F_2, c\#F_2\$F\#F\},$$
$$V_3 = T \cup \{Z'\}.$$

No communication between components is possible before any splicing, since $A_1 \cap V_j^* = \emptyset, j = 2, 3$; strings of A_2 are *passive*; $FF \in A_3$ is *passive*, and the rest of A_3 is *active*. Hence, splicing is possible only in the first and third components; $(a|X, X|Y) \models (aY, XX)$ in the first component and $(F_1c|F_1, |F_2) \models (F_1cF_2, F_1)$ in the third component. The string aY is communicated from component 1 to component 2, and in component 3, the string F_1 is a candidate for communication, since it is *inactive*. However, $F_1 \notin V_i^*, i = 1, 2, 3$, and hence remains in component 3.

In component 2, the string aY is spliced according to the rule $(a|Y, |bZ) \models (abZ, Y)$ and in component 3, the new splicings are $(F_1c|F_2, F|F) \models (F_1cF, FF_2)$ or $(F_1c|F_1, F_1c|F_2) \models (F_1cF_2, F_1cF_1)$. The string abZ in component 2 is *active*, whereas Y is *inactive*. Similarly, in component 3, the string F_1cF is *inactive*. Therefore, Y, F_1cF are candidates for communication in components 2,3. However, since $Y \notin V_1, V_3, F_1cF \notin V_1^*, V_2^*, V_3^*$, Y remains in component 2 and F_1cF in component 3. Continuing with abZ in component 2, we obtain $(ab|Z, Z'|Z') \models (abZ', Z'Z)$ or $(a|Y, a|bZ) \models (abZ, aY)$. Now the strings $abZ', Z'Z$ are *inactive* and therefore are candidates for communication. Of the two, abZ' is sent to component 3, while $Z'Z$ remains in component 2.

In component 3, abZ' is spliced as $(ab|Z', F_1|cF_1) \models (abcF_1, F_1Z')$. Now, F_1Z' is *inactive*; however since it does not belong to any V_i^*, it remains in component 3. The string $abcF_1$ is spliced as $(abc|F_1, |F_2) \models (abcF_2, F_1)$. Now, $abcF_2$ is *active*, and F_1 is *inactive*. F_1 remains in component 3 since it fails all filters, and we splice $abcF_2$. Some possible splicings are $(F_1c|F_1, abc|F_2) \models (F_1cF_2, abcF_1)$ or $(abc|F_1, abc|F_2) \models (abcF_2, abcF_1)$ or $(abc|F_2, F|F) \models (abcF, FF_2)$. All strings except $abcF$ are *active*, and $abcF$ is communicated to component 1.

In component 1, we have either the option of appending an aY to abc and thus continuing, or using $(abc|F, aY|) \models (abc, aYF)$. The string abc remains in component 1, and a copy is sent to components 2 and 3. Clearly, $L(\Gamma) = \{(abc)^n \mid n \geq 1\}$.

Theorem 2. $RE = LCDH_2$

Proof. Consider a type-0 grammar $G = (N, T, S, P)$. Let $N \cup T \cup \{B\} = \{D_1, \ldots, D_m\}$, where B is a new symbol. Since $N, T \neq \emptyset$, $m \geq 3$. Construct the LCDH system $\Gamma = (V, T, (A_1, R_1, V_1, e), (A_2, R_2, V_2, l))$, with

$V = N \cup T \cup \{X, Y, Z, Z', E_1, E_2, X_i, Y_i, X'_{2j}, Y'_{2j} \mid -1 \leq i \leq 2m, 1 \leq j \leq m\}$,

$A_1 = \{XBSY\} \cup \{ZvY \mid u \to v \in P\} \cup \{ZX'_{2i}Y'_{2i} \mid 1 \leq i \leq m\} \cup \{E_1E_2, XZ_0, Z_0Y\}$

$\quad \cup \{X_{2i}Z, ZY_{2i} \mid 1 \leq i \leq m\}$, and R_1 consists of the following rules:

Simulating rules of P :

1. $\#uY\$Z\#vY,\ u \to v \in P$,

Rotation : For $1 \leq i, j, k \leq m$,

2. $D_j\#D_iY\$ZX'_{2i}\#Y'_{2i}$,

3. $X\#D_jD_k\$ZX'_{2i}D_i\#Y$,

4. $\#XY\$Z\#X'_{2i}D_iD_j$,

Updation of Indices (Odd to even) :

5. $X_{2j+1}\#D_i\$X_{2j}\#Z,\ 0 \leq j \leq m, 1 \leq i \leq m$,

6. $D_i\#Y_{2j+1}\$Z\#Y_{2j},\ 0 \leq j \leq m, 1 \leq i \leq m$,

Going back to end markers X, Y, from X_0, Y_0

7. $X_0\#D_j\$X\#Z_0, 1 \leq j \leq m$,

8. $D_j\#Y_0\$Z_0\#Y, 1 \leq j \leq m$,

Possible Termination : For $D_j, D_k \in T, 1 \leq j, k \leq m$,

9. $D_j\#BY\$E_1\#E_2$,

10. $X\#D_k\$E_1\#BY$,

11. $E_1\#D_k\$\#E_1E_2$,

12. $D_j\#E_2\$E_1E_1E_2\#$,

$V_1 = N \cup T \cup \{B, X, Y, X_0, Y_0\} \cup \{X_{2i+1}, Y_{2i+1} \mid 0 \leq i \leq m-1\}$,

$A_2 = \{X_{2i}Z', Z'Y_{2i}, Z'Y_{2i-1}, X_{2i-1}Z', X_{-1}Z', Z'Y_{-1}\},\ 0 \leq i \leq m$, and

R_2 consists of rules

Initialize : For $1 \leq i, j \leq m$,

13. $X'_{2i}\#D_i\$X_{2i}\#Z'$,

14. $D_j\#Y'_{2i}\$Z'\#Y_{2i}$,

Updation of Indices (Even to odd) : For $1 \leq i, j \leq m$,

15. $X_{2i}\#D_j\$X_{2i-1}\$Z'$,

16. $D_j\#Y_{2i}\$Z'\#Y_{2i-1}$,

Removal of X_0, Y_0 : For $1 \leq j \leq m$,

17. $X_0\#D_j\$X_{-1}\#Z'$,

18. $D_j \# Y_0 \$ Z' \# Y_{-1}$.

$V_2 = N \cup T \cup \{B, X_{2i}, Y_{2i}, X'_{2j}, Y'_{2j} \mid 0 \le i \le m, 1 \le j \le m\}$

Let us examine the work of Γ. The underlying idea is to rotate and simulate. We start from the string $XBSY$ in component 1, and in component 2, there are no rules that can be applied with respect to strings in A_2. However since all strings in A_2 are *passive*, and since none of the strings in A_1 pass the filter V_2, there is no communication before any splicing. In the first component, we can simulate rules of P by using the rule 1, replacing suffixes. Since the new strings obtained as a result of rule 1 do not pass the filter V_2, and since there are no *inactive* strings in component 2, there is no communication between the components.

This can go on as long as rule 1 is applied. If we choose to rotate a symbol at any point of time, then we choose rule 2, giving $(Xw \mid D_i Y, ZX'_{2i} \mid Y'_{2i}) \models (XwY'_{2i}, ZX'_{2i}D_iY)$. Both of these strings $\notin V_2^*$, and hence cannot be communicated to component 2. We can choose next, $(X \mid D_j, ZX'_{2i}D_i \mid Y) \models (XY, ZX'_{2i}D_iD_jw_1Y'_{2i})$, provided $w = D_jw_1$. The two new strings obtained here also $\notin V_2^*$ and hence we continue in component 1. We can now use $(\mid XY, Z \mid X'_{2i}D_iDj) \models (X'_{2i}D_iD_jw_1Y'_{2i}, ZXY)$, and in this step, the string $X'_{2i}D_iD_jw_1Y'_{2i}$ is communicated to component 2. No string from component 2 is communicated to component 1.

In the next step, in component 2, we can use rules 13 or 14 to $X'_{2i}D_iD_jw_1Y'_{2i}$, resulting in $(X'_{2i}Z', X_{2i}D_iD_jw_1Y'_{2i})$ or $(X'_{2i}D_iD_jw_1Y_{2i}, Z'Y'_{2i})$. The strings $X'_{2i}Z', Z'Y'_{2i}$ in component 2 are *inactive*, and so are considered for communication. However, since they do not pass V_1, they remain in component 2. The strings $X'_{2i}D_iD_jw_1Y_{2i}$ or $X_{2i}D_iD_jw_1Y'_{2i}$ are *active* and so are not considered for communication. We can apply rule 14 or 15 to $X_{2i}D_iD_jw_1Y'_{2i}$ and rule 13 or 16 to $X'_{2i}D_iD_jw_1Y_{2i}$. In either case, we ultimately obtain the *inactive* string $X_{2i-1}D_iwY_{2i-1}$. Since $X_{2i-1}D_iwY_{2i-1} \in V_1^*$, it is sent to component 1.

Let $w' = D_iw$. In component 1, rules 5 and 6 are applicable to $X_{2i-1}w'Y_{2i-1}$. If we choose rule 5 first, we obtain $(X_{2i-1} \mid D_i, X_{2i-2} \mid Z) \models (X_{2i-1}Z, X_{2i-2}w'Y_{2i-1})$. Both these strings cannot be communicated to component 2, since they do not pass the filter. We continue with rule 6 to obtain $(D_k \mid Y_{2i-1}, Z \mid Y_{2i-2}) \models (X_{2i-2}w'Y_{2i-2}, ZY_{2i-1})$. We would obtain the same set of strings even if rule 6 is applied first. The string $X_{2i-2}w'Y_{2i-2}$ obtained after application of rules 5,6 is communicated to component 2, since it passes the filter.

In component 2, we now decrement the end markers using rules 15, 16. Observe that until both are used, we cannot communicate the intermediate string ($X_{2i-2}w'Y_{2i-3}$ or $X_{2i-3}w'Y_{2i-2}$), since it is *active*. The other strings obtained as a result of rules 15,16 are $X_{2i-2}Z', Z'Y_{2i-2}$, which cannot be communicated even though they are *inactive*, since they are not over V_1^*.

Continuing like this, a string $X_1w'Y_1$ is communicated to component 1. Now, using rules 5,6 as before we decrement X_1, Y_1 to X_0, Y_0. Note that before decrementing both X_1 and Y_1, we cannot communicate to component 2, since V_2 does not contain $X_{2i+1}, Y_{2i+1}, i \ge 0$. However, when we have $X_0w'Y_0$ in component 1, since V_1, V_2 contain X_0, Y_0, the string is communicated to component 2, and a copy is retained in component 1.

In component 1, the X_0 is replaced by X and Y_0 by Y by rules 7 and 8. Observe that the intermediate strings obtained (with X, Y_0 and X_0, Y as the end markers) cannot be communicated to component 2, since $X, Y \notin V_2$. But, we can start another simulation in component 1 using $Xw'Y$. Simultaneously, in component 2, rules 17,18 are applicable to $X_0 w' Y_0$. We do not consider the intermediate strings for communication since they are *active*. But, even after application of 17,18, the string we obtain, viz., $X_{-1} w' Y_{-1}$ cannot be communicated, since it is not over V_1.

Note that, the first time a rotation is done in component 1, the indices of the end markers will be the same, since rule 3 can be applied only after applying rule 2, thus obtaining the correct string $ZX'_{2i} D_i Y$. However, this is not the case for subsequent rotations. (since all strings $ZX'_{2i} D_i Y$ produced in previous steps will be available). In general, it is possible to obtain a string $X'_{2i} w' Y'_{2j}, i \neq j$ in component 1. We communicate this string to component 2, and, after a sequence of communications, we will end up with a string $X_0 w' Y_{2l}, l > 0$ or $X_{2k} w' Y_0, k > 0$. Let us examine how to handle this case.

Let us assume that we have the string $X_{2k} w' Y_0$ in component 1. Obviously, this is obtained after application of rules 5,6 in the two previous steps. This string is communicated to component 2 since it passes the filter V_2, without retaining a copy in component 1 ($X_{2k} w' Y_0 \notin V_1^*$). In component 2, rules 15,18 are applicable. This leads us to the intermediate strings $X_{2k-1} w' Y_0$ (15 applied first) or $X_{2k} w' Y_{-1}$ (18 applied first). In either case, both strings are *active*. We end up, in either case with $X_{2k-1} w' Y_{-1}$, which is *inactive*. But however, this string belongs to neither V_1^* nor V_2^* and so, remains in component 2, without contributing to the output.

Thus, we can continue a simulation iff we end up with $X_0 w'' Y_0$ in component 1, in which case, the copy sent to component 2 remains stuck there, but the copy in component 1 is useful by replacing X_0 by X and Y_0 by Y.

Let us now examine how a string over terminals can be generated, contributing to $L(\Gamma)$. Assume that we have in component 1, a string $XwBY$. We can choose to either rotate B using rule 2, or eliminate B using rule 9. Let us see what happens if rule 9 is chosen. We obtain $(Xw|BY, E_1|E_2) \models (XwE_2, E_1BY)$. Both these strings cannot be communicated, since they fail to pass the filter V_2. We can continue with rule 10, $(X|D_k, E_1|BY) \models (XBY, E_1 D_k w' E_2)$, provided $w = D_k w'$. Now, rule 11, $(E_1|D_k, |E_1 E_2) \models (E_1 E_1 E_2, D_k w' E_2)$ is used to remove E_1. This is followed by application of rule 12 removing E_2 and obtaining $D_k w'$. The only information we have about this string is that if $w' = w_1 D_j$ or $w = D_k w_1 D_j$, then $D_k, D_j \in T$. However, if this string is not over terminals, then it does not contribute to the language and is hence "lost". Thus, only terminal strings obtained starting from $XBSY$, which are rotated correctly every time (so that $X_0 w Y_0$ is obtained in component 1) can contribute to the language. □

Remark 1. To see how the above system communicates less, we will examine what happens if component 2 was eager in the above result. As long as no rotation takes place (for the first time) in component 1, there is no communication between components, irrespective of the nature of the individual components.

The number of strings communicated between components is the same (if component 2 is eager or lazy) even after rotation, in case, rotation takes place correctly, yielding X_0wY_0 in component 1. Now assume that rotation goes wrong, giving X_0wY_{2k} or $X_{2l}wY_0$, $k, l > 0$ in component 1. Either of these strings will be communicated to component 2. If component 2 was eager, then if rule 15 or 16 is chosen first, we get a string $X_{2l-1}wY_0$ or X_0wY_{2k-1}, which will be communicated to component 1, leading to wrong results. That means an extra communication is made, which also leads to wrong results in case component 2 was eager. But if component 2 is lazy, this communication will not be made, and the results also do not go wrong. The same is the case if X_0, Y_0 are not replaced in subsequent steps in component 1, when having X_0wY_0. ($X_0w'Y'_{2i}$ can be obtained in component 1, which will be communicated to component 2. Component 2 if eager, can then communicate $X_0w'Y_{2i-1}$ to component 1, and things go wrong).

4   Conclusion

We have improved the universality result of two-level distributed H systems, and conjecture that the result obtained is optimal. Likewise, by introducing laziness, we have proved that a better characterization of RE can be obtained, as compared to the result $CDH_3 = RE$ [5]. The power of $LCDH_2$, with both components being lazy, is open.

References

1. E. Csuhaj-Varju, L. Kari, Gh. Păun, Test tube distributed systems based on splicing, *Computers and AI*, 15, 5 (1996), 419–436.
2. Gh. Păun, Two-level distributed H systems, *Proc. of the Third Conf. on Developments in Language Theory*, Aristotle Univ. of Thessaloniki, 1997, 309-327.
3. Gh. Păun, DNA Computing: Distributed splicing systems, *Structures in Logic and Computer Science : A Selection of Essays in Honor of A. Ehrenfeucht*, LNCS 1261, 1997, 351–370.
4. Gh. Păun, G. Rozenberg, A. Salomaa, DNA Computing : New Computing Paradigms, Springer, 1998.
5. L. Priese, Y. Rogozhin, M. Margenstern, Finite H systems with 3 tubes are not predictable, *Pacific Symposium on Biocomputing*, Hawaii, 1998 (R. B. Altman, A. K. Dunker, L. Hunter, T. E. Klein, eds), World Sci, Singapore, 1998, 547-558.
6. C. Zandron, C. Ferretti, G. Mauri, A reduced distributed splicing system for RE languages, *New Trends in Formal Languages : Control, Cooperation, Combinatorics*, LNCS 1218, 1997, 346-366.

Intensive *In Vitro* Experiments of Implementing and Executing Finite Automata in Test Tube

Junna Kuramochi[1] and Yasubumi Sakakibara[2]

[1] Softbank BB Corporation, Japan
[2] Keio University, Department of Biosciences and Informatics,
3-14-1 Hiyoshi, Kohoku-ku, Yokohama, 223-8522, Japan
yasu@bio.keio.ac.jp

Abstract. We report our intensive *in vitro* experiments in which we have implemented and executed several finite-state automata in test tube. First, we employ the length-encoding technique proposed and presented in [4, 3] to implement finite automata in test tube. In the length-encoding method, the states and state transition functions of a target finite automaton are effectively encoded into DNA sequences, a computation (accepting) process of finite automata is accomplished by self-assembly of encoded complementary DNA strands, and the acceptance of an input string is determined by the detection of a completely hybridized double-strand DNA. Second, we design and develop practical laboratory protocols which combine several *in vitro* operations such as annealing, ligation, PCR, and streptavidin-biotin bonding to execute *in vitro* finite automata based on the length-encoding technique. We have carried laboratory experiments on various finite automata of from 2 states to 6 states for several input strings. To our knowledge, this is the first *in vitro* experiments that have succeeded to execute 6-states automaton in test tube.

1 Introduction

The finite-state automata (machines) are the most basic computational model in Chomsky hierarchy and are the start point to build universal DNA computers. Several works [1, 2, 3, 4] have attempted to develop finite automata *in vitro*. Benenson et al. [1] have successfully implemented the two state finite automata by the sophisticated use of the restriction enzyme (actually, *Fok*I) which cut outside of its recognition site in a double-stranded DNA. However, their method has some limitations for extending to more than 2 states. Yokomori et al. [4] have proposed a theoretical framework using length-encoding technique to implement finite automata on DNA molecules. Theoretically, the length-encoding technique has no limitations to implement finite automata of any larger states.

In this paper, we attempt to implement and execute finite automata of a larger number of states *in vitro*, and carry intensive laboratory experiments on various finite automata of from 2 states to 6 states for several input strings. To our knowledge, this is the first *in vitro* experiments that have succeeded to compute 6-states automaton in test tube.

A. Carbone and N.A. Pierce (Eds.): DNA11, LNCS 3892, pp. 193–202, 2006.

2   Methods

2.1   Length-Encoding Method to Implement Finite-State Automata

Let $M = (Q, \Sigma, \delta, q_0, F)$ be a (deterministic) finite automaton, where Q is a finite set of states numbered from 0 to k, Σ is an alphabet of input symbols, δ is a state-transition function such that $\delta : Q \times \Sigma \longrightarrow Q$, q_0 is the initial state, and F is a set of final states. We adopt the length-encoding technique [4] to encode each state in Q by the length of DNA subsequences.

For the alphabet Σ, we encode each symbol a in Σ into a single-strand DNA subsequence, denoted $e(a)$, of fixed length. For an input string w on Σ, we encode $w = x_1 x_2 \cdots x_m$ into the following single-strand DNA subsequence, denoted $e(w)$:

$$5'\text{-}\ e(x_1) \underbrace{X_1 X_2 \cdots X_k}_{k \text{ times}} e(x_2) \underbrace{X_1 X_2 \cdots X_k}_{k \text{ times}} \cdots e(x_m) \underbrace{X_1 X_2 \cdots X_k}_{k \text{ times}}\ \text{-}3',$$

where X_i is one of four nucleotides A, C, G, T, and the subsequences $X_1 X_2 \cdots X_k$ are used to encode $k + 1$ states of the finite automaton M. For example, when we encode a symbol '1' into a ssDNA subsequence GCGC and a symbol '0' into GGCC, and encode three states into TT, a string "1101" is encoded into the following ssDNA sequence:

$$5'\text{-}\ \overbrace{\text{GCGC}}^{1} \text{TT} \overbrace{\text{GCGC}}^{1} \text{TT} \overbrace{\text{GGCC}}^{0} \text{TT} \overbrace{\text{GCGC}}^{1} \text{TT}\ \text{-}3'$$

In addition, we append two supplementary subsequences at both ends for PCR primers and probes for affinity purifications with magnetic beads which will be used in laboratory protocol:

$$5'\text{-}\ \underbrace{S_1 S_2 \cdots S_s}_{\text{PCR primer}} e(x_1) X_1 X_2 \cdots X_k \cdots e(x_m) X_1 X_2 \cdots X_k \underbrace{Y_1 Y_2 \cdots Y_t}_{\text{probe}} \underbrace{R_1 R_2 \cdots R_u}_{\text{PCR primer}}\ \text{-}3'.$$

For a state-transition function from state q_i to state q_j with input symbol $a \in \Sigma$, we encode the state-transition function $\delta(q_i, a) = q_j$ into the following complementary single-strand DNA subsequence:

$$3'\text{-}\ \underbrace{\overline{X}_{i+1} \overline{X}_{i+2} \cdots \overline{X}_k}_{k-i \text{ times}} \overline{e(a)} \underbrace{\overline{X}_1 \overline{X}_2 \cdots \overline{X}_j}_{j \text{ times}}\ \text{-}5'$$

where $\overline{X}_i$ denotes the complementary nucleotide of X_i, and $\overline{y}$ denotes the complementary sequence of y. Further, we put two more complementary ssDNA sequences for the supplementary subsequences at both ends:

$$3'\text{-}\ \overline{S}_1 \overline{S}_2 \cdots \overline{S}_s\ \text{-}5', \qquad 3'\text{-}\ \underbrace{\overline{Y}_1 \overline{Y}_2 \cdots \overline{Y}_t \overline{R}_1 \overline{R}_2 \cdots \overline{R}_u}_{\text{biotinylated}}\ \text{-}5',$$

where the second ssDNA is biotinylated for streptavidin-biotin bonding.

Now, we put all those ssDNAs encoding an input string w and encoding state-transition functions and the supplementary subsequences of probes and PCR primers. Then, a computation (accepting) process of the finite automata M is accomplished by self-assembly among those complementary ssDNAs, and the acceptance of an input string w is determined by the detection of a completely hybridized double-strand DNA.

The main idea of length-encoding technique is explained as follows. Two consecutive valid transitions $\delta(h, a_n) = i$ and $\delta(i, a_{n+1}) = j$ are implemented by concatenating two corresponding encoded ssDNAs, that is,

$$3\text{'-} \underbrace{\text{AAA} \cdots \text{A}}_{k-h} \overline{e(a_n)} \underbrace{\text{AAA} \cdots \text{A}}_{i} \text{-5',}$$

and

$$3\text{'-} \underbrace{\text{AAA} \cdots \text{A}}_{k-i} \overline{e(a_{n+1})} \underbrace{\text{AAA} \cdots \text{A}}_{j} \text{-5'}$$

together make

$$3\text{'-} \underbrace{\text{AAA} \cdots \text{A}}_{k-h} \overline{e(a_n)} \underbrace{\text{AAA} \cdots \text{A}}_{k} \overline{e(a_{n+1})} \underbrace{\text{AAA} \cdots \text{A}}_{j} \text{-5'.}$$

Thus, the subsequence $\underbrace{\text{AAA} \cdots \text{A}}_{k}$ plays a role of "joint" between two consecutive state-transitions and it guarantees for the two transitions to be valid in M.

2.2 Designing Laboratory Protocols to Execute Finite Automata in Test Tube

In order to practically execute the laboratory experiments for the method described in the previous section, we design the following experimental laboratory protocol, which is also illustrated in Fig. 1:

0. Encoding: Encode an input string into a long ssDNA, and state-transition functions and supplementary sequences into short pieces of complementary ssDNAs.

1. Hybridization: Put all those encoded ssDNAs together into one test tube, and anneal those complementary ssDNAs to be hybridized.

2. Ligation: Put DNA "ligase" into the test tube and invoke ligations at temperature of 37 degree. When two ssDNAs encoding two consecutive valid state-transitions $\delta(h, a_n) = i$ and $\delta(i, a_{n+1}) = j$ are hybridized at adjacent positions on the ssDNA of the input string, these two ssDNAs are ligased and concatenated.

3. Denature and extraction by affinity purification: Denature double-stranded DNAs into ssDNAs and extract concatenated ssDNAs containing biotinylated probe subsequence by streptavidin-biotin bonding with magnetic beads.

4. Amplification by PCR: Amplify the extracted ssDNAs with PCR primers.

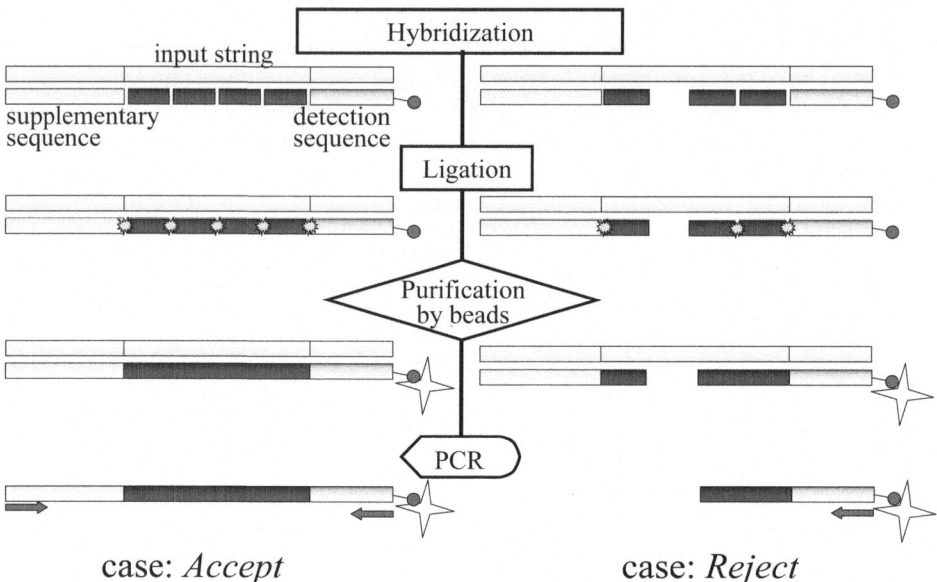

Fig. 1. The flowchart of laboratory protocol to execute *in vitro* finite automata which consists of five steps: hybridization, ligation, denature and extraction by affinity purification, amplification by PCR, and detection by gel-electrophoresis. The acceptance of the input string by the automata is the left case, and the rejection is the right case.

5. **Detection by gel-electrophoresis:** Separate the PCR products by length using gel-electrophoresis and detect a particular band of the full-length. If the full-length band is detected, that means a completely hybridized double-strand DNA is formed, and hence the finite automaton "accepts" the input string. Otherwise, it "rejects" the input string. In our laboratory experiments, we have used a "capillary" electrophoresis microchip-based system, called Bioanalyser 2100 (Agilent Technologies), in place of conventional gel-electrophoresis. The capillary electrophoresis is of higher resolution and more accurate than gel electrophoresis such as agarose gel.

Further, we have carefully designed DNA sequences encoding symbols, states, probes for affinity purification, PCR primers as follows. Two main factors for designing those DNA sequences are (1) Tm (melting temperature) to avoid mishybridizations and (2) to avoid forming secondary structures:

	DNA sequence	Tm
symbol '0'	GACGTTGGATGTGGG	50.165
symbol '1'	GCGTGTACGATGCAG	51.523
state	AAGCAGTTTT	23.641
probe	CTGGTTGCTTGTCCC	50.344
PCR primers	GCGTCTTGGTTGCTGAAATG	58.521
	CCGACTTCGTACGAGATTAG	55.481

3   Experiments

We have carried laboratory experiments on various finite automata of from 2 states to 6 states for several input strings.

3.1   2-States Automaton with Two Input Strings

Our experiments begin with a simple two-states automaton shown in Fig. 2 (left) with two input strings, (a) "1101" and (b) "1010". The language accepted by this automaton is $(10)^+$, and hence the automaton accepts the string 1010 and rejects the other string 1101.

The results of electrophoresis by Bioanalyser are displayed in the form of electropherogram (as shown in Fig. 3) which plots standard curve of migration time against DNA size where the x-axis is migration time and the y-axis is fluorescence intensity. They can also be displayed in gel-like image (as shown in Fig. 2 (right)). For these two input strings, the full-length DNA is of 190 bps (mer). Hence, if a band at position of 190 mer is detected in the result

Fig. 2. (Left:) A simple 2-states automaton used for this experiment. (Right:) The results of electrophoresis are displayed in gel-like image. Lane (a) is for the input string 1101, and lane (b) for 1010. Since the full-length band (190 mer) is detected only in lane (b), we determine the automaton accepts only the input string (b) 1010.

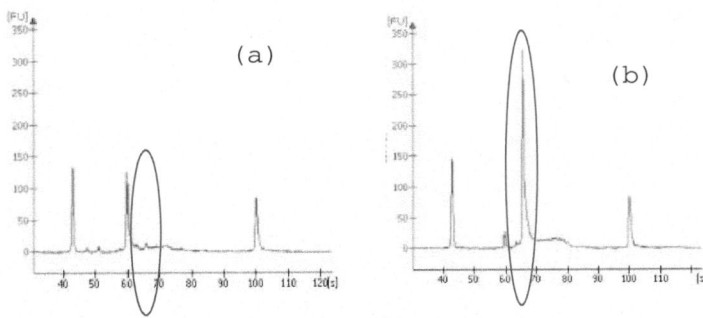

Fig. 3. The results of electrophoresis are displayed in the form of electropherogram where the vertical axis is fluorescence intensity. Peaks at the full-length position are marked with circles. Plot (a) is for the input string 1101, and plot (b) for 1010. A strong peak at the full-length position is detected only in plot (b), and hence we determine the automaton accepts the input string (b) 1010.

of electrophoresis, that means a completely hybridized double-strand DNA is formed, and hence the finite automaton "accepts" the input string.

Both figures 2 and 3 clearly show that our *in vitro* experiments have successfully identified the correct acceptance of this automaton for two input strings, and hence correctly executed the computation process of the automaton *in vitro*.

3.2   4-States Automaton with Three Input Strings

Our second experiment attempts 4-states automaton shown in Fig. 4 (upper left) for the three input strings (a) 1101, (b) 1110, and (c) 1010. This 4-states automaton accepts the language $(1(0\cup1)1)^* \cup (1(0\cup1)1)^*0$, and hence it accepts 1110 and 1010 and rejects 1101.

Fig. 4. (Left:) A 4-states automaton used for this experiment. (Right:) The results of electrophoresis are displayed in gel-like image. Lane (a) is for the input string 1101, lane (b) for 1110, and lane (c) for 1010. Since the full-length band (190 mer) is detected in lane (b) and (c), we determine the automaton accepts two input strings (b) 1110 and (c) 1010.

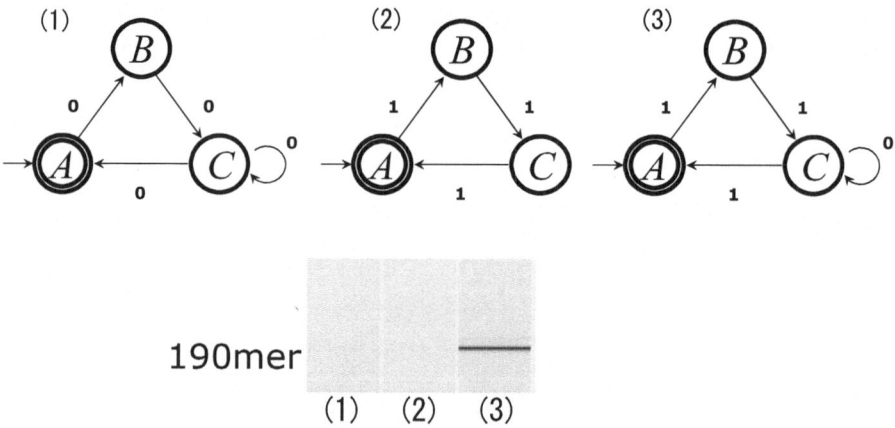

Fig. 5. (Upper:) Three different types of 3-states automata used for this experiment. (Lower:) The results of electrophoresis are displayed in gel-like image. Lane (1) is for the automaton (1), lane (2) for the automaton (2), and lane (3) for the automaton (3). Since the full-length band (190 mer) is detected only in lane (3), we determine the automaton (3) accepts the input string 1101.

The results are shown in Fig. 4 (upper right) in gel-like image. As in the first experiment, the full-length DNA is of 190 bps (mer). Bands at position of 190 mer is detected in lane (b) and lane (c). Hence, our *in vitro* experiments have successfully detected that the automaton accepts two input string (b) 1110 and (c) 1010.

3.3 Three 3-States Automata with One Input String

In this experiment, we execute three different types of 3-states automata shown in Fig. 5 (upper) for one input string "1101". The automaton (1) accepts the language 000*0, (2) accepts (111)*, and (3) accepts (110*1)*. Hence, the automaton (3) only accepts the input string 1101.

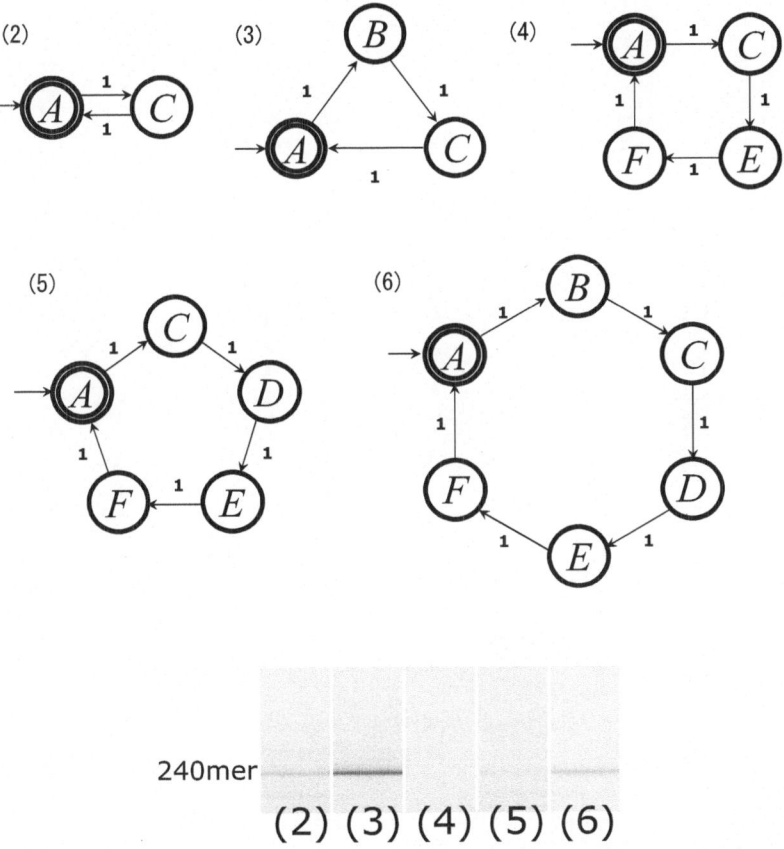

Fig. 6. (Upper:) Five different automata of from 2 states to 6 states used for this experiment. (Lower:) The results of electrophoresis are displayed in gel-like image. Lane (2) is for the automaton (2), (3) for (3), (4) for (4), (5) for (5), and (6) for (6). Since the full-length bands (240 mer) are detected in lane (2), (3) and (6), we determine the automata (2), (3) and (6) accepts the input string 111111.

The results are shown in Fig. 5 (lower) in gel-like image. Again, the full-length DNA is of 190 bps (mer). A band at position of 190 mer is detected only in lane (3). Hence, in our *in vitro* experiments, the automaton (3) has correctly accepted the input string 1101 and the automaton (1) and (2) have correctly rejected 1101.

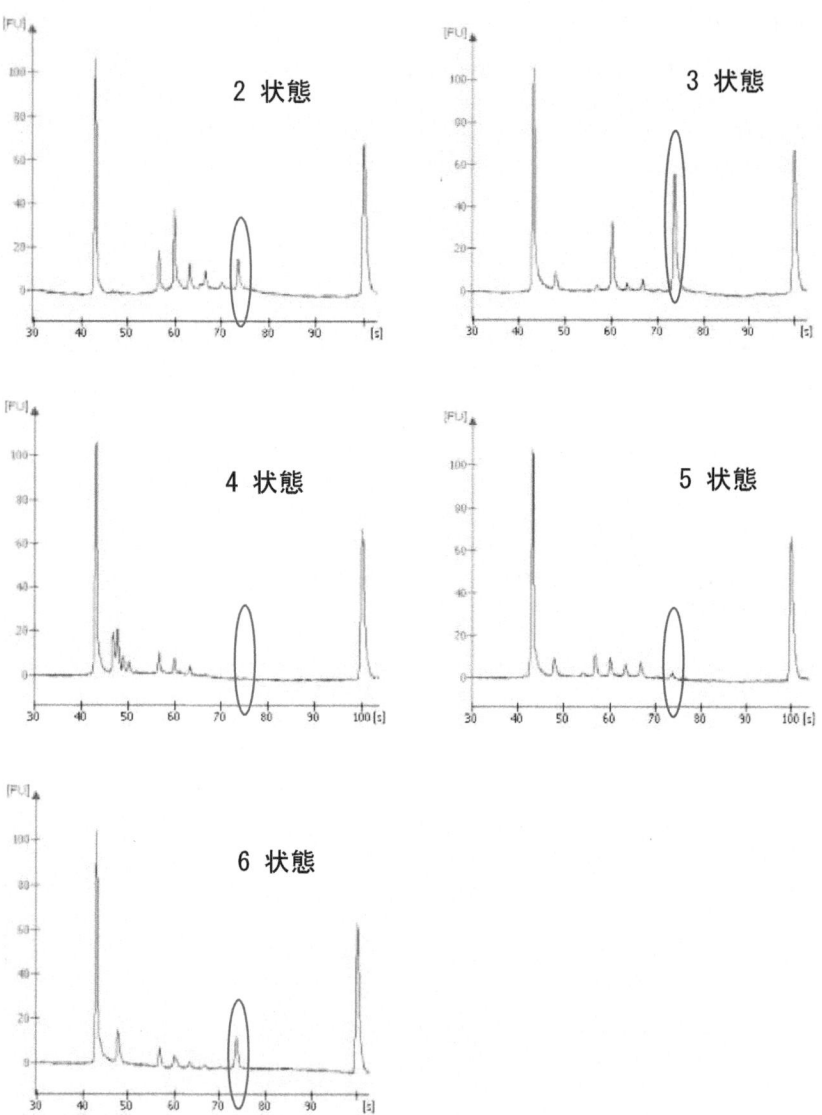

Fig. 7. The results of electrophoresis are displayed in the form of electropherogram. Peaks at the full-length position are detected in plot (2), (3) and (6).

3.4   From 2-States to 6-States Automata with One Input String "111111" of Length 6

Our final experiments are 5 different automata of from 2 states to 6 states shown in Fig. 6 (upper) for one input string "111111" of length 6. The automaton (2) accepts the language (11)*, that is, strings with even numbers of symbol '1', (3) accepts the language (111)*, strings repeating three times of 1s, (4) accepts the language (1111)*, strings repeating four times of 1s, (5) accepts the language (11111)*, strings repeating five times of 1s, (6) accepts the language (111111)*, strings repeating six times of 1s. Since 6 is a multiple of 2, 3 and 6, the automata (2), (3) and (6) accept the input string 111111 of length 6.

The results are shown in Fig. 6 (lower) in gel-like image and in Fig. 7 in the form of electropherogram. For the input string 111111, the full-length DNA is of 240 bps (mer). Bands at position of 240 mer are detected in lanes (2), (3) and (6) in Fig. 6, and strong peaks at the full-length position are also detected in plot (2), (3), and (6) in Fig. 7. Hence, in our *in vitro* experiments, the automaton (2), (3) and (6) have correctly accepted the input string 111111 and the automaton (4) and (5) have correctly rejected 111111.

4   Discussions

Since the full-length ssDNAs contain repeated patterns on DNA sequences, PCR amplifications with such templates produce many unexpected and unnecessary products which become obstacles for the precise detections using electrophoresis. The use of fluorescence tags is a possible solution for this problem.

An interesting future work is execute our *in vitro* automata for multiple input strings in parallel.

Acknowledgements

This work is supported in part by Grant-in-Aid for Scientific Research on Priority Area No. 14085205. This work was also performed in part through Special Coordination Funds for Promoting Science and Technology from the Ministry of Education, Culture, Sports, Science and Technology, the Japanese Government, and a grant of Keio Leading-edge Laboratory of Science and Technology (KLL) specified research projects.

References

1. Benenson, Y., T. Paz-Ellzur, R. Adar, E. Keinan, Z. Livneh, and E. Shapiro. Programmable and autonomous computing machine made of biomolecules. *Nature*, 414, 430–434, 2001.
2. Păun, Gh., G. Rozenberg, and A. Salomaa. *DNA Computing*. Springer-Verlag, Heidelberg, 1998.

3. Sakakibara, Y. and T. Hohsaka. In Vitro Translation-based Computations. *Proceedings of 9th International Meeting on DNA Based Computers*, Madison, Wisconsin, 175–179, 2003.
4. Yokomori, T., Y. Sakakibara, and S. Kobayashi. A Magic Pot : Self-assembly computation revisited. *Formal and Natural Computing*, LNCS 2300, Springer-Verlag, 418–429, 2002.

Development of an *In Vivo* Computer Based on *Escherichia coli*

Hirotaka Nakagawa[1], Kensaku Sakamoto[2], and Yasubumi Sakakibara[1]

[1] Keio University, Department of Biosciences and Informatics,
3-14-1 Hiyoshi, Kohoku-ku, Yokohama, 223-8522, Japan
yasu@bio.keio.ac.jp
[2] RIKEN Genomic Sciences Center,
1-7-22 Suehiro-cho, Tsurumi, Yokohama, 230-0045, Japan
sakamoto@gsc.riken.jp

Abstract. We present a novel framework to develop a programmable and autonomous *in vivo* computer using *E. coli*, and implement *in vivo* finite-state automata based on the framework by employing the protein-synthesis mechanism of *E. coli*. Our fundamental idea to develop a programmable and autonomous finite-state automata on *E. coli* is that we first encode an input string into one plasmid, encode state-transition functions into the other plasmid, and introduce those two plasmids into an *E. coli* cell by electroporation. Second, we execute a protein-synthesis process in *E. coli* combined with four-base codon techniques to simulate a computation (accepting) process of finite automata, which has been proposed for *in vitro* translation-based computations in [8]. This approach enables us to develop a programmable *in vivo* computer by simply replacing a plasmid encoding a state-transition function with others. Further, our *in vivo* finite automata are autonomous because the protein-synthesis process is autonomously executed in the living *E. coli* cell. We show some successful experiments to run an *in vivo* finite-state automaton on *E. coli*.

1   Introduction

The finite-state automata (machines) are the most basic computational model in Chomsky hierarchy and are the start point to build universal DNA computers. Several works have attempted to develop finite automata *in vitro*. However, there have been no experimental research works which attempt to build a finite automaton *in vivo*.

We have previously proposed a method using the protein-synthesis mechanism combined with four-base codon techniques to simulate a computation (accepting) process of finite automata *in vitro* [8] (a codon is normally a triplet of base, and different base triplets encode different amino acids in protein). The proposed method is quite promising and has several advanced features such as the protein-synthesis process is very accurate and overcomes mis-hybridization problem in the self-assembly computation and further offers an autonomous computation. Our aim was to extend this novel principle into a living system, by employing the *in vivo* protein-synthesis mechanism of *E. coli*. This *in vivo* computation

A. Carbone and N.A. Pierce (Eds.): DNA11, LNCS 3892, pp. 203–212, 2006.

possesses the following two novel features, not found in any previous biomolecular computer. First, an *in vivo* finite automaton is implemented in a living *E. coli* cell; it does not mean that it is executed simply by an incubation at a certain temperature. Second, this automaton increases in number very rapidly according to the bacterial growth; one bacterial cell can multiply to over a million cells overnight. The present study explores the feasibility of *in vivo* computation.

The main feature of our *in vivo* computer based on *E. coli* is that we first encode an input string into one plasmid, encode state-transition functions into the other plasmid, and transform *E. coli* cells with these two plasmids by electroporation. Second, we execute a protein-synthesis process in *E. coli* combined with four-base codon techniques to simulate a computation (accepting) process of finite automata, which has been proposed for *in vitro* translation-based computations in [8]. The successful computations are detected by observing the expressions of a reporter gene linked to mRNA encoding an input data. Therefore, when an encoded finite automaton accepts an encoded input string, the reporter gene, *lacZ*, is expressed and hence we observe a blue color. When the automaton rejects the input string, the reporter gene is not expressed and hence we observe no blue color. Our *in vivo* computer system based on *E. coli* is illustrated in Fig. 1.

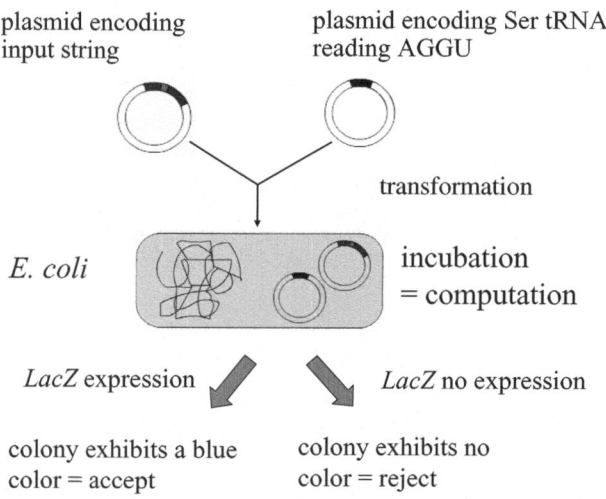

Fig. 1. The framework of our *in vivo* computer system based on *E. coli*

Thus, our *E. coli*-based computer enables us to develop a programmable and autonomous computer. To our knowledge, this is the first experimental development of *in vivo* computer and has succeeded to execute an finite-state automaton on *E. coli*.

2   Methods

2.1   A Framework of Programmable and Autonomous *In Vivo* Computer on *E. coli*

Two important issues on developing DNA-based computers are *programmable* and *autonomous*. We realize these two mechanisms by using the main features of our *in vivo* computer based on *E. coli*.

Programmable: The programmable means that a program is stored as a data (i.e., stored program computer) and any computation can be accomplished by just choosing a stored program. In DNA-based computers, it requires that a program is encoded into a molecule different from the main and fixed units of DNA computer, a molecule encoding programs can be stored and changed, and a change of molecules encoding programs accomplishes any computations.

The main features of our *in vivo* computer enable us to develop a programmable *in vivo* computer. We simply replace a plasmid encoding a state-transition function with other plasmid encoding a different state-transition function, and the *E. coli* cell transformed a new plasmid computes a different finite automaton.

Autonomous: The autonomous DNA computers mean that once we set a program and an input data and start a computation, the entire computational process is carried out without any operations from the outside. Our *in vivo* finite automata are autonomous in the sense that the protein-synthesis process which corresponds to a computation (accepting) process of an encoded finite automata is autonomously executed in a living *E. coli* cell and require no laboratory operations from the outside.

2.2   Simulating Computation Process of Finite Automata Using Four-Base Codons and Protein-Synthesis Mechanism

Sakakibara and Hohsaka [8] have proposed a method using the protein-synthesis mechanism combined with four-base codon techniques to simulate a computation (accepting) process of finite automata. An important objective of this paper is to execute the proposed method on *E. coli* in order to improve the efficiency of the method and further develop a programmable *in vivo* computer. We describe the proposed method using an example of simple finite automaton, illustrated in Fig. 3, which is of two states $\{s_0, s_1\}$, defined on one symbol '1', and accepts input strings with even numbers of symbol 1 and rejects input strings with odd numbers of 1s.

The input symbol '1' is encoded to the four-base subsequence AGGU and an input string is encoded into an mRNA by concatenating AGGU and A alternately and adding AAUAAC at the 3'-end. This one-nucleotide A in between AGGU is used to encode two states $\{s_0, s_1\}$, which is a same technique presented in [9]. For example, a string "111" is encoded into an mRNA:

AGGU A AGGU A AGGU AAAUAAC.
    1   1   1

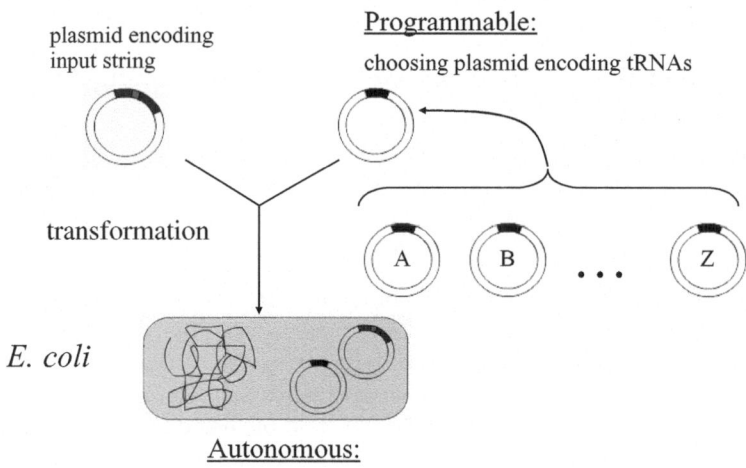

Fig. 2. A programmable and autonomous *in vivo* computer system based on *E. coli*

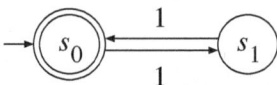

Fig. 3. A simple finite automaton of two states $\{s_0, s_1\}$, defined on one symbol '1', and accepting input strings with even numbers of symbol 1 and rejecting input strings with odd numbers of 1s

(This encoding will be replaced with other four-base encoding in real laboratory experiments because of the translation efficiency.) The four-base anticodon (3')UCCA(5') of tRNA encodes the transition rule $s_0 \xrightarrow{1} s_1$, that is a transition from state s_0 to state s_1 with input symbol 1, and the combination of two three-base anticodons (3')UUC(5') and (3')CAU(5') encodes the rule $s_1 \xrightarrow{1} s_0$. Further, the encoding mRNA is linked to *lacZ*-coding RNA subsequence as a reporter gene for the detection of successful computations. Together with these encodings and tRNAs containing four-base anticodon (3')UCCA(5'), if a given mRNA encodes an input string with odd numbers of symbol 1 , an execution of the *in vivo* protein-synthesis system stops at the stop codon, which implies that the finite automaton does not accept the input string, and if a given mRNA encodes even numbers of 1s, the translation goes through the entire mRNA and the detection of acceptance is found by the *blue* signal of *lacZ*. Examples of accepting processes are shown in Fig. 4: (Upper) For an mRNA encoding a string "1111", the translation successfully goes through the entire mRNA and translates the reporter gene of *lacZ* which emits the blue signal. (Lower) For an mRNA encoding a string "111", the translation stops at the stop codon UAA, does not reach to the *lacZ* region and produces no blue signal.

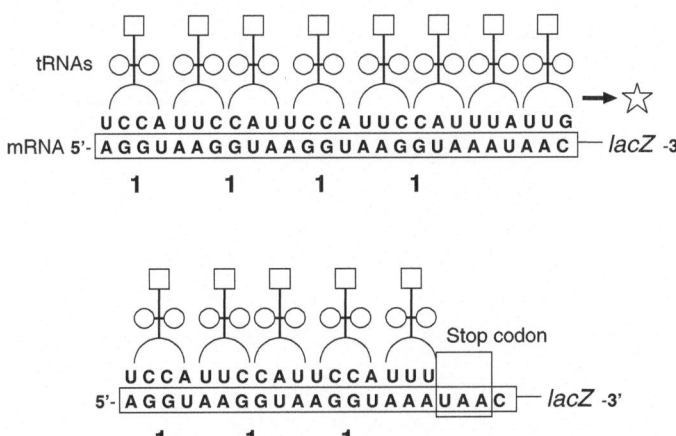

Fig. 4. Examples of accepting processes: (Upper) For an mRNA encoding a string "1111", the translation successfully goes through the mRNA and translates the reporter gene of *lacZ* emitting the blue signal. (Lower) For an mRNA encoding a string "111", the translation stops at the stop codon UAG, does not reach to the *lacZ* region and produces no blue signal.

If the competitive three-base anticodon (3')UCC(5') comes faster than the four-base anticodon (3')UCCA(5'), the incorrect translation (computation) immediately stops at the following stop codon UAA.

2.3 Designing Laboratory Protocols Using *E. coli*

In order to practically execute the laboratory experiments for our *in vivo* finite automata described in the previous sections, we have designed the following details of laboratory protocol. For the translation efficiency, we use tRNA with "UCCU" four-base anticodon in place of "UCCA".

(1) Construction of plasmid for tRNA with UCCU anticodon. The gene encoding a serine-inserting frameshift suppressor tRNA (designated as FS-Sup tRNA) [6] was generated by polymerase chain reaction (PCR) with four synthetic oligomers shown in Table 1. This PCR was performed using Pyrobest DNA polymerase (Takara Shuzo, Kyoto, Japan) and Gene Amp PCR System 2700 (ABI). The PCR product, after treated with MicroSpin Columns (QIAGEN), was digested with BamHI and Eco52I, and was then ligated into the corresponding sites of a derivative of pACYC184, by using Ligation kit ver.1 (Takara), to create plasmid pFSSuptRNA. This derivative of pACYC184 contains the lpp promoter before the BamHI site and the rrnC terminator after the Eco52I site. The use of these promoter and terminator for expressing tRNA in *E. coli* has been reported in [7]. *E. coli* MV1184 ElectroCells (Takara a) was transformed by electroporation with pFSSuptRNA and incubated in SOC medium at 37°C. The cells were then transferred onto LB Lennox plates (Nacarai) containing chloramphenicol (Wako) of 25 μg/ml to be inoculated at 37°C overnight. To extract

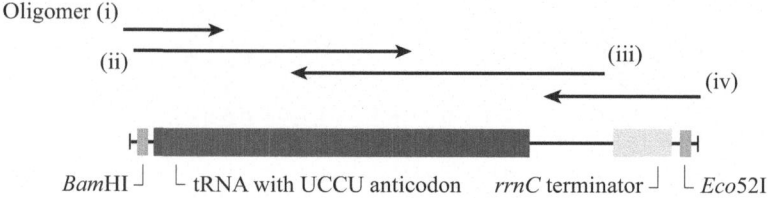

Fig. 5. Construction of FSSup tRNA with UCCU anticodon

Table 1. Oligomers for frameshift suppressor tRNA

Oligomer	Sequence
(i)	(5') CACACAGGATCCCCGTGGAGAGATGC (3')
(ii)	(5') GGATCCCCGTGGAGAGATGCCGGAGCGGCTGAACGGACCGGTCTTCCT
	AAACCGGAGTAGGGGCAAC (3')
(iii)	(5') GCTTTCGCTAAGGATCGTCGACTTTGGCGGAGAGAGGGGGATTTGAAC
	CCCCGGTAGAGTTGCCCCTACTCCGGTTTAG (3')
(iv)	(5') CACACACGGCCGTAAAAAAAATCCTTAGCTTTCGCTAAGGATCGTCG (3')

the plasmid, the cells from one colony were transplanted to LB Lennox medium of 1.5 ml containing chloramphenicol 25 μg/ml and cultured overnight at 37°C. Plasmid DNA from the cells was extracted by using QIAprep Spin Miniprep kit (Qiagen). Finally, the sequence of the FSSup tRNA gene was confirmed by sequencing using the standard dideoxy method.

(2) Plasmid carrying an encoded input string. DNA fragment carrying an encoded input string was made by annealing two oligomers, phosphorylated by T4 polynucleotide kinase (Toyobo), in an H buffer (Toyobo) with a thermal program of 95°C 2 mim followed by slow cooling to room temperature. The obtained fragment had overhanging bases at either end to be ligated into the PstI-XbaI sites pUC19 (Takara) (See Fig. 6). Amplification and sequence confirmation of this plasmid, pUC19 with the encoded input string, was performed as described in (1) except for a use of ampicillin (Nacarai) of 50 μg/ml in place of chloramphenicol.

(3) Cell preparation for electroporation. *E. coli* MV1184 with pFSSup-tRNA was cultured overnight in LB Lennox (3ml). This overnight culture was

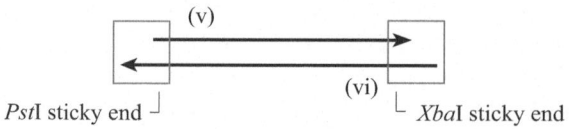

Fig. 6. Encoded input string

Table 2. Oligomers for an encoded input string

Oligomer	Sequence
(v)	(5') GC<u>AGGTA</u> ··· <u>AGGTA</u>AATAACACT (3')
	AGGTA×n
(vi)	(5') CTAGAGTGTTATT<u>TACCT</u> ··· <u>TACCT</u>GCTGCA (3')
	TACCT×n

added to LB Lennox (150 ml) containing chloramphenicol 25 μg/ml, and inoculated at 37°C until OD$_{595}$ becomes $0.6 \sim 0.8$. The fresh culture, thus prepared, was cooled on ice. Then, the culture was centrifuged at 5,000Xg for 15 min at 4°C, and then the supernatant was discarded and the pellet was re-suspended in cold water. This step of cell wash was repeated. The pellet thus obtained was suspended in 10 % glycerol, and was then centrifuged at 5,000Xg for 15 min at 4°C. After discarding the supernatant, the pellet was suspended in 10 % glycerol again. This cell suspension was applied to flash freezing with liquid nitrogen for store at -80°C.

E.coli MV1184 with pACYC184 instead of pFSSuptRNA was similarly treated for preparing cells for electroporation. MV1184 with pACYC184 was used for a control experiment.

(4) Calculation. The cells prepared in (3) were transformed with the plasmids carrying an encoded input string by electroporation. The transformed cells were added together with IPTG and X-Gal onto LB Lennox plates containing chloramphenicol of 25 μg/ml and ampicillin of 50 μg/ml. The cells were grown at 37°C overnight.

3 Experiments

We have done some laboratory experiments by following the laboratory protocols presented in Section 2.3 to execute the finite automaton shown in Fig. 3, which is of two states $\{s_0, s_1\}$, defined on one symbol '1', and accepts input strings with even numbers of symbol 1 and rejects input strings with odd numbers of 1s.

We tested our method for six input strings, "1", "11", "111", "1111", "11111", and "111111", to see whether the method correctly accepts the input string "11", "1111", "111111", and rejects the strings "1", "111", "11111".

The results are shown in Fig. 7. Blue-colored colonies which indicates the expression of *lacZ* reporter gene have been observed only in the plates for the input strings 11, 1111, and 111111. Therefore, our *in vivo* finite automaton has succeeded to correctly compute the six input strings, that is, it correctly accepts the input strings 11, 1111, 111111 of even numbers of symbol '1' and correctly rejects 1, 111, 11111 of odd number of 1s. To our knowledge, this is the first experimental development of *in vivo* computer and has succeeded to execute an finite-state automaton on *E. coli*.

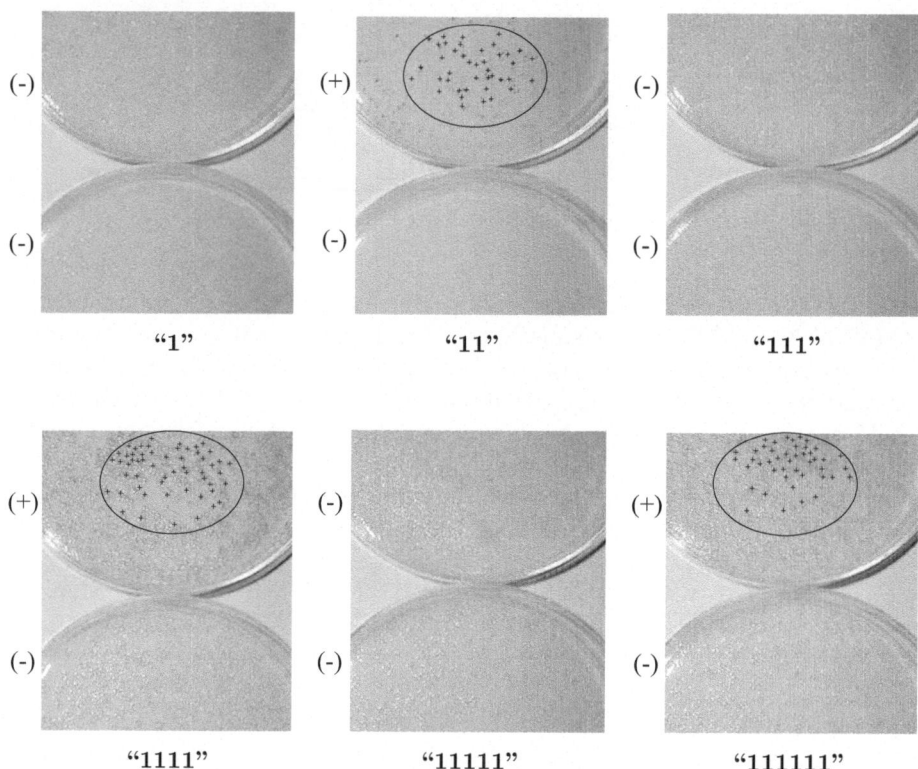

Fig. 7. Computation by the *E. coli* cells with plasmids of the input strings: 1, 11, 111, 1111, 11111, 111111. In each panel, the upper plate (part of a LB plate) shows the result in the presence of the suppressor tRNA with UCCU anticodon in the cell, while the lower plate shows the result of control experiment with no suppressor tRNA expressed. The signs (+) and (-) indicate the theoretical values about the expressions of *lacZ* reporter gene: (+) means that the cultured *E. coli* cells must express *lacZ* theoretically, and (-) means it must not express. Circles indicate the blue-colored colony expressing *lacZ*. Therefore, our *in vivo* finite automaton has correctly computed the six input strings, that is, it correctly accepts the input strings 11, 1111, 111111 of even numbers of symbol '1' and correctly rejects 1, 111, 11111 of odd number of 1s.

4   General Theory to Implement Finite Automata Using *n*-Base Codons

A general theory to implement any kinds of finite automata and any input strings on any alphabet is described as follows.

First, in theory, we assume that *n*-base codons (for arbitrary $n = 3, 4, 5, \ldots$), tRNAs containing the complementary *n*-base anticodons, and the *in vivo* protein-synthesis mechanism are available.

Next, we implement a finite automaton using *n*-base codons and some specific encodings. Let $M = (Q, \Sigma, \delta, q_0, F)$ be a (deterministic) finite automaton, where

Q is a finite set of states numbered from 0 to k, Σ is an alphabet of input symbols, δ is a state-transition function such that $\delta : Q \times \Sigma \longrightarrow Q$, q_0 is the initial state, and F is a set of final states.

For the alphabet Σ, we encode each symbol a in Σ into a DNA subsequence, denoted $e(a)$, of fixed length. For an input string w on Σ, we encode $w = x_1 x_2 \cdots x_m$ into the following DNA subsequence, denoted $e(w)$:

$$e(x_1) \underbrace{\mathtt{AA \ldots A}}_{k \text{ times}} e(x_2) \underbrace{\mathtt{AA \ldots A}}_{k \text{ times}} \ldots e(x_m) \underbrace{\mathtt{AA \ldots A}}_{k \text{ times}}$$

For the state-transition function from state q_i to state q_j with input symbol $a \in \Sigma$, we encode $\delta(q_i, a) = q_j$ into tRNA containing the following anticodon:

$$(3') \underbrace{\mathtt{UU \ldots U}}_{i \text{ times}} c(e(a)) \underbrace{\mathtt{UU \ldots U}}_{k-j \text{ times}} (5')$$

where $c(y)$ denotes the complementary sequence of y. Thus, we represent each state in Q by the length of DNA sequence. This is the same technique presented in [9]. Finally, we add some specific DNA subsequence containing stop codons at the 3'-end of the encoding sequence $e(w)$. This is for the *in vivo* protein-synthesis system to stop a translation if the finite automaton does not accept an input string.

It would be easy to see that the protein-synthesis mechanism of *E. coli* with these specific encodings of the input string and tRNAs containing the anticodons encoding the state-transition function will simulate the computation process of a target finite automaton.

In practice, several four-base anticodons such as AUCU, GGGA and GAUC are executable [6] and some five-base anticodons [1] have been proved in laboratory experiments.

5   Discussions

The presented experiments of our *in vivo* finite automata based on *E. coli* propose a kind of population computations in the following two senses: (1) While a computation by one single *E. coli* cell is not effective and accurate, a colony consisting of a large number of *E. coli* cells provides a reliable computation. (2) Since one bacterial cell can multiply to over a million cells overnight, our *in vivo* computation framework offers a massively parallel computation. Further, our *in vivo* finite automata have a quite distinguished feature that an *in vivo* finite automaton is implemented in a living *E. coli* cell; it is not implemented simply by an incubation at a certain temperature.

Acknowledgements

This work was also performed in part through Special Coordination Funds for Promoting Science and Technology from the Ministry of Education, Culture,

Sports, Science and Technology, the Japanese Government, and a grant of Keio Leading-edge Laboratory of Science and Technology (KLL) specified research projects.

References

1. Anderson, J. C., T. J. Magliery, and P. G. Schultz. Exploring the limits of codon and anticodon size. *Chemistry & Biology*, 9, 237–244, 2002.
2. Benenson, Y., T. Paz-Ellzur, R. Adar, E. Keinan, Z. Livneh, and E. Shapiro. Programmable and autonomous computing machine made of biomolecules. *Nature*, 414, 430–434, 2001.
3. Bishop, R. E., B. K. Leskiw, R. S. Hodges, C. M. Kay, and J. H. Weiner. The entericidin locus of *Escherichia coli* and its implications for programmed bacterial cell death. *Journal of Molecular Biology*, 280, 583–596, 1998.
4. Hohsaka, T., Y. Ashizuka, H. Taira, H. Murakami, M. Sisido. Incorporation of nonnatural amino acids into proteins by using various four-base codons in an *Escherichia coli* in vitro translation system. *Biochemistry*, 40, 11060–11064, 2001.
5. Hohsaka, T., Y. Ashizuka, H. Murakami, M. Sisido. Five-base codons for incorporation of nonnatural amino acids into proteins. *Nucleic Acids Research*, 29, 3646–3651, 2001.
6. Magliery, T. J., J. C. Anderson, and P. G. Schultz. Expanding the genetic code: selection of efficient suppressors of four-base codons and identification of "shifty" four-base codons with a library approach in *Escherichia coli*. *Journal of Molecular Biology*, 307, 755–769, 2001.
7. Normanly, J., J. M. Masson, L. G. Kleina, J. Abelson, and J. H. Miller. Construction of two Escherichia coli amber suppressor genes. Proceeding of the National Academy of Sciences USA, 83, 6548–6552, 1986.
8. Sakakibara, Y. and T. Hohsaka. In Vitro Translation-based Computations. *Proceedings of 9th International Meeting on DNA Based Computers*, Madison, Wisconsin, 175–179, 2003.
9. Yokomori, T., Y. Sakakibara, and S. Kobayashi. A Magic Pot : Self-assembly computation revisited. *Formal and Natural Computing*, LNCS 2300, Springer-Verlag, 418–429, 2002.

Control of DNA Molecules on a Microscopic Bead Using Optical Techniques for Photonic DNA Memory

Yusuke Ogura[1,3], Taro Beppu[1], Masahiro Takinoue[2,3],
Akira Suyama[2,3], and Jun Tanida[1,3]

[1] Graduate School of Information Science and Technology, Osaka University,
2-1 Yamadaoka, Suita, Osaka 565-0871, Japan
{ogura, t-beppu, tanida}@ist.osaka-u.ac.jp
[2] Graduate School of Arts and Sciences, The University of Tokyo,
3-8-1 Komaba, Meguro-ku, Tokyo 153-8902, Japan
suyama@dna.c.u-tokyo.ac.jp, takinoue@genta.c.u-tokyo.ac.jp
[3] Japan Science and Technology Agency (JST-CREST)

Abstract. This paper focuses on a photonic DNA memory, which is a DNA memory using optical techniques. Positional information of DNA is utilized for scaling up the address space of the DNA memory. Use of the optical techniques is useful in controlling positional addresses in parallel. We performed some experiments on control of the reactions of hairpin DNA molecules on a microscopic bead. Experimental results demonstrate that operations of writing and erasing of data DNA on a bead for the photonic DNA memory can be achieved by using optical techniques.

1 Introduction

Various computations can be implemented by use of parallelism and autonomous reactions of DNA. Storing of manipulated data improves the efficiency of the computations. This means that a DNA memory, which is a memory using DNA and its reactions, is considered to be fundamental to a wide range of applications of DNA computing[1, 2, 3].

The size of DNA is a nanometer scale. This is an important characteristic for constructing a valuable memory because showing the potential of the DNA memory as a high-capacity memory. For realizing the DNA memory, the capability to store huge data using a large amount of DNA is essential. In addition to this, it is required that one can access and use arbitrary data among the stored data at his disposal.

From this viewpoint, it is important to develop a method for addressing individual data to identify them. DNA molecules are distinguished depending on their base sequences; namely, the DNA molecules have their address information inherently, and addressing with DNA base sequences is possible. However, simple use of address information that relates to base sequences requires a hard task to design the sequences for making huge address space. In addition, it is difficult to control the DNA with such a variety of base sequences accurately.

A. Carbone and N.A. Pierce (Eds.): DNA11, LNCS 3892, pp. 213–223, 2006.

Use of information that is independent from base sequences is another possible strategy for scaling up the address space of a DNA memory. The positional information of DNA is considered to be usable information for the purpose. It it not necessary to control the positions of DNA molecules individually at a nanometer scale. For identifying the individual DNA molecules, addressing with base sequences is suitable. The positional information at a micrometer scale is more effective to use than that at a nanometer scale. Combining addresses relating to the base sequences and to the positional information provides large address space. For realizing this idea, methods should be developed for operating the DNA memory in the individuals of an array of micrometer-scale volumes.

We have been studying methods for manipulating DNA at a micrometer scale by the basis of optical techniques[4, 5]. For example, we demonstrated that multiple microscopic beads, on which many DNA molecules were attached, were simultaneously translated by optical manipulation that uses vertical-cavity surface-emitting laser (VCSEL) array sources. We also succeeded in controlling reactions of DNA with the resolution of a few micrometer by irradiating with a laser beam. These optical techniques for manipulating DNA are promising to realize addressing of the DNA memory using the positional information of DNA molecules. The characteristics of DNA and light can be effectively utilized by combining the method for addressing the DNA memory using optical techniques at a micrometer scale and the method using molecular reactions at a nanometer scale. The DNA memory is controlled in spatially parallel owing to the parallelism of light propagation. The reaction parallelism of DNA is exploited at a micrometer scale.

It is important to develop a method for addressing the DNA memory by using optical techniques not only from the viewpoint of DNA computing but also from the viewpoint of optical computing. Optical computing is a computational technique for parallel information processing that uses inherent property of light such as fast propagation, parallelism, and a large bandwidth. Many interesting results were obtained with various demonstration systems, which were associated with, for example, optical interconnection and digital optics[6]. However, the diffraction limit determines the resolution of the light and often restricts the density and capacity of information that is dealt with in an optical system. The precise alignment of optical devices is necessary for manipulating the light in diffraction limited systems.

The DNA memory that uses optical techniques, which we refer to as a photonic DNA memory, gives a practical solution for these problems. For example, the diffraction limit is approximately 1 μm in a typical optical system. On the other hand, a volume of 1 μm^3 of a DNA solution with density of 10 μM contains 6×10^3 DNA molecules. When DNA molecules each includes information of 1 bit, the amount of information of the volume is an order of 10^3 bits. This suggests that the photonic DNA memory has potential for dealing with information more than that is dealt with in the diffraction limited optical system. The difficult alignment of the optical system is avoidable because DNA molecules float in a volume of the solution and react autonomously.

In this paper, we focus on the photonic DNA memory, which uses optical techniques for addressing with positional information. We studied a method for controlling reactions of hairpin DNA on a microscopic bead by laser irradiation. Experimental results of the operations of writing and erasing on a bead are shown.

2   Photonic DNA Memory

In photonic DNA memory, DNA with a hairpin formation, which is referred to as hairpin DNA, is immobilized on the surface of a microscopic bead. The beads are used for translating DNA. The detail of translation is described later.

A solution of our DNA memory contains hairpin DNA, tag DNA, and anti-tag DNA. The base sequence of the tag DNA is completely or partially complement to that of the hairpin DNA. Anti-tag DNA that has a sequence complement to the tag DNA is immobilized on a substrate. The reason for using hairpin DNA is to achieve two stable states. If the temperature of the solution is decreased gradually, hairpin DNA and tag DNA molecules hybridize with each other. In contrast, if the temperature of the solution is decreased rapidly, the hairpin DNA forms the hairpin formation and does not hybridize with the tag DNA.

A tag DNA molecule includes address information of the DNA memory in its base sequence. When using tag DNA by itself as data, the state that the anti-tag DNA and the tag DNA construct dsDNA on the substrate is considered as the value "1", and the other state is considered as the value "0." One can, in contrast, append an additional DNA, genome DNA, proteins, and other molecules to the tag DNA as data. In this paper, we refer to tag DNA with or without an additional molecule as data DNA.

Let T_1 and T_1' be the melting temperature of the hairpin DNA and that of the dsDNA consisting of the tag DNA and the anti-tag DNA, respectively. Let $T_2(> T_1, T_1')$ denotes the melting temperature of dsDNA consisting of the hairpin DNA and the tag DNA.

Figure 1 shows the molecular reaction behavior of the DNA memory considered in this paper. The procedure for writing and erasing operations is as follows. At the initial condition, the temperature is set to $T_0(< T_1, T_1')$ and the data DNA molecules bind to the substrate by hybridization of the data DNA and anti-tag DNA. When the temperature of the solution is increased to higher than T_2, the data DNA is detached from the substrate and floats in the solution. The temperature is decreased to T_2, then the data DNA and the hairpin DNA hybridize. After the temperature is decreased to T_0, the DNA is stable. By the method, the data DNA can be read out from the substrate and written in to the bead.

When the temperature is decreased from higher than T_2 to T_0 rapidly, the hairpin DNA forms a hairpin formation, and cannot hybridize with the data DNA. As a result, the data DNA hybridizes with the anti-tag DNA, and it is immobilized to the substrate. This means that the data DNA can be read out from the bead and written in to the substrate. If hairpin DNA molecules that

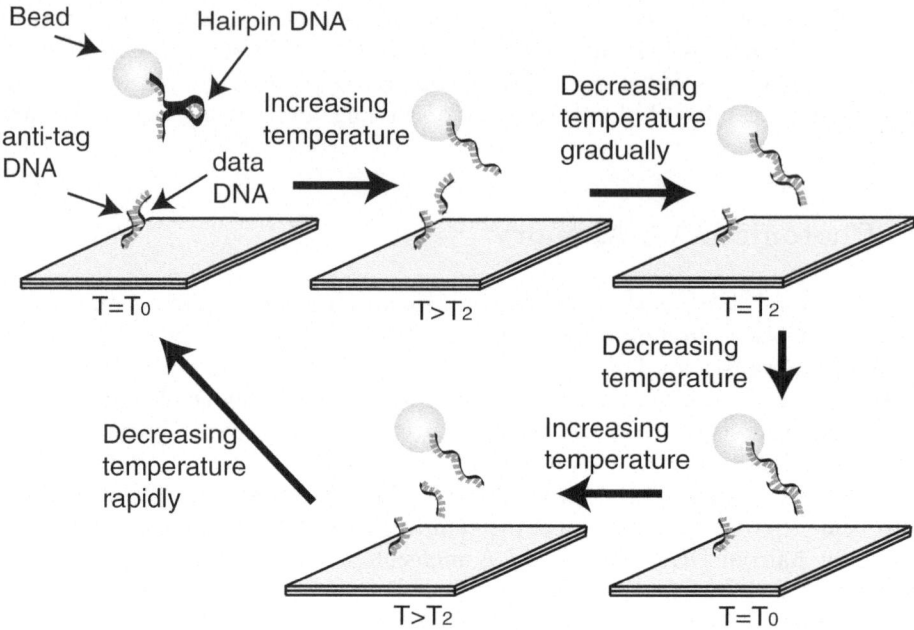

Fig. 1. The molecular reaction behavior of the photonic DNA memory

have different base sequences are immobilized to a bead, specific data DNA molecules are written in to the bead selectively due to addressing with base sequences.

Note that our scheme uses the substrate as a memory device and the positional addresses are defined on the substrate. The operations of writing and reading data DNA on beads are useful in storing a cluster of data DNA temporarily and necessary for translating the data from a position on the substrate to another position to process the data.

In the photonic DNA memory, the DNA memory is operated by using optical techniques. The scheme of a method for controlling molecular reactions of the photonic DNA memory is shown in Fig. 2. A solution containing microscopic beads is put on a substrate that is coated with a sort of material for light absorption. When the substrate is irradiated with a focused laser beam, the surface of the substrate is heated up owing to light absorption, and the temperature of the solution around irradiated area increases. By the basis of the phenomenon, the temperature of the solution can be controlled by changing the power of the beam used. The positional address of the photonic DNA memory can be used by changing the irradiating position.

It is possible to generate optical field patterns at a micrometer scale. Effective use of optical devices provides a method for generating various optical field patterns, so that the operations of the photonic DNA memory can be controlled in parallel. Operating the DNA memory at a local position using light is regarded as addressing based on positional information of the DNA memory. Different

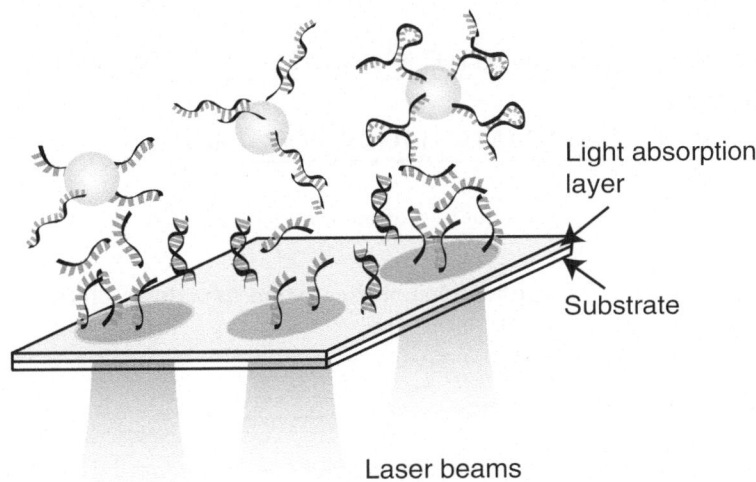

Fig. 2. The conceptual diagram of a method for controlling molecular reactions of the photonic DNA memory

positional addresses can be given at a pitch that is no more than several micrometer.

The bead with data DNA molecules can be translated by VCSEL array optical manipulation[4]. Optical manipulation is a non-contact manipulation method of an object using a radiation pressure force induced by the interaction between light and the object. A VCSEL array is high density array sources, the optical outputs of which can be controlled independently. Flexible manipulation for microscopic objects is achieved by control of spatial and temporal optical fields generated by the VCSEL array sources. The method is effective for parallel manipulation of multiple objects with compact hardware.

Use of the light in the DNA memory is effective in the following points.

1. It is possible to access to DNA memories that have different positional addresses in parallel.
2. Independent operations are executed for the DNA memories with the different positional addresses.
3. Physical interconnections are not required for flowing the data DNA.
4. Procedures of processing are programmable.

The photonic DNA memory can be applied, for example, to a programmable free-space micro-reactor array system. A variety of information is previously stored in individual reactors. Addressing with positional information of the DNA memory is performed by selecting reactors operated by optical field patterns. Operations of the DNA memory in the individual reactors are implemented by addressing with base sequences. The data DNA molecules are translated between reactors. The reactors are used as not only memories but also processing units and registers. The role of the individual reactors can be changed, so that the

Hairpin DNA
5'-biotin-ggacacggTGCAGTGTAAGCAACTATTGTCTccgtgtcc-3'

Data DNA
5'-GGACACGGAGACAATAGTTGCTTACACTGCA -3'

Fig. 3. The base sequences of hairpin DNA and data DNA

system is reconfigurable. Applications of the system include on-chip systems for genome analysis and DNA authentication.

3 · Experiments

We performed some experiments on writing and erasing of data DNA on a microscopic bead by using optical techniques. The operations on the bead are required to take data DAN selectively from the substate (DNA memory) and to use it in processing.

The base sequences of hairpin DNA and data DNA are shown in Fig. 3. In the hairpin DNA, the part of the sequence indicated with small letters is the part of a stem, and the part with capital letters forms a loop. The underlined letters of the hairpin DNA and the data DNA indicate a complementary part of the sequences. The detail of molecular reactions of the hairpin DNA and the data DNA is described in [7].

The hairpin DNA molecules which were modified with biotin at the 5'-end were mixed in a solution that contains polystyrene beads of 6 μm diameter coated with streptavidin. The hairpin DNA molecules were immobilized to the surface of the bead by biotin-streptavidin binding. The beads were extracted and put into a TE buffer solution. Fluorescence molecules (Molecular Probes, Alexa Fluor 546) were attached to the data DNA. The absorption and fluorescence emission maxima of the florescence molecules are 555 nm and 570 nm, respectively. The solution of the data DNA was mixed with the solution that contained beads with the hairpin DNA.

The substrate used in the experiments was a glass substrate that was coated with titanylphthalocyanine of $0.15 - \mu$m thickness as a layer of light absorption. The sample was irradiated from below with a beam that was generated from a semiconductor laser of a wavelength of 854 nm and focused with an objective lens (Olympus Corp., LUMPlan Fl 60×). A fluorescence microscope with a cooled CCD was used for observation.

We measured the optical power required for the operation of erasing on a bead. The data DNA and the hairpin DNA that was immobilized on the bead were annealed previously in a tube. The sample was put on the substrate and irradiated with the laser beam. An irradiation cycle consisted of the first phase of irradiating for 10 seconds and the second phase of capturing a fluorescence image for 2 seconds. This irradiation cycle was repeated. The power of the irradiation beam used was 1, 2, 3, 4, or 5 mW.

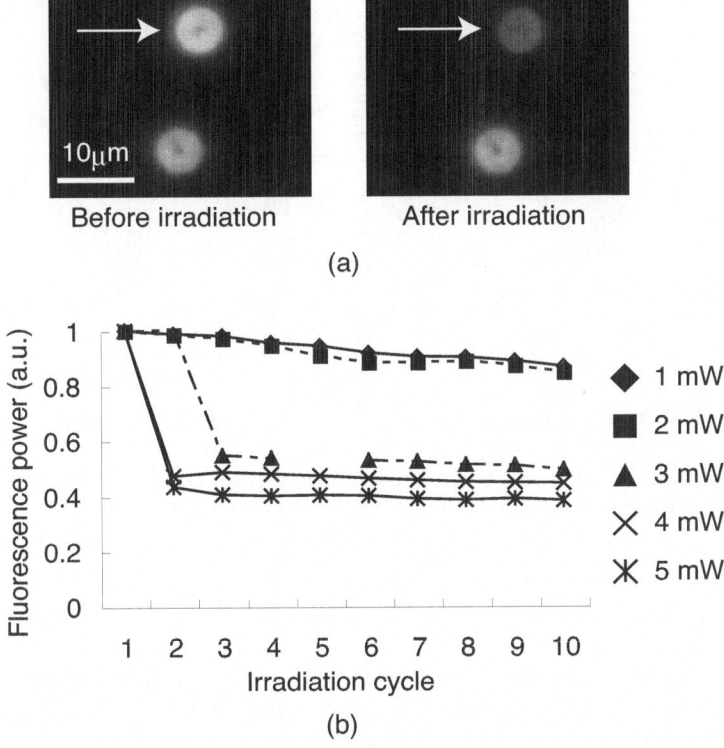

Fig. 4. (a) Fluorescence images captured before (left) and after irradiation (right). (b) The relationship between the number of irradiation cycles and fluorescence power.

If the data DNA molecules are immobilized to a bead, the fluorescence power observed around the bead is high because fluorescence molecules is concentrated on the bead. After denaturing, the data DNA dispersed in the solution, and the intensity around the bead decreases.

As an example, the fluorescence images captured before and after irradiating a bead with 5 mW for 10 seconds is shown in Fig. 4(a). Decrease of fluorescence intensity means that the data DNA was denatured by laser irradiation. Figure 4(b) shows the relationship between the number of irradiation cycles and fluorescence power. The fluorescence power was averaged values of 5 measurements. It can be seen from Fig. 4(b) that, with the irradiation power of 1 or 2 mW, the fluorescence power changes little. This result suggests that the temperature did not increase to the temperature required for denaturing. When the irradiation power was no less than 3 mW, in contrast, the fluorescence power decreased. We can conclude that the power of no less than 3 mW is required for erasing operation on the bead by denaturing the hairpin DNA and the tag DNA.

We investigated a suitable irradiating condition for writing the data DNA on a bead. The sample solution was prepared by mixing the solution of the beads including the hairpin DNA with the solution of the data DNA. The mixed

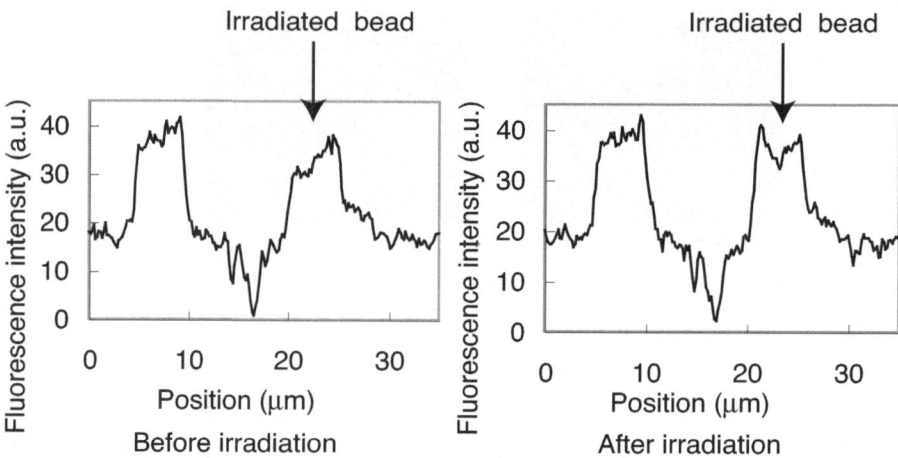

Fig. 5. Cross sections of a target bead of fluorescence images captured before (left) and after irradiating the bead with 2 mW for 30 seconds (right)

solution contained the enough data DNA for reaction. A suitable irradiation condition for hybridization was found by changing laser power, irradiation time, and other parameters.

As an example, we irradiated a target bead with a laser beam of 2 mW for 30 seconds and stopped irradiating. Cross sections of the bead of fluorescence images captured during this trial are shown in Fig. 5. The fluorescence intensity of the bead did not change, which means failure in writing.

In contrast, when a target bead was irradiated with the irradiation schedule shown in Fig. 6 (a), the obtained fluorescence images are shown in Fig. 6 (b). Figure 6 (c) shows cross sections of fluorescence images along the line indicated in Fig 6 (b). The background fluorescence intensity is removed to show these figures.

The fluorescence intensity of the irradiated bead increases obviously. This result indicates that the tag DNA hybridizes with the hairpin DNA on the bead. We succeeded in writing the data DNA on the bead using the optical technique. The fluorescence intensities of another beads around the target bead did not change, and a hybridization reaction can be controlled with the resolution of no more than $10\mu m$. If the reaction area is divided to many small areas of $10\mu m$ square, different positional addresses can be given to the individual small areas.

In the next experiment, we repeated writing and erasing operations on a bead. At the beginning, the data DNA molecules were not attached to a bead with the hairpin DNA. The following steps were repeated 3 times. Step 1: writing with the irradiation schedule shown in Fig. 6 (a), step 2: erasing by irradiating with the laser beam of 5 mW for 10 seconds.

Figure 7 shows the fluorescence power measured after the individual steps. The fluorescence power increases after step 1 and decreases after step 2. This is an expected result. Note that efficiencies of writing and erasing indicate almost the same values during 3 repetitions.

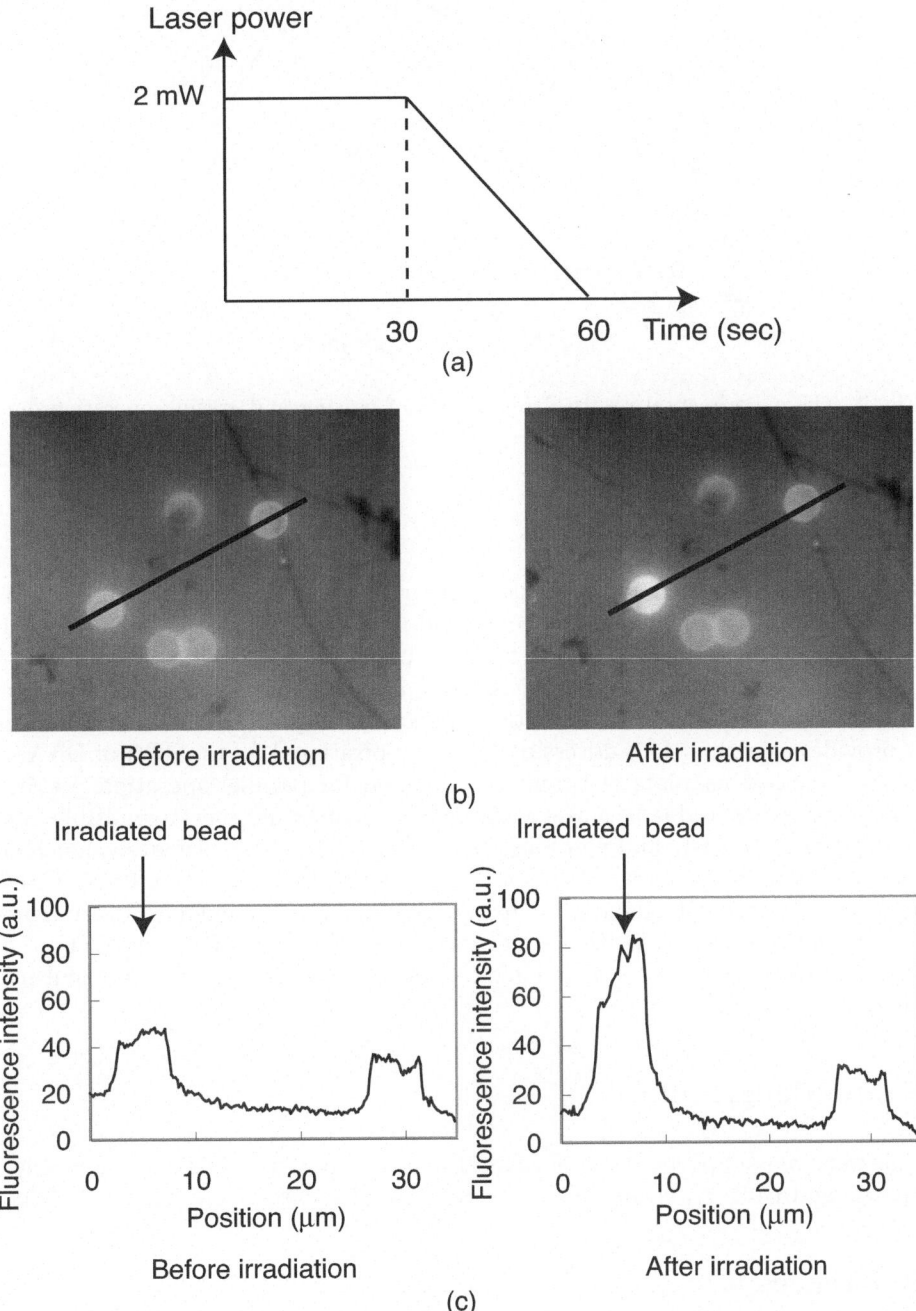

Fig. 6. Experimental results on writing data DNA on a bead. (a) Irradiation schedule, (b) fluorescence images, and (c) cross sections along the line shown in (b).

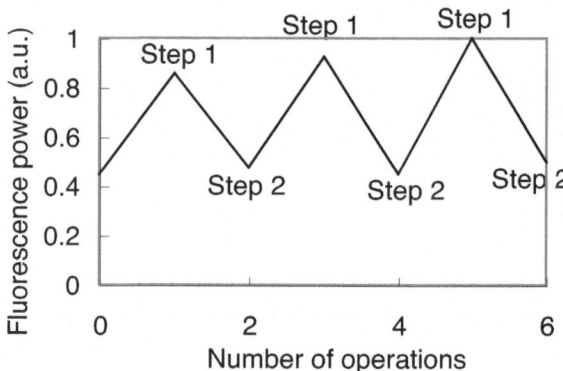

Fig. 7. The experimental result of repetitions of writing and erasing operation on a bead

4 Conclusions

We demonstrated that the operations of writing and erasing of data DNA on beads with hairpin DNA can be controlled by laser irradiation. The method is a fundamental technique for realizing the photonic DNA memory, which is a memory based on the nature of DNA and optical techniques. The use of optical techniques is effective to scale up the address space of a DNA memory because it provides a method for addressing based on positional information of DNA.

For practical use, lots of beams are required for parallel operation. VCSEL array sources are usable because, with the device, one can generate multiple laser beams simultaneously and modulate them independently. Fortunately, much effort is being made to increase the pixel number of VCSEL arrays, and the VCSEL arrays are expected to be applied to the photonic DNA memory. Future issues include optimization of writing conditions, transfer of data DNA molecules between a substrate and a bead, and demonstration of DNA memory using multiple kinds of data DNA.

Acknowledgments

This work was supported by JST CREST and the Ministry of Education, Science, Sports, and Culture, Grant-in-Aid for Scientific Research (A), 15200023, 2003.

References

1. Chen, J., Deaton, R., Wang, Y.: A DNA-based memory with *in vitro* Learning and associative recall. In: Chen, J., Reif, J. (eds.): DNA computing: 9th International Workshop on DNA Based Computers, DNA 9. Lecture Notes in Computer Science, Vol. 2943. Springer-Verlag, Berlin Heidelberg New York (2004) 145-156

2. Kameda, A., Yamamoto, M., Uejima, H., Hagiya, M., Sakamoto, K., Ohuchi, A.: Conformational addressing using the hairpin structure of single-strand DNA. In: Chen, J., Reif, J. (eds.): DNA computing: 9th International Workshop on DNA Based Computers, DNA 9. Lecture Notes in Computer Science, Vol. 2943. Springer-Verlag, Berlin Heidelberg New York (2004) 219-224

3. Takahashi, N., Kameda, A., Yamamoto, M., Ohuchi, A.: Aqueous computing with DNA hairpin-based RAM. In: Ferretti, C., Mauri, G., Zandron, C. (eds.): DNA computing: 10th International Workshop on DNA Computing, DNA 10. Lecture Notes in Computer Science, Vol. 3384. Springer-Verlag, Berlin Heidelberg New York (2005) 355-364

4. Ogura, Y., Kawakami, T., Sumiyama, F., Suyama, A., Tanida, J.: Parallel translation of DNA clusters by VCSEL array trapping and temperature control with laser illumination. In: Chen, J., Reif, J. (eds.): DNA computing: 9th International Workshop on DNA Based Computers, DNA 9. Lecture Notes in Computer Science, Vol. 2943. Springer-Verlag, Berlin Heidelberg New York (2004) 10-18

5. Ogura, Y., Sumiyama, F., Kawakami, T., Tanida, J.: Manipulation of DNA molecules using optical techniques for optically assisted DNA computing. In Dobisz, E.A., Eldada, L.A. (eds.): Nanoengineering: Fabrication, Properties, Optics, and Devices. Proceedings of SPIE, Vol. 5515. SPIE, Belligngham, WA (2004) 100-108

6. Tanida, J., Ichioka, Y.: Optical computing. In Brown, T.G., Creath, K., Kogelnik, H. (eds.): The Optics Encyclopedia, Vol. 3, Wiley-VCH, Berlin (2003) 1883-1902

7. Takinoue, M., Suyama, A.: Molecular reactions for a molecular memory based on hairpin DNA. Chem-Bio Infomatics Journal, **4** (2004) 93-100

Linearizer and Doubler: Two Mappings to Unify Molecular Computing Models Based on DNA Complementarity

Kaoru Onodera[1] and Takashi Yokomori[2]

[1] Mathematics Major, Graduate School of Education, Waseda University,
1-6-1 Nishiwaseda, Shinjyuku-ku, Tokyo 169-8050, Japan
kaoru@akane.waseda.jp
[2] Department of Mathematics, Faculty of Education and Integrated Arts
and Sciences, Waseda University, 1-6-1 Nishiwaseda, Shinjyuku-ku,
Tokyo 169-8050, Japan
yokomori@waseda.jp

Abstract. Two specific mappings called *doubler* f_d and *linearizer* f_ℓ are introduced to bridge two domains of languages. That is, f_d maps string languages into (double-stranded) molecular languages, while f_ℓ transforms in the other way around. Using these mappings, we give new characterizations for the families of sticker languages and of Watson-Crick languages, which leads to not only a unified view of the two families of languages but also a clarified view of the computational capability of the DNA complementarity. One of the results implies that any recursively enumerable language can be expressed as the projective image of $f_d(L)$ for a minimal linear language L.

1 Introduction

In the late 1990's history of theoretical research on molecular computing models, sticker systems have been proposed to model the behaviors of biomolecules with sticky ends and to investigate the computational capability of those molecules based on the biomolecular property of DNA complementary. On the other hand, almost in parallel a new type of machine model called Watson-Crick automaton was introduced and studied, which is taken as a finite state machine working on double-stranded molecules (rather than linear strings). Similarly, a sticker system was introduced as one of the generative systems by using the DNA complementarity. The above two systems have a great deal of potential to provide the promising models for DNA computings. One can find a huge amount of interesting results on a variety of families of these systems and automata in, e.g., [3].

The present paper concerns a new approach to unifying a great variety of these models of computation based on DNA complementarity. The purpose of this paper is twofold : One is to explore the computational power of annealing operations between complementary molecules in terms of notions in formal language theory. The other is to clarify the current (chaotic) landscape of a variety of existing computational models based on DNA complementarity, by providing a unified view of those models.

A. Carbone and N.A. Pierce (Eds.): DNA11, LNCS 3892, pp. 224–235, 2006.

For our purpose, we introduce two specific mappings "doubler f_d" and "linearizer f_ℓ" that can bridge the two worlds of *string* languages and of *double-stranded molecular* languages. Using these mappings, we will give new characterizations for the families of sticker languages and of Watson-Crick languages. For example, through the mapping f_d, we show that the difference between sticker systems and Watson-Crick automata is essentially reduced to the one between minimal linear and regular grammars, respectively.

2 Preliminaries

We assume the reader to be familiar with the rudiments on Watson-Crick finite automata and sticker systems as well as basic notions in formal language theory (see, e.g., [3, 4]).

For an alphabet V, $\rho \subseteq V \times V$ is a symmetric relation. We denote an element $(x_1, x_2) \in V^* \times V^*$ by $\binom{x_1}{x_2}$. Instead of using a notation $V^* \times V^*$, we often use $\binom{V^*}{V^*}$. For elements $\binom{x_1}{y_1}, \binom{x_2}{y_2} \in \binom{V^*}{V^*}$, by $\binom{x_1}{y_1}\binom{x_2}{y_2}$, we represent a double stranded molecule $(x_1 x_2, y_1 y_2) \in V^* \times V^*$.

Let $\begin{bmatrix} V \\ V \end{bmatrix}_\rho = \{ \binom{a}{b} \mid a, b \in V, \ (a, b) \in \rho \}$ and $WK_\rho(V) = \begin{bmatrix} V \\ V \end{bmatrix}_\rho^*$ (the set of all complete double stranded molecules over V including $\binom{\epsilon}{\epsilon}$).

For an element $\binom{a_1}{b_1}\binom{a_2}{b_2} \cdots \binom{a_n}{b_n} \in WK_\rho(V)$, we also write in the form $\begin{bmatrix} w_1 \\ w_2 \end{bmatrix}$, where $w_1 = a_1 a_2 \cdots a_n$, $w_2 = b_1 b_2 \cdots b_n$.

We define a set of incomplete molecules over V: $W_\rho(V) = L_\rho(V) \cup R_\rho(V) \cup LR_\rho(V)$, where

$$L_\rho(V) = \{ \boxed{\begin{array}{cc} x_1 & y_1 \\ \hline & y_2 \end{array}}, \ \boxed{\begin{array}{cc} & y_1 \\ \hline x_2 & y_2 \end{array}} \mid x_1, x_2 \in V^*, \begin{bmatrix} y_1 \\ y_2 \end{bmatrix} \in \begin{bmatrix} V \\ V \end{bmatrix}_\rho^* \},$$

$$R_\rho(V) = \{ \boxed{\begin{array}{cc} y_1 & z_1 \\ y_2 & \\ \hline \end{array}}, \ \boxed{\begin{array}{cc} y_1 & \\ y_2 & z_2 \\ \hline \end{array}} \mid z_1, z_2 \in V^*, \begin{bmatrix} y_1 \\ y_2 \end{bmatrix} \in \begin{bmatrix} V \\ V \end{bmatrix}_\rho^* \},$$

$$LR_\rho(V) = \{ \boxed{\begin{array}{ccc} x_1 & y_1 & z_1 \\ \hline & y_2 & \end{array}}, \ \boxed{\begin{array}{ccc} & y_1 & \\ \hline x_2 & y_2 & z_2 \end{array}}, \ \boxed{\begin{array}{cc} x_1 & y_1 \\ \hline & y_2 & z_2 \end{array}}, \ \boxed{\begin{array}{cc} & y_1 & z_1 \\ \hline x_2 & y_2 \end{array}} \mid$$
$$x_1, x_2, z_1, z_2 \in V^*, \begin{bmatrix} y_1 \\ y_2 \end{bmatrix} \in \begin{bmatrix} V \\ V \end{bmatrix}_\rho^+ \}.$$

Elements in $W_\rho(V)$ are called *bricks*.

[Sticker systems]
A *sticker system* is a 4-tuple $\gamma = (V, \rho, A, D)$, where V is a finite set of symbols, $\rho \subseteq V \times V$ is the complementary relation on V, $A \subseteq LR_\rho(V)$ is a finite set of axioms, and D is a finite set of elements in $W_\rho(V) \times W_\rho(V)$.

For $\gamma = (V, \rho, A, D)$ and $\alpha, \beta \in W_\rho(V)$, we write $\alpha \overset{d_\pi}{\Longrightarrow}_\gamma \beta$ (or simply $\alpha \Longrightarrow \beta$) if and only if $\beta = u\alpha v$, for some $d_\pi : (u, v) \in D$. That is, for example, in a graphical representation, it means

$$d_\pi : \left(\boxed{\begin{array}{cc} u_3 & u_2 \\ \bar{u}_2 & u_1 \end{array}} , \boxed{\begin{array}{cc} v_2 & v_3 \\ v_1 & \bar{v}_2 \end{array}} \right) = (u, v) \quad \text{and}$$

$$\alpha = \boxed{\begin{array}{ccc} \bar{u}_1 & \alpha_1 & \bar{v}_1 \\ & \bar{\alpha}_1 & \end{array}} \overset{d_\pi}{\Longrightarrow} \boxed{\begin{array}{ccccccc} u_3 & u_2 & \bar{u}_1 & \alpha_1 & \bar{v}_1 & v_2 & v_3 \\ \bar{u}_2 & u_1 & \bar{\alpha}_1 & & v_1 & \bar{v}_2 \end{array}} = \beta,$$

where u_1, u_2, v_1, v_2 and $\bar{u}_1, \bar{u}_2, \bar{v}_1, \bar{v}_2$ are complementary, respectively.

For any other types of bricks for d_π in γ, we similarly define $\overset{d_\pi}{\Longrightarrow}_\gamma$. We denote by $\Longrightarrow^*$ the reflexive and transitive closure of $\Longrightarrow$.

A set of molecules generated by γ called *molecular language* is defined by

$$LM(\gamma) = \{w \in WK_\rho(V) \mid x_1 \Longrightarrow^* w, \; x_1 \in A\}.$$

Furthermore, a *(string) language* $L(\gamma)$ *generated* by γ is a coding image of $LM(\gamma)$, i.e., the set of all *upper* components of the molecular language $LM(\gamma)$. The classes of molecular languages and of string languages generated by γ are denoted by $\mathcal{SL}_m$ and $\mathcal{SL}$, respectively.

[Watson-Crick finite automata]

A *Watson-Crick finite automaton* (abb. WK-automaton) is defined by the tuple

$$M = (V, \rho, Q, q_0, F, \delta).$$

V is an *(input) alphabet*, Q is a finite set of *states*, V and Q are disjoint alphabets. $\rho \subseteq V \times V$ is a symmetric relation. q_0 is the *initial state* in Q. $F \subseteq Q$ is the set of *final states*. $\delta : Q \times \binom{V^*}{V^*} \to \mathcal{P}(Q)$ is a *transition mapping* such that $\delta(q, \binom{x}{y})) \neq \phi$ only for finitely many triples $(s, x, y) \in Q \times V^* \times V^*$, where $\mathcal{P}(Q)$ is the set of all possible subsets of Q.

A transition in a WK-automaton can be defined as follows: For $\binom{x_1}{x_2}, \binom{u_1}{u_2}$, $\binom{y_1}{y_2} \in \binom{V^*}{V^*}$ with $\begin{bmatrix} x_1 u_1 y_1 \\ x_2 u_2 y_2 \end{bmatrix} \in WK_\rho(V)$, and $q_1, q_2 \in Q$, we write

$$\binom{x_1}{x_2} q_1 \binom{u_1}{u_2} \binom{y_1}{y_2} \Longrightarrow_M \binom{x_1}{x_2} \binom{u_1}{u_2} q_2 \binom{y_1}{y_2}$$

if and only if $\delta(q_1, \binom{u_1}{u_2})) \ni q_2$. We denote by $\Longrightarrow^*_M$ the reflexive and transitive closure of the relation $\Longrightarrow_M$. If there is no confusion, we use $\Longrightarrow$ instead of $\Longrightarrow_M$.

A *molecular language* and a *(string) language* over V recognized by M are defined by

$$LM(M) = \{\begin{bmatrix} w_1 \\ w_2 \end{bmatrix} \in WK_\rho(V) \mid q_0 \begin{bmatrix} w_1 \\ w_2 \end{bmatrix} \Longrightarrow_M^* \begin{bmatrix} w_1 \\ w_2 \end{bmatrix} q_f, \; q_f \in F\}$$

and $L(M)$ is the set of all *upper* components of the molecular language $LM(M)$.

A WK-automaton $M = (V, \rho, Q, q_0, F, \delta)$ is 1-*limited* if for any transition $\delta(q_1, \begin{pmatrix} x_1 \\ x_2 \end{pmatrix})) \ni q_2, |x_1 x_2| = 1$ holds.

Let WK_m and $1WK_m$ be the classes of molecular languages recognized by WK-automata and 1-limited WK-automata, resp. Further, WK and $1WK$ denote their corresponding string language classes.

Theorem 1. ([3]) $WK_m = 1WK_m$ (and $WK = 1WK$).

[External contextual grammars ([2])]
An *external contextual grammar* is a construct $G = (V, A, C)$, where V is an alphabet, $A \; (\subseteq V^*)$ is a finite set of axioms, C is a finite set of elements in $V^* \times V^*$. For $\alpha, \beta \in V^*$, we write $\alpha \Longrightarrow_G \beta$ if $\beta = u\alpha v$, for some $(u, v) \in C$.

An *external contextual language generated by* G is

$$L(G) = \{w \in V^* \mid x_1 \Longrightarrow_G^* w, \; x_1 \in A\}.$$

Let $\mathcal{EC}$ be the class of external contextual languages.

Theorem 2. ([2]) *It holds that* $\mathcal{EC} = \mathcal{MLIN}$ *(the class of minimal linear languages).*

[Twin-shuffle languages and their extensions]
Let V and $\bar{V} = \{\bar{a} \mid a \in V\}$ be alphabets. A *twin-shuffle language* over V is defined as

$$TS(V) = \bigcup_{x \in V^*} x \amalg \bar{x}, \quad \text{where}$$

$$x \amalg y = \{x_1 y_1 \cdots x_n y_n \mid x = x_1 \cdots x_n, \; y = y_1 \cdots y_n, \; n \geq 1, \; 1 \leq i \leq n, \; x_i, y_i \in V^*\}.$$

Consider alphabets V, $\bar{V}$ and V', where $V \cap V' = \phi$ and $\bar{V} \cap V' = \phi$. We define an *extended twin-shuffle language* over V and V' as follows :

$$ETS(V, V') = \{x_1 y_1 \cdots x_n y_n \mid n \geq 1, \text{ for } 1 \leq i \leq n, \; x_i \in TS(V), \; y_i \in V'^*\}.$$

3   Two Specific Mappings: Linearizer and Doubler

In order to materialize our goal of providing an unified view of WK-automata and sticker systems, we newly introduce two specific mappings : one is a mapping that linearizes a given molecular language (consisting of elements in $W_\rho(V)$) into its coded form of string language, and the other is the one that, given a string language, transforms into its double stranded version of molecular language.

[Linearizer mapping: f_ℓ]
In this paper, for an alphabet V let $\bar{V} = \{\bar{a} \mid a \in V\}$, and we assume that $\rho \subseteq V \times \bar{V}$ is a complementary symmetric relation and for any $a \in V$, $(a, \bar{a}) \in \rho$ and $\bar{\bar{a}} = a$. We first define a mapping f_ℓ to transform double strands into strings.

Let $\Sigma = V \cup \bar{V}$, then we introduce new notations : for $a \in \Sigma$, $\begin{pmatrix} a \\ \epsilon \end{pmatrix} = \hat{a}$, $\begin{pmatrix} \epsilon \\ a \end{pmatrix} = \check{a}$, $\begin{pmatrix} a \\ \bar{a} \end{pmatrix} = \tilde{a}$. Further, let $\hat{\Sigma} = \{\hat{a} \mid a \in \Sigma\}$, $\check{\Sigma} = \{\check{a} \mid a \in \Sigma\}$, $\tilde{\Sigma} = \{\tilde{a} \mid a \in \Sigma\}$. Now, we define the *linearizer mapping* f_ℓ:

$$f_\ell : W_\rho(V)^* \to (\hat{\Sigma} \cup \check{\Sigma})^* \; \tilde{\Sigma}^+ \; (\hat{\Sigma} \cup \check{\Sigma})^* \; \cup \; (\hat{\Sigma} \cup \check{\Sigma})^*,$$

which transforms double strands over Σ to single strands over $\hat{\Sigma} \cup \check{\Sigma} \cup \tilde{\Sigma}$.

For example, for double strands $u = \boxed{\begin{array}{cc} u_1 & u_2 \\ \hline \bar{u}_2 & u_3 \end{array}}$ (resp. $u = \boxed{\begin{array}{cc} u_2 & u_3 \\ \hline u_1 & \bar{u}_2 \end{array}}$), our intention is that $f_\ell(u) = \hat{u}_1 \tilde{u}_2 \check{u}_3$, (resp. $f_\ell(u) = \tilde{u}_1 \tilde{u}_2 \hat{u}_3$).

Formally, for a double strand $\begin{pmatrix} x_1 \\ x_2 \end{pmatrix} \begin{bmatrix} y_1 \\ y_2 \end{bmatrix} \begin{pmatrix} z_1 \\ z_2 \end{pmatrix}$, we define $f_\ell(\begin{pmatrix} x_1 \\ x_2 \end{pmatrix} \begin{bmatrix} y_1 \\ y_2 \end{bmatrix} \begin{pmatrix} z_1 \\ z_2 \end{pmatrix}) =$

$x\tilde{y}_1 z$, where $x = \begin{cases} \hat{x}_1 & \text{if } x_2 = \epsilon, \\ \check{x}_2 & \text{if } x_1 = \epsilon, \end{cases}$ $z = \begin{cases} \hat{z}_1 & \text{if } z_2 = \epsilon, \\ \check{z}_2 & \text{if } z_1 = \epsilon. \end{cases}$

For a double strand $\begin{pmatrix} a \\ \epsilon \end{pmatrix} = \hat{a}$ with $a \in V$, a double strand $\begin{pmatrix} \epsilon \\ \bar{a} \end{pmatrix} = \check{\bar{a}}$ is complementary, in the sense that $\begin{pmatrix} a \\ \epsilon \end{pmatrix} \begin{pmatrix} \epsilon \\ \bar{a} \end{pmatrix} = \begin{pmatrix} a \\ \bar{a} \end{pmatrix}$. Therefore, for $\hat{a}$ in $\hat{\Sigma}$ and $\check{a}$ in $\check{\Sigma}$, we consider a complementary relation defined by ψ as follows: $\psi(\hat{a}) = \check{b}$, $\psi(\check{a}) = \hat{b}$, where $(a, b) \in \rho$, i.e., $b = \bar{a}$. In $\hat{\Sigma}$ and $\check{\Sigma}$, we consider this complementary relation defined by ψ.

Thus, a twin-shuffle language $TS(\hat{\Sigma})$ is defined as follows:

$$TS(\hat{\Sigma}) = \bigcup_{x \in \hat{\Sigma}^*} x \;\text{Ш}\; \psi(x).$$

[Doubler mapping: f_d]
Conversely, we want to reconstruct a double strand from a string over $\hat{\Sigma} \cup \check{\Sigma} \cup \tilde{\Sigma}$.

Consider a string $y = y_1 a y_2 \in (\hat{\Sigma} \cup \check{\Sigma})^*$ with length n. For an alphabet $\hat{\Sigma}$, we say that a symbol a is $\hat{\Sigma}$-*occurrence at position i of y* with $1 \leq i \leq n$ if $|y_1|_{\hat{\Sigma}} = i - 1$ and a is in $\hat{\Sigma}$, where $|x|_V$ is the number of symbols in V in the string x.

Let y be a string in $TS(\hat{\Sigma})$ with length $2m \geq 2$. Consider a complete double strand of length m which satisfies the following conditions:

- if $a \in \Sigma$ is the i-th symbol with $1 \leq i \leq m$ in the upper strand, then $\hat{a}$ is $\hat{\Sigma}$-occurrence at position i of y.

- if $a \in \Sigma$ is the i-th symbol with $1 \le i \le m$ in the lower strand, then $\check{a}$ is $\check{\Sigma}$-occurrence at position i of y.

$$f_d(y) = \boxed{\begin{array}{c|c|c|c|c} a_1 & \cdots & \bar{a}_i & \cdots & a_m \\ \hline \bar{a}_1 & \cdots & a_i & \cdots & \bar{a}_m \end{array}} \quad y = \boxed{\check{\bar{a}}_1 \mid \hat{a}_1 \mid \cdots \mid \check{a}_i \mid \cdots\cdots \mid \hat{\bar{a}}_i \mid \cdots \mid \hat{a}_m \mid \check{\bar{a}}_m}$$

The double strand thus obtained is called the *doubler of y* and denoted by $f_d(y)$. In particular, for ϵ we define $f_d(\epsilon) = \begin{bmatrix} \epsilon \\ \epsilon \end{bmatrix}$. Note that for a string $y \notin TS(\hat{\Sigma})$, $f_d(y)$ is undefined. Then, $f_d(x) = \begin{bmatrix} \epsilon \\ \epsilon \end{bmatrix}$ implies that $x = \epsilon$.

For example, for strings $\hat{a}\check{\bar{b}}\hat{c}\check{a}\hat{b}\check{c}$, $\hat{a}\check{\bar{a}}\hat{\bar{b}}\hat{c}\check{b}\check{c}$, $\check{a}\hat{a}\hat{\bar{b}}\check{b}\check{c}\hat{c}$, in $(\hat{\Sigma} \cup \check{\Sigma})^*$,

$$f_d(\hat{a}\check{\bar{b}}\hat{c}\check{a}\hat{b}\check{c}) = f_d(\hat{a}\check{\bar{a}}\hat{\bar{b}}\hat{c}\check{b}\check{c}) = f_d(\check{a}\hat{a}\hat{\bar{b}}\check{b}\check{c}\hat{c}) = \begin{bmatrix} abc \\ \bar{a}b\bar{c} \end{bmatrix}.$$

Lemma 1. *For a string w in $(\hat{\Sigma} \cup \check{\Sigma})^*$, $f_d(w)$ is a complete double strand if and only if w is in $TS(\hat{\Sigma})$.*

We now want to extend the mapping f_d so as to apply to strings in $ETS(\hat{\Sigma}, \tilde{\Sigma})$.

Let $y = x_1 x_2 \cdots x_{2n}$ be a string in $ETS(\hat{\Sigma}, \tilde{\Sigma})$, where $n \ge 1$, and for $1 \le i \le n$, $x_{2i-1} \in TS(\hat{\Sigma})$, $x_{2i} \in \tilde{\Sigma}^*$. Then, a doubler mapping f_d is extended in the following manner.

- For a string $x_{2i} = \tilde{u}_{2i}$ in $\tilde{\Sigma}^*$, $f_d(x_{2i})$ is a complete double strand $\begin{bmatrix} u_{2i} \\ \bar{u}_{2i} \end{bmatrix}$.
- For a string x_{2i-1} in $TS(\hat{\Sigma})$, $f_d(x_{2i-1})$ is the same one as already defined.

In a graphical representation, this means the following :

$$f_d(y) = f_d(x_1)f_d(\tilde{u}_2)\cdots f_d(x_{2n-1})f_d(\tilde{u}_{2n}) = \boxed{f_d(x_1) \dfrac{u_2}{\bar{u}_2}} \cdots \boxed{f_d(x_{2n-1}) \dfrac{u_{2n}}{\bar{u}_{2n}}}$$

Note that for a string $y \notin ETS(\hat{\Sigma}, \tilde{\Sigma})$, $f_d(y)$ is undefined.

For example, for strings $\hat{a}\hat{\bar{b}}\check{a}\hat{b}\check{c}\tilde{d}\check{d}$, $\hat{a}\check{\bar{a}}\hat{\bar{b}}\hat{c}\check{b}\check{c}\tilde{d}$, $\tilde{a}\hat{\bar{b}}\check{c}\tilde{d}$ in $(\hat{\Sigma} \cup \check{\Sigma} \cup \tilde{\Sigma})^*$,

$$f_d(\hat{a}\hat{\bar{b}}\check{a}\hat{b}\check{c}\tilde{d}\check{d}) = f_d(\hat{a}\check{\bar{a}}\hat{\bar{b}}\hat{c}\check{b}\check{c}\tilde{d}) = f_d(\tilde{a}\hat{\bar{b}}\check{c}\tilde{d}) = \begin{bmatrix} a\bar{b}cd \\ \bar{a}b\bar{c}d \end{bmatrix}.$$

Note 1. f_d is different from ℓ_p (in [5]) in that ℓ_p has no $\check{\Sigma}$ for its alphabet, and f_d has more flexibility of y than ℓ_p to build up a double strand $f_d(y)$.

Lemma 2. *For a string w in $(\hat{\Sigma} \cup \check{\Sigma} \cup \tilde{\Sigma})^*$, $f_d(w)$ is a complete double strand if and only if w is in $ETS(\hat{\Sigma}, \tilde{\Sigma})$.*

4   Characterization Results in Terms of Doubler

In this section, by using the doubler mapping f_d, we characterize languages recognized by a Watson-Crick finite automaton and generated by a sticker system.

4.1   WK Molecular Languages Are f_d(Regular Languages)

Lemma 3. *For a Watson-Crick finite automaton M_W, there exists a finite automaton M such that $LM(M_W) = f_d(L(M)) = \{f_d(w) \mid w \in L(M)\}$.*

Proof. We may consider a 1-limited WK-automaton $M_W = (\Sigma, \rho, Q, q_0, F, \delta_W)$. Then, construct a finite automaton $M = (\hat{\Sigma} \cup \check{\Sigma}, Q, q_0, F, \delta)$ *derived from M_W* as follows: For $\delta_W(q_i, x) \ni q_j$ in M_W, construct $\delta(q_i, f_\ell(x)) \ni q_j$ in M.

It suffices to show that $\begin{bmatrix} z_1 \\ z_2 \end{bmatrix}$ is in $LM(M_W)$ if and only if there exists a string $w \in L(M)$ such that $f_d(w) = \begin{bmatrix} z_1 \\ z_2 \end{bmatrix}$.

Assume that $\begin{bmatrix} z_1 \\ z_2 \end{bmatrix}$ is in $LM(M_W)$ and there exists a transition,

$$\begin{pmatrix} u_1 \cdots u_i \\ v_1 \cdots v_i \end{pmatrix} q_i \begin{pmatrix} u_{i+1} \\ v_{i+1} \end{pmatrix} \begin{pmatrix} u_{i+2} \cdots u_n \\ v_{i+2} \cdots v_n \end{pmatrix} \Longrightarrow_{M_W} \begin{pmatrix} u_1 \cdots u_i \\ v_1 \cdots v_i \end{pmatrix} \begin{pmatrix} u_{i+1} \\ v_{i+1} \end{pmatrix} q_{i+1} \begin{pmatrix} u_{i+2} \cdots u_n \\ v_{i+2} \cdots v_n \end{pmatrix},$$

where $n \geq 1$, $\begin{bmatrix} u_1 \cdots u_n \\ v_1 \cdots v_n \end{bmatrix} = \begin{bmatrix} z_1 \\ z_2 \end{bmatrix}$, $0 \leq i \leq n$, and $\begin{pmatrix} u_j \\ v_j \end{pmatrix} \in \begin{pmatrix} \Sigma \\ \epsilon \end{pmatrix} \cup \begin{pmatrix} \epsilon \\ \Sigma \end{pmatrix}$, for $1 \leq j \leq n$, and $q_n \in F$.

From the way of constructing δ, for each $0 \leq i \leq n$, there exists a transition $\delta(q_i, b_{i+1}) \ni q_{i+1}$ in M, where $b_{i+1} = f_\ell(\begin{pmatrix} u_{i+1} \\ v_{i+1} \end{pmatrix})$. Then, there exists a transition $\delta(q_0, b_1 \cdots b_n) \ni q_n$ in M.

Since $\begin{bmatrix} z_1 \\ z_2 \end{bmatrix}$ is the complete double strand, the i-th symbol in the upper strand and the i-th symbol in the lower strand are complementary. Therefore, for the string $w = b_1 \cdots b_n$, $\hat{\Sigma}$-occurrence at position i of w and $\check{\Sigma}$-occurrence at position i of w are complementary. Then, it holds that $b_1 \cdots b_n \in \hat{u}_1 \cdots \hat{u}_n \amalg \check{v}_1 \cdots \check{v}_n$, which leads to that $f_d(b_1 \cdots b_n) = \begin{bmatrix} z_1 \\ z_2 \end{bmatrix}$.

Conversely, assume that a string w is in $L(M)$ such that $f_d(w) = \begin{bmatrix} z_1 \\ z_2 \end{bmatrix}$.

Let $w = b_1 \cdots b_{2n}$, where $n \geq 1$, for $1 \leq i \leq 2n$, $b_i \in \hat{\Sigma} \cup \check{\Sigma}$, then from Lemma 1, w is in $TS(\hat{\Sigma})$. Let $w \in \hat{u}_1 \cdots \hat{u}_n \amalg \check{v}_1 \cdots \check{v}_n$.

There exists a transition $\delta(q_0, b_1 \cdots b_{2n}) \ni q_{2n}$ with $q_{2n} \in F$. From the way of constructing δ, for a transition $\delta(q_i, b_{i+1}) \ni q_{i+1}$ in M, there exists a transition $\delta_W(q_i, b'_{i+1}) \ni q_{i+1}$, where $0 \leq i \leq 2n - 1$, $b_{i+1} = f_\ell(b'_{i+1})$.

Then, there exists a transition in M_W, $q_0 \begin{pmatrix} u_1 \cdots u_n \\ v_1 \cdots v_n \end{pmatrix} \Longrightarrow^*_{M_W} \begin{pmatrix} u_1 \cdots u_n \\ v_1 \cdots v_n \end{pmatrix} q_f$,

where q_f is in F. Since w is in $TS(\hat{\Sigma})$, for each $1 \leq i \leq n$, u_i and v_i are complementary, which means that $\begin{bmatrix} u_1 \cdots u_n \\ v_1 \cdots v_n \end{bmatrix} \in WK_\rho(V)$. $\qquad\square$

Lemma 4. *For a finite automaton $M = (\hat{\Sigma} \cup \check{\Sigma}, Q, q_0, F, \delta)$, there exists a Watson-Crick finite automaton $M_W = (\Sigma, \rho, Q, q_0, F, \delta_W)$ such that $LM(M_W) = f_d(L(M)) = \{f_d(w) \mid w \in L(M)\}$.*

Proof (sketch). For a finite automaton $M = (\hat{\Sigma} \cup \check{\Sigma}, Q, q_0, F, \delta)$, construct a WK-automaton $M_W = (\Sigma, \rho, Q, q_0, F, \delta_W)$ as follows:

For a transition $\delta(q_i, \hat{a}) \ni q_j$ in M, construct $\delta_W(q_i, \begin{pmatrix} a \\ \epsilon \end{pmatrix}) \ni q_j$ in M_W.

For a transition $\delta(q_i, \check{a}) \ni q_j$ in M, construct $\delta_W(q_i, \begin{pmatrix} \epsilon \\ a \end{pmatrix}) \ni q_j$ in M_W.

It suffices to show that a complete double strand $\begin{bmatrix} z_1 \\ z_2 \end{bmatrix}$ is in $LM(M_W)$ if and only if there exists a string $w \in L(M)$ such that $f_d(w) = \begin{bmatrix} z_1 \\ z_2 \end{bmatrix}$, which is proved in a manner similar to the above lemma. Thus, we can prove the equation $LM(M_W) = \{f_d(w) \mid w \in L(M)\}$. □

From Lemmas 3 and 4, we have the following theorem.

Theorem 3. *A molecular language L is in $\mathcal{WK}_m$ if and only if there exists a regular language R such that $L = f_d(R)$.*

4.2   Sticker Molecular Languages Are f_d(Minimal Linear Languages)

We slightly extend f_d to f'_d as follows : For $w = xyz$ in $(\hat{\Sigma} \cup \check{\Sigma} \cup \tilde{\Sigma})^*$ such that only $f_d(y)$ is well-defined and x, z are in $\hat{\Sigma}^* \cup \check{\Sigma}^*$, define $f'_d(w) = x' f_d(y) z'$, where x' (z') represents that x (z) forms an "upper stand" if x (z) is in $\hat{\Sigma}^*$ or "lower one" otherwise.

Lemma 5. *For a sticker system γ_W, there exists an external contextual grammar G such that $LM(\gamma_W) = f_d(L(G)) = \{f_d(w) \mid w \in L(G)\}$.*

Proof. For a sticker system $\gamma_W = (\Sigma, \rho, A_W, D_W)$, we define an external contextual grammar $G = (\hat{\Sigma} \cup \check{\Sigma} \cup \tilde{\Sigma}, A, C)$ *derived from* γ_W as follows:

For (u, v) in D_W, construct $(f_\ell(u), f_\ell(v))$ in C. Let $A = \{f_\ell(\alpha) \mid \alpha \in A_W\}$.

It suffices to show that for any $\alpha' = \begin{pmatrix} \alpha'_1 \\ \alpha'_2 \end{pmatrix}$ in $LR_\rho(\Sigma)$, there exists a computation $\begin{pmatrix} \alpha'_1 \\ \alpha'_2 \end{pmatrix} \Longrightarrow^n_{\gamma_W} \begin{bmatrix} z_1 \\ z_2 \end{bmatrix}$ if and only if there exists a computation $\alpha'' \Longrightarrow^n_G z$, where $f'_d(\alpha'') = \alpha'$ and $f_d(z) = \begin{bmatrix} z_1 \\ z_2 \end{bmatrix}$.

We will prove this by the induction on n.

Base step : $(n = 0)$ It trivially holds. *Induction step* : There exists a computation

$$\begin{pmatrix} \alpha'_1 \\ \alpha'_2 \end{pmatrix} \Longrightarrow_{\gamma_W} \begin{pmatrix} u \\ u' \end{pmatrix} \begin{pmatrix} \alpha'_1 \\ \alpha'_2 \end{pmatrix} \begin{pmatrix} v \\ v' \end{pmatrix} \Longrightarrow^n_{\gamma_W} \begin{bmatrix} z_1 \\ z_2 \end{bmatrix}$$

iff (by inductive hypothesis and the way of constructing C) there uniquely exist $(f_\ell(\begin{pmatrix} u \\ u' \end{pmatrix}), f_\ell(\begin{pmatrix} v \\ v' \end{pmatrix}))$ in C such that

$$f_\ell(\alpha') \Longrightarrow_G f_\ell(\begin{pmatrix} u \\ u' \end{pmatrix}) f_\ell(\alpha') f_\ell(\begin{pmatrix} v \\ v' \end{pmatrix})) = w \quad \text{and} \quad w \Longrightarrow_G^n z,$$

where $f_d'(w) = \begin{pmatrix} u \\ u' \end{pmatrix} \begin{pmatrix} \alpha_1' \\ \alpha_2' \end{pmatrix} \begin{pmatrix} v \\ v' \end{pmatrix}$ and $f_d(z) = \begin{bmatrix} z_1 \\ z_2 \end{bmatrix}$

iff there exists

$$f_\ell(\alpha') \Longrightarrow_G f_\ell(\begin{pmatrix} u \\ u' \end{pmatrix}) f_\ell(\alpha') f_\ell(\begin{pmatrix} v \\ v' \end{pmatrix})) \Longrightarrow_G^n z, \quad \text{and} \quad f_d(z) = \begin{bmatrix} z_1 \\ z_2 \end{bmatrix}.$$

Considering $\alpha' = \alpha \in A_W$, we have that

$$\alpha \Longrightarrow_{\gamma_W}^n \begin{bmatrix} z_1 \\ z_2 \end{bmatrix} \text{ if and only if } f_\ell(\alpha) \Longrightarrow_G^n z, \text{ where } f_d(z) = \begin{bmatrix} z_1 \\ z_2 \end{bmatrix}. \qquad \square$$

An external contextual grammar $G = (\hat{\Sigma} \cup \check{\Sigma} \cup \tilde{\Sigma}, A, C)$ is said to be *restricted* if (1) for any (u, v) in C, u and v are in $(\hat{\Sigma} \cup \check{\Sigma})^* \tilde{\Sigma}^+ (\hat{\Sigma} \cup \check{\Sigma})^* \cup (\hat{\Sigma} \cup \check{\Sigma})^*$, and (2) $A \subset (\hat{\Sigma} \cup \check{\Sigma})^* \tilde{\Sigma}^+ (\hat{\Sigma} \cup \check{\Sigma})^*$.

Let r-$\mathcal{EC}$ be the class of languages generated by restricted external contextual grammars.

Lemma 6. *For a given restricted external contextual grammar $G = (\hat{\Sigma} \cup \check{\Sigma} \cup \tilde{\Sigma}, A, C)$, there exists a sticker system $\gamma_W = (\Sigma, \rho, A_W, D_W)$ such that $LM(\gamma_W) = \{f_d(w) \mid w \in L(G)\}$.*

Proof (sketch). For a given G above, we construct a sticker system $\gamma_W = (\Sigma, \rho, A_W, D_W)$ as follows : For (x, y) in C, construct $(f_d'(x), f_d'(y))$ in D_W. Let $A_W = \{f_d'(\alpha) \mid \alpha \in A\}$.

We can prove that $\begin{bmatrix} z_1 \\ z_2 \end{bmatrix}$ is in $LM(\gamma_W)$ if and only if there exists a string $w \in L(G)$ such that $f_d(w) = \begin{bmatrix} z_1 \\ z_2 \end{bmatrix}$, in a manner similar to the above lemma, which implies the equation $LM(\gamma_W) = \{f_d(w) \mid w \in L(G)\}$. $\qquad \square$

From Lemmas 5 and 6, we have the following theorem.

Theorem 4. *A molecular language L is in $\mathcal{SL}_m$ if and only if there exists an external contextual language R in r-$\mathcal{EC}$ such that $L = f_d(R)$.*

4.3 Characterizing Recursively Enumerable Languages by f_d

Based on the doubler mapping f_d and a projection, we first introduce a mapping f_{pr}. For the projection $pr_T : V_2^* \to T^*$, we define $f_{pr} : (\hat{\Sigma} \cup \check{\Sigma} \cup \tilde{\Sigma})^* \to T^*$ as follows: $f_{pr}(w) = pr_T(x)$, where $f_d(w) = \begin{bmatrix} x \\ x' \end{bmatrix}$ for some $x' \in \Sigma^*$. Then, using the class $\mathcal{EC}$ and f_{pr}, we have the following characterization of recursively enumerable languages.

Theorem 5. *For a recursively enumerable language L, there exists an external contextual grammar G such that $f_{pr}(L(G)) = L$.*

Proof. It is well known that for any recursively enumerable language $L \subseteq T^*$, there exist two ϵ-free morphisms h_1, h_2, a regular language R, and a projection pr_T such that $L = pr_T(h_1(EQ(h_1, h_2)) \cap R)$. Let $M = (Q, \Gamma_2, \delta, q_0, F)$ be a finite automaton such that $L(M) = R$. Let $h_1, h_2 : V_1^* \to V_2^*$.

We construct an external contextual grammar $G = (\hat{\Sigma} \cup \check{\Sigma} \cup \tilde{\Sigma}, A, C)$:

1. Let $(w_a, u_a v_a)$ be in C, where $u_a = \hat{b}_1 \cdots \hat{b}_n$, with $h_1(a) = b_1 \cdots b_n$,
$$v_a = \check{c}_1 \cdots \check{c}_m, \text{ with } h_2(a) = c_1 \cdots c_m,$$
for any q_1 in Q, $w_a = \hat{\tilde{q}}_{n+1} \tilde{\#} \check{\tilde{q}}_n \; \hat{q}_n \tilde{\#} \check{\tilde{q}}_{n-1} \cdots \hat{q}_3 \tilde{\#} \check{\tilde{q}}_2 \; \hat{q}_2 \tilde{\#} \check{\tilde{q}}_1$,
   if $\delta(q_i, b_i) \ni q_{i+1}$ for each $1 \le i \le n$, $q_{n+1} \notin F$,
   $w_a = \tilde{q}_{n+1} \tilde{\#} \check{\tilde{q}}_n \; \hat{q}_n \tilde{\#} \check{\tilde{q}}_{n-1} \cdots \hat{q}_3 \tilde{\#} \check{\tilde{q}}_2 \; \hat{q}_2 \tilde{\#} \check{\tilde{q}}_1$,
   if $\delta(q_i, b_i) \ni q_{i+1}$ for each $1 \le i \le n$, $q_{n+1} \in F$.

The strings u_a, v_a are used to check the equality for the homomorphisms h_1 and h_2. The string w_a corresponds to the brick

q_{n+1}	#q_n#	$\cdots q_3$#q_2#	
	#$\bar{q}_n$#$\bar{q}_{n-1}$	$\cdots$ #$\bar{q}_2$#	$\bar{q}_1$

which is used to check whether a string is in $R = L(M)$.

2. Let $\Sigma = \Gamma_Q \cup \Gamma_2 \cup \Gamma_\#$ for $\Gamma_Q = Q \cup \bar{Q}$, $\Gamma_2 = V_2 \cup \bar{V}_2$, $\Gamma_\# = \{\#, \bar{\#}\}$ and let $A = \{\hat{q}_0 \tilde{\#}\}$.

We will show the equality $f_{pr}(L(G)) = L$. Assume that w is in $f_{pr}(L(G))$, then there exists a string w' in $L(G)$ such that $f_{pr}(w') = w$. From the definition of C, $w' = w_1 \hat{q}_0 \tilde{\#} w_2$, where $w_1 \in (\hat{\Gamma}_Q \cup \check{\Gamma}_Q \cup \{\tilde{\#}\})^*$, $w_2 \in (\hat{\Gamma}_2 \cup \check{\Gamma}_2)^*$. Since $f_{pr}(w')$ is defined, w' is in $ETS(\hat{\Sigma}, \tilde{\Sigma})$. Further, $w_1 \in ETS(\hat{\Gamma}_Q, \check{\Gamma}_Q \cup \{\tilde{\#}\})$, $w_2 \in TS(\hat{\Gamma}_2)$. Then, there must exist a string $z = a_1 \cdots a_m \in \Gamma_1^*$ which satisfies the following two conditions : for $h_1(z) = b_1 \cdots b_{m'}$ with $m' \ge 1$,

- $w_1 = \check{\tilde{q}}_{m'} \tilde{\#} \check{\tilde{q}}_{m'-1} \; \hat{q}_{m'-1} \tilde{\#} \check{\tilde{q}}_{m'-2} \cdots \hat{q}_2 \tilde{\#} \check{\tilde{q}}_1 \; \hat{q}_1 \tilde{\#} \check{\tilde{q}}_0$, where for $0 \le i \le m' - 1$, $\delta(q_i, b_{i+1}) \ni q_{i+1}$, $q_{m'} \in F$. This implies that $b_1 \cdots b_{m'}$ is in R.
- $w_2 \in \hat{b}_1 \cdots \hat{b}_{m'} \; \text{⧢} \; \check{b}_1 \cdots \check{b}_{m'}$ This implies that $h_1(z) = h_2(z)$.

Therefore, $b_1 \cdots b_{m'}$ is in $h_1(EQ(h_1, h_2)) \cap R$, then from the definition of f_{pr}, $f_{pr}(w') = pr_T(b_1 \cdots b_{m'}) \in L$.

Conversely, assume that w is in L. Then, there exists a string z such that $z \in h_1(EQ(h_1, h_2))$, $z \in R$ and $pr_T(z) = w$. Then, there exists a string $z' = a_1 \cdots a_m$ such that $z' \in EQ(h_1, h_2)$ and $h_1(z') = h_1(a_1) \cdots h_1(a_m) = b_1 \cdots b_{m'} = h_2(a_1) \cdots h_2(a_m)$ for $m' \ge 1$.

For z', there exists a derivation $\hat{q}_0 \tilde{\#} \Longrightarrow_G^* w_1 \hat{q}_0 \tilde{\#} w_2$ in G such that $w_2 = \hat{h}_1(a_1) \check{h}_2(a_1) \cdots \hat{h}_1(a_m) \check{h}_2(a_m)$, where $\hat{h}_1(a_i) = \hat{b}_{i1} \cdots \hat{b}_{ik}$ for $h_1(a_i) = b_{i1} \cdots b_{ik}$, $\check{h}_2(a_i) = \check{c}_{i1} \cdots \check{c}_{i\ell}$ for $h_2(a_i) = c_{i1} \cdots c_{i\ell}$. Then, $w_2 \in \hat{b}_1 \cdots \hat{b}_{m'} \; \text{⧢} \; \check{b}_1 \cdots \check{b}_{m'}$.

At the same time, since $\delta(q_0, z) \ni q_f$ with $q_f \in F$, from the way of constructing C, $w_1 = \check{\tilde{q}}_{m'} \tilde{\#} \check{\tilde{q}}_{m'-1} \; \hat{q}_{m'-1} \tilde{\#} \check{\tilde{q}}_{m'-2} \cdots \hat{q}_2 \tilde{\#} \check{\tilde{q}}_1 \; \hat{q}_1 \tilde{\#} \check{\tilde{q}}_0$, where for $0 \le i \le m' - 1$, $\delta(q_i, b_{i+1}) \ni q_{i+1}$, $q_{m'} \in F$. Therefore, $w_1 \hat{q}_0 \tilde{\#} w_2$ is in $ETS(\hat{\Sigma}, \tilde{\Sigma})$. Then, from the definition of f_{pr}, we have $f_{pr}(w_1 \hat{q}_0 \tilde{\#} w_2) = w$. $\square$

4.4 WK Molecular Languages Are Proj(Sticker Molecular Languages)

Let us define a *double-strand projection* (abb. d-projection) *d-pr* on double strands as follows : For z in $\begin{bmatrix} V_1 \cup V_2 \\ V_1 \cup V_2 \end{bmatrix}$, $d\text{-}pr(z) = z$ if z is in $\begin{bmatrix} V_1 \\ V_1 \end{bmatrix}$ and $d\text{-}pr(z) = \begin{bmatrix} \epsilon \\ \epsilon \end{bmatrix}$ otherwise.

Lemma 7. *For any Watson-Crick finite automaton M, there exists a sticker system γ such that $LM(M) = d\text{-}pr(LM(\gamma))$.*

Proof. (sketch) From Theorem 1, we may consider a 1-limited WK-automaton $M = (\Sigma, \rho, Q, q_0, F, \delta)$. Based on M, construct a sticker system $\gamma = (\Sigma \cup \Gamma_Q \cup \Gamma_\#, \rho_Q, A, D)$ as follows : $\Gamma_Q = Q \cup \bar{Q}$, $\Gamma_\# = \{\#, \bar{\#}\}$. For a transition $\delta(q_i, \begin{pmatrix} x_1 \\ x_2 \end{pmatrix}) \ni q_j$ in M with $q_j \notin F$, construct $(\begin{pmatrix} q_j \\ \epsilon \end{pmatrix} \begin{bmatrix} \# \\ \bar{\#} \end{bmatrix} \begin{pmatrix} \epsilon \\ \bar{q_i} \end{pmatrix}, \begin{pmatrix} x_1 \\ x_2 \end{pmatrix})$ in D. For a transition $\delta(q_i, \begin{pmatrix} x_1 \\ x_2 \end{pmatrix}) \ni q_f$ with $q_f \in F$, construct $(\begin{bmatrix} q_f \\ \bar{q_f} \end{bmatrix} \begin{bmatrix} \# \\ \bar{\#} \end{bmatrix} \begin{pmatrix} \epsilon \\ \bar{q_i} \end{pmatrix}, \begin{pmatrix} x_1 \\ x_2 \end{pmatrix})$ in D. Finally, let $A = \{ \begin{pmatrix} q_0 \# \\ \epsilon \# \end{pmatrix} \}$. Consider a d-projection $d\text{-}pr$ on the alphabet $\begin{bmatrix} \Sigma \\ \bar{\Sigma} \end{bmatrix}$. From the way of constructing γ, by the induction on the length of a computation, we can prove that for $q_f \in F$, $q_0 \begin{pmatrix} w_1 \\ w_2 \end{pmatrix} \Longrightarrow_M^* \begin{pmatrix} w_1 \\ w_2 \end{pmatrix} q_f$ if and only if there exists a computation $\begin{pmatrix} q_0 \# \\ \epsilon \# \end{pmatrix} \Longrightarrow_\gamma^* \begin{bmatrix} q_f \\ \bar{q_f} \end{bmatrix} \begin{bmatrix} z_1 \\ z_2 \end{bmatrix} \begin{bmatrix} \# \\ \# \end{bmatrix} \begin{pmatrix} w_1 \\ w_2 \end{pmatrix}$, where $\begin{bmatrix} z_1 \\ z_2 \end{bmatrix} \in \begin{bmatrix} \#Q \\ \#\bar{Q} \end{bmatrix}^+$. Finally, from the definition of $d\text{-}pr$, it holds that a complete double strand $\begin{bmatrix} w_1 \\ w_2 \end{bmatrix}$ is in $LM(M)$ if and only if there exists a complete double strand $\begin{bmatrix} q_f \\ \bar{q_f} \end{bmatrix} \begin{bmatrix} z_1 \\ z_2 \end{bmatrix} \begin{bmatrix} \# \\ \# \end{bmatrix} \begin{bmatrix} w_1 \\ w_2 \end{bmatrix} \in LM(\gamma)$ such that $d\text{-}pr(\begin{bmatrix} q_f \\ \bar{q_f} \end{bmatrix} \begin{bmatrix} z_1 \\ z_2 \end{bmatrix} \begin{bmatrix} \# \\ \# \end{bmatrix} \begin{bmatrix} w_1 \\ w_2 \end{bmatrix}) = \begin{bmatrix} w_1 \\ w_2 \end{bmatrix}$. $\square$

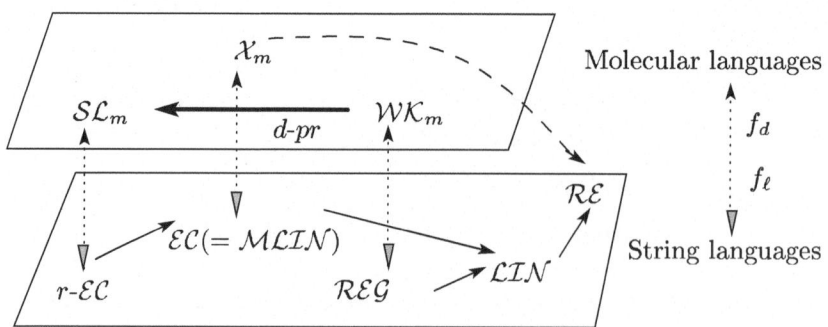

Fig. 1. Landscape of Double-Decker Families of Languages

5   Conclusion

By introducing two specific mappings called "doubler f_d" and "linearizer f_ℓ", we have given new characterization results for the families of sticker languages and of Watson-Crick languages which lead to not only an unified view of the two families of languages but also a clarified view of the computational capability of the DNA complementarity. From Theorems 1 and 5, we have the result $\mathcal{RE} = f_{pr}(\mathcal{MLIN})$, which seems shed some new insights into computations in comparison to the existing ones such as $\mathcal{RE} = dgsm(\mathcal{SL})$ or $\mathcal{RE} = coding(\mathcal{WK})$ (in [3]).

Acknowledgements

This work is supported in part by Grant-in-Aid for Scientific Research on Priority Area no.14085205, Ministry of Education, Culture, Sports, Science and Technology of Japan.

References

1. Hoogeboom, H.J. and Vugt, N.V. : Fair sticker languages, *Acta Informatica*, **37**, 213–225 (2000).
2. Păun, Gh. : *Marcus contextual grammar*. Kluwer Academic Publishers (1997).
3. Păun, Gh., Rozenberg, G. and Salomaa, A. : *DNA Computing. New Computing Paradigms.*, Springer (1998).
4. Rozenberg, G. and Salomaa, A. (Eds.) : *Handbook of Formal Languages*, Springer (1997).
5. Salomaa, A. : *Turing, Watson-Crick and Lindenmayer : Aspects of DNA Complementarity*, In *Unconventional Models of Computation*, Auckland, Springer, 94–107 (1998).
6. Sakakibara, Y. and Kobayashi, S. : *Sticker systems with complex structures. Soft Computing*, **5**, 114–120 (2001).
7. Vliet, R. van, Hoogeboon, H.J. and Rozenberg, G. : *Combinatorial Aspects of Minimal DNA Expressions*, Pre-proc. In *Tenth International Meeting on DNA Computing*, Univ. of Milano-Bicocca, Italy, 84–96 (2004).

Analysis and Simulation of Dynamics in Probabilistic P Systems⋆

Dario Pescini[1], Daniela Besozzi[2],
Claudio Zandron[1], and Giancarlo Mauri[1]

[1] Università degli Studi di Milano-Bicocca,
Dipartimento di Informatica, Sistemistica e Comunicazione,
Via Bicocca degli Arcimboldi 8, 20126 Milano, Italy
{pescini, zandron, mauri}@disco.unimib.it
[2] Università degli Studi di Milano,
Dipartimento di Informatica e Comunicazione,
Via Comelico 39, 20135 Milano, Italy
besozzi@dico.unimi.it

Abstract. We introduce dynamical probabilistic P systems, a variant where probabilities associated to the rules change during the evolution of the system, as a new approach to the analysis and simulation of the behavior of complex systems. We define the notions for the analysis of the dynamics of these systems and we show an application for the investigation of the properties of the Brusselator (a simple scheme for the Belousov-Zhabothinskii reaction).

1 Introduction

P systems [8] are a class of distributed and parallel computing devices, inspired by the structure and the functioning of cells. The basic model consists of a cell-like membrane structure, composed by several compartments where multisets of objects evolve according to given rules, in a nondeterministic and maximally parallel manner. A computation device is obtained starting from an initial configuration and letting the system evolve. In the following, we assume that the reader is familiar with the basic notions and the terminology underlying P systems. We refer, for details, to [9]. Updated information about P systems can be found at http://psystems.disco.unimib.it/.

Many research studies around P systems concentrates on computational power aspects. In this paper, we propose a new approach for the investigation and the application of P systems, which consists in interpreting them as tools for the description and the analysis of the *dynamical* behavior of complex systems. A similar approach is considered also in [3, 10, 12], where different methods are used to investigate several biological and chemical processes, among which one can find the Belousov-Zhabothinskii reaction. As said, membrane systems are inspired from the functioning of the cell, hence it is natural to consider them for

---

⋆ Work supported by the Italian Ministry of University (MIUR), under project PRIN-04 "Systems Biology: modellazione, linguaggi e analisi (SYBILLA)".

A. Carbone and N.A. Pierce (Eds.): DNA11, LNCS 3892, pp. 236–247, 2006.

modelling different cellular processes and natural living systems, with the final goal of producing new tools and acquiring useful information for the scientists (mainly, biologists) working on the modelled system. Some first steps in this direction have already been made, see [4] for various applications.

Since we are interested in describing the evolution of a complex system, and since changes of many different conditions can have direct influence on the reaction parameters and behavior, the basic non-deterministic model of P systems is not suitable to describe these kind of processes. Indeed, many efforts have been recently done to introduce the notion of probability in P systems. The first definition of a probabilistic P system appeared in [7], where probabilities are assigned to evolution rules, an initial probability distribution is defined in each region, and vectors related to each rule specify which rules can be applied at the next step. Though, two assumptions are made which seem quite unnatural from a biological point of view: priority relations among rules are used and, above all, probability values are initially assigned and never change during a computation, which corresponds to a *static* nature of the system. In [6] some more proposals for approaching probabilistic P systems are suggested: priority relations are no more formally considered, though they are implicitly included in computations, since one does not consider a stochastic application of rules. Lately, P systems with probabilistic rules have been also applied for the investigation of cellular phenomena and structures, such as respiration and photosynthesis processes in [2], mechanosensitive channels in [1].

In order to overcome the limitations outlined above, we propose a new version of probabilistic P systems, where probability values are *dynamically* assigned to evolution rules, according to the form of the current multiset. Moreover, the application of rules is stochastic (we will talk about *evolution* instead of *computation*).

The paper is structured as follows. In Section 2 we give the formal definition of dynamical probabilistic P systems, in Section 3 we introduce some notions which will then be used to analyze, via software tools, the behavior of such systems. In Section 4 we show an application to the Brusselator, a well known and simplified theoretical scheme which describes the Belousov-Zhabotinskii reaction (BZ, in short). Finally, in Section 5 we present the conclusion and give some perspective for future work.

2   Dynamical Probabilistic P Systems

In this section we give the definition of a probabilistic P system, where the probabilities associated to the rules vary during the evolution of the system. The method for evaluating probabilities and the way the system works are explained in details. Then, we extend the definition to consider families of P systems of this type, whose members differ among each other for the choice of some parameters, but not for the main structure.

We assume the reader to be familiar with the basic notions and notations of P systems [9]. Some prerequisites about multisets are here recalled.

Let V be an alphabet, we denote by V^* the set of all strings over V, by λ the empty string, and by $V^+ = V^* \backslash \{\lambda\}$ the set of non-empty strings. A multiset over V is a map $M : V \to \mathbb{N}$, where $M(a)$ is the multiplicity of any symbol $a \in V$, $\mathbb{N}$ is the set of natural numbers. A multiset M over $V = \{a_1, \ldots, a_l\}$ can be explicitly represented by the string $x = a_1^{M(a_1)} a_2^{M(a_2)} \ldots a_l^{M(a_l)}$, for all $a_i \in V$ such that $M(a_i) \neq 0$, and by all its possible permutations. By interpreting a multiset in the corresponding form of a string x, we can denote by $|x|$ its *length* and by $|x|_a$ the number of occurrences of a symbol a in x. The set of symbols from V occurring in x is denoted by $alph(x)$. Moreover, to every string $x \in V^*$ we can associate the *Parikh vector* $\Psi_V(x) = (|x|_{a_1}, |x|_{a_2}, \ldots, |x|_{a_l})$.

Definition 1. A *dynamical probabilistic P system* (DPP, in short) of degree n is a construct $\Pi = (V, O, \mu, M_0, \ldots, M_{n-1}, R_0, \ldots, R_{n-1}, E, I)$ where:

- V is the alphabet of the system, $O \subseteq V$ is the set of *analyzed symbols*;
- μ is a membrane structure consisting of n membranes labelled with the numbers $0, \ldots, n-1$. The skin membrane is labelled with 0;
- $M_i, i = 0, \ldots, n-1$, is the multiset over V initially present inside membrane i;
- $R_i, i = 0, \ldots, n-1$, is a finite set of evolution rules associated with membrane i. An evolution rule is of the form $r : u \xrightarrow{k} v$, where u is a multiset over V, v is a string over $V \times (\{here, out\} \cup \{in_j \mid 1 \leq j \leq n-1\})$ and $k \in \mathbb{R}^+$ is a constant associated to the rule;
- $E = \{V_E, M_E, R_E\}$ is called the *environment*, it consists of an alphabet $V_E \subseteq V$, a *feeding multiset* M_E over V_E and a finite set of *feeding rules* R_E of the type $r : u \to (v, in_0)$, for u, v multisets over V_E;
- $I \subseteq \{0, \ldots, n-1\} \cup \{\infty\}$ is the set of labels of the *analyzed regions* (the label ∞ corresponds to the environment).

The alphabet O and the set I specify which symbols and regions (environment included) are of peculiar importance in Π, namely those elements whose evolution will be actually analyzed and simulated.

Definition 2. Let Π be a DPP. We call the *parameters* of Π the set $\mathcal{P}$ consisting of: (1) the multisets $M_0, \ldots, M_{n-1}, M_E$ initially present in μ and in E, (2) the constants associated to all rules in $R_0, \ldots, R_{n-1}$.

Note that the alphabets V, O, V_E, the membrane structure μ, the form of the rules in $R_0, \ldots, R_{n-1}, R_E$ and the set I of analyzed regions do not belong to the set of parameters of Π. We call these components the *main structure* of Π. We can now extend Definition 1 and consider a *family* of DPPs, where the main structure is equal for all members of the family, while the parameters can change from member to member.

Definition 3. A *family* of DPPs is defined as $\mathcal{F} = \{(\Pi, \mathcal{P}_i) \mid \Pi$ is a DPP and $\mathcal{P}_i$ is the set of parameters of $\Pi, i \geq 1\}$.

Hence, given any two elements $(\Pi, \mathcal{P}_1), (\Pi, \mathcal{P}_2) \in \mathcal{F}$, it holds $\mathcal{P}_1 \neq \mathcal{P}_2$ for the choice of (all or some) values in $\mathcal{P}_1$ and $\mathcal{P}_2$. For instance, one can choose to analyze the same DPP with some different settings of initial conditions, such as different initial multisets and/or different rule constants (this can be useful when not all of them are previously known) and/or different feeding multisets.

In the following, we will talk about the *evolution*, not computation, of a DPP, since we are not interested in generating languages but in simulating biological or chemical systems. The family $\mathcal{F}$ describes a general model of the biological or chemical system of interest and, for any choice of the parameters, we can investigate the evolution of the corresponding fixed DPP.

A fixed initial configuration of Π depends on the choice of $\mathcal{P}$, hence it consists of the multisets initially present inside the membrane structure, the chosen rule constants and the feeding multiset, which is given as an input to the skin membrane from the environment at each step of the evolution by applying the feeding rules. Different strategies in the feeding process can be used: for instance, one can use the feeding rules to keep at a constant value the concentrations of chemicals involved in a certain reaction (see Section 4 for an application of this strategy to the BZ), or to increase the concentrations of substances mimicking the biological transport from the extracellular space. We assume that, as long as the system evolves, the environment contains as many symbols as they are needed to continuously feed the system.

At each step of the evolution, all applicable rules are simultaneously applied and all occurrences of the left-hand sides of the rules are consumed, hence the parallelism is maximal at both levels of objects and of rules. For simplicity, in this paper we assume that the system evolves according to a universal clock, that is, all membranes and the application of all rules are synchronized. The applied rules are chosen according to the probability values dynamically assigned to them; the rules with the highest normalized probability value will be more frequently tossed. In simulations, the tossing process is obtained by means of a random number generator, as described below. If some rules compete for objects and have the same probability values, then objects are nondeterministically assigned to those rules.

The probability associated to each rule in any set R_i, $i = 0, \ldots, n-1$, is a function of its constant and of the current multiset occurring in membrane i, and it is evaluated as follows. Let $V = \{a_1, \ldots, a_l\}$, M_i be the multiset inside membrane i, $r : u \xrightarrow{k} v$ a rule in R_i; let $u = a_1^{\alpha_1} \ldots a_s^{\alpha_s}$, $alph(u) = \{a_1, \ldots, a_s\}$ and $H = \{1, \ldots, s\}$. To obtain the actual normalized probability p_i of applying r with respect to all other rules that are applicable in membrane i at the same step, we need to evaluate the non-normalized probability $\widetilde{p}_i(r)$ of r, which depends on the constant associated to r and on the left-hand side of r, namely:

$$\widetilde{p}_i(r) = \begin{cases} 0 & \text{if } M_i(a_h) < \alpha_h \text{ for some } h \in H \\ k \cdot \prod_{h \in H} \dfrac{M_i(a_h)!}{\alpha_h!(M_i(a_h) - \alpha_h)!} & \text{if } M_i(a_h) \geq \alpha_h \text{ for all } h \in H \end{cases} \qquad (1)$$

that is, whenever the current multiset inside membrane i contains *all* occurrences of *all* symbols appearing in the left-hand side of rule r (second case in Equation (1)), then $\widetilde{p}_i(r)$ is dynamically defined according to the current multiset inside membrane i: we choose α_h copies of each symbol a_h among all its $M_i(a_h)$ copies currently available in the membrane itself. In other words, we consider all possible distinct combinations of the symbols appearing in $alph(u)$. Thus, $\widetilde{p}_i(r)$ corresponds to the probability of having a collision among reactant objects, which are considered undistinguishable.

If $R_i = \{r_1, \ldots, r_m\}$, the normalized probability of any rule r_j is

$$p_i(r_j) = \frac{\widetilde{p}_i(r_j)}{\sum_{j=1}^{m} \widetilde{p}_i(r_j)}. \tag{2}$$

In the simulations, the parallel application of the rules is done by splitting one parallel step into several sequential sub-steps. It is possible to separate each single parallel step into two stages, exploiting the fact that the probability distribution and the applicability of the rules are functions only of the left-hand side of the rules and their constants. In the first stage objects are assigned to rules by means of a random number generator, while in the second one the multiset is updated using a stored trace of the rules previously tossed. It should be pointed out that, during the first stage, the probability distribution of the rules has to be kept constant, otherwise the application of the rules would become sequential. A detailed description of he simulation algorithm will appear elsewhere.

Remark 1. A different probability distribution over rules could be obtained by using the classical *rate law* of Chemistry, though the approach used in Equation (1) is more accurate from the combinatorial point of view (see also [5], where a similar approach is considered). Indeed, at high concentrations (multiplicities) the two approaches are undistinguishable, but at lower ones our choice is preferable since it accounts for the exact number of all possible tuples of evolving objects.

3 Analysis of the Dynamics in DPP

In this section we introduce some notions that will be used for the analysis of the behavior of a DPP via software tools, whose complete description and functioning will appear in a forthcoming paper. The final goal is to introduce an appropriate definition of the phase space, thus creating a bridge between P systems and well known tools from the Physics of dynamical systems. Usually, the evolution of a physical system is completely determined by means of the motion equations, a set of differential equations inferred by the system properties. In the case of P systems this role should be accomplished by the evolution rules, which create a one-to-one mapping between the application of each rule and the relative displacement of the system in the phase space.

First of all, to keep trace of the system evolution we extend the definition of the alphabet $V = \{a_1, \ldots, a_l\}$ of Π by introducing the parameter *time*, that is, we define the space $\widetilde{V} := V \times \mathbb{N} = V \times \{\text{time}\}$.

Definition 4. Let $M = \{a_1^{\alpha_1} \ldots a_l^{\alpha_l}\}$ be a multiset over V, where $\alpha_i \geq 0$ for all $h = 1, \ldots, l$. We call a *t-multiset* the structure $M = \{a_1^{\alpha_1}, \ldots, a_l^{\alpha_l}, t\} \in \widetilde{V}$.

By abuse of notation, we will denote both the multiset over V and the t-multiset in $\widetilde{V}$ with the same symbol M, being it clear when one considers also the time component or not. To represent a t-multiset in the space $\widetilde{V}$ we define its position relatively to the t-multiset $O = \{0, \ldots, 0\}$ of $\widetilde{V}$ (the first l components of O are the null multiplicities of the symbols from V). We need also to extend the notion of Parikh vector to the space $\widetilde{V}$ as $\Psi_{\widetilde{V}}(M) = (\alpha_1, \ldots, \alpha_l, t)$. This is necessary if we want to distinguish among two multisets having the same total numbers of symbols but different multiplicities for (at least) one symbol from V.

Definition 5. The *position* of a t-multiset $M \in \widetilde{V}$ is the vector $\overrightarrow{M} = \Psi_{\widetilde{V}}(M)$. The vector $\overrightarrow{O} = \Psi_{\widetilde{V}}(O)$ is called the *origin* of $\widetilde{V}$.

From Definition 5 it follows that the positions of t-multisets $\overrightarrow{O}$ and $\overrightarrow{M}$ are vectors in the space $\mathbb{N}^{l+1}$. The next step is to introduce a scalar product in $\mathbb{N}^l$, to naturally define the notion of distance between t-multisets, thus giving the structure of an euclidean space to $\mathbb{N}^l$. By convention, in the following we will always denote the components of a generic position $\overrightarrow{M}_i$ as the $l + 1$-tuple $(\alpha_{i,1}, \alpha_{i,2}, \ldots, \alpha_{i,l}, t_i)$.

Definition 6. Let $\overrightarrow{M}_i, \overrightarrow{M}_j$ be two positions in $\mathbb{N}^l \times \mathbb{N}$. The *distance* between $\overrightarrow{M}_i, \overrightarrow{M}_j$ is a function $d : \mathbb{N}^{l+1} \times \mathbb{N}^{l+1} \longrightarrow \mathbb{R}^+$ defined as

$$d^2(\overrightarrow{M}_i, \overrightarrow{M}_j) = \sum_{k=1}^{m}(\alpha_{i,k} - \alpha_{j,k})^2 \ . \tag{3}$$

Note that the two positions $\overrightarrow{M}_i, \overrightarrow{M}_j$ in Definition 6 need not to be necessarily one the evolution of the other (that is, the multiset inside the same membrane taken into different time steps). In fact, given a family $\mathcal{F}$ of DPP and two positions $\overrightarrow{M}_i, \overrightarrow{M}_j$, the following cases may hold: (*i*) $\overrightarrow{M}_i, \overrightarrow{M}_j$ occur in distinct time steps, in the same membrane of the same DPP with equal setting $\mathcal{P}$; (*ii*) $\overrightarrow{M}_i, \overrightarrow{M}_j$ occur in distinct or equal time steps, in different membranes of the same DPP with equal setting $\mathcal{P}$; (*iii*) $\overrightarrow{M}_i, \overrightarrow{M}_j$ occur in distinct or equal time steps, in the same membrane of the same DPP with different settings $\mathcal{P}_1, \mathcal{P}_2$; (*iv*) $\overrightarrow{M}_i, \overrightarrow{M}_j$ occur in distinct or equal time steps, in different membranes of the same DPP with different settings $\mathcal{P}_1, \mathcal{P}_2$. That is, we might be interested in looking at the multiset occurring inside a membrane during its evolution, or comparing two multisets of different membranes of the same DPP (in equal or different time steps), or else two multisets inside the same (or even a different) membrane but analyzed in two different evolutions of the *family* of the DPP. In each of the four cases, the distance gives information about "how far" the states in the two trajectories are (that is, the t-multisets in the two evolutions).

In particular, given any couple of positions $\overrightarrow{M_i}, \overrightarrow{M_j}$ of the same DPP (for the same or different set of fixed parameters $\mathcal{P}$), we can say that they are *simultaneous* if they exist at the same time step. This concept can be useful mainly when one considers a membrane structure with degree $n > 1$, where many multisets are co-evolving.

Definition 7. Let $\overrightarrow{M_i}, \overrightarrow{M_j}$ be two positions in $\mathbb{N}^{l+1}$. The *displacement* between $\overrightarrow{M_i}, \overrightarrow{M_j}$ is a function $\overrightarrow{u} : \mathbb{N}^{l+1} \times \mathbb{N}^{l+1} \longrightarrow \mathbb{Z}^l$ defined as

$$\overrightarrow{u}(\overrightarrow{M_i}, \overrightarrow{M_j}) = (\alpha_{i,1} - \alpha_{j,1}, \ldots, \alpha_{i,l} - \alpha_{j,l}) . \tag{4}$$

Note that the displacement can be either a positive or negative value, and it tells how the system "moves"; in details, it tells how the multiplicities in the positions $\overrightarrow{M_j}$ differ from those in $\overrightarrow{M_i}$. Hence, it gives more information than the distance, since it also considers the direction of the variation. Indeed, it is also possible to construct the *versor* $\widehat{u} : \mathbb{N}^{l+1} \times \mathbb{N}^{l+1} \longrightarrow \mathbb{R}^l$ of the displacement which only gives the information about the direction of $\overrightarrow{u}$:

$$\widehat{u}(\overrightarrow{M_i}, \overrightarrow{M_j}) = \left(\frac{\alpha_{i,1} - \alpha_{j,1}}{d(\overrightarrow{M_i}, \overrightarrow{M_j})}, \ldots, \frac{\alpha_{i,l} - \alpha_{j,l}}{d(\overrightarrow{M_i}, \overrightarrow{M_j})} \right) . \tag{5}$$

Note that $\overrightarrow{u} = \widehat{u} \cdot d$, by definition.

The last step before arriving to the definition of the phase space consists in defining the velocity, which carries on the information about the time the displacement between two t-multisets (in the same DPP, with equal initial settings) needs to take place. That is, it tells how fast the evolution from one state of the DPP to the other is.

Definition 8. Let $\overrightarrow{M_i}, \overrightarrow{M_j}$ be two positions with $t_i \neq t_j$ occurring inside the same membrane of a DPP (for a fixed choice of the parameters). The *average velocity* between $\overrightarrow{M_i}, \overrightarrow{M_j}$ is a function $\overrightarrow{v} : \mathbb{N}^{l+1} \times \mathbb{N}^{l+1} \longrightarrow \mathbb{R}^l$ defined as

$$\overrightarrow{v}(\overrightarrow{M_i}, \overrightarrow{M_j}) = \left(\frac{\alpha_{i,1} - \alpha_{j,1}}{t_i - t_j}, \ldots, \frac{\alpha_{i,l} - \alpha_{j,l}}{t_i - t_j} \right) . \tag{6}$$

When $t_i - t_j = 1$, which is the minimal time increment allowed in P systems, then the average velocity $\overrightarrow{v}(\overrightarrow{M_i}, \overrightarrow{M_j})$ becomes the "instantaneous" velocity between time steps t_j and $t_i = t_j + 1$, that we denote by $\overrightarrow{v}(\overrightarrow{M_j})$. Note that if $\overrightarrow{M_i}$ is the position evolved from $\overrightarrow{M_j}$ in the same membrane, then the instantaneous velocity gives the variation of that multiset in a single time step.

We are now ready to define the phase space for a DPP, which is constructed as the cartesian product of the phase spaces of all membranes in the DPP. Let $\overrightarrow{M^i} = (\alpha_1, \ldots, \alpha_l, t)$ be the position of the t-multiset inside membrane i at time t, and let $\overrightarrow{v}(\overrightarrow{M^i}) = (v_1, \ldots, v_l)$ be its instantaneous velocity.

Definition 9. We call a *phase point* of $\overrightarrow{M^i}$ the vector $\overrightarrow{\varphi_t^i} = (\alpha_1, \ldots, \alpha_l, v_1, \ldots, v_l) \in \mathbb{N}^l \times \mathbb{R}^l$, for any fixed $t \in \mathbb{N}$.

The phase point represents the state of membrane i at any given time t. The evolution of the multiset in membrane i can be described by the *phase curve*, which is a function $\overrightarrow{\varphi}^i : \mathbb{N} \longrightarrow \mathbb{N}^l \times \mathbb{R}^l$ such that $\overrightarrow{\varphi}^i(t) = \overrightarrow{\varphi}_t^i$.

The space $\Phi^i \subseteq \mathbb{N}^l \times \mathbb{R}^l$ is the set of all the points $\overrightarrow{\varphi}_t^i$ corresponding to an evolution of the multiset inside any membrane.

Definition 10. Let Π be a DPP of degree n, for some $n \geq 1$. The space $\Phi^i \subseteq \mathbb{N}^l \times \mathbb{R}^l$ is called the phase space of the membrane i, $\Phi^E \subseteq \mathbb{N}^l \times \mathbb{R}^l$ is the phase space of the environment. The space $\Phi_\Pi = \Phi^0 \times \cdots \times \Phi^{n-1} \times \Phi^E \subseteq (\mathbb{N}^l \times \mathbb{R}^l)^{n+1}$ is called the *phase space* of the DPP.

Hence, the phase space of a DPP describes the evolution of the whole system, with respect to both the change of all multisets and the passing of time. Actually, in analyzing the behavior of a given DPP, we will be interested in considering only the phase space restricted to the regions specified in the set I (see Definition 1). Similarly, only the evolution of symbols from O will be analyzed for the multisets present in the regions appearing in I.

4   Case Study: The Belousov-Zhabotinskii Reaction

The BZ chemical reaction is considered the prototype oscillator and exhibits an extraordinary variety of temporal and spatial phenomena. Its oscillating behavior is one of the most widely studied, both theoretically and experimentally, thus making this reaction a suitable workbench for the capabilities of DPP. Its basic mechanism consists in the oxidation of malonic acid, in acid medium, by bromate ions and catalyzed by cerium, which has two states. The sustained periodic oscillations are observed in the cerium ions. The Brusselator is a simplified theoretical scheme introduced in [11] to explain the nonlinear oscillating behavior, and after that was carefully studied in, e.g., [13]. Despite the fact that it is physically unrealistic, as it involves a trimolecular state, it is recognized to be the skeleton for the explanation of the oscillating behavior in chemical reactions. Moreover, it has a very simple description: $A \xrightarrow{k_1} X, B + X \xrightarrow{k_2} Y + D, 2X + Y \xrightarrow{k_3} 3X, X \xrightarrow{k_4} E$.

In this section we describe the Brusselator in terms of DPP and we show the analysis and some results obtained from the simulations. Indeed, in order to describe a chemical or a biological system evolving over time, a kind of rule able to react to the variation of occurrences of symbols (that is, concentrations of substances) is needed. For this purpose, we believe that the dynamical probabilistic rules are really suitable, so we consider the DPP defined as $\Pi_{BZ} = (V, O, \mu, M_0, R_0, E_{BZ}, 0)$ where

- $V = \{A, B, X, Y\}$, $O = \{X, Y\}$;
- $\mu = [_0 \;]_0$;
- $M_0 = \{A^{m_1} B^{m_2} X^{m_3} Y^{m_4}\}$;
- R_0 consists of the rules

$$r_1 : A \xrightarrow{k_1} X$$

$$r_2 : BX \xrightarrow{k_2} Y$$

$$r_3 : XXY \xrightarrow{k_3} XXX$$

$$r_4 : X \xrightarrow{k_4} \lambda$$

for some $k_1, \ldots, k_4 \in \mathbb{R}^+$;

– the environment E_{BZ} is given by the alphabet {A,B}, the multiset $M_{E_{BZ}} = \{A^{n_1}, B^{n_2}\}$, for some $n_1, n_2 \in \mathbb{N}$, and the feeding rules $R_{E_{BZ}} = \{r_5 : A \longrightarrow (A, in_0), r_6 : B \longrightarrow (B, in_0)\}$.

Note that, with respect to the original equations in the Brusselator, we choose not to consider the chemicals D and E since they are not relevant for the system evolution. According to Definition 2, the set of parameters of Π_{BZ} is $\mathcal{P}_{BZ} = \{m_1, \ldots, m_4, k_1, \ldots, k_4, n_1, n_2\}$. A family $\mathcal{F}_{BZ}$ can be given by considering different values for the elements in $\mathcal{P}_{BZ}$.

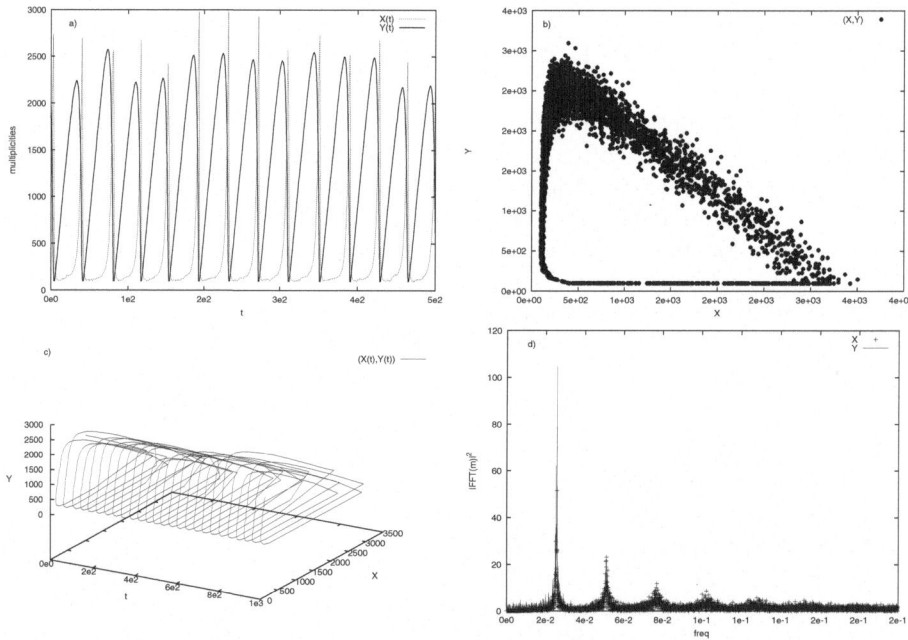

Fig. 1. Quasi-periodic cycle

The simulations based on the DPP approach have shown all the dynamical behaviors which characterize the continuously stirred BZ (see for example [3, 5, 13]); here we present the quasi periodic oscillations (in Figure 1, for $\mathcal{P}_{BZ}^{qp} = \{100, 100, 1000, 2000, 50, 0.5, 5 \cdot 10^{-5}, 5, 100, 100\}$) and the attractor (in Figure 2, for $\mathcal{P}_{BZ}^{att} = \{100, 100, 1000, 2000, 1, 1, 1, 1, 100, 100\}$). A fading transition from one to the other is possible by tuning the parameters in $\mathcal{P}_{BZ}$. Since

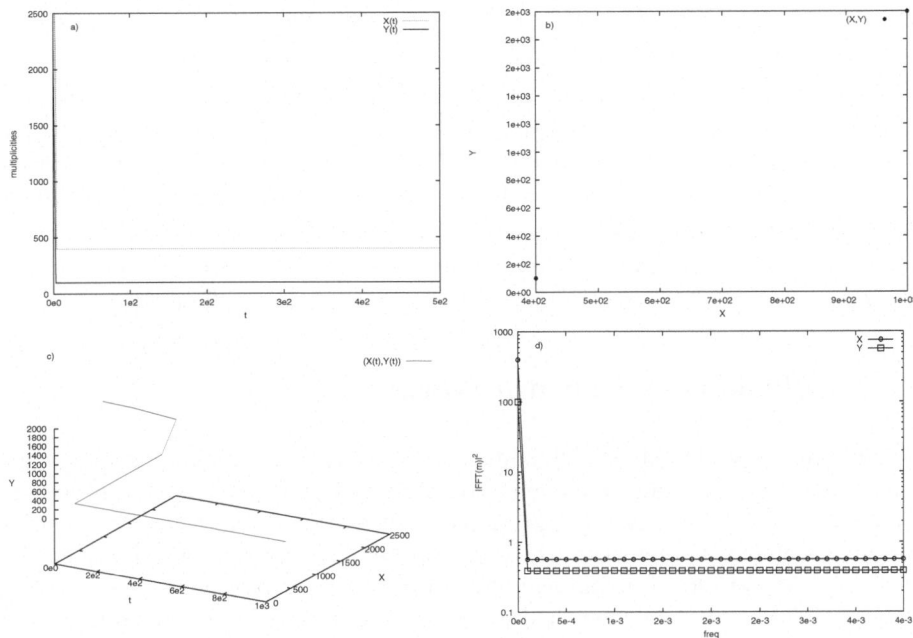

Fig. 2. Attractor

in the literature about the Brusselator the phase plane has been widely identified with the X-Y plane, our attention is focused on the dynamic of these symbols. A first characterization of the system dynamic can be obtained by looking directly to the temporal evolution of the two variables: Fig.1.(a) and Fig.2.(a) allow to discriminate the quasi periodic oscillation of the first case from the attracted dynamic of the second one. Fig.1.(b) and Fig.2.(b) show the phase space of membrane 0: in the first case we obtain a limit cycle, in the second case only the initial multiset (point at right-up corner) and the attractor (point at left-bottom corner) can be displayed. Fig.1.(c) and Fig.2.(c) show the evolution of multiplicities of X and Y; the projection on $X - Y$ plane of these pictures obviously correspond to Fig.1.(b), Fig.2.(b), respectively. Finally, Fig.1.(d) and Fig.2.(d) show the spectra: in the first case, the spectrum shows the highest peak, corresponding to the principal oscillation frequency, and some other harmonics, plus the stochastic contribute which is spread all over the other frequencies; in the second case (where the Y axis is in logarithmic scale), the spectrum corresponds to a δ of Dirac centered in the 0 frequency (the height of δ is equal to the mean value of the multiplicities of X and Y), since this is the Fourier transform of a constant (in time) signal.

Remark 2. To make clear the definitions of Section 3, we give some examples by extracting three t-multisets from the simulated evolution of $(\Pi_{BZ}, \mathcal{P}_{BZ}^{qp})$. Chosen the t-multisets $M_{39} = \{100, 100, 1921, 1029, 39\}$, $M_{40} = \{100, 100, 2701, 262, 40\}$, $M_{53} = \{100, 100, 109, 1055, 53\}$, their positions are $\overrightarrow{M}_{39} = (100, 100, 1921, 1029,$

39), $\vec{M}_{40} = (100, 100, 2701, 262, 40)$, $\vec{M}_{53} = (100, 100, 109, 1055, 53)$. The distance between M_{53} and M_{39} is $d(\vec{M}_{53}, \vec{M}_{39}) = (0 + 0 + (-1812)^2 + 26^2)^{1/2} \approx$ 1812.19, while the displacement is $\vec{u}(\vec{M}_{53}, \vec{M}_{39}) = (0, 0, -1812, 26)$. The versor associated to this displacement is $\hat{u}(\vec{M}_{53}, \vec{M}_{39}) = (0, 0, -\frac{1812}{1812.19}, \frac{26}{1812.19}) \approx$ $(0, 0, -0.99, 0.0014)$, which says that the predominant direction of the motion is along the X axes (that is, the highest variation occurs for the multiplicities of the symbol X). The average velocity $\vec{v}(\vec{M}_{53}, \vec{M}_{39}) = (0, 0, -\frac{1812}{14}, \frac{26}{14}) \approx$ $(0, 0, -129.43, 1.86)$ is quite different from the instantaneous one, which is $\vec{v}(\vec{M}_{39}) = (0, 0, 780, -767)$ (evaluated between time steps 39 and 40).

5   Conclusions and Future Work

In this paper we introduced dynamical probabilistic P systems as a new approach for describing and analyzing complex biological or chemical processes. We also sketched some novel definitions, such as timed-multisets, the position and displacement of a multiset, the phase space of a P system, which are needed for the investigations of dynamical properties of the system of interest.

In particular, we applied such system to the analysis of well-known Belousov-Zhabotinskii reaction, showing that we can simulate the behavior of chemical oscillator reactions. Indeed, the interaction of two or more oscillating systems is of interest for many biological processes and systems, as it constitutes an important factor to keep alive an organism or a complex system constituted by several sub-components of different types.

The future work will consist in a further deep investigation of our model, both from a theoretical and an experimental point of view, e.g., by considering also non-synchronized evolutions, as well as in its use for the analysis of complex cellular processes. For instance, we are currently applying dynamical probabilistic P systems and the tools here introduced to the analysis of the role of protein p53 in cell growth arrest and apoptosis.

References

1. I.I. Ardelean, D. Besozzi, M.H. Garzon, G. Mauri, S. Roy, *P system models for mechanosensitive channels*, in [4].
2. I.I. Ardelean, M. Cavaliere, Modelling biological processes by using a probabilistic P system software, *Natural Computing*, 2 (2003), 173-197.
3. L. Bianco, F. Fontana, G. Franco, V. Manca, *P systems for biological dynamics*, in [4].
4. G. Ciobanu, G. Păun, M.J. Pérez-Jiménez eds., *Applications of Membrane Computing*, Springer–Verlag, Berlin, in press.
5. D.T. Gillespie, Exact stochastic simulation of coupled chemical reactions, *Journ. Phys. Chem.*, 81 (1977), 2340-2361.
6. A. Obtułowicz, G. Păun, (In search of) probabilistic P systems, *BioSystems*, 70 (2003), 107-121.

7. M. Madhu, Probabilistic rewriting P systems, *Int. J. Found. Comput. Sci.*, 14, 1 (2003), 157-166.
8. G. Păun, Computing with membranes, *Journal of Computer and System Sciences*, 61, 1 (2000), 108-143.
9. G. Păun, *Membrane Computing. An introduction.* Springer–Verlag, Berlin, 2002.
10. M.J. Pérez-Jiménez, F.J. Romero-Campero, Modelling EGFR signalling cascade using continuous membrane systems, *Pre-Proceedings of CMSB* (G. Plotkin ed.), Edinburgh, 3-5 April 2005, 118-129.
11. I. Prigogine, R. Lefever, Symmetry breaking instabilities in dissipative systems. II, *Journ. Chem. Phys.*, 48 (1968), 1695-1700.
12. Y. Suzuki, H. Tanaka, Abstract rewriting systems on multisets and their application for modelling complex behaviours, *Rovira i Virgili Univ. Tech. Rep. 26* (M. Cavaliere, C. Martín-Vide, G. Păun, eds.), Brainstorming Week on Membrane Computing, Tarragona 2003, 313-331.
13. J.J. Tyson, Some further studies of nonlinear oscillations in chemical systems, *Journ. Chem. Phys.*, 58 (1973), 3919-3930.

Experimental Validation of DNA Sequences for DNA Computing: Use of a SYBR Green I Assay

Wendy K. Pogozelski[1,*], Matthew P. Bernard[1,**],
Salvatore F. Priore[1,***], and Anthony J. Macula[2,†]

[1] Department of Chemistry, SUNY Geneseo, Geneseo, NY 14454
pogozels@geneseo.edu
[2] Department of Mathematics, SUNY Geneseo, Geneseo, NY 14454
macula@geneseo.edu

Abstract. In developing hybridization-based DNA computing methods, DNA codes must be created that behave as predicted; otherwise computing errors can result. Here we describe the experimental validation of two DNA codes, each containing 16-nucleotide strands designed to hybridize only with their complements and not with themselves or with any other strands in the set. Code I was constructed simply to restrict potential for cross-hybridized (CH) secondary structure. Code II was constructed using the software *SynDCode*, incorporating nearest-neighbor thermodynamics and generalizations of the Levenshtein edit distance. Every combination of strands was tested for potential to mispair, both in individual pairings and in pools. Since the strands are designed to be linked together in a long bit string, we also tested end-to-end junctions of Code II strands. Hybridization was examined by measuring fluorescence as a function of temperature in the presence of SYBR Green I, a dye whose fluorescence increases exponentially when bound to double-stranded DNA. This method shows promise as a means for rapid experimental validation of large numbers of sequences.

1 Introduction

The success of hybridization-based architectures for DNA computing depends on the predictability of the behavior of the DNA. DNA strands must have far greater affinity for their canonical Watson-Crick base-paired reverse complements than for any other strand in the set. Sequences must be carefully designed to avoid secondary structures such as loops (both symmetrical and non-symmetrical) and to avoid misalignments, non-Watson-Crick base pairs and other mismatches.

* Partially supported by FA8750-04-2-0218, Air Force Research Laboratory, AFRL, IFTC, Rome, NY and NSF- UBM 0436298.
** Partially supported by FA8750-04-2-0218, AFRL, IFTC, Rome, NY and SUNY Geneseo Foundation.
*** Partially supported supported by NSF-UBM 0436298 and SUNY Geneseo Foundation.
† Partially supported by AFOSR F30602-03-C-0059 and NSF-UBM 0436298.

A. Carbone and N.A. Pierce (Eds.): DNA11, LNCS 3892, pp. 248–256, 2006.

Furthermore, DNA codes must be designed with the goal of finding a fixed temperature that is well below the melting point of properly-paired duplexes and well above the melting point of mispaired duplexes. Lastly, properly-paired sequences in a code should have similar melting points.

The affinity of one DNA strand for another and the stability of the resultant helix are dependent on several factors. The greatest contribution, according to both mathematical models and empirical verification, is the vertical stacking (mainly $\pi - \pi$ interactions) of adjacent base pairs [1]. Therefore, the identities of the nearest-neighbor bases are crucially important, as they determine the extent of stacking [2]. The nearest-neighbor model has been extended for heteroduplex stability to include parameters for the interactions that arise with mismatches [3], [4] These models were used in designing a DNA code in which potential mispairing was minimized.

Here we show how the reliability of these DNA strands can be validated experimentally. We describe a method employing the dye SYBR Green I and a Sequence Detection System, also known as a Real-time PCR thermalcycler. This instrument contains a light source, various filters, a 96-well platform, a programmable heating/cooling apparatus and a fluorescence detector capable of monitoring SYBR Green absorption and emission. While there are some limitations to this method, it is fast and suitable for testing large numbers of sequences.

SYBR Green I is a nonsymmetric positively-charged cyanine dye whose fluorescence emission at 510–520 nm increases markedly in the presence of double-stranded DNA. Recently, a structure was proposed for SYBR Green on the basis of nuclear magnetic resonance and mass spectrometry experiments [5]. (See Figure 1). The compound's chemical name, which is proprietary, was reported on the basis of these experiments to be to be [2-[N-(3-dimethylaminopropyl) -N-propylamino]-4-[2,3-dihydro-3-methyl-(benzo-1,3-thiazol-2-yl)-methylidene] - 1 - phenyl -quinolinium.

In this study, we exploited SYBR Green's binding and fluorescent properties to distinguish between DNA strands likely to form mismatches, and strands with suitably large preference for their reverse complements.

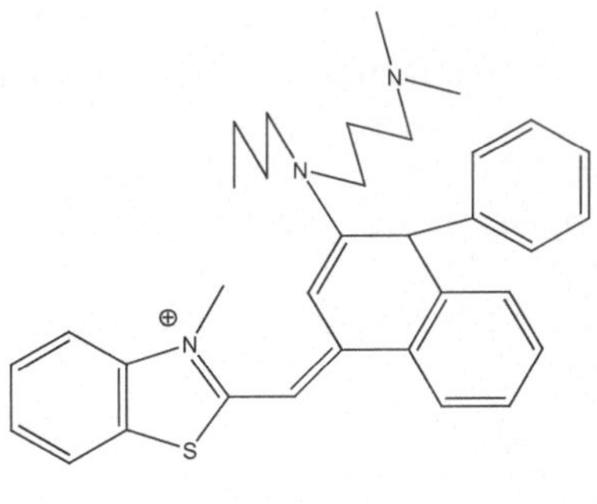

Fig. 1. Proposed structure of SYBR Green I

2   Methods

2.1   Creation of DNA Strands

We first created two codes (I and II) comprised of primary strands of length 16 and their Watson-Crick complements. Code I is actually a subset of a code of 26 pairs constructed so that, apart from the 26 proper Watson-Crick pairs (here referred to as WC pairs), no other pair of different or identical strands would have a common subsequence of length greater than nine. In terms of the well-known insertion-deletion distance (very similar to the sequence alignment Levenshtein edit distance), each pair of crosshybridized strands (here referred to as CH strands) requires at least 14 insertions and/or deletions to transform one into the other. Because the code is closed under complementation, these equivalent conditions imply that any mispaired duplex can have at most nine complementary base pairings in any secondary structure without pseudoknots. Code I was constructed simply to reduce crosshybridizations by severely restricting the potential for CH secondary structure. The only other enhancement to the properties of Code I was the condition that the subword GGGG not appear in any strand. The method did not use the state-of-the-art nearest-neighbor thermodynamic model for DNA hybridization. A subset of 13 pairs of this code was tested simply as an initial study of our validation method. The advantage of using this suboptimal code was that it gave us a few sequences that we knew had very different melting temperatures and would probe the limits of our assay.

Code II is a subcode of a complemented DNA code constructed by the software tool *SynDCode* containing 233 WC pairs. *SynDCode* incorporates the nearest-neighbor thermodynamics and generalizations of the Levenshtein edit distance. Specifics are described [6],[7].

1. Each WC duplex $x : \overline{x}$ has ΔG_{hyb} between -16 kcal/mol and -20 kcal/mol.
2. GGG does not appear in any codeword.
3. Each CH duplex $x : y$ has ΔG_{hyb} greater than -10 kcal/mol.
4. Each CH duplex $x : y$ has at most eight stacked pairs (2-stems) in any secondary structure.
5. Each CH duplex $x : y$ has at most four stacked triples (3-stems) in any secondary structure. Moreover, if x is a codeword and z is the 16-mer in the center (bases 9-24) of any junction strand (of which x is not a part), then the same conditions 2-5 hold for the CH duplex $x : z$. This is if:
   a. GGG does not appear in any junction strand.
   b. Each CH duplex $x : z$ has its ΔG_{hyb} greater than -10 kcal/mol.
   c. Each CH duplex $x : z$ has at most eight stacked pairs (2-stems) in any secondary structure
   d. Each CH duplex $x : z$ has at most four stacked triples (3-stems) in any secondary structure.

Code II was also designed to take into account the sequences that would result when strands are ligated together. In other words, the strands were designed with the extra constraint to minimize mispairing that could potentially occur in junctions of sequences. For this purpose, junction oligonucleotides were constructed that represented the latter half (3'-end) of one strand and the first half (5'-end) of the sequence it would be joined to if two primary strands were ligated together. Strands could be ligated with either a primary strand or a complement. We tested 32 pairs (16 strands and 16 complements) as well as the 64 junction strands that would arise from the ligation of the various strands and complements.

The DNA oligonucleotides were synthesized using phosphoramidite chemistry and were desalted (InVitrogen). It should be noted that we tested more highly-purified strands (HPLC-purified) and found that the additional purification did not alter our experimental results in any observable way. Because these strands are short, truncation of the strands is not likely to be a problem. Lyophilized oligonucleotides were dissolved in 10 mM Tris buffer/1 mM EDTA for a concentration of 1 μg/μL (0.48 M). All water used in dilutions and buffers was distilled and deionized via a Millipore purification system.

2.2   Fluorescence Measurements

Immediately prior to insertion into the sequence detection system, strands were heated to 90 °C in a standard thermalcycler to remove secondary structure. Samples were then slowly cooled to 25 °C to allow duplexes to form. Upon reaching 25 °C, samples were pipetted into a 96-well plate (Applied Biosystems) and placed in an Applied Biosystems Model 7000 Sequence Detection System.

In testing Code I, every possible combination of the 26 DNA strands was made. To do so, each well of the 96-well plates consisted of 0.5 μg of each oligonucleotide, 1X SYBR Green I Master Mix that included a buffer and SYBR Green (Applied Biosystems), and enough distilled deioinized water for a 50 μL volume. The Master Mix included a passive reference for standardization of the fluorescence. It was important to keep the concentration of SYBR Green constant, since excess SYBR Green can quench the DNA-mediated fluorescence [8]. The actual concentration of SYBR Green is proprietary, regardless of who manufactures it, but Zipper et al. have estimated that most preparations of 10000X are approximately 10 mg/mL5.

Fluorescence emission was monitored at 520 nm over a 35 °C temperature window. Measurements were made by slowly increasing the temperature to 60–70 °C over a period of several minutes. The software collected raw fluorescence data (relative to the passive reference) and plotted it as a function of temperature. In addition, the data were converted into melting curves by plotting the negative derivative for fluorescence vs. temperature (-dF/dT vs. T). Data were exported to Microsoft Excel for additional analysis. The maximum of each derivative curve corresponds to the melting temperature (T_m) of the duplex. We show derivative plots in Figures 2 and 3 and a raw fluorescence plot in Figure 4.

Code II was tested in a slightly different manner. All strands were examined in the "pooled" format—that is; all strands were pooled in a single well to test for CH strands in the presence of all potentially competing strands. The approach was the following: one well contained all strands except the perfect reverse complement of the strand being tested. In this way we could see whether or not SYBR Green would bind to potential CH duplexes that would form in the absence of the WC duplex. In another well, the same strands were pipetted, but this time the strand able to form a WC duplex was included. This procedure allowed us to measure the fluorescence when the complement was present and compare it with the fluorescence that would be detected without the complement. Monitoring fluorescence in the absence of a known complement was an extra-stringent test of cross-hybridization. A similar protocol was followed in testing junction sequences.

In testing Code II, we obtained plots of both raw fluorescence and the derivative of the fluorescence, but here we report only the former, since we are less interested in the T_m.

3   Results and Discussion

3.1   Use of Fluorescence to Monitor Hybridization

SYBR Green I shows greatly increased fluorescence when bound to double-stranded DNA. The change in fluorescence as a function of temperature is low at very low temperatures, shows a maximum at the melting temperature (T_m) of the sequence, and returns to low values at higher temperatures. A typical illustration of the change in fluorescence beginning at the T_m can be seen in Figure 2. This figure shows the sequence 5'-AGAAACGGACTAGTGG-3' being tested for hybridization with its complement as well as against each of the other sequences of the code. The change in fluorescence(-dF/dT) is plotted as a function of temperature. Each curve represents a different pair combination of sequences. The magnitude of the change in fluorescence for the test sequence binding to its complement is far greater than that of any other combination. Moreover, the temperature which corresponds to the maximum in the curve, the melting temperature or T_m is far greater for the WC duplex than for any other combination. This observation indicates that the test sequence with its complement is a far more stable than any other combination of strands. The code strand thus shows very little potential to cross-hybridize with other strand in the set other than its WC complement, even when the WC complement is absent from the mixture.

An example of a strand that *does* cross-hybridize with other strands in the code or pairs to itself is shown in Figure 3. In this plot, the sequence 5'-$A_4T_8A_4$-3' shows the highest fluorescence when combined with the complement but the strand can form competing duplex regions that bind SYBR Green. In particular, it binds with itself. This strand is clearly not acceptable for DNA computing. Although it is obvious from inspection that this strand would form CH duplexes, it is nonetheless useful to show the type of results that a CH duplex would yield using this experimental validation method.

Analysis of the experimental results for all sequences of Code I led to the following conclusion. If the derivative fluorescence data of a CH duplex was more than 10% that of the Watson-Crick duplex, then that particular sequence

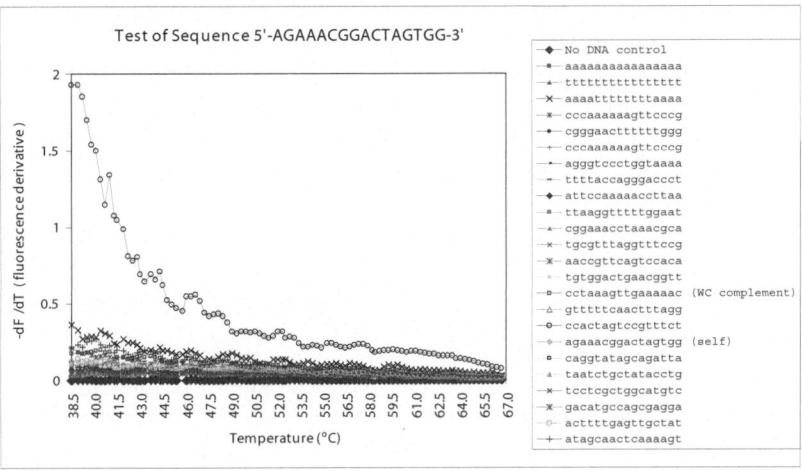

Fig. 2. Rate of change of fluorescence as a function of temperature for the sequence 5'-AGAAACGGACTAGTGG-3' with its perfect complement as well as with itself, and with every other sequence in the set. The maximum corresponds to the T_m.

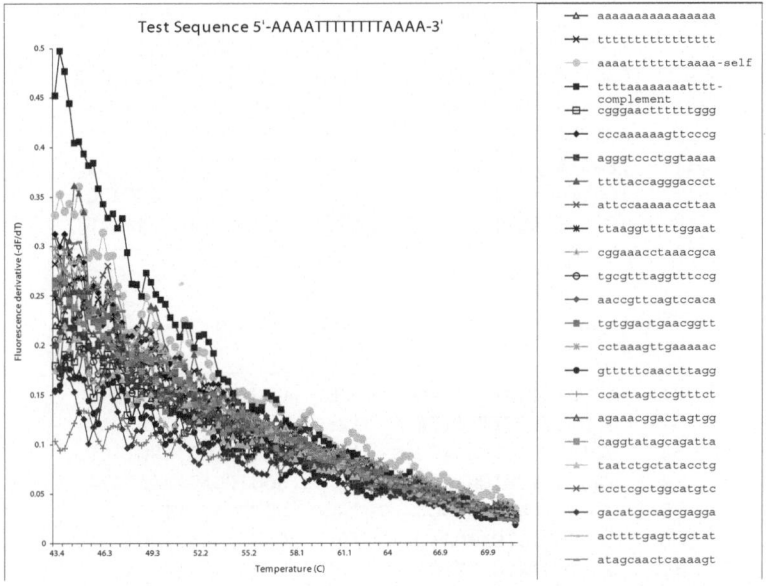

Fig. 3. Example of a sequence (5'-$A_4T_8A_4$-3') that forms competing CH duplexes with itself and with other sequences in the code

and its complement should be omitted from the final DNA code. The resulting collection of nine complementary pairs of sequences has the property that each forms a Watson-Crick duplex that is at least ten times as favorable as any of the

162 potential cross-hybridized duplexes. The revised set of sequences is shown in Table 1 (without parentheses).

We then tested a subcode of the more rigorously-designed Code II. These experiments were performed a bit differently in that the strands were pooled rather than tested in pair combinations. All of the strands of Code II were found to be acceptable by the above test; they showed far greater affinity for their reverse complement than for other members of the set or for self-binding.

Since we envision an architecture in which these individual strands will ultimately be linked together to form much longer bit strings, we also needed to test whether the junctions of these sequences might be able to mispair. One could

Table 1. Subset of Code I. This code included some known mismatches to test our assay. The codewords in parentheses showed cross-hybridization. "S" indicates primary strand; "C" indicates complement.

S1.	(AAAAAAAAAAAAAAAA)	C1.	(TTTTTTTTTTTTTTTT)
S2.	(AAAATTTTTTTTAAAA)	C2.	(TTTTAAAAAAAATTTT)
S3.	CGGGAACTTTTTTGGG	C3.	CCCAAAAAAGTTCCCG
S4.	(AGGGTCCCTGGTAAAA)	C4.	(TTTTACCAGGGACCCT)
S5.	ATTCCAAAAACCTTAA	C5.	TTAAGGTTTTTGGAAT
S6.	CGGAAACCTAAACGCA	C6.	TGCGTTTAGGTTTCCG
S7.	AACCGTTCAGTCCACA	C7.	TGTGGACTGAACGGAA
S8.	(CGCGGGCCCACCAATT)	C8.	(AATTGGTGGGCCCGCG)
S9.	CCTAAAGTTGAAAAAC	C9.	GTTTTTCAACTTTAGG
S10.	CCACTAGTCCGTTTCT	C10.	AGAAACGGACTAGTGG
S11.	CAGGTATAGCAGATTA	C11.	TAATCTGCTATACCTG
S12.	TCCTCGCTGGCATGTC	C12.	GACATGCCAGCGAGGA
S13.	ACTTTTGAGTTGCTAT	C13.	ATAGCAACTCAAAAGT

Table 2. Subset of Code II Designed with *SynDCode*

S1.	AGGCTAAAGTTATCAC	C1.	GTGATAACTTTAGCCT
S2.	GTCTTCGTTTTTTTCA	C2.	TGAAAAAAACGAAGAC
S3.	GCAAGCGACCAATACT	C3.	AGTATTGGTCGCTTGC
S4.	TACCTTTTCTCGACGC	C4.	GCGTCGAGAAAAGGTA
S5.	CTCAATAAAATGCGCG	C5.	CGCGCATTTTATTGAG
S6.	CGTTGCACTCAAGATC	C6.	GATCTTGAGTGCAACG
S7.	GACTGGAATGTTTTGT	C7.	ACAAAACATTCCAGTC
S8.	GGATGCAGGTTGATTA	C8.	TAATCAACCTGCATCC
S9.	AAGCCTTAGAAGAGAG	C9.	CTCTCTTCTAAGGCTT
S10.	TTTCTGTGGCACTGGT	C10.	ACCAGTGCCACAGAAA
S11.	TGTGTGTCCGATGAGA	C11.	TCTCATCGGACACACA
S12.	TTAAAAGACGTTGGTT	C12.	AACCAACGTCTTTTAA
S13.	TACGCTAATCGGTAAG	C13.	CTTACCGATTAGCGTA
S14.	TGGAGGAACTACCGGA	C14.	TCCGGTAGTTCCTCCA
S15.	CCATAGCTGAGTTCTT	C15.	AAGAACTCAGCTATGG

imagine that the new sequence created when two strands are ligated end-to-end might potentially form CH duplexes. For example, while strand S1 might not hybridize to any other strand in the code except its perfect complement C1, we needed to consider whether or not S1/S2 (strands S1 and S2 ligated together) or S1/C2 (strands S1 and C2 ligated together) would have affinity for other oligonucleotides in the mixture. Therefore, we created 56 strands of 32 nucleotides in length that represented junctions of sequences. We studied the fluorescence of these junction strands both in the presence and absence of their complements. The complement would be expected to bind to only half of the junction strand. For example, if testing junction strand S14/C15, we would expect strand S15 (the complement of C15) to bind to the latter (3') portion (the C15 portion) of the junction strand.

Typical results for a successful experiment to test junction sequences are shown in Figure 4. The graph shows the junction sequence S14/C15 being tested against all other strands in the code. Here we show raw fluorescence rather than derivative fluorescence. All of the primary strands (S1–S15) and complementary strands (C1–C15) show baseline fluorescence when pooled, indicating that there is no appreciable duplex formation among these strands. When the junction sequence S14/C15 (bearing half of S14 and half of C15 as illustrated in Figure 4) is added to the pool of strands C1–C15, there is no additional fluorescence, indicating that the junction strand does not cross-hybridize to any strand in

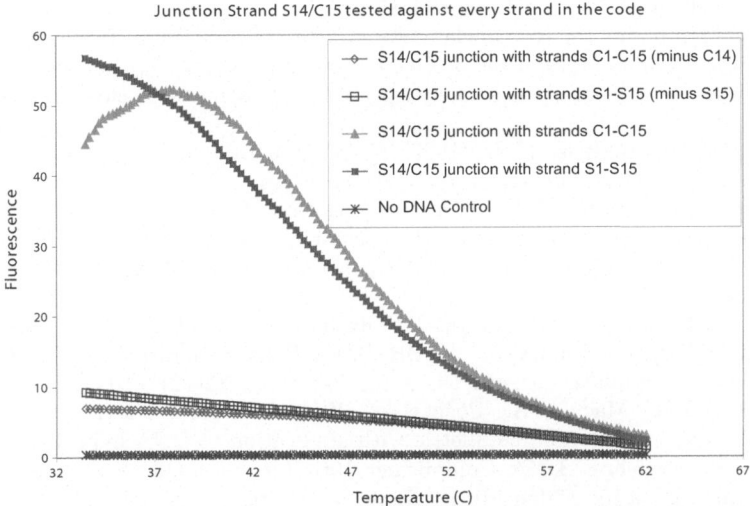

Fig. 4. Example of fluorescence experiment testing a junction sequence against all other strands in the DNA code. Closed squares and triangles (the topmost curves) show fluorescence of the duplex created when half the complementary strand is present. The open symbols show fluorescence when known complements are omitted. The low fluorescence for these bottom curves indicates little or no hybridization.

this set. In mixing the junction sequence S14/C15 with the complement "S" pool of strands (S1–S15), we had to be a little more careful and take extra steps. We knew that junction sequence S14/C15 would hybridize with C14 and S15. Therefore, we tested binding to the "S" pool both in the presence and absence of these complements. Fluorescence was barely above baseline when all the "S" strands (minus S14 and S15) were mixed with the junction C14/C15 and rose dramatically when S15 was added. This is the behavior expected for sequences that do not cross-hybridize. Therefore, these sequences appear to be well-behaved enough to be used in DNA computing. All of the strands in Code II were found to be acceptable.

4   Conclusions

The method as described here does have a few disadvantages; for example, the Sequence Detection System is automated and doesn't permit much user control. Users are limited to measuring fluorescence within a 35 °C temperature range, and the rate of data sampling cannot be altered. However, the method is fast, suitable for screening large numbers of strands, and is easy, robust, and effective.

References

1. Borer, P.N., Gengler, B., Tinoco, I., Jr. (1974) Stability of ribonucleic acid double-stranded helices. J. Mol. Biol. 86, 843–853.
2. Freier, S.M., Sugimoto, N., Sinclair, A., Alkema, D., Neilson, T., Kierzek, R. et al. (1986). Stability of XGCGCp- GCGCYp- and XGCGCYp helixes: an empirical estimate of the energetics of hydrogen bonds in nucleic acids. Biochemistry, 25- 3214–3219.
3. Allawi, H.T., SantaLucia, J., Jr. (1997) Thermodynamics and NMR of internal G-T mismatches in DNA. Biochemistry, 36, 10581–10594.
4. McDowell, J.A., Turner, D.H. (1996) Investigation of the structural basis for thermodynamic stabilities of tandem GU mismatches: solution structure of (rGAG-GUCUC)2 by two-dimensional NMR and simulated annealing, Biochemistry, 35, 14077–14089.
5. Zipper, H., Brunner, H., Bernhage, J., Vitzthum, F. Investigations on DNA intercalation and surface binding by SYBR Green I, its structure determination and methodological implications. Nucleic Acids Res. 2004, 32(12), e103.
6. D'yachkov A.G., Macula, A., Pogozelski, W.K., Renz, T.E., Rykov, V., Torney, D.C. An insertion-deletion like metric with application to DNA hybridization thermodynamic modeling, DNA Computing: 10th International Workshop on DNA Computing, DNA10, Milan, Italy, June 7–10, 2004, Revised Selected Papers, Springer-Verlag, LNCS, Volume 3384, 90–103, 2005.
7. Bishop, M., Macula, A.J., Pogozelski, W.K., Renz, T.E., Rykov, V.V., SynDCode: Cooperative DNA code generating software, DNA 11 Conf. Preproceedings, London, Ont. June 6–9, 2005 .
8. Lipsky, R.H., Mazzanti, C.M., Rudolph, J.G., Xu, K., Vyas, G., Bozak, D., Radel, M.Q., Goldman, D. (2001) DNA melting analysis for detection of single nucleotide polymorphisms, Clinical Chemistry, 47, 635–644.

Complexity of Graph Self-assembly in Accretive Systems and Self-destructible Systems*

John H. Reif[1], Sudheer Sahu[1], and Peng Yin[1]

Department of Computer Science, Duke University,
Box 90129, Durham, NC 27708-0129, USA
{reif, sudheer, py}@cs.duke.edu

Abstract. Self-assembly is a process in which small objects autonomously associate with each other to form larger complexes. It is ubiquitous in biological constructions at the cellular and molecular scale and has also been identified by nanoscientists as a fundamental method for building nano-scale structures. Recent years see convergent interest and efforts in studying self-assembly from mathematicians, computer scientists, physicists, chemists, and biologists. However most complexity theoretic studies of self-assembly utilize mathematical models with two limitations: 1) only attraction, while no repulsion, is studied; 2) only assembled structures of two dimensional square grids are studied. In this paper, we study the complexity of the assemblies resulting from the cooperative effect of repulsion and attraction in a more general setting of graphs. This allows for the study of a more general class of self-assembled structures than the previous tiling model. We define two novel assembly models, namely the accretive graph assembly model and the self-destructible graph assembly model, and identify one fundamental problem in them: the sequential construction of a given graph, referred to as Accretive Graph Assembly Problem (AGAP) and Self-Destructible Graph Assembly Problem (DGAP), respectively. Our main results are: (i) AGAP is **NP**-complete even if the maximum degree of the graph is restricted to 4 or the graph is restricted to be planar with maximum degree 5; (ii) counting the number of sequential assembly orderings that result in a target graph (#AGAP) is #**P**-complete; and (iii) DGAP is **PSPACE**-complete even if the maximum degree of the graph is restricted to 6 (this is the first **PSPACE**-complete result in self-assembly). We also extend the accretive graph assembly model to a stochastic model, and prove that determining the probability of a given assembly in this model is #**P**-complete.

1 Introduction

Self-assembly is the ubiquitous process in which small objects associate autonomously with each other to form larger complexes. For example, atoms can self-assemble into molecules; molecules into crystals; cells into tissues, *etc*. Recently, self-assembly has also been explored as a powerful and efficient mechanism for constructing synthetic molecular scale objects with nano-scale features. This approach is particularly fruitful in DNA based nanoscience, as exemplified by the diverse set of DNA lattices made from

* The work is supported by NSF ITR Grants EIA-0086015 and CCR-0326157, NSF QuBIC Grants EIA-0218376 and EIA-0218359, and DARPA/AFSOR Contract F30602-01-2-0561.

A. Carbone and N.A. Pierce (Eds.): DNA11, LNCS 3892, pp. 257–274, 2006.

self-assembled branched DNA molecules (DNA tiles) [9, 15, 23, 25, 30, 43, 44]. Another nanoscale example is the self-assembly of peptide molecules [8]. Self-assembly is also used for mesoscale construction, for example, via the use of capillary forces [29] or magnetic forces [1] to provide attraction and repulsion between mesoscale tiles and other objects.

Building on classical Wang tiling models [40] dating back to 1960s, Rothemund and Winfree [31] in 2000 proposed an elegant discrete mathematical model for complexity theoretic studies of self-assembly known as the *Tile Assembly Model*. In this model, DNA tiles are treated as oriented unit squares (*tiles*). Each of the four sides of a tile has a glue with a positive integral strength. Assembly occurs by accretion of tiles iteratively to an existing assembly, starting with a distinguished *seed* tile. A tile can be "glued" to a position in an existing assembly if the tile can fit in the position such that each pair of abutting sides of the tile and the assembly have the same glue and the total strength of the glues is greater than or equal to the *temperature*, a system parameter. Research in this field largely focuses on studying the complexity of and algorithms for (uniquely and terminally) producing assemblies with given properties, such as shape. It has been shown that the construction of $n \times n$ squares has a program size complexity (the minimum number of distinct types of tiles required) of $\Theta(\frac{\log n}{\log \log n})$ [3, 31]. The upper bound is obtained by simulating a binary counter and the lower bound by analyzing the Kolmogorov complexity of the tiling system. The model was later extended by Adleman *et al.* to include the time complexity of generating specified assemblies [3]. Later work studies various topics, including combinatorial optimization, complexity problems, fault tolerance, and topology changes, in the standard Tile Assembly Model as well as some of its variants [4, 6, 10, 11, 12, 13, 14, 19, 27, 33, 34, 35, 36, 37, 38, 41, 42].

Though substantial progress has been made in recent years in the study of self-assembly using the above tile assembly model, which captures many important aspects of self-assembly in nature and in nano-fabrications, the complexity of some other important aspects of self-assembly requires further study:

- Only attraction, while no repulsion, is studied. However, repulsive forces often occur in self-assembly. For example, there is repulsion between hydrophobic and hydrophilic tiles [7, 29]; between tiles labeled with magnetic pads of the same polarity [1]; and there is also static electric repulsion in molecular systems, *etc.*. Indeed, the study of repulsive forces in the self-assembly system was posed as an open question by Adleman and colleagues in [3]. Though there has been previous work on the kinetics of such systems [20], no complexity theoretic study has been directed towards such systems.
- Tile Assembly Model captures well assembled structures of two dimensional square grids, but are not well adaptable to study assemblies of general graph structure. However, many molecular self-assemblies using DNA and other materials involve the assembly of more diverse graph-like structures in both two and three dimensions. Pioneer work in modeling DNA self-assembly as graphs include [16, 17, 18, 32]. In particular, Jonoska et al studied the computational capacity of the self-assembly of realistic DNA graphs and showed that 3SAT and 3-colorability problems can be solved in constant laboratory steps in theory [16, 17, 18]. In addition, Seeman's group have experimentally constructed

topoisomers of self-assembled DNA graphs [32]. Klavins showed how to produce desired topology of self-assembled structures with planar graph structure using graph grammars [21, 22].

In this paper, we study the cooperative effect of repulsion and attraction on the complexity of the self-assembly system in a graph setting. This approach thus allows the study of a more general class of assemblies.

We distinguish two systems, namely the *accretive system* and the *self-destructible system*. In an accretive system, an assembled component cannot be removed from the assembly. In contrast, in the self-destructible system, a previously assembled component can be "actively" removed from the assembly by the repulsive force exerted by another newly assembled component. In other words, the assembly can (partially) *destruct* itself. We define the *accretive graph assembly model* for the former and the *self-destructible graph assembly model* for the latter.

We first define an accretive assembly model and study a fundamental problem in this model: the sequential construction of a given graph, referred to as Accretive Graph Assembly Problem (AGAP). Our main result for this model is that AGAP is **NP**-complete even if the maximum degree of vertices in the graph is restricted to 4; the problem remains **NP**-complete even for planar graphs (planar AGAP or PAGAP) with maximum degree 5. We also prove that the problem of counting the number of sequential assembly orderings that lead to a target graph (#AGAP) is #**P**-complete. We further extend the AGAP model to a stochastic model, and prove that determining the probability of a given assembly (stochastic AGAP or SAGAP) is #**P**-complete.

If we relax the assumption that an assembled component always stays in the assembly, repulsive force between assembled components can cause self-destruction in the assembly. Self-destruction is a common phenomenon in nature, at least in biological systems. One renowned example is apoptosis, or programmed cell death [39]. Programmed cell death can be viewed as a self-destructive behavior exercised by a multi-cellular organism, in which the organism actively kills a subset of its constituent cells to ensure the normal development and function of the whole system. It has been shown that abnormalities in programmed cell death regulation can cause a diverse range of diseases such as cancer and autoimmunity [39]. It is also conceivable that self-destruction can be exploited in self-assembly based nano-fabrication: the components that serve to generate intermediate products but are unnecessary or undesirable in the final product should be actively removed.

To the best of our knowledge, our self-destructible graph assembly model is the first complexity theoretic model that captures and studies the fundamental phenomenon of self-destruction in self-assembly systems. Our model is different from previous work on reversible tiling systems [2, 5]. These previous studies use thermodynamic or stochastic techniques to investigate the reversible process of tile assembly/disassembly: an assembled tile has a probability of "falling" off the assembly in a kinetic system. In contrast, our self-destructible system models the behavior of a self-assembly system that "actively" destructs part of itself.

To model the self-destructible systems, we define a self-destructible graph assembly model, and consider the problem of sequentially constructing a given graph, referred to

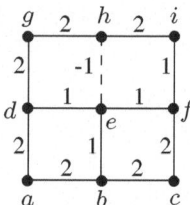

Fig. 1. An example of graph assembly in the accretive model

as the Self-Destructible Graph Assembly Problem (DGAP). We prove that DGAP is **PSPACE**-complete even if the graph is restricted to have maximum degree 6.

The rest of the paper is organized as follows. We first define the accretive graph assembly model and the AGAP problem in Section 2. In this model, we first show the **NP**-completeness of AGAP and PAGAP (planar AGAP) in Section 3 and then show the #**P**-completeness of SAGAP (stochastic AGAP) in Section 4. Next, we define the self-destructible graph assembly model and the DGAP problem in Section 5 and show the **PSPACE**-completeness of DGAP in Section 6. We close with a discussion of our results in Section 7.

2   Accretive Graph Assembly Model

Let $\mathbb{N}$ and $\mathbb{Z}$ denote the set of natural numbers and the set of integers, respectively. A *graph assembly system* is a quadruple $\mathcal{T} = \langle G = (V, E), v_s, w, \tau \rangle$, where $G = (V, E)$ is a given graph with vertex set V and edge set E, $v_s \in V$ is a distinguished *seed vertex*, $w : E \rightarrow \mathbb{Z}$ is a weight function (corresponding to the glue function in the standard tile assembly model [31]), and $\tau \in \mathbb{N}$ is the *temperature* of the system (intuitively temperature provides a tunable parameter to control the stability of the assembled structure). In contrast to the canonical tile assembly model in [31], which allows only positive edge weight, we allow both positive and negative edge weights, with positive (resp. negative) edge weight modeling the attraction (resp. repulsion) between the two vertices connected by this edge. We will see that this simple extension makes the assembly problem significantly more complex.

Roughly speaking, given a graph assembly system $\mathcal{T} = \langle G, v_s, w, \tau \rangle$, G is *sequentially constructible* if we can attach all its vertices one by one, starting with the seed vertex; a vertex x can be assembled if the *support* to it is equal to or greater than the system temperature τ, where support is the sum of the weights of the edges between x and its assembled neighbors.

Figure 1 gives an example. Here the temperature is set to 2. If h gets assembled before e, then the whole graph can get assembled: an example assembly ordering can be $a \prec b \prec c \prec d \prec f \prec g \prec h \prec i \prec e$. In contrast, if vertex e gets assembled before h, the graph cannot be assembled: h can be assembled only if it gets support from *both* g and i; while i cannot get assembled without the support from h.

Formally, given a graph assembly system $\mathcal{T} = \langle G, v_s, w, \tau \rangle$, G is *sequentially constructible* if there exists an ordering of *all* the vertices in V, $\mathcal{O}_\mathcal{T} = (v_s = v_0 \prec v_1 \prec$

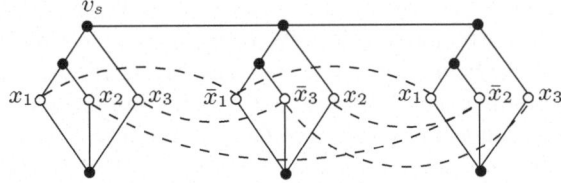

Fig. 2. A graph construction corresponding to an AGAP reduction from 3SAT formula $(x_1 \lor x_2 \lor x_3) \land (\bar{x}_1 \lor \bar{x}_3 \lor x_2) \land (x_1 \lor \bar{x}_2 \lor x_3)$. An edge between two literal vertices is depicted as a dashed arch and assigned weight -1; all other edges have weight 2.

$v_2 \prec \cdots \prec v_{n-1})$ such that $\sum_{v_j \in N_G(v_i), j < i} w(v_i, v_j) \geq \tau, 0 < i \leq n - 1$, where $N_G(v_i)$ denotes the set of vertices adjacent to v_i in G. The ordering $\mathcal{O}_{\mathcal{T}}$ is called an *assembly ordering* for G. $\sigma_{\mathcal{O}}(v_i) = \sum_{v_j \in N_G(v_i), j < i} w(v_i, v_j)$ is called the *support* of v_i in ordering $\mathcal{O}$. When the context is clear, we simply use $\mathcal{O}$ and $\sigma(v_i)$ to denote assembly ordering and support, respectively.

We define the *accretive graph assembly problem* as follows,

Definition 1. Accretive Graph Assembly Problem (AGAP): *Given a graph assembly system $\mathcal{T} = \langle G, v_s, w, \tau \rangle$ in the accretive model, determine whether there exists an assembly ordering $\mathcal{O}$ for G.*

The above model is *accretive* in the sense that once a vertex is assembled, it cannot be "knocked off" by the subsequent assembly of any other vertex. If we relax this assumption, we will obtain a self-destructible model, which is described in Section 5.

3   AGAP and PAGAP Are NP-Complete

3.1   4-DEGREE AGAP Is NP-Complete

Lemma 1. AGAP *is in* **NP**.

Proof. Given an assembly ordering of the vertices, sequentially check whether each vertex can be assembled. This takes polynomial time.   □

Recall that the **NP**-complete 3SAT problem asks: Given a Boolean formula ϕ in conjunctive normal form with each clause containing 3 literals, determine whether ϕ is satisfiable [26]. 3SAT remains **NP**-complete for formulas in which each variable appears at most three times, and each literal at most twice [26]. We will reduce this restricted 3SAT to AGAP to prove AGAP is **NP**-hard.

Lemma 2. AGAP *is* **NP**-*hard.*

Proof. Given a 3SAT formula ϕ where each variable appears at most three times, and each literal at most twice, we will construct below an accretive graph assembly system $\mathcal{T} = \langle G, v_s, w, \tau \rangle$ for ϕ. We will then show that the satisfiability problem of ϕ can be reduced (in logarithmic space) to the sequential constructibility problem of G in $\mathcal{T}$.

For each clause in ϕ, construct a *clause gadget* as in Figure 2. For each literal, we construct a *literal vertex* (colored white). We further add *top vertices* (black) above and *bottom vertices* (black) below the literal vertices. We next take care of the structure of formula ϕ as follows. Connect all the clause gadgets sequentially via their top vertices as in Figure 2; connect two literal vertices if and only if they correspond to two complement literals. This produces graph G. Designate the leftmost top vertex as the seed vertex v_s. We next assign weight -1 to an edge between two literal vertices and weight 2 to all the other edges. Finally, set the temperature $\tau = 2$. This completes the construction of $\mathcal{T} = \langle G, v_s, w, \tau \rangle$.

The following proposition implies the lemma.

Proposition 1. *There is an assembly ordering $\mathcal{O}$ for $\mathcal{T}$ if and only if ϕ is satisfiable.*

$\Rightarrow$

First we show that if ϕ can be satisfied by truth assignment T, then we can derive an assembly ordering $\mathcal{O}$ based on T.

Stage 1. Starting from the seed vertex, assemble all the top vertices sequentially. This can be easily done since each top vertex will have support 2, which is greater than or equal to $\tau = 2$, the temperature.

Stage 2. Assemble all the literal vertices assigned $true$. Since two $true$ literals cannot be complement literals, no two literal vertices to be assembled at this stage can have a negative edge between them. Hence all these $true$ literal vertices will receive a support 2 ($\geq \tau = 2$).

Stage 3. Assemble all the bottom vertices. Note that truth assignment T satisfies ϕ implies that every clause in ϕ has at least one $true$ literal. Thus every clause gadget in G has at least one literal vertex (a $true$ literal vertex) assembled in stage 2, which in turn allows us to assemble the bottom vertex in that clause gadget.

Stage 4. Assemble all the remaining literal vertices (the $false$ literal vertices). Observe that any remaining literal vertex v has support 4 from its already assembled neighboring top vertex and bottom vertex and that v can have negative support at most -2 from its assembled literal vertex neighbors (recall that each literal vertex can have at most two literal vertex neighbors since each variable appears at most three times in ϕ). Hence the total support for v will be at least 2 ($\geq \tau$).

$\Leftarrow$

Suppose that there exists an assembly ordering $\mathcal{O}$, then we can derive a satisfying truth assignment T for ϕ. For each literal vertex, assign its corresponding literal $true$ if it appears in $\mathcal{O}$ before *all* of its literal vertex neighbors (this assures no two complement literals are both assigned $true$); otherwise assign it $false$.

To show T satisfies ϕ, we only need to show every clause contains at least one $true$ literal. For contradiction, suppose there exists a clause gadget A with three $false$ literal vertices, where v is the literal vertex assembled first. However, v cannot be assembled: it has support 2 from the top vertex; no support from the bottom vertex (v gets assembled first and hence the bottom vertex in A cannot be assembled before v); at least -1 negative support from one of its literal vertex neighbors (v is assigned $false$); the total support of v is thus at most 1, less than temperature $\tau = 2$. Contradiction. Hence T must satisfy ϕ. $\square$

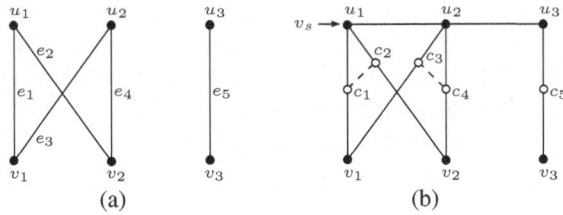

Fig. 3. (a) and (b) show an example bipartite graph B and the corresponding graph G used in the proof of Lemma 4, respectively. In (b), c_i's denote connector vertices (colored white); u_1 is the seed vertex. The weight of an edge connecting two connector vertices (dashed line) is -4; the weight of any other edge is 2.

We note that the technique of translating 3SAT formula into graph structure by modeling variables as vertices and connecting complement literals is a classical technique [26], and has also been used powerfully in other different graph self-assembly context [18].

The following theorem follows immediately from Lemma 1 and Lemma 2.

Theorem 1. AGAP *is* **NP**-*complete.*

Let k-DEGREE AGAP be the AGAP in which the largest degree of any vertex in graph G is k. Observe that the largest degree of any vertex in the graph construction in the proof of Lemma 2 is 4. Hence we have

Corollary 1. *4*-DEGREE AGAP *is* **NP**-*complete.*

3.2 5-DEGREE PAGAP Is NP-Complete

We next study the planar AGAP (PAGAP) problem, where the graph G in the assembly system $\mathcal{T}$ is planar. Here, we show PAGAP is **NP**-hard by a reduction from the **NP**-hard planar three-satisfiability problem (P3SAT) [24]. The reduction is in similar spirit as that in the proof of Lemma 1. For lack of space, we skip the proof and only state our results.

Theorem 2. PAGAP *is* **NP**-*complete.*

Corollary 2. *5*-DEGREE PAGAP *is* **NP**-*complete.*

4 #AGAP and SAGAP Are #P-Complete

4.1 #AGAP Is #P-Complete

We now consider a more general version of AGAP: given an accretive graph assembly system $\mathcal{T} = \langle G, v_s, w, \tau \rangle$ and a *target vertex set* $V_t \subseteq V$, determine if there exists an ordering $\tilde{\mathcal{O}}(V, V_t)$ of V such that V_t is assembled after we *attempt* assembling each vertex $v \in V$ sequentially according to $\tilde{\mathcal{O}}$. Vertex v will be assembled if there is enough support; otherwise it will not. $\tilde{\mathcal{O}}$ is called the *assembly ordering* of V for V_t. When the context is clear, we simply call it assembly ordering for V_t and denote it by $\tilde{\mathcal{O}}$. Note that the assembly ordering $\tilde{\mathcal{O}}$ is an ordering on all the vertices in V, but we only care about the assembly of the target vertex set V_t: the assembly of vertices in $V \setminus V_t$ is neither required nor prohibited. For $V_t = V$, the general AGAP is then precisely the standard AGAP. The problem of counting the number of assembly orderings for $V_t \subseteq V$ under this general AGAP model is referred to as #AGAP.

Lemma 3. #AGAP *is in* #P.

We next show #AGAP is #P-hard, using a reduction from the #P-complete problem PERMANENT, the problem of counting the number of perfect matchings in a bipartite graph [26].

Lemma 4. #AGAP *is* #P*-hard.*

Proof. Given a bipartite graph $B = (U, V, E)$ with two partitions of vertices U and V and edge set E, where $U = \{u_1, \ldots, u_n\}$, $V = \{v_1, \ldots, v_n\}$, and $E = \{e_1, \ldots, e_m\}$ (recall that by definition of bipartite graph, there is no edge between any two vertices in U and no edge between any two vertices in V), we construct an assembly system $\mathcal{T} = \langle G, v_s, w, \tau \rangle$. First, we derive graph G by adding vertices and edges to B (see Figure 3 for an example): on each edge e_k, add a splitting *connector vertex* c_k; add an edge (dashed line) between two connector vertices if they share a same neighbor in U; connect u_i and u_{i+1} for $i = 1, \ldots, n-1$. Next, assign weight -4 to an edge between two connector vertices; assign weight 2 to all the other edges. Finally, designate u_1 as the seed vertex v_s, and set the temperature $\tau = 2$. The target vertex set V_t is $U \bigcup V$.

A crucial property of G is that the assembly of one connector vertex c will make all of c's connector vertex neighbors unassemblable, due to the negative edge connecting c and its neighbors. Thus, starting from a vertex $u \in U$, only one connector vertex and hence only one $v \in V$ can be assembled. For a concrete example, see Figure 3 (b): starting from u_1, if we sequentially assemble c_1 and v_1, vertex c_1 will render c_2 unassemblable, and hence the assembly sequence $u_1 \prec c_2 \prec v_2$ is not permissible.

We first show that if there is no perfect matching in B, there is no assembly ordering for $U \bigcup V$. If there is no perfect matching in B, there exists $S \subseteq V$ s.t. $|N(S)| < |S|$ (Hall's theorem), where $N(S) \subseteq U$ is the set of neighboring vertices to the vertices in S in *original* graph B. However, as argued above, one vertex in U can lead to the assembly of at most one vertex in V. Thus $|N(S)| < |S|$ implies that at least one vertex in S remains unassembled. Hence, no assembly ordering exists that can assemble all vertices in $U \bigcup V$.

Next, when there exists perfect matching(s) in B, we can show that each perfect matching in B corresponds to a *fixed* number of assembly orderings for $U \bigcup V$. First note that the total number of vertices in graph G is $2n + m$ (recall that m is the number of edges in B and hence the number of connector vertices in G), giving a total $s = (2n + m)!$ permutations. We divide s by the following factors to get the number of assembly orderings for $U \bigcup V$.

1. For every matching edge e_k between $u \in U$ and $v \in V$, we have to follow the strict order $u \prec c_k \prec v$, where c_k is the connector vertex on e_k. This is ensured by our construction as argued above. There are altogether n such matching edges. So we need to further divide s by $(3!)^n$.

2. For the n vertices in U, we have to follow the strict order of assembling the vertices from left to right, and hence we need to divide s by $n!$.

3. Denote by d_i the degree of u_i in graph B. For the d_i connector vertices corresponding to the d_i edges incident on u_i, the connector vertex corresponding to the matching edge must be assembled first, and thus, we need to further divide s by $\Pi_{i=1}^n d_i$.

Putting together 1), 2), and 3), we have that each perfect matching in B corresponds to $\frac{(2n+m)!}{(3!)^n (n!)(\Pi_{i=1}^n d_i)}$ assembly orderings for $U \bigcup V$ in G. □

Lemma 3 and Lemma 4 imply

Theorem 3. #AGAP *is #P-complete.*

4.2 SAGAP Is #P-Complete

An intimately related question to counting the total number of assembly orderings is the problem to calculate the probability of assembling a target structure in a stochastic setting. We next extend the accretive graph self-assembly model to stochastic accretive graph self-assembly model. Given a graph $G = (V, E)$, where $|V| = n$, starting with the seed vertex v_s, what is the probability that the target vertex set $V_t \subseteq V$ gets assembled if anytime any unassembled vertex can be picked with equal probability? This problem is referred to as stochastic AGAP (SAGAP).

Since any unassembled vertex has equal probability of being selected and the assembly has to start with the seed vertex, the total number of possible orderings are $(n - 1)!$. Then SAGAP asks precisely how many of these $(n - 1)!$ orderings are assembly orderings for the target vertex set V_t. Thus, #AGAP can be trivially reduced to SAGAP, and the reduction is obviously a logarithmic space parsimonious reduction. We immediately have

Theorem 4. SAGAP *is #P-complete.*

5 Self-destructible Graph Assembly Model

The assumption in the above accretive model is that once a vertex is assembled, it cannot be "knocked off" by the later assembly of another vertex. Next, we relax this assumption and obtain a more general model: the *self-destructible graph assembly model*. In this model, the incorporation of a vertex a that repulses an already assembled vertex b can make b unstable and hence "knock" b off the assembly. This phenomenon renders the assembly system an interesting dynamic property, namely (partial) self-destruction.

The *self-destructible graph assembly system* operates on a *slot graph*. A slot graph $\tilde{G} = (S, E)$ is a set of "slots" S connected by edges $E \subseteq S \times S$. Each "slot" $s \in S$ is associated with a set of vertices $V(s)$. During the assembly process, a slot s is either empty or is occupied by a vertex $v \in V(s)$. A slot s occupied by a vertex v is denoted as $\langle s, v \rangle$.

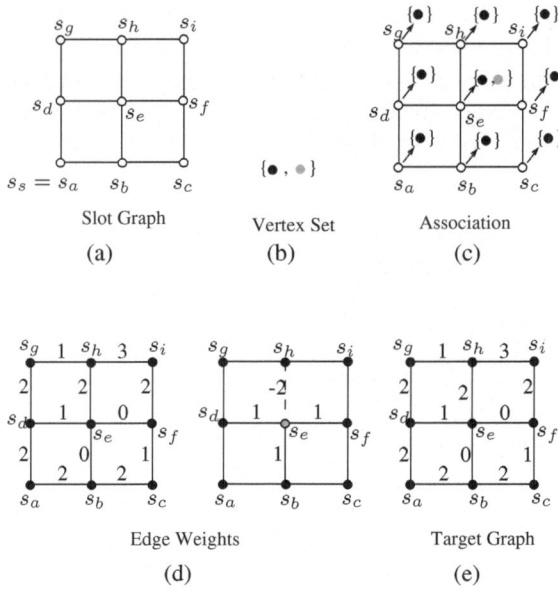

Fig. 4. An example self-destructible graph assembly system

A *self-destructible graph assembly system* is defined as $T = \langle \tilde{G} = (S, E), V, M, w,$ $\langle s_s, v_s \rangle, \tau \rangle$, where $\tilde{G} = (S, E)$ is a given slot graph with slot set S and edge set $E \subseteq S \times S$; $V = \bigcup_{s \in S} V(s)$ is the set of vertices; the association rule $M \subseteq S \times V$ is a binary relation between S and V, which maps each slot s to its associated vertex set $V(s)$ (note that the sets $V(s)$ are *not* necessarily disjoint); for any edge $(s_a, s_b) \in E$, we define a weight function $w : V(s_a) \times V(s_b) \rightarrow \mathbb{Z}$ (here a weight is determined cooperatively by an edge (s_a, s_b) and the two vertices occupying s_a and s_b); $\langle s_s, v_s \rangle$ is a distinguished *seed slot* s_s occupied by vertex v_s; $\tau \in \mathbb{N}$ is the *temperature* of the system. The size of a self-destructible assembly system is the bit representation of the system.

A *configuration* of $\tilde{G}$ is a function $A : S \rightarrow V \bigcup \{empty\}$, where *empty* indicates a slot being un-occupied. For ease of exposition, a configuration is alternatively referred to as a *graph*, denoted as G. When the context is clear, we simply refer to a slot occupied by a vertex as a *vertex*, for readability.

Given the above self-destructible graph assembly system, we aim at assembling a target graph, *i.e.* reaching a target configuration, G_t, starting with the seed vertex $\langle s_s, v_s \rangle$ and using the following *unit assembly operations*. In each unit operation, we temporarily attach a vertex v to the current graph G and obtain a graph G', and then repeat the following procedure until no vertex can be removed from the assembly: inspect all the vertices in current graph G'; find the vertex v' with the smallest *support*, *i.e.* the sum of the weights of edges between v' and its assembled neighbors, and break the ties arbitrarily (note that v' can be v); if the support to v' is less than τ, remove v'. This procedure ensures that when a vertex that repulses its assembled neighbors is incorporated in the existing assembly, all the vertices whose support drops below system temperature will be removed. However, in the case when a vertex to be attached exerts no repulsive force to its already assembled neighbors, the above standard unit assembly operation can be simplified as follows: a vertex can be assembled if the total support it receives from its assembled neighbors is equal to or greater than the system temperature τ – this is exactly the same as the operation in the accretive graph assembly model.

Figure 4 gives a concrete example of a self-destructible graph assembly system $\mathcal{T} = \langle \tilde{G} =(S, E), V, M, w, \langle s_s, v_s \rangle, \tau \rangle$. Here, slot s_a is designated as the distinguished seed slot s_s and temperature τ is set to 2. Figure 4 (a) depicts the slot graph $\tilde{G} = (S, E)$, where $S = \{s_a, s_b, s_c, s_d, s_e, s_f, s_g, s_h, s_i\}$, $E = \{(s_a, s_b), (s_b, s_c), (s_a, s_d), (s_b, s_e), (s_c, s_f), (s_d, s_e), (s_e, s_f), (s_d, s_g), (s_e, s_h), (s_f, s_i), (s_g, s_h), (s_h, s_i)\}$. Figure 4 (b) gives the vertex set $V = \{black, grey\}$. Figure 4 (c) shows the association rule M: $V(s_e) = \{black, grey\}$; $V(s) = \{black\}$, for $s \in S \setminus s_e$. Figure 4 (d) illustrates w. A numerical value indicates the weight of an edge incident to two occupied slots. The left panel of Figure 4 (d) describes the cases when both slots incident to an edge are occupied by black vertices; the right panel describes the case when slot s_e is occupied by a grey vertex but its neighboring slot is occupied by a black vertex. For example, the weight for edge (s_e, s_h), when both s_e and s_h are occupied with black vertices, is 2; when s_e is occupied by a grey vertex and s_h by a black vertex, the weight is -2. The negative weight is indicated by a dashed edge. Figure 4 (e) depicts the target graph (configuration) G_t, where each the slot in S is occupied by a black vertex, *i.e.* $A(s) = black$ for any $s \in S$.

Now we are ready to define the Self-Destructible Graph Assembly Problem (**DGAP**).

Definition 2. Self-destructible Graph Assembly Problem (DGAP): *Given a self-destructible graph assembly system* $\mathcal{T} = \langle G = (S, E), V, M, w, \langle s_s, v_s \rangle, \tau \rangle$ *and a target graph (configuration)* G_t, *determine whether there exists a sequence of assembly operations such that* G_t *can be assembled starting from* $\langle s_s, v_s \rangle$.

6   DGAP Is PSPACE-Complete

Theorem 5. DGAP *is* **PSPACE**-*complete.*

Proof. Recall that the **PSPACE**-complete problem IN-PLACE ACCEPTANCE is as follows: given a deterministic Turing machine (TM for short) U and an input string x, does U accept x without leaving the first $|x| + 1$ symbols of the string [26]? We reduce IN-PLACE ACCEPTANCE to DGAP using a direct simulation of a deterministic TM U on x with self-destructible graph assembly in **PSPACE**.

The proof builds on 1) a classical technique for simulating TM using self-assembly of square tiles [28, 31], which takes exponential space for deciding **PSPACE**-complete languages; and 2) our new cyclic gadget, which helps the classical TM simulation to reuse space and thus achieve a **PSPACE** simulation. We will first reproduce the classical simulation; next introduce our modification to the classical simulation; then describe our cyclic gadget; finally integrate the cyclic gadget with the modified TM simulation to obtain a **PSPACE** simulation and thus conclude the proof.

Classical TM simulation. The classical scheme uses the assembly of vertices on a 2D square grid to mimic a TM's transition history [28, 31]. Consecutive configurations of TM are represented by successive horizontal rows of assembled-vertices.

Given a TM $U(Q, \Sigma, \delta, q_0)$, where Q is a finite set of states, Σ is a finite set of symbols, δ is the transition function, and $q_0 \in Q$ is the initial state, we construct a self-destructible assembly system $\mathcal{T} = \langle G = (S, E), V, M, w, \langle s_s, v_s \rangle, \tau \rangle$ as follows. The slot graph $G = (S, E)$ is an infinite 2D square grid; each node of the grid corresponds to a slot $s \in S$. A vertex $v \in V$ is represented as a quadruple $v = \langle a, b, c, d \rangle$, where a, b, c, and d are referred to as the North, East, South, and West '*glues*' (see Figure 5). Each glue x is associated with an integral strength $g(x)$. More specifically, we construct the following vertices:

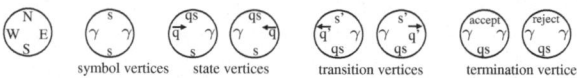

symbol vertices   state vertices   transition vertices   termination vertices

Fig. 5. Vertices used in the basic TM simulation

- For each $s \in \Sigma$, construct a *symbol vertex* $\langle s, \gamma, s, \gamma \rangle$, where γ is a special symbol $\notin \Sigma$.
- For each $\langle q, s \rangle \in Q \times \Sigma$, construct *state vertices* $\langle \langle q, s \rangle, \gamma, s, \overrightarrow{q} \rangle$ and $\langle \langle q, s \rangle, \overleftarrow{q}, s, \gamma \rangle$.
- For each transition $\langle q, s \rangle \to \langle q', s', \mathrm{L} \rangle$ (resp. $\langle q, s \rangle \to \langle q', s', \mathrm{R} \rangle$), where L (resp. R) is the head moving direction "Left" (resp. "Right"), construct a *transition vertex* $\langle s', \gamma, \langle q, s \rangle, \overleftarrow{q'} \rangle$ (resp. $\langle s', \overrightarrow{q'}, \langle q, s \rangle, \gamma \rangle$).
- For transition $\langle q, s \rangle \to$ ACCEPT (resp. REJECT), construct a *termination vertex* $\langle \mathrm{ACCEPT}, \gamma, \langle q, s \rangle, \gamma \rangle$ (resp. $\langle \mathrm{REJECT}, \gamma, \langle q, s \rangle, \gamma \rangle$).

The glue strength $g(\langle q, s \rangle)$ is set to 2; all other glue strengths are 1. Mapping relation M: every vertex in V can be mapped to every slot in S. We next describe weight function $V \times V \times E \to \mathbb{Z}$. Consider two vertices $v_1 = \langle a, b, c, d \rangle$ and $v_2 = \langle a', b', c', d' \rangle$ connected by edge e, if e is horizontal and v_1 lies to the East (resp. West) of v_2, the weight function is $g(b', d)$ (resp. $g(b, d')$); if e is vertical and v_1 lies to the North (resp. South) of v_2, the weight function is $g(c, a')$ (resp. $g(a, c')$); where $g(x, y) = g(x)$ (resp. 0) if $x = y$ (resp. $x \neq y$). In other words, the edge weight for two neighboring vertices is the strength of the abutting glues, if the abutting glues are the same; otherwise it is 0.

It is straightforward to show that the assembly of the vertices in V on the slot graph $G = (S, E)$ simulates the operation of the TM U. Figure 6 (a) gives a concrete example to illustrate the simulation process as in [31]. Here we assume the bottom row in the assembly in Figure 6 (a) is pre-assembled.

Our modified TM simulation. We add two modifications to the classical simulation and obtain the scheme in Figure 6 (b): 1) a set of vertices are added to assemble an input row (bottom row in the figure) and 2) a dummy column is added to the leftmost of the assembly. For the construction, see the self-explanatory Figure 6 (b). The leftmost bottom vertex is the seed vertex and a thick line indicates a weight 2 edge. The reason for adding the dummy column is as follows. The glue strength $g(\langle q, s \rangle)$ is 2 in Figure 6 (a); this is necessary to initiate the assembly of a new row and hence a transition to next configuration. However, due to a subtle technical point explained later (in the part "Integrating cyclic gadget with TM simulation"), we cannot allow weight 2 edge(s) in a column unless all the edges in this column have weight 2. So we add the leftmost dummy column of vertices connected by weight 2 edges, and this enables us to set $g(\langle q, s \rangle) = 1$ and thus avoid weight 2 edge other than those in the dummy column. The modified scheme simulates a TM on input x with the head initially residing at s_0 and never moving to the left of s_0. The assembly proceeds from bottom to top; within each row, it starts from the leftmost dummy vertex and proceeds to the right (note the difference in the assembly sequence in Figure 6 (a) and (b), as indicated by the thick grey arrows).

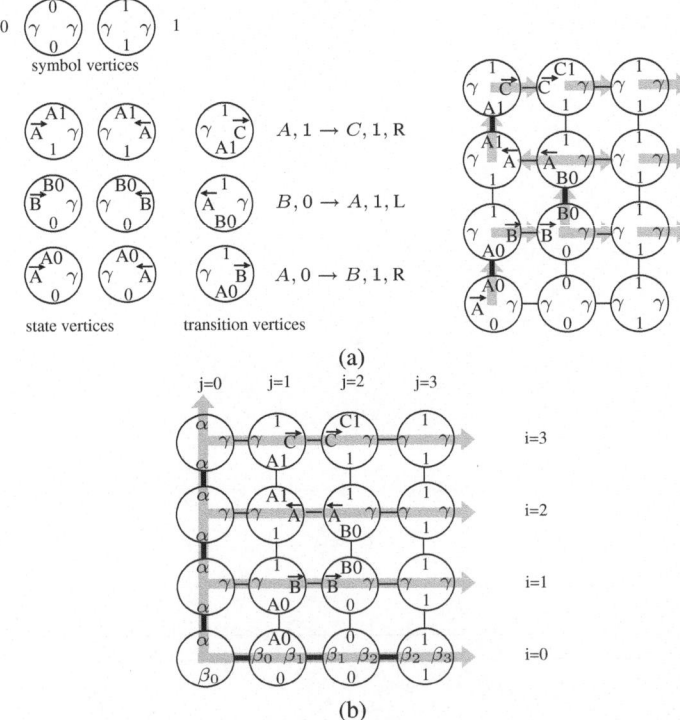

Fig. 6. (a) An example classical simulation of a Turing machine $U(Q, \Sigma, \delta, q_0)$, where $Q = \{A, B, C\}$; $\Sigma = \{0, 1\}$; transition function δ is shown in the figure; $q_0 = A$. The top of the left panel shows two symbol vertices; below are some example transition rules and the corresponding state vertices and transition vertices. The right panel illustrates the simulation of U on input 001 (simulated as the bottom row, which is assumed to be preassembled), according to the transition rules in the figure; the head's initial position is on the leftmost vertex. Each transition of U adds a new row. (b) Our modified scheme. The leftmost bottom vertex is the seed vertex. The leftmost column is the dummy column. In both (a) and (b), a thick line indicates a weight 2 edge; a thin line indicates weight 1; thick grey arrows indicate the assembly sequence.

Our cyclic gadget. The above strategy to simulate TM by laying out its configurations one above another can result in a graph with height exponential in the size of the input ($|x|$): the height of the graph is precisely the number of transitions plus one. A crucial observation is that once row i is assembled, row $i - 1$ is no longer needed: row i holds sufficient information for assembling row $i + 1$ and hence for the simulation to proceed. Thus, we can evacuate row $i - 1$ and reuse the space to assemble a future row, say row $i + 2$. Using this trick, we can shrink the number of rows from an exponential number to a constant. The self-destructible graph assembly model can provide us with precisely this power. To realize this power of evacuating and reusing space, we construct a *cyclic gadget*, shown in Figure 7 (a). The gadget contains three kinds of vertices: the *computational vertices* (a, b, and c) that carry out the actual simulation of the Turing machine; the *knocking vertices* (x, y, and z) that serve to knock off the computational vertices and thus release the space; the *anchor vertices* (x', y', and z') that anchor the knocking vertices. Edge weights are labeled in the figure.

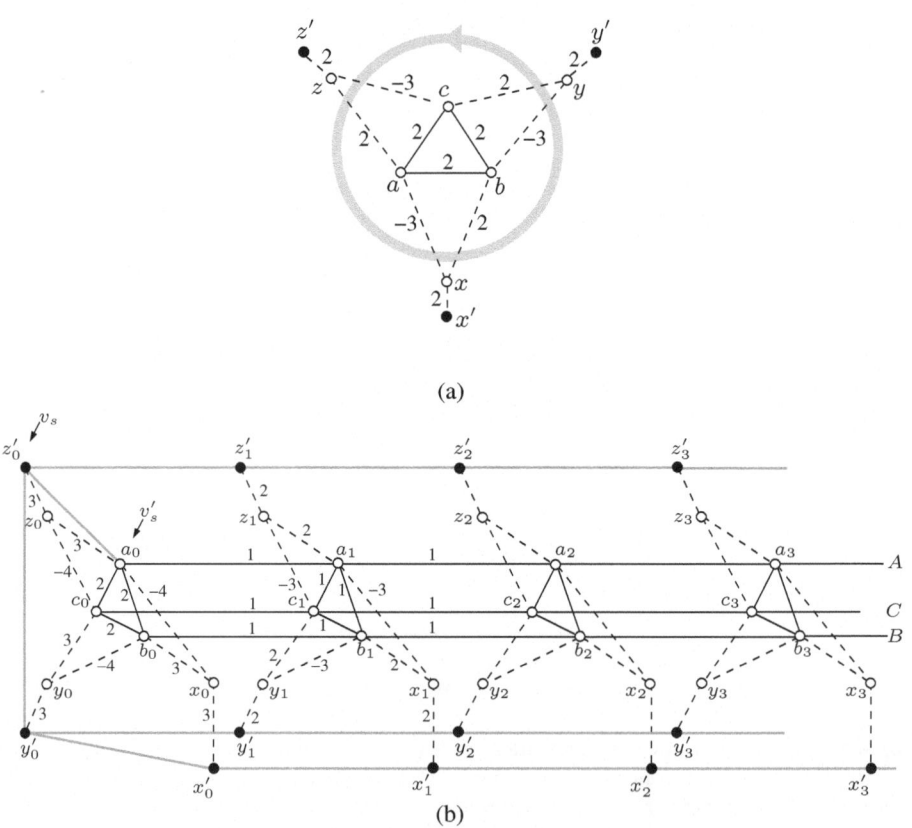

(a)

(b)

Fig. 7. (a) The construction and operation of our cyclic gadget. The counterclockwise grey cycle indicates the desired sequence of events. (b) The integrated scheme. Grey edges have weight 2. Unlabeled black edges have weight 1. v_s indicates the seed vertex; z_0 is the seed slot. v'_s indicates a distinguished computational "seed".

For ease of exposition, we introduce a little more notation. The event in which a new vertex b is attached to a pre-assembled vertex a is denoted as $a \cdot b$; the event in which a knocks off b is denoted as $a \dashv b$.

We next describe the operation of the cyclic gadget. We require that anchor vertices x', y', and z' and computational vertex a are pre-assembled. The anchor vertices and computational vertices will keep getting assembled and then knocked off in a counterclockwise fashion. First, b is attached to a (event $a \cdot b$). Then x is attached to b (event $b \cdot x$). At this point, x has total support 1 from b, x', and a (providing support 2, 2, and -3, respectively); a has total support -1 from b and x (providing support 2 and -3, respectively). Since the temperature is 2, x will knock off a ($x \dashv a$). Next, we have $b \cdot c$ followed by $c \cdot y$. At this point, y has total support 1 from c and y'; b has total support 1 from x and c. Therefore, either $y \dashv b$ or $b \dashv y$ can happen, but $y \dashv b$ is in the desired counterclockwise direction. Next, we will have cycles of (reversible) events. In summary, the following sequence of events occur, providing the desired cyclicity:

$$a \cdot b, b \cdot x, x \dashv a; b \cdot c, c \cdot y, y \dashv b; (c \cdot a, a \dashv x, a \cdot z, z \dashv c; a \cdot b, b \dashv y, b \cdot x, x \dashv a;$$
$$b \cdot c, c \dashv z, c \cdot y, y \dashv b)^*;$$

The steps in the $()$ will keep repeating. Note that the steps in the $()$ are reversible, which will facilitate our reversible simulation of a Turing machine below.

Integrating cyclic gadget with TM simulation. We next integrate the cyclic gadget with the modified TM simulation in Figure 6 (b). In the resulting scheme, we obtain a reversible simulation of a deterministic TM on a slot graph of constant height, by evacuating old rows and reusing the space: row i is evacuated after the assembly of row $i + 1$, providing space for the assembly of row $i + 3$.

Figure 7 (b) illustrates the integrated scheme. Slot rows A, B, and C correspond to rows $i = 3r$, $i = 3r + 1$, and $i = 3r + 2$ in Figure 6 (b), respectively. Let $|x| = n$. A is a sequence of slots $A = [a_0, a_1, \ldots, a_{n+1}]$; similarly, $B = [b_0, b_1, \ldots, b_{n+1}]$ and $C = [c_0, c_1, \ldots, c_{n+1}]$ as in Figure 7 (b). Slots a_0, b_0, and c_0 are dummy slots (corresponding to the dummy column in Figure 6 (b)). For each a_j, b_j, and c_j, we construct a cyclic gadget by introducing slots x_j, y_j, z_j, x'_j, y'_j, and z'_j.

Slot z'_0 is designated as the seed slot s_s and one of its associated vertices as the seed vertex v_s and the temperature is again set to 2.

The edge weights are shown in the figure. We emphasize that the weight for an edge between two computational vertices (vertices in A, B, and C) u and v is set to the glue strength if u and v have the same glue on their abutting sides; otherwise it is 0. This is consistent with the scheme in Figure 6 (b) and helps to ensure the proper operation of the computational assembly. In contrast, the weight for any other edge is always set to the value shown in Figure 7 (b), regardless of the actual computational vertices present in the slots in A, B, and C; this ensures the proper operation of the cyclic gadget.

There are some subtle technical points regarding edge weight assignment. First, the weight for the edge connecting vertices $v_s = z_0$ and v'_s is 2; while the weight for an edge connecting z'_0 and subsequent vertices other than v'_s that occupy slot a_0 is 0. This ensures the correct operation of the cyclic gadget for the dummy slots. Second, the assembly of the first row (input row) involves computational vertices with glue strength 2 (rather than 1) and hence weight 2 edges between neighboring vertices in this row. However *no* modification on the edge weight of the edges incident to the knocking vertices and anchor vertices is required to accommodate this edge weight difference: the initial step $(a \cdot b, \; b \cdot x, \; x \dashv a)$ is irreversible and it is straightforward to check that $x \dashv a$ can occur successfully. Third, except for the edges connecting dummy vertices, no weight 2 edge exists between the computational vertices after the evacuation of the input row. This is essential for upper bounding the number of vertices associated with each slot: otherwise, an exponential number of knocking vertices and anchor vertices would be required.

The assembly proceeds as follows. First, the frame of anchoring vertices (subgraph with grey edges) will be assembled, starting from the seed vertex at z'_0. The seed vertex at z'_0 will pull in a distinguished computational vertex v'_s (corresponding to the seed vertex in Figure 6 (b)) at slot a_0, and v'_s subsequently initiates the assembly of the input row (corresponding to the bottom row in Figure 6 (b)). Then the computational vertices will assemble, simulating the process shown in Figure 6 (b). Meanwhile, the cyclic gadget functions along each layer of a_j, b_j, and c_j (corresponding to column j in Figure 6 (b)), effecting the reusing of space. More specifically, vertices corresponding to those in rows $i = 3r$, $i = 3r + 1$, and $i = 3r + 2$ in Figure 6 (b) will be assembled in A, B, and C respectively. Similar to the process in Figure 7 (a), row $i + 1$ gets assembled with the support from row i, and subsequently pulls in knocking vertices, which knock off row i and thus evacuate space for future row $i + 3$ to assemble. Within a row, the vertices are knocked off sequentially from left to right, starting with the dummy vertex.

Concluding the proof. We set the target graph G_t as a *complete* row of vertices containing ACCEPT termination vertex $\langle \text{ACCEPT}, \gamma, \langle s, q \rangle, \gamma \rangle$. Then G_t can be assembled if and only if TM U accepts x. We insist G_t to be a complete row of vertices (occupying $s_0, s_1, \ldots, s_{|x|+1}$, where $s \in \{a, b, c\}$) to avoid false positives. Note the size of the slot graph used in the proof is polynomial in the size of the input $|x|$ and hence our simulation is in **PSPACE**.   □

Corollary 3. *6*-DEGREE DGAP *is* **PSPACE**-*complete.*

7   Conclusion

In this paper, we define two new models of self-assembly and obtain the following complexity results: 4-DEGREE AGAP is **NP**-complete; 5-DEGREE PAGAP is **NP**-complete; #AGAP and SAGAP are #**P**-complete; 6-DEGREE DGAP is **PSPACE**-complete. One immediate open problem is to determine the complexity of these problems with lower degrees. In addition, it would be nice to find approximation algorithms for the optimization version of the **NP**-hard problems. Note AGAP can be solved in polynomial time if only positive edges are permitted in graph G, using a greedy heuristic. In contrast, when negative edges are allowed, for each negative edge $e = (v_1, v_2)$, we need to decide the relative order for assembling v_1 and v_2. Thus k negative edges will imply 2^k choices, and we have to find out whether any of these 2^k choices can result in the assembly of the target graph. This is the component that makes the problem hard.

References

1. http://mrsec.wisc.edu/edetc/selfassembly/.
2. L. Adleman. Towards a mathematical theory of self-assembly. Technical Report 00-722, University of Southern California, 2000.
3. L. Adleman, Q. Cheng, A. Goel, and M.D. Huang. Running time and program size for self-assembled squares. In *Proceedings of the thirty-third annual ACM symposium on Theory of computing*, pages 740–748. ACM Press, 2001.
4. L. Adleman, Q. Cheng, A. Goel, M.D. Huang, D. Kempe, P.M. de Espans, and P.W.K. Rothemund. Combinatorial optimization problems in self-assembly. In *Proceedings of the thirty-fourth annual ACM symposium on Theory of computing*, pages 23–32. ACM Press, 2002.
5. L. Adleman, Q. Cheng, A. Goel, M.D. Huang, and H. Wasserman. Linear self-assemblies: Equilibria, entropy, and convergence rate. In *Sixth International Conference on Difference Equations and Applications*, 2001.
6. G. Aggarwal, M.H. Goldwasser, M.Y. Kao, and R.T. Schweller. Complexities for generalized models of self-assembly. In *Proceedings of 15th annual ACM-SIAM Symposium on Discrete Algorithms (SODA)*, pages 880–889. ACM Press, 2004.
7. N. Bowden, A. Terfort, J. Carbeck, and G.M. Whitesides. Self-assembly of mesoscale objects into ordered two-dimensional arrays. *Science*, 276(11):233–235, 1997.
8. R.F. Bruinsma, W.M. Gelbart, D. Reguera, J. Rudnick, and R. Zandi. Viral self-assembly as a thermodynamic process. *Phys. Rev. Lett.*, 90(24):248101, 2003 June 20.
9. N. Chelyapov, Y. Brun, M. Gopalkrishnan, D. Reishus, B. Shaw, and L. Adleman. DNA triangles and self-assembled hexagonal tilings. *J. Am. Chem. Soc.*, 126:13924–13925, 2004.
10. H.L. Chen, Q. Cheng, A. Goel, M.D. Huang, and P.M. de Espanes. Invadable self-assembly: Combining robustness with efficiency. In *Proceedings of the 15th annual ACM-SIAM Symposium on Discrete Algorithms (SODA)*, pages 890–899, 2004.
11. H.L. Chen and A. Goel. Error free self-assembly using error prone tiles. In *DNA Based Computers 10*, pages 274–283, 2004.
12. Q. Cheng, A. Goel, and P. Moisset. Optimal self-assembly of counters at temperature two. In *Proceedings of the first conference on Foundations of nanoscience: self-assembled architectures and devices*, 2004.
13. M. Cook, P.W.K. Rothemund, and E. Winfree. Self-assembled circuit patterns. In *DNA Based Computers 9*, volume 2943 of *LNCS*, pages 91–107, 2004.
14. K. Fujibayashi and S. Murata. A method for error suppression for self-assembling DNA tiles. In *DNA Based Computing 10*, pages 284–293, 2004.
15. Y. He, Y. Chen, H. Liu, A.E. Ribbe, and C. Mao. Self-assembly of hexagonal DNA two-dimensional (2D) arrays. *J. Am. Chem. Soc.*, 127:12202–12203, 2005.
16. N. Jonoska, S.A. Karl, and M. Saito. Three dimensional DNA structures in computing. *BioSystems*, 52:143–153, 1999.
17. N. Jonoska and G.L. McColm. A computational model for self-assembling flexible tiles. *Unconventional Computing*. To appear, 2005.
18. N. Jonoska, P. Sa-Ardyen, and N.C. Seeman. Genetic programming and evolvable machines. *Computation by Self-assembly of DNA Graphs*, 4.
19. M. Kao and R. Schweller. Reduce complexity for tile self-assembly through temperature programming. In *Proceedings of 17th annual ACM-SIAM Symposium on Discrete Algorithms (SODA)*. To appear. ACM Press, 2006.
20. E. Klavins. Toward the control of self-assembling systems. In *Control Problems in Robotics*, volume 4, pages 153–168. Springer Verlag, 2002.
21. E. Klavins. Directed self-assembly using graph grammars. In *Foundations of Nanoscience: Self Assembled Architectures and Devices*, Snowbird, UT, 2004.

22. E. Klavins, R. Ghrist, and D. Lipsky. Graph grammars for self-assembling robotic systems. In *Proceedings of the International Conference on Robotics and Automation*, 2004.

23. T.H. LaBean, H. Yan, J. Kopatsch, F. Liu, E. Winfree, J.H. Reif, and N.C. Seeman. The construction, analysis, ligation and self-assembly of DNA triple crossover complexes. *J. Am. Chem. Soc.*, 122:1848–1860, 2000.

24. D. Lichtenstein. Planar formulae and their uses. *SIAM J. Comput.*, 11(2):329–343, 1982.

25. J. Malo, J.C. Mitchell, C. Venien-Bryan, J.R. Harris, H. Wille, D.J. Sherratt, and A.J. Turberfield. Engineering a 2D protein-DNA crystal. *Angew. Chem. Intl. Ed.*, 44:3057–3061, 2005.

26. C.M. Papadimitriou. *Computational complexity*. Addison-Wesley Publishing Company, Inc., 1st edition, 1994.

27. J.H. Reif, S. Sahu, and P. Yin. Compact error-resilient computational DNA tiling assemblies. In *Proc. 10th International Meeting on DNA Computing*, pages 248–260, 2004.

28. R.M. Robinson. Undecidability and non periodicity of tilings of the plane. *Inventiones Math*, 12:177–209, 1971.

29. P.W.K. Rothemund. Using lateral capillary forces to compute by self-assembly. *Proc. Natl. Acad. Sci. USA*, 97(3):984–989, 2000.

30. P.W.K. Rothemund, N. Papadakis, and E. Winfree. Algorithmic self-assembly of DNA sierpinski triangles. *PLoS Biology 2 (12)*, 2:e424, 2004.

31. P.W.K. Rothemund and E. Winfree. The program-size complexity of self-assembled squares (extended abstract). In *Proceedings of the thirty-second annual ACM symposium on Theory of computing*, pages 459–468. ACM Press, 2000.

32. P. Sa-Ardyen, N. Jonoska, and N.C. Seeman. Self-assembling DNA graphs. *Lecture Notes in Computer Science*, 2568:1–9, 2003.

33. S. Sahu, P. Yin, and J.H. Reif. A self assembly model of time-dependent glue strength. In *Proc. 11th International Meeting on DNA Computing*, pages 113–124, 2005.

34. R. Schulman, S. Lee, N. Papadakis, and E. Winfree. One dimensional boundaries for DNA tile self-assembly. In *DNA Based Computers 9*, volume 2943 of *LNCS*, pages 108–125, 2004.

35. R. Schulman and E. Winfree. Programmable control of nucleation for algorithmic self-assembly. In *DNA Based Computers 10*, LNCS, 2005.

36. R. Schulman and E. Winfree. Self-replication and evolution of DNA crystals. In *The 13th European Conference on Artificial Life (ECAL)*, 2005.

37. D. Soloveichik and E. Winfree. Complexity of compact proofreading for self-assembled patterns. In *Proc. 11th International Meeting on DNA Computing*, pages 125–135, 2005.

38. D. Soloveichik and E. Winfree. Complexity of self-assembled shapes. In *DNA Based Computers 10*, LNCS, 2005.

39. A. Strasser, L. O'Connor, and V.M. Dixit. Apoptosis signaling. *Annu. Rev. Biochem.*, 69:217–245, 2000.

40. H. Wang. Proving theorems by pattern recognition ii. *Bell Systems Technical Journal*, 40: 1–41, 1961.

41. E. Winfree. Self-healing tile sets. Draft, 2005.

42. E. Winfree and R. Bekbolatov. Proofreading tile sets: Error correction for algorithmic self-assembly. In *DNA Based Computers 9*, volume 2943 of *LNCS*, pages 126–144, 2004.

43. E. Winfree, F. Liu, L.A. Wenzler, and N.C. Seeman. Design and self-assembly of two-dimensional DNA crystals. *Nature*, 394(6693):539–544, 1998.

44. H. Yan, T.H. LaBean, L. Feng, and J.H. Reif. Directed nucleation assembly of DNA tile complexes for barcode patterned DNA lattices. *Proc. Natl. Acad. Sci. USA*, 100(14): 8103–8108, 2003.

Designing Nucleotide Sequences for Computation: A Survey of Constraints

Jennifer Sager and Darko Stefanovic

Department of Computer Science, University of New Mexico,
MSC01 1130, 1 University of New Mexico,
Albuquerque, NM 87131
{sagerj, darko}@cs.unm.edu

Abstract. We survey common biochemical constraints useful for the design of DNA code words for DNA computation. We define the DNA Code Constraint Problem and cover biochemistry topics relevant to DNA libraries. We examine which biochemical constraints are best suited for DNA word design.

1   Introduction

Most DNA[1] computation models assume that computation is error-free. For example, Adleman [2] and Lipton [3] used randomly generated DNA strings in their experiments because they assumed that errors due to false positives are rare. However, it has been experimentally shown that randomly generated codes are inadequate for accurate DNA computation as the size of the problem grows [4], since a poorly chosen set of DNA strands can cause errors. Therefore, for many types of DNA computers, it may be practical or even necessary to create a 'library' or 'pool' of DNA word codes suitable for computation.[2]

A properly constructed library will help to minimize errors so that DNA computation is more practical, reliable, scalable, and less costly in terms of materials and laboratory time. However, the construction of a library is non-trivial for two reasons. First, there are 4^N unique DNA strings of length N; thus the number of candidate molecules grows exponentially in the length of the DNA string. Second, the constraints used to find a library are complex since they are subject to the laws of biochemistry as well as the specific algorithm and computation style. Given an algorithm for a type of DNA computer, the DNA Code Constraint Problem is to find a set of constraints that the DNA strands must satisfy to minimize the number of errors due to the choice of DNA strands. The constraints are determined by the physical reality of performing the algorithm in the laboratory and the specific algorithm and computation style. We examine the biomolecular constraints typically used to choose a set of DNA strings suitable for computation.

---

[1] Even though we describe most of the constraints in terms of DNA, RNA computers also exist (for an example see [1]) and all of the constraints described here are also relevant to RNA.

[2] For an overview of library design see [5]. For a survey of algorithms that have been used to solve the DNA/RNA Code Design Problem see [6].

A. Carbone and N.A. Pierce (Eds.): DNA11, LNCS 3892, pp. 275–289, 2006.

2 Positive and Negative Design

Even though there are many types of DNA computers, most share similar biochemical requirements because they use the same fundamental biochemical processes for computation. The fundamental computation step for most DNA computers occurs through the bonding (hybridization) and unbonding (denaturation) of oligonucleotides (short strands of DNA). For background information about DNA chemistry, see Appendix A.

Creating an error-free library typically requires that planned hybridizations and denaturations (between a word and its Watson-Crick complement) do occur and unplanned hybridizations and denaturations (between all other combinations of code words and their complements) do not occur. The former situation is referred to as the *positive design problem* while the latter is referred to as the *negative design problem* [6, 7].

The positive design problem requires that there exists a sequence of reactions that produces the desired outputs, starting from the given inputs. Thus, positive design attempts to "optimize affinity for the target structure" [7]. These reactions must occur within a reasonable amount of time for feasible concentrations. Usually the strands must satisfy a specified secondary structure criterion (e.g., the strand must have a desired secondary structure or have no secondary structure at all). Since a strand is typically identified by hybridization with its perfect Watson-Crick complement, the positive design problem requires that each Watson-Crick duplex is stable. In addition, for computation styles that use denaturation, the positive design problem often requires all of the strands in the library to have similar melting temperatures, or melting temperatures above some threshold. In short, positive design tries to maximize hybridization between perfect complements.

The negative design problem requires that: (1) no strand has undesired secondary structure such as hairpin loops, (2) no string in the library hybridizes with any other string in the library, and (3) no string in the library hybridizes with the complement of any other string in the library. Thus, negative design attempts to "optimize specificity for the target structure" [7]. Unplanned hybridizations can cause two types of potential errors: false positives and false negatives. False negatives occur when all (except an undetectable amount) of DNA that encodes a solution is hybridized in unproductive mismatches. Since mismatched strands are generally less stable than perfectly matched strands, false negatives can be controlled by adjusting strand concentrations. Deaton experimentally verified the occurrence of false positives, which happen when a mismatched hybridization causes a strand to be incorrectly identified as a solution [4]. False positives can be prevented by ensuring that all unplanned hybridizations are unstable. In short, the negative design problem tries to minimize non-specific hybridization.

Positive design often uses GC-content and energy minimization as heuristics (see below). Negative design uses combinatorial methods (such as Hamming distance, reverse complement Hamming distance, shifted Hamming distance, and sequence symmetry minimization), and thermodynamic methods (such as minimum free energy). Constraints which incorporate both positive and negative design are probability, average incorrect nucleotides, energy gap, probability gap, and energy minimization in combination with sequence symmetry minimization. The best-performing models for designing single-strand secondary structure use simultaneous positive and negative design, and significantly outperform either method alone; however, kinetic constraints must be considered separately since low free energy does not necessarily imply fast folding [7]. We believe that this same principle holds for designing hybridizations between multiple strands.

2.1 Secondary Structure of Single Strands

Most DNA computation styles need strands with no secondary structure (i.e., no tendency to hybridize with itself). There are, on the other hand, cases where specific secondary structures are desired, such as for deoxyribozyme logic gates [8]; Figure 1 shows the desired structure. Even there, structures different from the desired must be eliminated.

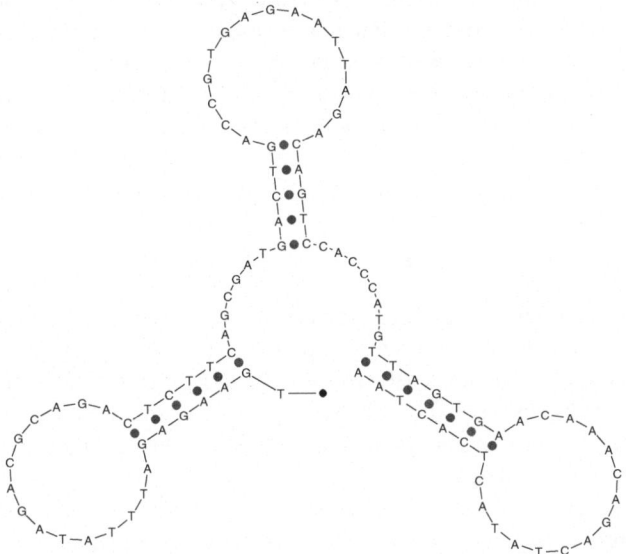

Fig. 1. Example of secondary structure in Stojanovic and Stefanovic's DNA automaton [8] as computed by Mfold [9, 10, 11] using 140 mM Na^+ and 2 mM Mg^{++} at 25°C. The strand has three hairpin loops, which is the desired secondary structure. Here ΔG is -12.3 kcal/mol.

There are several heuristics that are used to prevent secondary structure. Sometimes, repeated substrings and complementary substrings within a single strand which are non-overlapping and longer than some minimum length are forbidden in order to prevent stem formation. This heuristic is often called *sequence symmetry minimization* [12, 7] or *substring uniqueness* [13]. Another heuristic is to forbid particular substrings; these *forbidden substrings* are usually strings known to have undesired secondary structure. For example, sequences containing GGGGG should be avoided because they may form the four-stranded G4-DNA structure [14, 15].[3] Alternatively, strands are designed using

---

[3] For more information about alternative base pairing structures see [16].

only a *three-letter alphabet* (A, C, T for DNA and A, C, U for RNA) to eliminate the potential for GC pairs which could cause unwanted secondary structure [17].

In order to design a strand with a desired secondary structure, the nucleotides at positions which bond together must be complementary. This simple approach can be improved by also requiring the strands to satisfy some free-energy-based criteria, such as those described below from Dirks et al. [7].

The *minimum free energy* constraint, which can be calculated in $O(N^3)$ time for structures with no pseudoknots [18], is used to choose sequences such that the target structure has the minimum free energy. However, since this method is negative design, it does not ensure the absence of other structures that the sequence is likely to form. Algorithms also exist to determine whether a set of strands are structure free, where a set of sequences is considered to be structure free if the minimum free energy of every strand in the set is greater than or equal to zero [19, 20, 21].

The *energy minimization* constraint is used to choose sequences which have a low free energy in the target structure, but not necessarily the minimum free energy. To design strands with this constraint, first generate a random string s that satisfies the complementary requirements of the desired secondary structure. For each step (Dirks used 10^6 steps), choose a random one-point mutation. Let s' be the sequence with this random one-point mutation (and a mutation in the corresponding base required by the structure constraint, if any). Accept the mutation by replacing s with s' if:

$$e^{-\frac{\Delta G(s') - \Delta G(s)}{RT}} \geq \rho$$

where $\rho \in [0, 1]$ is a random number drawn from a uniform distribution, $\Delta G(s)$ is the free energy of the sequence in secondary structure s, and $\Delta G(s')$ is the free energy of the sequence in secondary structure s' (the free energy of a given structure can be calculated in $O(N)$ time). Thus, this equation always accepts any mutations which result in no change or a decrease in free energy, and accepts with some probability any mutations which increase the free energy.

Sequences can also be chosen which maximize the *probability* of sampling the target structure. The probability $p(s)$ that every nucleotide in the sequence exactly matches the target structure s at thermodynamic equilibrium is calculated by:

$$p(s) = \frac{1}{Q} e^{-\frac{\Delta G(s)}{RT}}$$

where $\Delta G(s)$ is the free energy of the sequence in secondary structure s. The partition function, Q, is:

$$Q = \sum_{s \in \Omega} e^{-\frac{\Delta G(s)}{RT}}$$

where Ω is the set of all secondary structures that the sequence can form in equilibrium. If s^* is the target secondary structure and $p(s^*) \approx 1$, then the sequence has a high affinity and high specificity for s^*. An optimal dynamic programming algorithm calculates $p(s^*)$ for structures with no pseudoknots in $O(N^3)$ time [22]; $p(s^*)$ for secondary structures with pseudoknots can be calculated in $O(N^5)$ time [23].

Additionally, sequences can be chosen to minimize the *average number of incorrect nucleotides*, $n(s)$, over all equilibrium secondary structures Ω. The structure matrix S_s for a given sequence of length N in structure s is:

$$S_s[i,j] = \begin{cases} 1, \text{ if base } i \text{ is paired with base } j \text{ in } s \\ 0, \text{ otherwise} \end{cases}$$

$$S_s[i,N+1] = \begin{cases} 1, \text{ if base } i \text{ is unpaired in } s \\ 0, \text{ otherwise} \end{cases}$$

where $1 \leq i \leq N$ and $1 \leq j \leq N$. The probability matrix P_s is:

$$P_s[i,j] = \sum_{s \in \Omega} p(s)S_s[i,j]$$

where $1 \leq i \leq N$ and $1 \leq j \leq N+1$. When $1 \leq j \leq N$, $P_s[i,j]$ is the probability of forming a base pair between the nucleotides at position i and j. $P_s[i,N+1]$ is the probability that base i is unpaired. $n(s)$ is the average number of incorrect nucleotides over the equilibrium ensemble of secondary structures Ω. If s^* is the target structure then:

$$n(s^*) = N - \sum_{i=1}^{N} \sum_{j=1}^{N+1} P_s[i,j]S_{s^*}[i,j]$$

$n(s^*)$ can be calculated in $O(N^3)$ time in structures with no pseudoknots and $O(N^5)$ in structures with pseudoknots.

Dirks et al. determined that the best-performing models are probability, average incorrect nucleotides, and energy minimization in combination with sequence symmetry minimization for the substrings that are not constrained by the desired secondary structure. The models with medium performance are the negative design methods (minimum free energy, and sequence symmetry minimization alone). The worst performing model is energy minimization (a positive design method). Surprisingly, minimum free energy performs similarly to sequence symmetry minimization; these results show that free energy measurements do not guarantee good design. An effective search must use both positive and negative design methods.

2.2 Secondary Structure of Multiple Strands

The strength of a perfectly matched duplex, a positive constraint, is often estimated by either: (1) the type of hydrogen bonds, AT vs. GC, expressed as the percentage of nucleotides that are G and C bases in a strand or duplex, which is known as *GC-content*; or (2) the amount of free energy released from the formation of the hydrogen bonds and the phosphodiester bonds that hold together adjacent nucleotides in a strand. The latter model is known as the nearest-neighbor model.

Since GC base pairs are held together by three hydrogen bonds while AT base pairs are held together by only two hydrogen bonds, double-stranded DNA with a high GC content is *often* more stable than DNA with a high AT content. Many DNA library searches require each strand to have a 50% GC-content to balance the requirement of stable matched hybridizations for identification purposes with the requirements of denaturation. The GC-content heuristic is simple to calculate; only the length and the number of GC bases are needed, where the length refers to the number of nucleotide base pairs. However the nearest-neighbor heuristic is more accurate than the GC-content heuristic because the nearest neighbor base stacking energies account for more of the change in free energy than the energy of the hydrogen bonding between nucleotide bases.

Requiring all pairs of strings in the library to have at least a given minimum *Hamming distance* (i.e., the number of characters in corresponding places which differ between two strings), is intended to satisfy the negative requirement that no pair of strings in the library should hybridize. A variation of this idea is the *reverse complement Hamming distance* which is the number of corresponding positions which differ in the complement of s_1 and the reverse of s_2. This constraint is used to reduce the false positives that occur from hybridization between a word and the reverse of another word in the library.

The advantage of Hamming distance (and its variations) is its theoretical simplicity and the vast body of extant work in coding theory. Many bounds have been calculated on the optimal size of codes with various Hamming-distance-based constraints [24]. Many early DNA library search algorithms used Hamming distance as a constraint to develop combinatorial algorithms based on the results from coding theory. However, Hamming distance alone is an insufficient constraint.

One problem with Hamming-distance-based heuristics is that this measure assumes that position i of the first string is aligned with position i of the second string. However, since duplexes can be formed with dangling ends and loops, this is not the only possible alignment. Various *Hamming distance slides*, *substring uniqueness* [13], partial words [25], and H-measure [26] constraints have been developed to fix the alignment problem. Another problem with heuristics based on Hamming distance is that the percentage of matching base pairs necessary to form a duplex is not necessarily known. Melting temperature can be used to approximate what the minimum Hamming distance should be[4], however, for a given temperature and word set, there can be significant variation in the required minimum distance.

Now that accurate free-energy information is available for all but the most complicated secondary structures (e.g., branching loops), the nearest-neighbor model is a much more accurate method to use than the constraints based on Hamming distance. It has also been experimentally determined for a sequence A of length n and a sequence B of length m that minimum free energy is a superior constraint to BP, where

$$BP = min(n,m) - min_{-m < k < n} H(A, \sigma^k(\overline{B}))$$

where $H(*, *)$ is the Hamming distance, $\overline{B}$ is the reverse complement of B, and σ^k is the shift rightward when $k > 0$ or leftward when $k < 0$ [14][5]. One way of using free-energy-based calculations as a constraint to prevent mismatched duplexes is to maximize the

---

[4] Deaton estimates the melting temperature of mismatched duplexes by decreasing $1°C$ per 1% mismatch between oligonucleotides [4]. Since this calculation is outdated (see Section 3), if this heuristic is used for a library search, it is recommended that the melting temperature should be estimated from free energy calculations.

[5] BP is equivalent to the H-measure constraint if $n = m$.

gap between the free energy of the weakest specific hybridization and the free energy of strongest nonspecific hybridization, which we refer to as the *energy gap*; this approach was used by Penchovsky [27]. A metric also exists which calculates the maximum number of stacked base pairs in any secondary structure; a thermodynamic weighting of this metric gives an upper bound on the free energy of duplex formation [28]. The probability, $p(s^*)$, measurement could also be applied to duplexes. A reasonable heuristic would be to maximize the gap between the lowest probability of the desired specific hybridizations and the highest probability of undesired non-specific hybridizations, which we refer to as the *probability gap*. Algorithms exist which calculate the probability, $p(s^*)$, for all possible combinations of single and double stranded foldings between a pair of strands [29]. Various equilibrium thermodynamic approaches have been used [30, 31, 32, 33, 34]. Computational incoherence, ξ, predicts the probability of error hybridization per-structure based on statistical thermodynamics [30, 35].

The physically-based models can be divided into categories based on the level of chemical detail [36]. Techniques which model single molecules include molecular mechanics models such as Monte Carlo minimum free energy simulations and molecular dynamics which models the change of the system with time. Techniques which average system behavior, or mass action approaches, are less accurate but more computationally feasible. Molecular mechanics (which models the movement of the system to the lowest energy), chemical kinetics, melting temperature, and statistical thermodynamics are all mass action approaches.

3 Melting Temperature

Melting temperature is typically used as a constraint in DNA paradigms that use multiple hybridization and denaturation steps to identify the answer, for an example see [1]. When DNA is heated, the hydrogen bonds that bind two bases together tend to break apart, and the strands tend to separate from each other. The probability that a bond will break increases with temperature. This probability can be described by the melting temperature, which is the temperature in equilibrium at which 50% of the oligonucleotides are hybridized and 50% of the oligonucleotides are separated. Since temperature control is often used to help denature the strands in intermediate steps, it is advantageous for these paradigms to require all of the strands in the library to have similar melting temperatures, or melting temperatures above some threshold.

The melting temperature of a perfectly matched duplex can be roughly estimated from the 2–4 rule [5], which predicts the melting temperature as twice the number of AT base pairs plus 4 times the number of GC base pairs. Another rough estimate of the change in melting temperature due to mismatched duplexes can also be obtained by decreasing the melting temperature of a corresponding matched duplex by 1°C per 1% mismatch; unfortunately, the inaccuracy is typically greater than 10°C [37]. Neither method is recommended. A better method is to use the nearest-neighbor model regardless of whether the duplex is perfectly matched or mismatched. This method produces more accurate results because melting temperature is closely related to free energy. Melting temperature has been used to characterize the hybridization potential of a duplex [38, 39], but this measure cannot be used to predict whether two strands are bound at a given temperature since the melting temperatures of different duplexes do not necessarily correspond to relative rankings of stability.

4 Reaction Rates

Once the structure of candidate strands is known, the next logical question to ask is how fast do these reactions occur and what concentration is needed. Kinetics deals with the rate of change of reactions. For some implementations of DNA computers, the rate of the reaction could be an additional search constraint. System-level simulation software has been described for this purpose [40].

5 DNA Prediction Software

There exists many software packages that predict DNA/RNA structure, thermodynamics, or kinetics. A few well-know structure prediction software packages are: Dynalign [41], MFold [9], NUPACK [23, 42], RNAsoft [43], RNAstructure [44], and the Vienna Package [45]. RNA free energy nearest neighbor parameters are available from the Turner Group [44]. Some software packages which calculate thermodynamics are: HyTher [46, 10, 47], BIND [38], MELTING [48], MELTSIM [49], and MeltWin [50]. Kinfold [51] simulates kinetics. EdnaCo [52] and Visual OMP (Oligonucleotide Modeling Platform; DNA Software Inc.) [53] simulate biochemical protocols *in silico*. In addition, there are many library design software packages such as: DNA Design Toolbox [54], DNASequenceCompiler [13], DNASequenceGenerator [13], NACST/Seq [55], NucleicPark [34], PERMUTE [1], PUNCH [56], SCAN [39], SEQUIN [12], SynDCode [28, 57, 58], and TileSoft [59].

6 Conclusion

Structure prediction can be separated into two problems. The first is to understand how DNA folds in nature. The second is to understand how computers should fold DNA strands to obtain the structure. Since nature has the advantage of parallel processing and the proximity of the molecules in space, the way nature finds the solution to the folding problem should not necessarily be the same as the way a computer finds the solution.

Early algorithms to find DNA word sets focused on the Hamming distance constraint or variations thereof to achieve a theoretical abstraction of the constraints, which allowed the use of combinatorial algorithms and proofs of completeness (i.e., that the size of the pool is optimal or near optimal). However, in the process the constraints are simplified so much that they no longer accurately predict DNA structure. Current algorithms tend to use a more complex combination of the constraints. However, since these constraints are difficult to abstract, more recent programs resort to genetic algorithms, random search, exhaustive search, and local stochastic search algorithms.

Thermodynamics, melting temperature and kinetics are best at predicting DNA structure, reaction rates and reaction temperatures. However, calculating these measures can be costly. According to the requirements mentioned for the negative design problem, checking that a library of size M meets specifications requires $O(M^2)$ string comparisons, where each comparison of a pair of strings of length N is potentially polynomial in N. Thus, the weaker combinatorial and heuristic predictors could be used to quickly filter a candidate set of library molecules, and then the free energy model could be used to more accurately check this set. If this approach is adopted, the correlation between these alternative heuristics and free energy measurements should be explored. Alternatively, free energy or probability approximation algorithms could be used. This approach has the advantage that techniques from randomized algorithm analysis could be used to prove the correctness of the approximation.

Research in DNA libraries has two main goals: (1) to further understand DNA chemistry, and (2) to understand search techniques useful for constructing sets of DNA codes. Although there is a growing consensus that DNA computers will never be as practical or as fast as conventional computers, biological computers have the advantage that their style of computation is closer to natural processes. Deaton states that the process of converting an algorithm into a biomolecular systems "is as difficult [i.e., NP-hard or harder] as the combinatorial optimization problems they are intended to solve" [60]. However, successful research in DNA libraries will help to reduce errors in DNA computation and may discover new information about how DNA interacts with itself. Although current DNA computers are simplistic in comparison to natural biochemical processes, DNA computation may help to develop alternative theories for how cells work or could have evolved [61]. In addition, research in DNA design also pertains to DNA nanotechnology, PCR-based applications, and DNA arrays. Breakthroughs in this field will add to the current knowledge of DNA chemistry as well as DNA computers.

Acknowledgments

We are grateful to Milan Stojanovic for his advice and encouragement, and to the anonymous reviewers for their extensive comments. This material is based upon work supported by the National Science Foundation (grants CCR-0219587, CCR-0085792, EIA-0218262, EIA-0238027, and EIA-0324845), Sandia National Laboratories, Microsoft Research, and Hewlett-Packard (gift 88425.1). Any opinions, findings, and conclusions or recommendations expressed in this material are those of the author(s) and do not necessarily reflect the views of the sponsors.

References

1. Dirk Faulhammer, Anthony R. Cukras, Richard J. Lipton, and Laura F. Landweber. Molecular computation: RNA solutions to chess problems. *Proceedings of the National Academy of Sciences of the USA (PNAS)*, 97(4):1385–1389, February 2000. The PERMUTE Program is available at http://www.pnas.org/cgi/content/full/97/4/1385/DC1.
2. Leonard M. Adleman. Molecular computation of solutions to combinatorial problems. *Science*, 266(5187):1021–1024, November 1994.
3. Richard J. Lipton. DNA solution of hard computational problems. *Science*, 268:542–545, April 1995.

4. Russell J. Deaton, Randy C. Murphy, Max Garzon, Donald R. Franceschetti, and S. E. Stevens, Jr. Good encodings for DNA-based solutions to combinatorial problems. In Landweber and Baum [62], pages 247–258.

5. Arwen Brenneman and Anne E. Condon. Strand design for bio-molecular computation. Technical report, University of British Columbia, March 2001.

6. Giancarlo Mauri and Claudio Ferretti. Word design for molecular computing: A survey. In Junghuei Chen and John H. Reif, editors, *DNA Computing: 9th International Workshop on DNA-Based Computers, DNA 2003 (University of Wisconsin: Madison, WI)*, volume 2943 of *Lecture Notes in Computer Science*, pages 37–47. Springer, 2004.

7. Robert M. Dirks, Milo Lin, Erik Winfree, and Niles A. Pierce. Paradigms for computational nucleic acid design. *Nucleic Acids Research*, 32(4):1392–1403, 2004.

8. Milan N. Stojanovic and Darko Stefanovic. A deoxyribozyme-based molecular automaton. *Nature Biotechnology*, 21(9):1069–1074, September 2003.

9. Michael Zuker. Mfold web server for nucleic acid folding and hybridization prediction. *Nucleic Acids Research*, 31(13):3406–3415, 2003. Mfold is available at http://www.bioinfo.rpi.edu/applications/mfold

10. John SantaLucia, Jr. A unified view of polymer, dumbbell, and oligonucleotide DNA nearest-neighbor thermodynamics. *Proceedings of the National Academy of Sciences of the USA (PNAS)*, 95:1460–1465, 1998.

11. Nicolas Peyret. *Prediction of Nucleic Acid Hybridization: Parameters and Algorithms*. PhD thesis, Wayne State University, Dept. of Chemistry, 2000.

12. Nadrian C. Seeman. *De Novo* design of sequences for nucleic acid structural engineering. *Journal of Biomolecular Structure & Dynamics*, 8(3):573–581, 1990.

13. Udo Feldkamp, Hilmar Rauhe, and Wolfgang Banzhaf. Software tools for DNA sequence design. *Genetic Programming and Evolvable Machines*, 4(2):153–171, June 2003.

14. Fumiaki Tanaka, Atsushi Kameda, Masahito Yamamoto, and Azuma Ohuchi. Specificity of hybridization between DNA sequences based on free energy. In Carbone et al. [63], pages 366–375.

15. Dipankar Sen and Walter Gilbert. Formation of parallel four-stranded complexes by guanine-rich motifs in DNA and its implications for meiosis. *Nature*, 334(6180):364–366, July 1988.

16. Nadrian C. Seeman. It started with Watson and Crick, but it sure didn't end there: Pitfalls and possibilities beyond the classic double helix. *Natural Computing: an international journal*, 1(1):53–84, 2002.

17. Kalim U. Mir. A restricted genetic alphabet for DNA computing. In Landweber and Baum [62].

18. Michael Zuker and Patrick Stiegler. Optimal computer folding of large RNA sequences using thermodynamics and auxiliary information. *Nucleic Acids Research*, 9(1):133–148, 1981.

19. Mirela Andronescu, Danielle Dees, Laura Slaybaugh, Yinglei Zhao, Anne Condon, Barry Cohen, and Steven Skiena. Algorithms for testing that sets of DNA word designs avoid unwanted secondary structure. In Hagiya and Ohuchi [64], pages 182–195.

20. Satoshi Kobayashi. Testing structure freeness of regular sets of biomolecular sequences. In Ferretti et al. [65], pages 395–404.

21. Atsushi Kijima and Satoshi Kobayashi. Efficient algorithm for testing structure freeness of finite set of biomolecular sequences. In Carbone et al. [63], pages 278–288.

22. John S. McCaskill. The equilibrium partition function and base pair binding probabilities for RNA secondary structure. *Biopolymers*, 29(6-7):1105–1119, May-Jun 1990.

23. Robert M. Dirks and Niles A. Pierce. A partition function algorithm for nucleic acid secondary structure including pseudoknots. *Journal of Computational Chemistry*, 24(13):1664–1677, October 2003. NUPACK is available at http://www.acm.caltech.edu/~niles/software.html.

24. Amit Marathe, Anne E. Condon, and Robert M. Corn. On combinatorial DNA word design. *Journal of Computational Biology*, 8(3):201–220, 2001.
25. Peter Leupold. Partial words for DNA coding. In Ferretti et al. [65].
26. Max Garzon, P. Neathery, Russell J. Deaton, Randy C. Murphy, Donald R. Franceschetti, and S. E. Stevens, Jr. A new metric for DNA computing. In *Proceedings 2nd Genetic Programming Conference*, pages 472–478, 1997.
27. Robert Penchovsky and Jorg Ackermann. DNA library design for molecular computation. *Journal of Computational Biology*, 10(2):215–229, 2003.
28. Arkadii G. D'yachkov, Anthony J. Macula, Wendy K. Pogozelski, Thomas E. Renz, Vyacheslav V. Rykov, and David C. Torney. A weighted insertion-deletion stacked pair thermodynamic metric. In Claudio Ferretti, Giancarlo Mauri, and Claudio Zandron, editors, *DNA Computing: 10th International Workshop on DNA-Based Computers, DNA 2004 (University of Milano-Bicocca: Milan, Italy)*, volume 3384 of *Lecture Notes in Computer Science*, pages 90–103. Springer, 2005. SynDCode is available at http://cluster.ds.geneseo.edu:8080/ParallelDNA/.
29. Roumen A. Dimitrov and Michael Zuker. Prediction of hybridization and melting for double-stranded nucleic acids. *Biophysical Journal*, 87:215–226, July 2004.
30. John A. Rose, Russell J. Deaton, Donald R. Franceschetti, Max Garzon, and S. E. Stevens, Jr. A statistical mechanical treatment of error in the annealing biostep of DNA computation. In *Special program in GECCO-99*, pages 1829–1834, June 1999.
31. John A. Rose and Russell J. Deaton. The fidelity of annealing-ligation: A theoretical analysis. In Anne Condon and Grzegorz Rozenberg, editors, *DNA Computing: 6th International Workshop on DNA-Based Computers, DNA 2000 (Leiden Center for Natural Computing: Leiden, The Netherlands)*, volume 2054 of *Lecture Notes in Computer Science*. Springer, 2001.
32. John A. Rose, Russell J. Deaton, Masami Hayiya, and Akira Suyama. The fidelity of the tag-antitag system. In Jonoska and Seeman [66].
33. John A. Rose, Russell J. Deaton, Masami Hagiya, and Akira Suyama. An equilibrium analysis of the efficiency of an autonomous molecular computer. *Physical Review E*, 65(021910), 2002.
34. John A. Rose, Masami Hagiya, and Akira Suyama. The fidelity of the tag-antitag system II: Reconcilation with the stringency picture. In *Proceedings of the Congress on Evolutionary Computation*, pages 2749–2749, 2003. NucleicPark is available at http://hagi.is.s.u-tokyo.ac.jp/johnrose/ and http://engronline.ee.memphis.edu/molec/demos.htm.
35. John A. Rose, Russell J. Deaton, Donald R. Franceschetti, Max Garzon, and S. E. Stevens, Jr. Hybridization error for DNA mixtures of N species. http://engronline.ee.memphis.edu/molec/Misc/ci.pdf, 1999.
36. John A. Rose and Akira Suyama. Physical modeling of biomolecular computers: Models, limitations, and experimental validation. *Natural Computing*, 3(4):411–426, 2004.
37. John SantaLucia, Jr. and Donald Hicks. The thermodynamics of DNA structural motifs. *Annual Review of Biophysics Biomolecular Structure*, 33:415–40, June 2004.
38. Alexander J. Hartemink and David K. Gifford. Thermodynamic simulation of deoxyoligonucleotide hybridization for DNA computation. In Harvey Rubin and David Harlan Wood, editors, *Preliminary Proceedings of DNA Based Computers III, DIMACS Workshop 1997 (University of Pennsylvania: Philadelphia, PA)*, pages 15–25, 1997.
39. Alexander J. Hartemink, David K. Gifford, and Julia Khodor. Automated constraint-based nucleotide sequence selection for DNA computation. In Lila Kari, Harvey Rubin, and David Harlan Wood, editors, *DNA Based Computers IV, DIMACS Workshop 1998 (University of Pennsylvania: Philadelphia, PA)*, *Biosystems*, volume 52, issues 1-3, pages 227–235. Elsevier, October 1999.

40. Akio Nishikawa, Masayuki Yamamura, and Masami Hagiya. DNA computation simulator based on abstract bases. *Soft Computing*, 5(1):25–38, 2001.
41. David H. Mathews and Douglas H. Turner. Dynalign: An algorithm for finding the secondary structure common to two RNA sequences. *Journal of Molecular Biology*, 317(217):191–203, 2002.
42. Robert M. Dirks and Niles A. Pierce. An algorithm for computing nucleic acid base-pairing probabilities including pseudoknots. *Journal of Computational Chemistry*, 25:1295–1304, 2004.
43. Mirela Andronescu, Rosalia Aguirre-Hernandez, Anne Condon, and Holger H. Hoos. RNA-soft: a suite of RNA secondary structure prediction and design software tools. *Nucleic Acids Research*, 31(13):3416–3422, 2003. RNAsoft is available at `http://www.rnasoft.ca/`.
44. David H. Mathews, Matthew D. Disney, Jessica L. Childs, Susan J. Schroeder, Michael Zucker, and Douglas H. Turner. Incorporating chemical modification constraints into a dynamic programming algorithm for prediction of RNA secondary structure. *Proceedings of the National Academy of Sciences of the USA (PNAS)*, 101(19):7287–7292, May 2004. The free energy nearest neighbor parameters are available at `http://rna.chem.rochester.edu/`, RNAstructure is available at `http://128.151.176.70/RNAstructure.html`.
45. Ivo Ludwig Hofacker. Vienna RNA secondary structure server. *Nucleic Acids Research*, 31(13):3429–3431, 2003. Vienna Package is available at `http://www.tbi.univie.ac.at/~ivo/RNA/`.
46. Nicolas Peyret, Pirro Saro, and John SantaLucia, Jr. HyTher server. HyTher Version 1.0 is available at `http://ozone2.chem.wayne.edu/`.
47. Nicolas Peyret, P. Ananda Seneviratne, Hatim T. Allawi, and John SantaLucia, Jr. Nearest-neighbor thermodynamics and NMR of DNA sequences with internal A-A, C-C, G-G, and T-T mismatches. *Biochemistry*, 38:3468–3477, 1999.
48. Nicolas Le Novère. MELTING, computing the melting temperature of nucleic acid duplex. *Bioinformatics*, 17(12):1226–1227, 2001. Melting is available at `http://www.ebi.ac.uk/~lenov/meltinghome.html`.
49. Richard D. Blake, Jeffrey W. Bizzaro, Jonathan D. Blake, G. R. Day, S. G. Delcourt, J. Knowles, Kenneth A. Marx, and John SantaLucia, Jr. Statistical mechanical simulation of polymeric DNA melting with MELTSIM. *Bioinformatics*, 15(5):370–375, 1999.
50. MeltWin. MeltWin is available at `http://www.meltwin.com/`.
51. Christoph Flamm, Walter Fontana, Ivo L. Hofacker, and Peter Schuster. RNA folding at elementary step resolution. *RNA*, 6:325–338, 2000. Kinfold is available at `http://www.www.tbi.univie.ac.at/~xtof/RNA/Kinfold/`.
52. Max Garzon, Russell J. Deaton, John A. Rose, L. Lu, and Donald R. Franceschetti. Soft molecular computing. *Proc. DNA5-99 Workshop, AMS DIMACS Series in Theoretical Computer Science*, 54:91–100, 2000. EdnaCo is available at `http://zorro.cs.memphis.edu/~cswebadm/csweb/research/pages/bmc/` or `http://engronline.ee.memphis.edu/molec/demos.htm`.
53. Visual OMP (Oligonucleotide Modeling Platform), DNA Software, Inc. Visual OMP is available at `http://www.dnasoftware.com`.
54. The DNA and Natural Algorithms Group. DNA design toolbox. DNA Design Toolbox is available at `http://www.dna.caltech.edu/DNAdesign/`.
55. Dongmin Kim, Soo-Yong Shin, In-Hee Lee, and Byoung-Tak Zhang. NACST/Seq: A sequence design system with multiobjective optimization. In Hagiya and Ohuchi [64], pages 242–251.
56. Adam J. Ruben, Stephen J. Freeland, and Laura F. Landweber. PUNCH: An evolutionary algorithm for optimizing bit set selection. In Jonoska and Seeman [66], pages 150–160.

57. Morgan Bishop, Anthony J. Macula, Wendy K. Pogozelski, Thomas E. Renz, and Vyach-eslav V. Rykov. SynDCode: Cooperative DNA code generating software. In Carbone et al. [63], page 391.

58. Wendy K. Pogozelski, Matthew P. Bernard, Salvatore F. Priore, and Anthony J. Macula. Experimental validation of DNA sequences for DNA computing: Use of a SYBR green assay. In Carbone et al. [63], pages 322–331.

59. Peng Yin, Bo Guo, Christina Belmore, Will Palmeri, Erik Winfree, Thomas H. LaBean, and John H. Reif. Tilesoft: Sequence optimization software for designing DNA secondary structures. http://www.cs.duke.edu/~reif/paper/peng/TileSoft/TileSoft.pdf, January 2004.

60. Russell J. Deaton and Max Garzon. Thermodynamic constraints on DNA-based computing. In Gheorghe Păun, editor, *Computing with Bio-Molecules*, pages 138–152. Springer-Verlag, Singapore, 1998.

61. Warren D. Smith. DNA computers in vitro and vivo. In Richard J. Lipton and Eric B. Baum, editors, *DNA Based Computers, DIMACS Workshop 1995 (Princeton University: Princeton, NJ)*, volume 27 of *Series in Discrete Mathematics and Theoretical Computer Science*, pages 121–185. American Mathematical Society, 1996.

62. Laura F. Landweber and Eric B. Baum, editors. *DNA Based Computers II, DIMACS Workshop 1996 (Princeton University: Princeton, NJ)*, volume 44 of *Series in Discrete Mathematics and Theoretical Computer Science*. American Mathematical Society, 1999.

63. Alessandra Carbone, Mark Daley, Lila Kari, Ian McQuillan, and Niles Pierce, editors. *Preliminary Proceedings of the 11th International Workshop on DNA-Based Computers, DNA 2005 (University of Western Ontario: London, Ontario, Canada)*, June 2005.

64. Masami Hagiya and Azuma Ohuchi, editors. *DNA Computing: 8th International Workshop on DNA-Based Computers, DNA 2002 (Hokkaido University: Sapporo, Japan)*, volume 2568 of *Lecture Notes in Computer Science*. Springer, 2003.

65. Claudio Ferretti, Giancarlo Mauri, and Claudio Zandron, editors. *Preliminary Proceedings of the 10th International Workshop on DNA-Based Computers, DNA 2004 (University of Milano-Bicocca: Milan, Italy)*, 2004.

66. Nataša Jonoska and Nadrian C. Seeman, editors. *DNA Computing: 7th International Workshop on DNA-Based Computers, DNA 2001 (University of South Florida: Tampa, FL)*, volume 2340 of *Lecture Notes in Computer Science*. Springer, 2002.

67. Peter Schuster. Counting and maximum matching of RNA structures. Preprint, http://www.tbi.univie.ac.at/~pks accessed on 2/1/2005, January 2004.

68. James D. Watson, Nancy H. Hopkins, Jeffrey W. Roberts, Joan Argetsinger Steitz, and Alan M. Weiner. *Molecular Biology of the Gene*. Benjamin/Cummings, Menlo Park, CA, fourth edition, 1988.

69. Mitsuhiro Kubota and Masami Hagiya. Minimum basin algorithm: An effective analysis technique for dna energy landscapes. In Ferretti et al. [65], pages 202–213.

70. Ignacio Tinoco, Jr., Kenneth Sauer, James C. Wang, and Joseph D. Puglisi. *Physical Chemistry: Principles and Applications in Biological Sciences*. Prentice Hall, fourth edition, 2002.

71. Bruce Alberts, Alexander Johnson, Julian Lewis, Martin Raff, Keith Roberts, and Peter Walter. *Molecular Biology of the Cell*. Garland, New York, fourth edition, 2002.

72. Peter Schuster, Peter F. Stadler, and Alexander Renner. RNA structures and folding: from conventional to new issues in structure predictions. *Current Opinion in Structural Biology*, 7(2):229–235, April 1997.

73. Douglas H. Turner, Naoki Sugimoto, and S. M. Freier. RNA structure prediction. *Annual Review of Biophysics and Biophysical Chemistry*, 17:167–192, June 1988.

A   Appendix: DNA Structure

This section provides background information on DNA chemistry that pertains to DNA word design.

A single strand of DNA is a sequence of nucleotides. Each nucleotide contains a sugar (deoxyribose or ribose), a phosphate group, and one of four bases, adenine (A), thymine (T), guanine (G), or cytosine (C). RNA is composed similarly except that thymine is replaced by the closely related uracil (U). Hybridization or annealing occurs when a sequence of nucleotides bonds to the nucleotides of another sequence, starting from the 5' end (the ribose end) of one sequence and the 3' end (the phosphate end) of the other sequence. The nucleotides only form stable bonds in certain combinations: A hydrogen-bonds to T or U, and G hydrogen-bonds to C. Thus A is the Watson-Crick complement of T/U, and G is the Watson-Crick complement of C. In addition, the "wobble pair", G and U, can form weak bonds. The Watson-Crick complement of a strand is obtained by first reversing it, and then complementing each base.

DNA (and RNA) can fold back upon itself into loops or other irregular complex twisted shapes. The remaining subsections can be a combination of different types of loop structures, which are single-stranded sections bounded by bonded base pairs (stem sections). Loops can be classified into several categories, Figure 2. A hairpin loop is a loop with a single stem. Internal loops are loops with single bases on both sides of the stem. Bulging loops are loops with single bases on only one side of the stem. Loops with three or more stems are called branching loops.

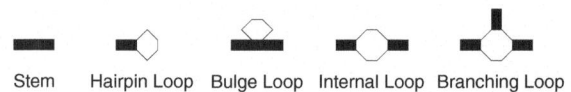

Stem   Hairpin Loop   Bulge Loop   Internal Loop   Branching Loop

Fig. 2. DNA loops. Solid areas represent double stranded sections. Lines represent single stranded sections.

Structure calculations attempt to predict which reactions will occur (i.e., which bonds will form and which will break). The tendency of the atoms in a molecule to bond together is referred to as the molecule's stability. Stability is affected by the sequence of bases, as well as environmental factors such as temperature, pH, the time given to allow reaction to complete, salt concentration, and the concentrations of the chemical components. Temperature is the most significant of these environmental factors. The DNA folding problem refers to the prediction of the structure and folding energy of a given sequence. There is an exponential number (approximately 1.8^N) of possible secondary structures for a sequence of length N [37, 67]. The inverse of this problem is the selection of a sequence with a given structure.

The stability of a DNA structure is a result of the change in free energy owing to bonding. The simplest explanation of free energy is that "free energy is energy that has the ability to do work" [68]. When a spontaneous reaction occurs at constant temperature and pressure, there is a decrease in free energy. This decrease in free energy is equal to the maximum amount of work that the system can do on its surroundings.

Conversely, for a non-spontaneous reaction, the free energy is the amount of work that must be done to cause the reaction to occur. The change in free energy is denoted ΔG. If $\Delta G < 0$, the reaction is spontaneous in the forward direction. If $\Delta G = 0$, the reaction is at equilibrium. If $\Delta G > 0$, the reaction is spontaneous in the reverse direction. When a bond between atoms forms, stronger bonds produce bigger changes in free energy; consequently, atoms that bond strongly together are more likely to exist in bonded form.

The most widely used method to estimate the free energy of DNA is the *nearest neighbor model*, which predicts the free energy of a duplex as the sum of the free energy of each nearest neighbor pair plus a few correction factors. The model is valid for single strands, Watson-Crick complementary duplexes, and mismatched duplexes. It can be adjusted for various temperature, pH, and salt conditions. Nearest neighbor parameters have been measured for several different types of nearest neighbors including matched pairs, internal mismatched pairs, dangling ends, internal loops, hairpin loops, and bulge loops. However, the fastest algorithms assume that the structure has no pseudoknots. (A pseudoknot is an occurrence of two pairs of bonded bases at positions (i,k) and (j,l) such that $i < j < k < l$.) Probabilistic measurements of free energy can also be derived from the nearest neighbor model to predict the most likely structure. Algorithms also exist which predict the energy landscape of the structures that a strand can form [69].

Thus, DNA is more stable when it has lower free energy and in most cases it will fold into the structure that has the minimum free energy. However, this structure is not necessarily the most likely structure to form. In fact, the equilibrium structure may not be a single structure at all; "what actually occurs, on the time scale of most enzymatic reactions relevant for biological function, is rather an ensemble of related structures interchanging more or less rapidly with one another" [22]. For example, the structure of the DNA of the bacterial virus T4 has several forms in solution including a tight coil and an extended form [70].

For more comprehensive information about DNA chemistry, see [68,71]. For a summary of nearest-neighbor thermodynamics see [37]. For more information about nucleotide structures see [72, 67]. For more information about structure prediction algorithms see [73].

A Self-assembly Model of Time-Dependent Glue Strength*

Sudheer Sahu, Peng Yin, and John H. Reif

Department of Computer Science, Duke University,
Box 90129, Durham, NC 27708-0129, USA
{sudheer, py, reif}@cs.duke.edu

Abstract. We propose a self-assembly model in which the glue strength between two juxtaposed tiles is a function of the time they have been in neighboring positions. We then present an implementation of our model using strand displacement reactions on DNA tiles. Under our model, we can demonstrate and study catalysis and self-replication in the tile assembly. We then study the tile complexity for assembling shapes in our model and show that a thin rectangle of size $k \times N$ can be assembled using $O(\frac{\log N}{\log \log N})$ types of tiles.

1 Introduction

Self-assembly is a ubiquitous process in which small objects self-organize into larger and complex structures. Examples in nature are numerous: atoms self-assemble into molecules, molecules into cells, cells into tissues, and so on. Recently, self-assembly has also been demonstrated as a powerful technique for constructing nano-scale objects. For example, a wide variety of DNA lattices made from self-assembled branched DNA molecules (DNA tiles) [9, 19, 21, 22, 40, 42, 43] have been successfully constructed. Peptide self-assembly provides another nanoscale example [8]. Self-assembly is also used for mesoscale constructions using capillary forces [7, 26] or magnetic forces [1].

Mathematical studies of tiling dates back to 1960s, when Wang introduced his tiling model [36]. The initial focus of research in this area was towards the decidability/undecidability of the tiling problem [25]. A revival in the study of tiling was instigated in 1996 when Winfree proposed the simulation of computation [41] using self-assembly of DNA tiles.

In 2000, Rothemund and Winfree [28] proposed the *abstract Tile Assembly Model*, a mathematical model for theoretical studies of self-assembly. This model was later extended by Adleman *et al.* to include the time complexity of generating specified assemblies [3]. Later work includes combinatorial optimization, complexity problems, fault tolerance, and topology changes, in the abstract Tile Assembly Model as well as in some of its variants [4, 5, 6, 10, 11, 12, 13, 14, 17, 18, 20, 23, 24, 27, 29, 31, 32, 34, 35, 38, 39].

In this paper, we use the term *standard model* to refer to the above *abstract Tile Assembly Model* proposed by Winfree. For detailed description of the *standard model*, see [28].

---

* The work is supported by NSF ITR Grants EIA-0086015 and CCR-0326157, NSF QuBIC Grants EIA-0218376 and EIA-0218359, and DARPA/AFSOR Contract F30602-01-2-0561.

A. Carbone and N.A. Pierce (Eds.): DNA11, LNCS 3892, pp. 290–304, 2006.

Roughly speaking, a tile in the standard model is a unit square where each side of the square has a glue from a set Σ associated with it. In this paper we use the terms *pad* and side of the tile interchangeably. Formally, a tile is an ordered quadruple $(\sigma_n, \sigma_e, \sigma_s, \sigma_w) \in \Sigma^4$, where σ_n, σ_e, σ_s, and σ_w represent the *northern*, *eastern*, *southern*, and *western* side glues of the tile, respectively. Σ also contains a special symbol *null*, which is a zero-strength glue. T denotes the set of all tiles in the system. A tile cannot be rotated. So, $(\sigma_1, \sigma_2, \sigma_3, \sigma_4) \neq (\sigma_2, \sigma_3, \sigma_4, \sigma_1)$. Also defined are various projection functions $n : T \rightarrow \Sigma$, $e : T \rightarrow \Sigma$, $s : T \rightarrow \Sigma$, and $w : T \rightarrow \Sigma$, where $n(\sigma_1, \sigma_2, \sigma_3, \sigma_4) = \sigma_1$, $e(\sigma_1, \sigma_2, \sigma_3, \sigma_4) = \sigma_2$, $s(\sigma_1, \sigma_2, \sigma_3, \sigma_4) = \sigma_3$, and $w(\sigma_1, \sigma_2, \sigma_3, \sigma_4) = \sigma_4$.

A glue strength function $g : \Sigma \times \Sigma \rightarrow \mathbb{R}$ determines the glue strength between two abutting tiles. $g(\sigma, \sigma') = g(\sigma', \sigma)$ is the strength between two tiles that abut on sides with glues σ and σ'. If $\sigma \neq \sigma'$, $g(\sigma, \sigma') = 0$; otherwise it is a positive value. It is also assumed that $g(\sigma, null) = 0$, $\forall \sigma \in \Sigma$. In the tile set T, there is a special *seed* tile s. There is a system parameter to control the assembly known as *temperature* and denoted as τ. All the ingredients described above constitute a *tile system*, a quadruple $\langle T, s, g, \tau \rangle$. A *configuration* is a snapshot of the assembly. More formally, it is the mapping from $\mathbb{Z}^2$ to $T \bigcup \{EMPTY\}$ where $EMPTY$ is a special tile $(null, null, null, null)$, indicating a tile is not present. For a configuration C, a tile $A = (\sigma_n, \sigma_e, \sigma_s, \sigma_w)$ is attachable at position (i, j) iff $C(i, j) = EMPTY$ and $g(\sigma_e, w(C(i, j + 1))) + g(\sigma_n, s(C(i + 1, j))) + g(\sigma_w, e(C(i, j - 1))) + g(\sigma_s, n(C(i - 1, j))) \geq \tau$.

Assembly takes place sequentially starting from a seed tile s at a known position. For a given tile system, any assembly that can be obtained by starting from the *seed* and adding tiles one by one, is said to be *produced*. An assembly is called to be *terminally produced* if no further tiles can be added to it. The *tile complexity* of a shape S is the size of the smallest tile set required to uniquely and terminally assemble S under a given assembly model. One of the well-known results is that the tile complexity of self-assembly of a square of size $N \times N$ in standard model is $\Theta(\frac{\log N}{\log \log N})$ [3, 28].

Adleman introduced a reversible model [2], and studied the kinetics of the reversible linear self-assemblies of tiles. Winfree also proposed a kinetic assembly model to study the kinetics of the self-assembly [37]. Apart from these basic models, various generalized models of self-assembly are also studied [6, 16]: namely, multiple temperature model, flexible glue model, and q-tile model.

Though all these models contribute greatly towards a good understanding of the process of self-assembly, there are still a few things that could not be easily explained or modeled (for example, the process of catalysis and self-replication in tile assembly). Recently, Schulman and Winfree show self-replication using the growth of DNA crystals [33], but their system requires shear forces to separate the replicated units. In this paper we propose a new model, in which catalysis and self-replication is possible without external intervention. In this new model, which is built on the basic framework of *abstract Tile Assembly Model*, the glue strength between different glues is dependent on the time for which they have remained together.

The rest of the paper is organized as follows. First we define our model formally in Section 2. We then put forth a method to physically implement such a system in Section 3. Then we present the processes of catalysis and self-replication in tile assembly in our model in Sections 4 and 5, respectively. In Section 6, we discuss the tile complexity of assembly of various shapes. We conclude with the discussion of our results and future research directions in Section 7.

2   Time-Dependent Glue Model

We propose a Time-dependent Glue Model, which is built on the framework described above. In this model, the glue-strength between two tiles is dependent upon the time for which the two tiles have remained together.

Let τ be the temperature of the system. Tiles are defined as in *standard model*. However, in our model, glue strength function g is defined as $g : \Sigma \times \Sigma \times \mathbb{R} \to \mathbb{R}$.

In $g(\sigma, \sigma', t)$ the argument t is the time for which two sides of the tiles with glue-labels σ and σ' have been juxtaposed. For every pair (σ, σ'), the value $g(\sigma, \sigma', t)$ increases with t up to a maximum limit and then takes a constant value determined by σ and σ'. We define the time when g reaches this maximum as *time for maximum strength*, $tms : \Sigma \times \Sigma \to \mathbb{R}$. Note $g(\sigma, \sigma', t) = g(\sigma, \sigma', tms(\sigma, \sigma'))$ for $t > tms(\sigma, \sigma')$.

We also have a function *minimum interaction time* defined as $mit : \Sigma \times \Sigma \to \mathbb{R}$.

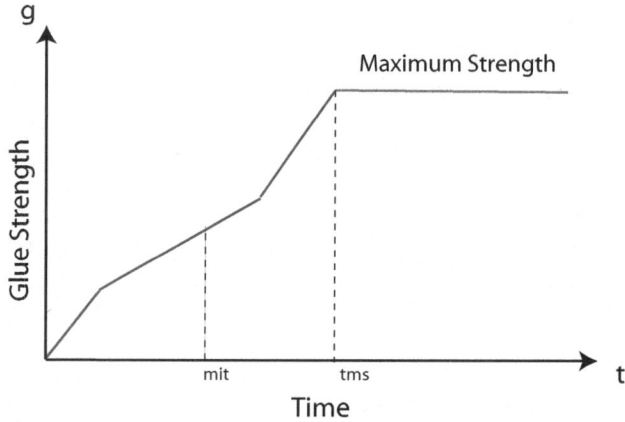

Fig. 1. Figure illustrates the concept of time-dependent glue strength, minimum interaction time, and time for maximum strength

For every pair (σ, σ'), a function $mit(\sigma, \sigma')$ is defined as the minimum time for which the two tiles with abutting glue symbols σ and σ' stay together. If $g(\sigma, \sigma', mit(\sigma, \sigma'))$ $\geq \tau$, the two tiles will stay together; otherwise they will separate if there is no other force holding them in their abutting positions. An example of glue-strength function is shown in Figure 1. Intuitively speaking, *mit* serves as the minimum time required by the pads to decide whether they want to separate or remain joined. We further define $mit(\sigma, null) = 0$, $tms(\sigma, null) = 0$, and $g(\sigma, null, t) = 0$.

Next we give the justification and estimation of *mit* for a pair (σ, σ') of glues. Let $g(\sigma, \sigma', t)$ be the glue strength function. For more realistic estimation of *mit*, consider a physical system in which, in addition to association, dissociation reactions also occur.

Let $p(b)$ be the probability of dissociation when the bond strength is b, and $f(t)$ be the probability that no dissociation takes place in the time interval $[0 \dots t]$. Then,

$$f(t + \delta t) = f(t) \cdot (1 - p(g(t + \delta t))) \cdot \delta t,$$
$$\frac{f(t + \delta t)}{f(t)} = (1 - p(g(t + \delta t))) \cdot \delta t.$$

The probability that the dissociation takes place between time t and $t + \delta t$ is given by $f(t) \cdot p(g(t + \delta t)) \cdot \delta t$. Since *mit* is defined as the time for which two glues are expected to remain together once they come in contact, its expected value is:

$$E[mit] = \lim_{\delta t \to 0} \sum_{t=0}^{\infty} t \cdot f(t) \cdot p(g(t + \delta t)) \cdot \delta t,$$

where $p(b)$ can be determined using Winfree's kinetic model [37]. Hence, based on the knowledge of glue strength function it is possible to determine the expected minimum interaction time for a pair (σ, σ'). For simplicity, we will use the expected value of *mit* as the actual value of *mit* for a pair of glues (σ, σ').

Next we illustrate the time-dependent model with an example of the addition of a single tile to an aggregate. When a position (i, j) becomes available for the addition of a tile A, it will stay at (i, j) for a time interval t_0, where $t_0 = \max \{mit(e(A), w(C(i, j+1))), mit(n(A), s(C(i+1, j))), mit(w(A), e(C(i, j-1))), mit(s(A), n(C(i-1, j)))\}$. Recall that our model requires that if two tiles ever come in contact, they will stay together till the minimum interaction time of the corresponding glues.

After this time interval t_0, if $g(e(A), w(C(i, j + 1)), t_0) + g(n(A), s(C(i + 1, j)), t_0) + g(w(A), e(C(i, j - 1)), t_0) + g(s(A), n(C(i - 1, j)), t_0) < \tau$, A will detach; otherwise, A will continue to stay at position (i, j).

We describe in the next section a method to implement our model of time-dependent glue strength with DNA tiles.

3   Implementation of Time-Dependent Glue Model

If the hydrogen bonds between the bases in two hybridizing DNA strands build up sequentially, the total binding force between the two strands will increase with time up to the complete hybridization, which will provide a simple way of obtaining time-dependent glue strength between DNA tiles. However, even if we assume that the hybridization of two complementary DNA strands is instantaneous, we can design a multi-step binding mechanism to implement the idea of time-dependent glue strength, which exploits the phenomenon of strand displacement.

Figure 2 (a) illustrates the process of strand displacement in which strand B displaces strand C from strand A. Figure 2 (b) illustrates one step during this process. At any time either the hybridization of B with A (and hence dehybridization of C from A) or hybridization of C with A (and hence dehybridization of B from A) can proceed with 1/2 probability. Hence, we can model the strand displacement process as a random walk, with forward direction corresponding to hybridization between B

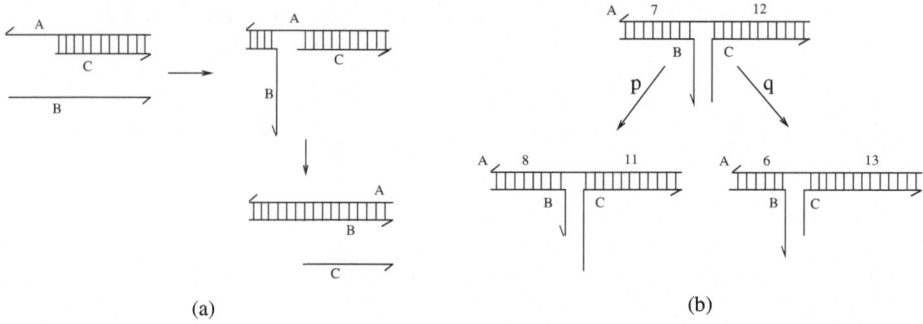

Fig. 2. Figure (a) illustrates the process of strand displacement. Figure (b) shows a single step of strand-displacement as single step of random walk. In (b), the numbers represent the number of DNA base pairs.

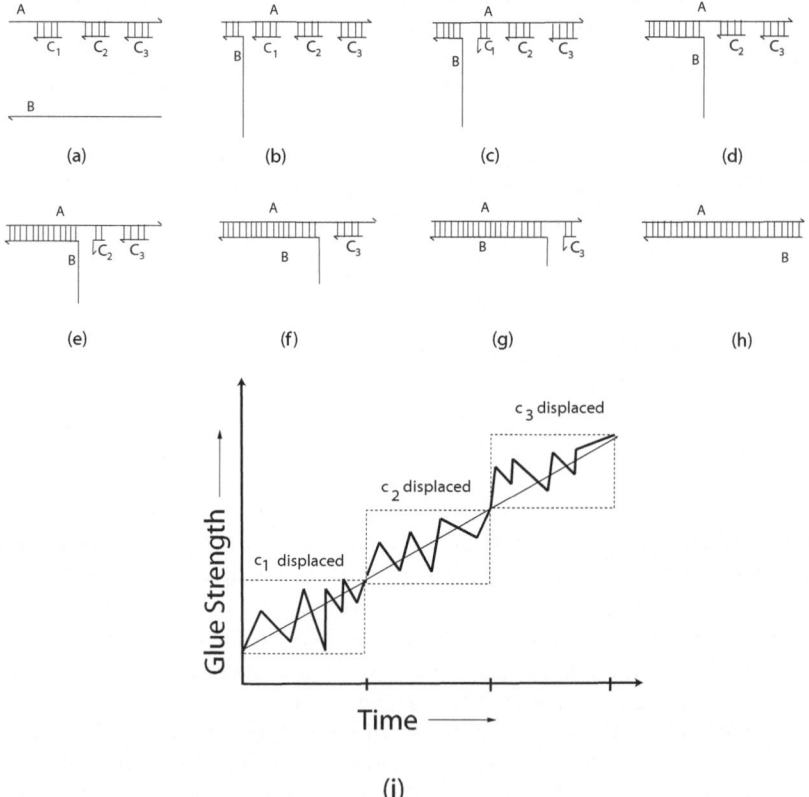

Fig. 3. Figures (a) to (h) illustrate a mechanism by which strand displacement reaction is used to implement time-dependent glue between two pads. They show step by step removal of C_i's by B from A. In Figure 3 (i) an imaginary graph illustrates the variation of glue-strength between A and B w.r.t. time.

and A, and backward direction corresponding to hybridization between C and A. To simplify the model, we can assume that the step length in this random walk is 1 base pair long. Hence, if the length of C is n bases, the expected number of steps required for B to replace C is n^2 [15].

Next we describe the design of the pads of DNA tiles with time dependent glue using the above mechanism of strand displacement.

To make the glue between pad A and pad B time-dependent, we need a construction similar to the one in Figure 3 (a). Strand representing pad A has various smaller strands (C_i's, called *protector strands*) hybridized to it as shown in Figure 3 (a). Strand B will displace these protector strands C_i sequentially.

The variable *tms* here will be the time required for B to displace all the C_i's. In the case when there are k different small strands C_i of length n_i attached to A, *tms* is $\sum_{i=1}^{k} n_i^2$.

Figure 3 gives the step by step illustration of the above process. The variation of glue strength between A and B is shown in Figure 3 (i). By controlling the length of various C_i's (*i.e.* $n_1, n_2, \ldots, n_k$), we can control the glue-strength function g for a pair of tile-pads (or glues). Thus, we have shown a method to render the DNA tiles the characteristic of time-dependent glue strength.

An interesting property is that the individual strand displacement of B against C_i is modeled as a random walk, but the complete process described above can be viewed as *roughly* monotonic. As shown in Figure 3 (i), the strength of the hybridization between strand A and strand B increases in a roughly monotonic fashion with the removal of every C_i. However during the individual competition between B and C_i, the increase is not monotonic.

4   Catalysis

Catalysis is the phenomenon in which an external substance facilitates the reaction of other substances, without itself being used up in the process. The following question was posed by Adleman [2]: can we model the process of catalysis in self-assembly of tiles? In this section, we present a model for catalysis in self-assembly of tiles using time-dependent glue model. Now consider a supertile $\mathcal{X}$ (composed of two attached tiles C and D) and two single tiles A and B as shown in Figure 4 (a). We describe below how $\mathcal{X}$ can serve as a catalyst for the assembly of A and B. Assume $t_0 = mit(e(A), w(B))$ such that $g(e(A), w(B), t_0)$ is less than the temperature τ. Let $mit(s(A), n(C)) = mit(s(B), n(D)) = t_1 > t_0$. Also assume $g(s(A), n(C), t_1) + g(s(B), n(D), t_1) < \tau$ and $g(e(A), w(B), t_1) \geq \tau$.

The graph in Figure 4 (b) illustrates an example set of required conditions for the glue strength functions in the system.

To show that $\mathcal{X}$ acts as a catalyst, we first show that without $\mathcal{X}$ stable $A \cdot B$ can not form. Next we show that $A \cdot B$ will form when $\mathcal{X}$ is present and $\mathcal{X}$ will be recovered unchanged after the formation of $A \cdot B$.

Without $\mathcal{X}$ in the system, A and B can only be held in neighboring positions for time $t_0 = mit(e(A), w(B))$, since $g(e(A), w(B), t_0) < \tau$. Hence, at t_0, A and B will fall apart.

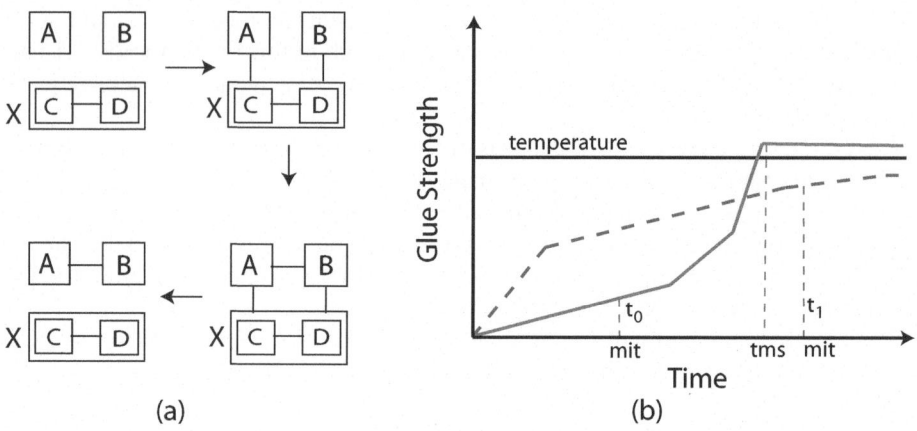

Fig. 4. Figure (a) shows catalyst $\mathcal{X}$ with the tiles C and D catalyzes the formation of $A \cdot B$. (b) shows the conditions required for catalysis in terms of the glue strength function. Solid line shows the plot of $g(e(A), w(B), t)$ and dashed line shows the plot of $g(s(A), n(C), t) + g(s(B), n(D), t)$.

However, in the presence of $\mathcal{X}$, the situation changes. Supertile $\mathcal{X}$ has two neighboring tiles C and D. Tiles A and B attach themselves to C and D as shown in Figure 4 (a). Since we let $mit(s(A), n(C)) = mit(s(B), n(D)) = t_1 > t_0$, tiles A and B are held in the same position for time t_1. By our construction, as shown in Figure 4 (b), the following two events will occur at time t_1:

- At t_1, the glue strength between A and B is $g(e(A), w(B), t_1) \geq \tau$ and hence A and B will be glued together. That is, in the presence of $\mathcal{X}$, A and B remain together for a longer time, producing stably glued $A \cdot B$.
- At t_1, the total glue strength between $A \cdot B$ and $\mathcal{X}$ is $g(s(A), n(C), t_1) + g(s(B), n(D), t_1) < \tau$, and the glued $A \cdot B$ will fall off $\mathcal{X}$. $\mathcal{X}$ is recovered unchanged from the reaction and the catalysis is complete. Now $\mathcal{X}$ is ready to catalyze other copies of A and B.

Note that if only A (resp. B) comes in to attach with C (resp. D), it will fall off at the end of time $mit(s(A), n(C))$ (resp. $mit(s(B), n(D))$). If assembled $A \cdot B$ comes in, it will also fall off, at time t_1. These two reactions are futile reactions, and do not block the desired catalysis reaction. However, as the concentration of $A \cdot B$ increases and the concentration of unattached A and B decreases, the catalysis efficiency of $\mathcal{X}$ will decrease due to the increased probability of the occurrence of futile reaction between $A \cdot B$ and $C \cdot D$.

5   Self-replication

Self-replication process is one of the fundamental process of nature, in which a system creates copies of itself. We discuss below an approach to model self-replication using the time-dependent glue model.

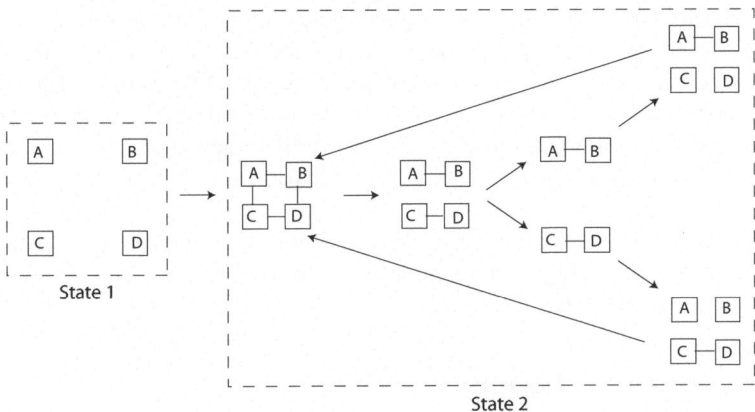

Fig. 5. A schematic of self-replication

Our approach is built on the above described process of catalysis: a product $A \cdot B$ catalyzes the formation of $C \cdot D$, which in turn catalyzes the formation of $A \cdot B$. And hence an exponential growth of self-replicated $A \cdot B$ and $C \cdot D$ takes place.

More precisely, let $t_0 < t_1$, consider tiles A, B, C, and D, such that :

$$mit(e(A), w(B)) = mit(e(C), w(D)) = t_0,$$
$$mit(s(A), n(C)) = mit(s(B), n(D)) = t_1,$$
$$g(e(A), w(B), t_0) = g(e(C), w(D), t_0) < \tau,$$
$$g(e(A), w(B), t_1) = g(e(C), w(D), t_1) > \tau,$$
$$g(s(A), n(C), t_1) + g(s(B), n(D), t_1) < \tau.$$

A system containing these four types of tiles has two states:

State 1. If there is no template $A \cdot B$ or $C \cdot D$ in the system, no assembled super-tile exists since no two tiles can be held together long enough to form strong enough glue between them such that they become stably glued. Since $mit(e(A), w(B)) = mit(e(C), w(D)) = t_0$ and $g(e(A), w(B), t_0) = g(e(C), w(D), t_0) < \tau$, neither stable $A \cdot B$ nor stable $C \cdot D$ can form. Similarly, $mit(s(A), n(C)) = mit(s(B), n(D)) = t_1$, $g(s(A), n(C), t_1) < \tau$, and $g(s(B), n(D), t_1) < \tau$ implies that neither stable $A \cdot C$ nor stable $B \cdot D$ can form.

State 2. In contrast, if there is an initial copy of stable $A \cdot B$ in the system, self-replication occurs as follows. $A \cdot B$ serves as catalyst for the formation of $C \cdot D$, and $C \cdot D$ and $A \cdot B$ separate from each other at the end of the catalysis period, as described in Section 4; in turn, $C \cdot D$ serves as catalyst for the formation of $A \cdot B$. Thus we have a classical self-replication system: one makes a copy of itself via its complement. The number of the initial template $(A \cdot B)$ and its complement $(C \cdot D)$ grows exponentially in such system.

Hence, if the system is in state 1, it needs a triggering activity (formation of an stable $A \cdot B$ or $C \cdot D$) to go into state 2. Once the system is in state 2, it starts the self-replication process. Figure 5 illustrates the process of self-replication in the assembly of tiles.

If the system is in state 1, then the triggering activity (formation of an stable $A \cdot B$ or $C \cdot D$) can take place only if A, B, C, D co-position themselves so that the east side of A faces the west side of B and the south side of A faces the north side of C, and at the same time the south side of B faces the north side of D. In such a situation, A and C will remain abutted till time t_1, B and D will remain abutted till time t_1, and A and B (and C and D) might also remain together for time t_1, producing stable $A \cdot B$ and stable $C \cdot D$. And this will bring the system to state 2. But such copositioning of 4 tiles is a very low probability event. Thus a very low probability event can perturb a system in state 1 and triggers tremendous changes by bringing the system to state 2 where self-replication occurs.

6 Tile Complexity Results

In the standard model, the tile complexity of assembling an $N \times N$ square is $\Theta(\frac{\log N}{\log \log N})$ [3, 28]. It is also known that the upper bound on the tile complexity of assembling a $k \times N$ rectangle in the standard model is $O(k + N^{1/k})$ and that the lower bound on tile complexity of assembling a $k \times N$ rectangle is $\Omega(\frac{N^{1/k}}{k})$ [6]. For small values of k this lower-bound is asymptotically larger than $O(\frac{\log N}{\log \log N})$. Here we claim that, in our model, as in the multi-temperature model defined in [6], a $k \times N$ rectangle can be self-assembled using $O(\frac{\log N}{\log \log N})$ types of tiles, even for small values of k. The proof technique follows the same spirit as in [6].

Theorem 1. *In time-dependent glue model, the tile complexity of self-assembling a $k \times N$ rectangle for an arbitrary integer $k \geq 2$ is $O(\frac{\log N}{\log \log N})$.*

Proof. The tile complexity of self-assembling a $k \times N$ rectangle is $O(N^{\frac{1}{k}} + k)$ for the standard model [6]. In time dependent glue model, we can use the similar idea as in [6] to reduce the tile complexity of assembling thin rectangles. For given k and N, build a $j \times N$ rectangle with $j > k$ such that the glues among the first k rows become strong after their *mit (minimum interaction time)*, while the glues among the last $j - k$ rows do not become as strong. As such, these $j - k$ rows, referred to as *volatile rows*, will fall apart after certain time and produce the target $k \times N$ rectangle.

The tile set required to accomplish this construction is shown in Figure 6, which is similar to the one used in [6]. For more detailed illustration of this tile set, refer to [6]. First, a j-digit m-base counter is assembled as follows. Starting from the west edge of the seed tile, a chain of length m is formed in the first row using m chain tiles. At the same time tiles in the seed column also start assembling. It should be noted that first k tiles in the seed column have sufficient glue-strength and they are stable. Now starting from their west edges, the 0 normal tiles start filling the $m - 1$ columns in the upper rows. Then the hairpin tiles H_1^P and H_1^R assemble in the second row, which causes the assembly of further m chain tiles in the first row, and the assembly of 1 normal tiles in the second row (and 0 normal tiles in the upper rows) in the next section of m columns. Generally speaking, whenever a C_{m-1} chain tile is assembled in the first row, probe tiles in the upper rows are assembled until reaching a row that does not contain an $m - 1$ normal tile. In such a row, the appropriate hairpin tiles are assembled and this further propagates the assembly of return probe tiles downwards until the first row is

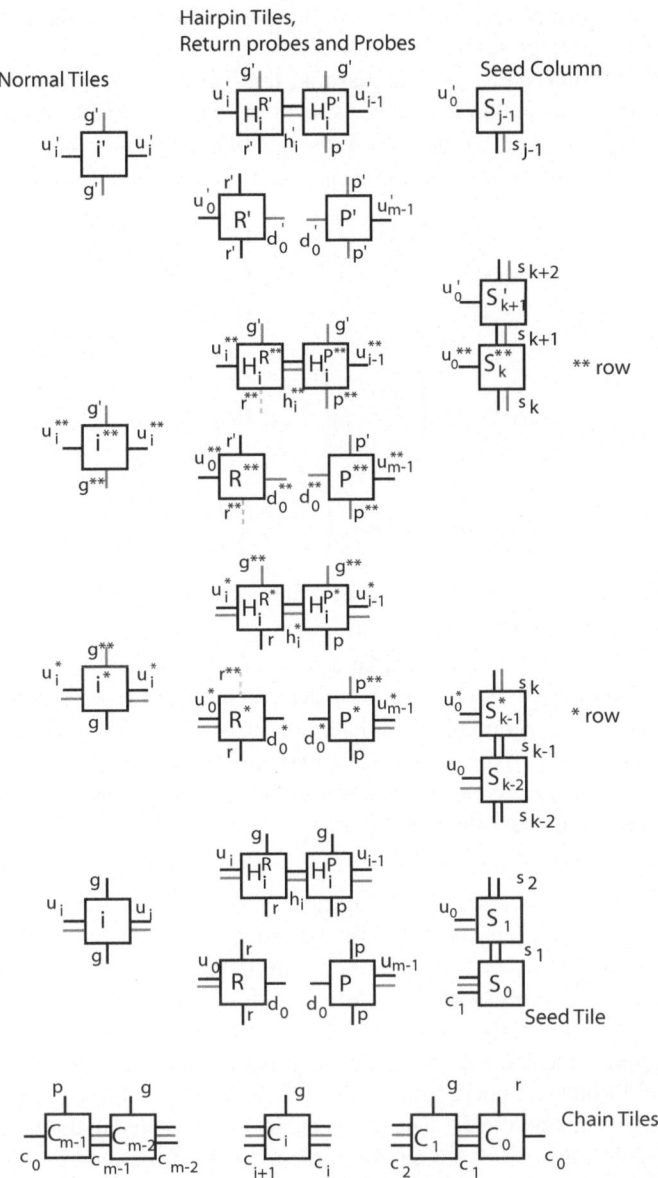

Fig. 6. Tile set to construct a $k \times N$ rectangle using only $O(N^{1/j} + j)$ tiles. The glue strength functions of gray, dashed, and black glues are defined in the proof.

reached, where a C_0 chain tile gets assembled. This again starts an assembly of a chain of length m. The whole process is repeated until a $j \times m^j$ rectangle is assembled.

Next we describe our modifications which are required for the $j - k$ upper volatile rows to get disassembled after the complete assembly of the $j \times m^j$ rectangle. First of all we need to have a special $(k+1)$-th row (∗∗ row), which will assemble to the north of the k-th row (∗ row), as shown in Figure 6.

The operating temperature $\tau = 2$. Assume that for all glue-types, $mit = t_0$ and $tms = t_1$. There are three kinds of glues shown in Figure 6: black, gray, and dashed. Assume that the glue-strength function for a single black glue is $g_{\text{black}}(t)$, a single gray glue is $g_{\text{gray}}(t)$, and a single dashed glue is $g_{\text{dashed}}(t)$. They are defined as

$$g_{\text{black}}(t) = \begin{cases} \frac{4t}{5t_0} & t < t_0 \\ \frac{4}{5} + \frac{t-t_0}{5(t_1-t_0)} & t_0 \le t < t_1 \\ 1 & t \ge t_1 \end{cases}$$

$$g_{\text{gray}}(t) = \begin{cases} \frac{2t}{5t_0} & t < t_0 \\ \frac{2}{5} + \frac{t-t_0}{10(t_1-t_0)} & t_0 \le t < t_1 \\ \frac{1}{2} & t \ge t_1 \end{cases}$$

$$g_{\text{dashed}}(t) = \begin{cases} \frac{2t}{5t_0} & t < t_0 \\ \frac{2}{5} & t \ge t_0 \end{cases}$$

Multiple glues shown on the same side of a tile in Figure 6 are additive. For example, the glue strength between C_i and C_{i+1} ($0 \le i \le m - 2$) is $2g_{\text{black}}(t) + g_{\text{gray}}(t)$.

This system will start assembling like a base $N^{1/j}$ counter of j digits, as briefed above and detailed in [3, 6]. It will first construct a rectangle of size $j \times N$ using $N^{1/j} + j$ type of tiles. Once the rectangle is complete, the tile on the north-west corner will start the required disassembly of the upper $(j - k)$ volatile rows, which results in the formation of a $k \times N$ rectangle. We call these two phases *Assembly phase* and *Disassembly phase* respectively, and describe them below.

Assembly Phase

In the Assembly Phase, we aim at constructing a $j \times N$ rectangle. In the time dependent model, the assembly proceeds as in the standard model until the assembly of P^* tile in the k-th row (∗ row). At this point, an $H^{R^{**}}$ tile is required to get assembled. However, when the $H^{R^{**}}$ tile is assembled in the $(k + 1)$-th row, the total support on $H^{R^{**}}$ from its east neighbor is only $\frac{4}{5} + \frac{2}{5} < 2$ at the end of mit. Thus $H^{R^{**}}$ must obtain additional support; otherwise it will get disassembled, blocking the desired assembly process. The additional support comes both from its south neighbor and its west neighbor. (1) On the south front, tile R^* can arrive and be incorporated in the k-th row (∗ row) of the assembly. It holds $H^{R^{**}}$ for another time interval of mit and provides a support of $\frac{2}{5}$. Further note that during this second interval, an R tile can be assembled in the $(k - 1)$-th row, and the R^* tile in the k-th row will then have support 2 at mit and hence stay attached. In addition, tile R has support 2 at mit, so it will also stay attached. Regarding $H^{R^{**}}$, the end result is that it receives an additional *stable* support $\frac{2}{5}$ from its south neighbor. However, the maximum support from both the south and the east is at most $1 + \frac{1}{2} + \frac{2}{5}$, which is still less than $\tau = 2$. Fortunately, additional rescue comes from the west. (2) On the west front, an i^{**} tile can get attached to $H^{R^{**}}$, and stabilize it by raising its total support above 2. However, this support is unstable, or *volatile*, in the sense that i^{**} itself needs additional support from its own west and south neighbors to stay attached. If this support can not come in time, that is, before *mit*, i^{**} will get disassembled, in turn causing the disassembly of $H^{R^{**}}$. The key observation here is that this assembly/disassembly is a reversible dynamic process: the disassembly may stop and start going backwards (*i.e.* assembling again) at any point. Thus in a dynamic, reversible fashion, the target structure of the Assembly Phase, namely the $j \times N$ rectangle, can be eventually constructed.

The above added complication is due to the fact that we require the $H^{R^{**}}$ tiles in the $(k+1)$-th row to get a total support of < 2 from the south and the east. This is crucial because during the subsequent Disassembly Phase (as we describe next) the desired disassembly can only carry through if the total support of each volatile tile from the south and the east is < 2.

Disassembly Phase

In the Disassembly Phase, we will remove the $j - k$ volatile rows, and reach the final target structure, a $k \times N$ rectangle. Once the $j \times N$ rectangle is complete, the tile T at the north-west corner (P' tile in the j-th row) initiates the disassembly. When the *mit* of the glue-pairs between tile T and its neighbors is over, tile T will get detached because the total glue strength that it has accumulated is $\frac{4}{5} + \frac{2}{5} < \tau = 2$. Note that, unlike the above case for $H^{R^{**}}$, no additional support can come from the west for tile T since T is the west-most tiles. As such, T is doomed to get disassembled. With T gone, T's east neighbor will get removed next, since it now has a total glue strength $\leq 1 + \frac{1}{2} < \tau$. Similarly, all the tiles in this row will get removed one by one, followed by the removal of the tiles in the next row (south row). Such disassembly of the tiles continues until we are left with the target rectangle of size $k \times N$, whose constituent tiles, at this stage, all have a total glue strength no less than $\tau = 2$, and hence stay stably attached.

Note that, similar as in the Assembly Phase, the volatile tiles that just got removed might come back. But again, ultimately they will have to *all* fall off (after the *mit*), and produce the desired $k \times N$ rectangle.

Concluding the Proof

We can construct a $k \times N$ rectangle using $O(N^{1/j} + j)$ type of tiles (where $j > k$). As in [6], it can be reduced to $O(\frac{\log N}{\log \log N})$ by choosing $j = \frac{\log N}{\log \log N - \log \log \log N}$. $\square$

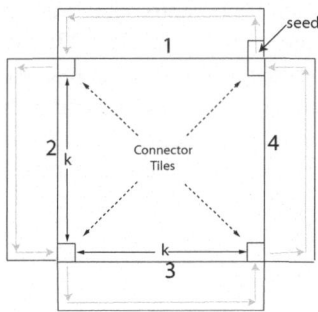

Fig. 7. Direction of the gray arrow shows the direction of construction of a square with a hole, starting from the indicated seed

Thin rectangles can serve as building blocks for the construction of many other interesting shapes. One example is a square of size $N \times N$ with a large square hole of size $k \times k$ ($k \sim N$). Under the standard model, the lower bound can be shown to be $\Omega(\frac{(k)^{\frac{2}{N-k}}}{N-k})$ by a lower bound argument similar to the one in [6]. Note that as $N - k$ decreases, *i.e.* the square hole in the square increases, the lower bound increases. In the case when $N - k$ is smaller than $\frac{\log N}{\log \log N - \log \log \log N}$, the lower bound is more than $\frac{\log N}{\log \log N}$. In the case when $N - k$ is a small constant, the complexity is almost N^c, where c is some constant < 1. However, in time-dependent model, the tile complexity of this shape can be reduced to $O(\frac{\log k}{\log \log k})$ even for small values of $N - k$, using our thin rectangle construction.

The basic idea is quite simple. We sequentially grow four different $(\frac{N-k-2}{2}) \times (k + 2)$ rectangles that will make up the major part of the square's sides. To enable the sequential growth of these rectangles, we introduce four *connector tiles* that concatenate them. After the completion of one rectangle the connector tile will assemble and provide basis for the assembly of the subsequent rectangle. Finally, some more constant type of tiles will be introduced to fill in the gaps at the four corners this $N \times N$ square, and the gap between two subsequent connector tiles, producing the target $N \times N$ square with a $k \times k$ hole.

The upper bound on the number of tiles is exactly the same as the upper bound for constructing the four thin rectangles, which is $O(\frac{\log k}{\log \log k})$.

7 Discussion and Future Work

In this paper, we define a model in which the glue strength between tiles depends upon the time they have been abutting each other. Under this model, we demonstrate and analyze catalysis and self-replication, and show how to construct a thin $k \times N$ rectangle using $O(\frac{\log N}{\log \log N})$ tiles. The upper bound on assembling a thin rectangle is obtained by applying similar assembly strategy as in the multi-temperature model [6]. Thus, an interesting question is whether the multi-temperature model can be simulated using our time-dependent model. We also want to further investigate if under our model the lower bound of $\Omega(\frac{\log N}{\log \log N})$ for the tile complexity of an $N \times N$ square can be further improved.

Another interesting direction is to study the kinetics of the catalysis and self-replication analytically. Winfree's kinetic model [37] can be used to study them, but the challenge here is that the rate constant for the dissociation for a particular species varies with time because of changing glue strengths of its bonds. This makes the analytical study hard. However, these catalytic and self-replicating systems can be modeled as a continuous time markov chain, and studied using computer simulation to obtain empirical results.

References

1. http://mrsec.wisc.edu/edetc/selfassembly/.
2. L. Adleman. Towards a mathematical theory of self-assembly. Technical Report 00-722, University of Southern California, 2000.

3. L. Adleman, Q. Cheng, A. Goel, and M.D. Huang. Running time and program size for self-assembled squares. In *Proceedings of the thirty-third annual ACM symposium on Theory of computing*, pages 740–748. ACM Press, 2001.
4. L. Adleman, Q. Cheng, A. Goel, M.D. Huang, D. Kempe, P.M. de Espans, and P.W.K. Rothemund. Combinatorial optimization problems in self-assembly. In *Proceedings of the thirty-fourth annual ACM symposium on Theory of computing*, pages 23–32. ACM Press, 2002.
5. L. Adleman, J. Kari, L. Kari, and D. Reishus. On the decidability of self-assembly of infinite ribbons. In *Proceedings of the 43rd Symposium on Foundations of Computer Science*, pages 530–537, 2002.
6. G. Aggarwal, M.H. Goldwasser, M.Y. Kao, and R.T. Schweller. Complexities for generalized models of self-assembly. In *Proceedings of 15th annual ACM-SIAM Symposium on Discrete Algorithms (SODA)*, pages 880–889. ACM Press, 2004.
7. N. Bowden, A. Terfort, J. Carbeck, and G.M. Whitesides. Self-assembly of mesoscale objects into ordered two-dimensional arrays. *Science*, 276(11):233–235, 1997.
8. R.F. Bruinsma, W.M. Gelbart, D. Reguera, J. Rudnick, and R. Zandi. Viral self-assembly as a thermodynamic process. *Phys. Rev. Lett.*, 90(24):248101, 2003 June 20.
9. N. Chelyapov, Y. Brun, M. Gopalkrishnan, D. Reishus, B. Shaw, and L. Adleman. DNA triangles and self-assembled hexagonal tilings. *J. Am. Chem. Soc.*, 126:13924–13925, 2004.
10. H.L. Chen, Q. Cheng, A. Goel, M.D. Huang, and P.M. de Espanes. Invadable self-assembly: Combining robustness with efficiency. In *Proceedings of the 15th annual ACM-SIAM Symposium on Discrete Algorithms (SODA)*, pages 890–899, 2004.
11. Q. Cheng and P.M. de Espanes. Resolving two open problems in the self-assembly of squares. Technical Report 03-793, University of Southern California, 2003.
12. Q. Cheng, A. Goel, and P. Moisset. Optimal self-assembly of counters at temperature two. In *Proceedings of the first conference on Foundations of nanoscience: self-assembled architectures and devices*, 2004.
13. M. Cook, P.W.K. Rothemund, and E. Winfree. Self-assembled circuit patterns. In *DNA Based Computers 9*, volume 2943 of *LNCS*, pages 91–107, 2004.
14. K. Fujibayashi and S. Murata. A method for error suppression for self-assembling DNA tiles. In *DNA Based Computing 10*, pages 284–293, 2004.
15. B.D. Hughes. *Random Walks and Random Environments, Vol. 1: Random Walks*. New York: Oxford University Press, 1995.
16. M. Kao and R. Schweller. Reduce complexity for tile self-assembly through temperature programming. In *Proceedings of 17th annual ACM-SIAM Symposium on Discrete Algorithms (SODA) (to appear)*. ACM Press, 2006.
17. E. Klavins. Directed self-assembly using graph grammars. In *Foundations of Nanoscience: Self Assembled Architectures and Devices*, Snowbird, UT, 2004.
18. E. Klavins, R. Ghrist, and D. Lipsky. Graph grammars for self-assembling robotic systems. In *Proceedings of the International Conference on Robotics and Automation*, 2004.
19. T.H. LaBean, H. Yan, J. Kopatsch, F. Liu, E. Winfree, J.H. Reif, and N.C. Seeman. The construction, analysis, ligation and self-assembly of DNA triple crossover complexes. *J. Am. Chem. Soc.*, 122:1848–1860, 2000.
20. M.G. Lagoudakis and T.H. LaBean. 2-D DNA self-assembly for satisfiability. In *DNA Based Computers V*, volume 54 of *DIMACS*, pages 141–154. American Mathematical Society, 2000.
21. D. Liu, M. Wang, Z. Deng, R. Walulu, and C. Mao. Tensegrity: Construction of rigid DNA triangles with flexible four-arm dna junctions. *J. Am. Chem. Soc.*, 126:2324–2325, 2004.
22. C. Mao, W. Sun, and N.C. Seeman. Designed two-dimensional DNA holliday junction arrays visualized by atomic force microscopy. *J. Am. Chem. Soc.*, 121:5437–5443, 1999.
23. J.H. Reif, S. Sahu, and P. Yin. Compact error-resilient computational DNA tiling assemblies. In *Proc. 10th International Meeting on DNA Computing*, pages 248–260, 2004.

24. J.H. Reif, S. Sahu, and P. Yin. Complexity of graph self-assembly in accretive systems and self-destructible systems. In *Proc. 11th International Meeting on DNA Computing*, pages 101–112, 2005.
25. R.M. Robinson. Undecidability and non periodicity of tilings of the plane. *Inventiones Math*, 12:177–209, 1971.
26. P.W.K. Rothemund. Using lateral capillary forces to compute by self-assembly. *Proc. Natl. Acad. Sci. USA*, 97(3):984–989, 2000.
27. P.W.K. Rothemund. *Theory and Experiments in Algorithmic Self-Assembly*. PhD thesis, University of Southern California, 2001.
28. P.W.K. Rothemund and E. Winfree. The program-size complexity of self-assembled squares (extended abstract). In *Proceedings of the thirty-second annual ACM symposium on Theory of computing*, pages 459–468. ACM Press, 2000.
29. P. Sa-Ardyen, N. Jonoska, and N.C. Seeman. Self-assembling DNA graphs. *Lecture Notes in Computer Science*, 2568:1–9, 2003.
30. S. Sahu, P. Yin, and J.H. Reif. A self-assembly model of DNA tiles with time-dependent glue strength. Technical Report CS-2005-04, Duke University, 2005.
31. R. Schulman, S. Lee, N. Papadakis, and E. Winfree. One dimensional boundaries for DNA tile self-assembly. In *DNA Based Computers 9*, volume 2943 of *LNCS*, pages 108–125, 2004.
32. R. Schulman and E. Winfree. Programmable control of nucleation for algorithmic self-assembly. In *DNA Based Computers 10*, LNCS, 2005.
33. R. Schulman and E. Winfree. Self-replication and evolution of DNA crystals. In *The 13th European Conference on Artificial Life (ECAL)*, 2005.
34. D. Soloveichik and E. Winfree. Complexity of compact proofreading for self-assembled patterns. In *Proc. 11th International Meeting on DNA Computing*, pages 125–135, 2005.
35. D. Soloveichik and E. Winfree. Complexity of self-assembled shapes. In *DNA Based Computers 10*, LNCS, 2005.
36. H. Wang. Proving theorems by pattern recognition ii. *Bell Systems Technical Journal*, 40: 1–41, 1961.
37. E. Winfree. Simulation of computing by self-assembly. Technical Report 1998.22, Caltech, 1998.
38. E. Winfree. Self-healing tile sets (draft). 2005.
39. E. Winfree and R. Bekbolatov. Proofreading tile sets: Error correction for algorithmic self-assembly. In *DNA Based Computers 9*, volume 2943 of *LNCS*, pages 126–144, 2004.
40. E. Winfree, F. Liu, L.A. Wenzler, and N.C. Seeman. Design and self-assembly of two-dimensional DNA crystals. *Nature*, 394(6693):539–544, 1998.
41. E. Winfree, X. Yang, and N.C. Seeman. Universal computation via self-assembly of DNA: Some theory and experiments. In L.F. Landweber and E.B. Baum, editors, *DNA Based Computers II*, volume 44 of *DIMACS*, pages 191–213. American Mathematical Society, 1999.
42. H. Yan, T.H. LaBean, L. Feng, and J.H. Reif. Directed nucleation assembly of DNA tile complexes for barcode patterned DNA lattices. *Proc. Natl. Acad. Sci. USA*, 100(14): 8103–8108, 2003.
43. H. Yan, S.H. Park, G. Finkelstein, J.H. Reif, and T.H. LaBean. DNA-templated self-assembly of protein arrays and highly conductive nanowires. *Science*, 301(5641):1882–1884, 2003.

Complexity of Compact Proofreading
for Self-assembled Patterns

David Soloveichik and Erik Winfree

Department of CNS and CS, California Institute of Technology
{dsolov, winfree}@caltech.edu

Abstract. Fault-tolerance is a critical issue for biochemical computation. Recent theoretical work on algorithmic self-assembly has shown that error correcting tile sets are possible, and that they can achieve exponential decrease in error rates with a small increase in the number of tile types and the scale of the construction [24, 4]. Following [17], we consider the issue of applying similar schemes to achieve error correction without any increase in the scale of the assembled pattern. Using a new proofreading transformation, we show that compact proofreading can be performed for some patterns with a modest increase in the number of tile types. Other patterns appear to require an exponential number of tile types. A simple property of existing proofreading schemes – a strong kind of redundancy – is the culprit, suggesting that if general purpose compact proofreading schemes are to be found, this type of redundancy must be avoided.

1 Introduction

The Tile Assembly Model [22, 23] formalizes a generalized crystal growth process by which an organized structure can spontaneously form from simple parts. This model considers the growth of two dimensional "crystals" made out of square units called tiles. Typically, there are many types of tiles that must compete to bind to the crystal. A new tile can be added to a growing complex if it binds strongly enough. Each of the four sides of a tile has an associated bond type that interacts with matching sides of other tiles that have already been incorporated. The assembly starts from a specified seed assembly and proceeds by sequential addition of tiles. Tiles do not get used up since it is assumed there is an unbounded supply of tiles of each type. This model has been used to theoretically examine how to use self-assembly for massively parallel DNA computation [21, 26, 16, 13], for creating objects with programmable morphogenesis [10, 1, 2, 20], for patterning of components during nanofabrication of molecular electronic circuits [6], and for studying self-replication and Darwinian evolution of information-bearing crystals [18, 19]. Fig. 1 illustrates two different patterns and the corresponding tile systems that self-assemble into them. Both patterns are produced by similar tile systems using only two bond types, four tile types, simple boolean rules and similar seed assemblies (the L-shaped boundaries).

Confirming the physical plausibility and relevance of the abstraction, several self-assembling systems have been demonstrated using DNA molecules as tiles,

A. Carbone and N.A. Pierce (Eds.): DNA11, LNCS 3892, pp. 305–324, 2006.
© Springer-Verlag Berlin Heidelberg 2006

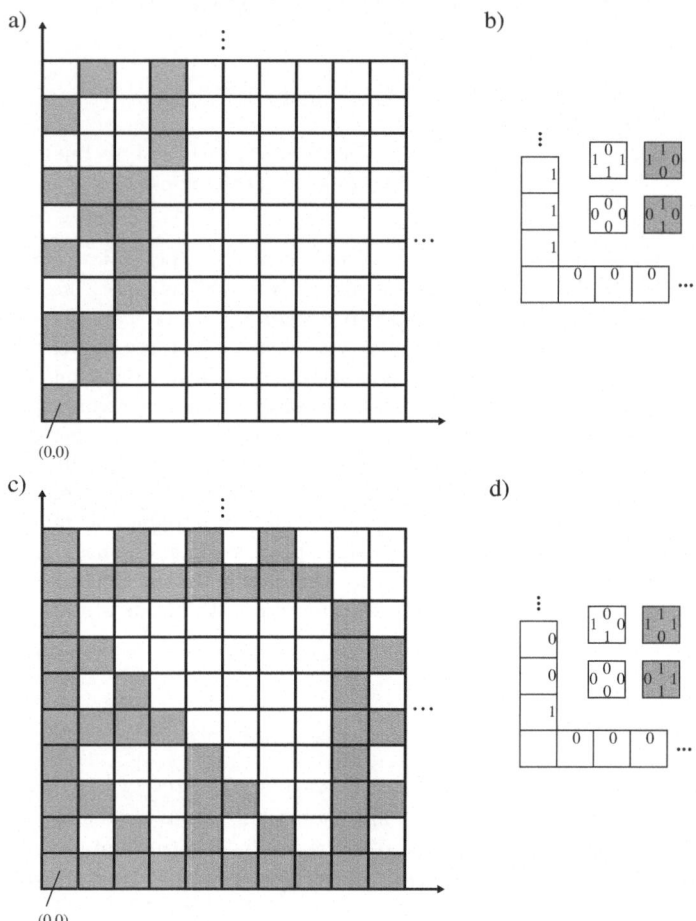

Fig. 1. (a) A binary counter pattern and (b) a tile system constructing it. (c) A Sierpinski pattern and (d) a tile system constructing it. In this formalism, identically-labeled sides match and tiles cannot be rotated. Tiles may attach to the growing assembly only if at least two sides match, i.e., if two bonds can form. Mismatches neither help nor hinder assembly. Note that the tile choice at each site is deterministic for these two tile sets.

including both periodic [25, 15, 12] and algorithmic patterns [14, 9, 3]. A major stumbling block to making algorithmic self-assembly practical is the error rate inherent in any stochastic biochemical implementation. Current implementations seem to suffer error rates of 1% to 15% [9, 3]. This means that on average every eighth to hundredth tile that is incorporated does not correctly bond with its neighbors. Once such a mistake occurs, the erroneous information can be propagated to tiles that are subsequently attached. Thus, a single mistake can result in a drastically different pattern being produced. With this error rate, structures of size larger than roughly 100 tiles cannot be assembled reliably.

There are generally two ways to improve the error-robustness of the assembly process. First, the physics of the process can be modified to achieve a lower probability of the incorporation of incorrect tiles into the growing complex. The second method, which we pursue here, is to use some logical properties of the tiles to perform error correction.

Proofreading tile sets for algorithmic self-assembly were introduced by Winfree and Bekobolatov [24]. The essential idea was to make use of a redundant encoding of information distributed across k tiles, making isolated errors impossible: to continue growth, errors must appear in multiples of k. Thanks to the reversible nature of crystallization, growth from erroneous tiles stalls and the erroneous tiles subsequently dissociate, allowing another chance for correct growth. Using this approach, a large class of tile sets can be transformed into more robust tile sets that assemble according to the same logic.

However (a) the proofreading tile sets produce assemblies k times larger than the original tile sets, involving k^2 times as many tiles; and (b) the improvement in error rates did not scale well with k in simulation. Chen and Goel [4] developed *snaked proofreading tile sets* that generalize the proofreading construction in a way that further inhibits growth on crystal facets. They were able to prove, with respect to a reversible model of algorithmic self-assembly, that error rates decrease exponentially with k, and thus to make an $N \times N$ pattern required only $k = \Omega(\log N)$. This provides a solution for (b), although the question of optimality remains open. Reif et al [17] raised the question of whether more compact proofreading schemes could be developed, and showed how to transform the two tiles sets shown in Fig. 1 to obtain lower error rates without any sacrifice in scale. However, Reif et al did not give a general construction that works for any original tile set, and did not analyze how the number of tile types would scale if the construction were to be generalized to obtain greater degrees of proofreading. Thus, question (a) concerning whether this can be improved in general and at what cost remained open.

The question of compactness is particularly important when self-assembly is used for molecular fabrication tasks, in which case the scale of the final pattern is of direct and critical importance. Furthermore, the question of scale is a fundamental issue for the theory of algorithmic self-assembly. In the error-free case, disregarding scale can drastically change the minimal number of tile types required to produce a given shape [20]; some shapes can be assembled from few tile types at a small scale, while other shapes can *only* be assembled from few tile types at a large scale. Examining whether proofreading can be performed without sacrificing scale is both of practical significance and could lead to important theoretical distinctions.

If it is the case that some patterns can't be assembled with low error rates at the original scale using a concise tile set, while for other patterns compact proofreading can be done effectively, then we would be justified in calling the former intrinsically fragile, and the latter intrinsically robust. Any such distinctions should be independent of any particular proofreading scheme. Indeed, we here show that this is true (in a certain sense), and we give a combinatorial criterion that distinguishes fragile patterns from robust patterns. As examples, we show that the two patterns discussed in Reif et al's work on compact proofreading [17] and shown in Fig. 1 are fundamentally different, in that (within a wide class of potential proofreading schemes considered here) the cost of obtaining reliable assembly at the same scale becomes dramatically different as lower error rates are required.

1.1   The Abstract Tile Assembly Model

This section informally summarizes the abstract Tile Assembly Model (aTAM). See [8, 20] for a formal treatment. Self assembly occurs on a $\mathbb{Z} \times \mathbb{Z}$ grid of unit square locations, on which unit-square **tiles** may be placed under specific conditions. Each tile has **bond types** on its north, east, south and west sides. A finite set of **tile types** defines the set of possible tiles that can be placed on the grid. Tile types are oriented and therefore a rotated version of a tile type is considered to be a different tile type. A single tile type may be used an arbitrary number of times. A **configuration** is a set of tiles such that there is at most one tile in every location $(i, j) \in \mathbb{Z} \times \mathbb{Z}$. Two adjacent tiles **bond** if their abutting sides have matching bond types. Further, each bond type forms bonds of a specific strength, called its **interaction strength**. In this paper the three possible strengths of bonds are $\{0, 1, 2\}$. A new tile can be added to an empty spot in a configuration if and only if the sum of its interaction strengths with its neighbors reaches or exceeds some parameter τ. The tile systems shown in this paper use $\tau = 2$, i.e., at least a single strong (strength 2) or two weak (strength 1) bonds are needed to secure a tile in place.

For the purposes of this paper, a tile system consists of a finite set of tile types T with specific interaction strengths associated with each bond type, and a start configuration. Whereas a configuration can be any arrangement of tiles, we are interested in the subclass of configurations that can result from a self-assembly process. Thus, an **assembly** is a configuration that can result from the start configuration by a sequence of additions of tiles according to the above rules at $\tau = 1$ or $\tau = 2$ (i.e., it is connected). A τ-**stable** assembly is one that cannot be split into two parts without breaking bonds with a total strength of at least τ. **Deterministic** tile systems are those whose assemblies can incorporate at most 1 tile type at any location at any time.

1.2   The Kinetic Tile Assembly Model and Errors

The Kinetic Tile Assembly Model (kTAM) augments the abstract Tile Assembly Model with a stochastic model of self-assembly dynamics, allowing calculation of error rates and the duration of self-assembly. Following [23, 24] we make the following assumptions. First, the concentration of each tile type in solution is held constant throughout the self-assembly process, and the concentrations of all tile types are equal. We assume that for every tile association reaction there is a corresponding dissociation reaction (and no others). We further assume that the rate of addition (**forward rate** f) of any tile type at any position of the perimeter of the growing assembly is the same. Specifically, $f = k_f e^{-G_{mc}}$ where k_f is a constant that sets the time scale, and G_{mc} is the logarithm of the concentration of each tile type in solution. The rate that a tile falls off the growing assembly (**reverse rate** r_b) depends exponentially on the number of bonds that must be broken. Specifically, $r_b = k_f e^{-bG_{se}}$ where b is the total interaction strength with which the tile is attached to the assembly, and G_{mc} is the unit bond free energy, which may depend, for example, on temperature.

We assume the following concerning f and r_b. Following [23] we let $f \approx r_2$ for a $\tau = 2$ system since it provides the optimal operating environment [23]. Further, we assume f (and therefore r_2) can be arbitrarily chosen in our model by changing G_{mc} and G_{se}, for example by changing tile concentrations and temperature. (In practice, there are limits to how much these parameters can be changed.) However, k_f is assumed to be a physical constant not under our control.

In the kTAM, the $\tau = 2$ tile addition requirement imposed by the abstract Tile Assembly Model is satisfied only with a certain probability: assuming $f \approx r_2$ so $r_1 \gg f$, if a tile is added that bonds only with strength 1, it falls off very quickly as it should in the aTAM with $\tau = 2$. Tiles attached with strength 2 stick much longer, allowing an opportunity for other tiles to attach to them. Once a tile is bonded with total strength 3, it is very unlikely to dissociate (unless surrounding tiles fall off first).

Following [4], the fundamental kind of error we consider here is an **insufficient attachment**. At threshold $\tau = 2$, an insufficient attachment occurs when a tile attaches with strength 1, but before falling off, another tile attaches next to it, resulting in a 2-stable assembly. Since insufficient attachments are the only kind of error we analyze in this paper, we'll use "error" and "insufficient attachment" interchangeably.

Chen and Goel [4] make use of a simplification of the kTAM that captures the essential behavior while being more tractable for rigorous proofs. Under the conditions where $f = r_2$, the self-assembly process is dominated by tiles being added with exactly 2 bonds and tiles falling off via exactly 2 bonds. The **locking** kTAM model assumes that these are the only possible single-tile events. That is, $r_b = 0$ for $b \geq 3$, and tiles never attach via a single strength-1 bond. Additionally, insufficient attachments are modeled in the locking kTAM as atomic events, in which two tiles are added simultaneously at any position in which an insufficient attachment can occur. Specifically, any particular pair of tile types that can create an insufficient attachment in the kTAM is added at a rate $f_{err} = O(e^{-3G_{se}})$. (This is asymptotically the rate that insufficient attachments occur in kTAM [4].) Thus the total rate of insufficient attachments at a particular location is $Q f_{err}$, where Q is the number of different ways (with different tile types) that an insufficient attachment can occur there. We don't absorb Q into the $O(\cdot)$ notation because we will be considering tile sets with an increasing number of tile types that can cause errors. Note that Q can be bounded by the square of the total number of tile types. These insufficient attachments are the sole cause of errors during growth.[1] Growth during which no insufficient attachments occur we call (reversible) $\tau = 2$ **growth**.

1.3 Quarter-Plane Patterns

The output of the self-assembly process is usually considered to be either the shape of the uniquely produced terminal assembly [10, 1, 2, 20] or the pattern produced if we focus on the locations of certain types of tiles [24, 4, 17, 6]. Here we will focus on self-assembling of **quarter-plane patterns**. A quarter-plane pattern (or just **pattern** for short) **P** is an assignment of symbols from a finite alphabet of "colors" to points on the quarter plane ($\mathbb{Z}^+ \times \mathbb{Z}^+$ by convention). A deterministic tile system can be thought to construct a pattern in the sense that there is some function (not necessarily a bijection) mapping tile types to colors such that tiles in any produced assembly correctly map to corresponding colors of the pattern. As the assembly grows, a larger and larger portion of the pattern gets filled. There are patterns that cannot be deterministically constructed by any tile system (e.g., uncomputable ones), but for the purposes of this paper we consider patterns constructible from deterministic tile systems where all bond strengths are 1 and the seed assembly (defining the boundary conditions) is an infinite L shape that is eventually periodic, with its corner on the origin. See Fig. 1 for two examples. Such tile system we'll call **quarter plane tile systems** and the patterns produced by them the **constructible quarter-plane patterns**. These systems include a wide variety of patterns, including the Sierpinski pattern, the binary counter pattern, the Hadamard pattern [6], and patterns containing the space-time history of arbitrary 1D block cellular automata and Turing machines. Note that by including the infinite seed assembly we are avoiding the issue of nucleation, which requires distinct error correcting techniques [18].

2   Making Self-assembly Robust

The kinetic Tile Assembly Model predicts that for any quarter plane tile system, arbitrarily small error rates can be achieved by increasing G_{mc} and G_{se}, but at the cost of decreasing the overall rate of assembly. Specifically, the worst case

---

[1] Another error, with respect to the aTAM, that can occur in the original kTAM is when a tile attached by strength 3 (or more) falls off. Why do we feel comfortable neglecting this error in the locking kTAM, especially since as a function of G_{se}, both r_3 and f_{err} are both $O(e^{-3G_{se}})$? One reason is that in practice the dissociation of tiles held to the assembly with strength 3 does not seem to cause the problems that insufficient attachments induce, in tile sets that we have simulated and examined: no incorrect tiles are immediately introduced, often the correct tile will quickly arrive to repair the hole, and if an incorrect tile fills the hole, further growth may be impossible, usually allowing time for the incorrect tile to fall off. A second reason is that as the number of tile types increases (i.e., with more complex patterns or more complex proofreading schemes), Qf_{err} becomes arbitrarily large, while r_3 stays constant. Nonetheless, a more satisfying treatment would not make these approximations and would address the original kTAM directly.

analysis (which assumes that after any single error, assembly can be continued by valid $\tau = 2$ growth) predicts that the relationship between per tile error rate ε and the rate of assembly r (layers per second) approximately satisfies $r \propto \varepsilon^2$ [23]. This is rather unsatisfactory since, for example, decreasing the error rate by a factor of 10 necessitates slowing down self-assembly by a factor of 100.

Rather than talking about the relationship between the per tile error rate and the total rate of self-assembly, following [4] one can ask how long it takes to produce the correct $N \times N$ initial portion of the pattern with high probability. To produce this initial portion correctly with high probability, we need the per-tile error rate to be $\varepsilon = O(N^{-2})$ to ensure that no mistake occurs. This implies that $r = O(N^{-4})$ for worst case tile sets. This informal argument suggests that the time to produce the $N \times N$ square is $\Omega(N^5)$. This is unsatisfactory, because the same assembly can be grown in time $O(N)$ in the aTAM augmented with rates [1], and thus the cost of errors appears to be considerable.

Despite this pessimistic argument, certain kinds of tile systems can achieve better error rate/rate of assembly tradeoffs. Indeed, the reversibility of the self-assembly process can help. Some tile systems have the property that upon encountering an error, unless many more mistakes are made, the self-assembly process stalls. Stalling gives time for the incorrectly incorporated tiles to be eventually replaced by the correct ones in a random walk process, so long as not too many incorrect tiles have been added.

Exploiting this observation, several schemes have been proposed for converting arbitrary quarter plane tile systems into tile systems producing a *scaled-up* version of the same pattern, resulting in better robustness to error. The initial proposal due to Winfree and Bekbolatov [24] suggests replacing each tile type of the original tile system with k^2 tile types, with unique internal strength-1 bonds (Fig. 2(a)). Such proofreading assemblies have the property that for a block corresponding to a single tile in the old system to get completed, either no

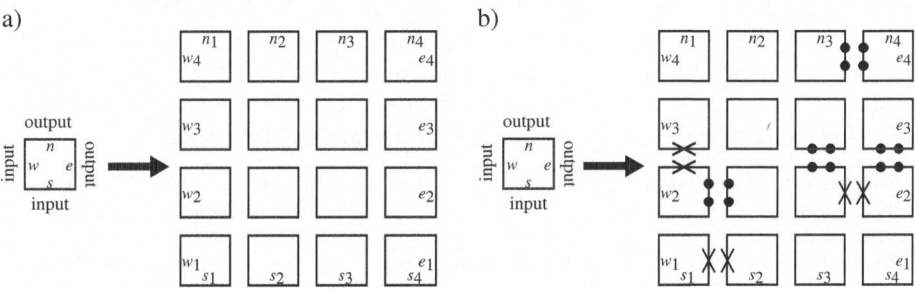

Fig. 2. Winfree and Bekbolatov (a) and Chen and Goel (b) proofreading transformations using 4×4 blocks. Each tile type is replaced with k^2 tile types that fit together to form the block as shown. Strength 2 bonds are indicated with 2 dots. Strength 0 bonds are indicated with a cross. All unlabeled (internal) bond types are unique (within the block and between blocks.) The placement of weak and strong bonds is dependent upon the orientation of growth, which in this case is to the north-east, since for quarter plane tile systems the input is always received from the west and south sides.

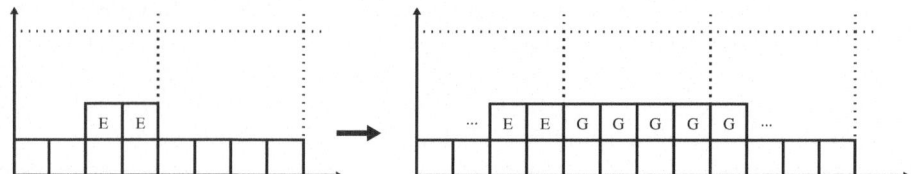

Fig. 3. The Winfree and Bekbolatov proofreading scheme is susceptible to single facet nucleation errors. If an insufficient attachment results in the two E tiles shown, then subsequent $\tau = 2$ growth (G) can continue indefinitely to the right. Thus many incorrect tiles can be added following a single facet nucleation error even if the block that E is in does not get completed. The dotted lines indicate block boundaries (for 4×4 blocks). Note that most of the incorrect tiles are attached with strength 3; therefore, they do not easily fall off, except at the left and the right sides.

mistakes, or at least k mistakes must occur. However, this scheme suffers from the problem that the self-assembly process after a single insufficient attachment can still result in a large number of incorrect tiles that must later be removed, spanning the length of the assembly. Consider the situation depicted in Fig. 3. If the insufficient attachment illustrated occurs (The first E is added with interaction strength 1, but before it dissociates, a tile attaches to it on the right with interaction strength 2), the incorrect information can be propagated indefinitely to the edge of the assembly by subsequent $\tau = 2$ tile additions.

Currently the only scheme that provably achieves a guaranteed level of proofreading is due to Chen and Goel [4] using the locking kTAM model. Their proofreading scheme, called *snaked proofreading*, is similar to the Winfree and Bekbolatov system, but additionally controls the order of self-assembly within each block by using strength-0 and strength-2 bonds, making sure that not too many incorrect tiles can be added by $\tau = 2$ growth after an insufficient attachment. In particular, the strength-0 bonds ensure that unless most of the block gets completed, self-assembly stalls. Fig. 2(b) shows their 4×4 construction; see their paper for the general construction for arbitrary block size.[2] They can attain a polynomial decrease in the error rate with only a logarithmic increase in k. Specifically the formal results they obtain are the following:[3][4]

Theorem 1 (theorem 4.2 of [4]). *For any constant $p < 1$, the $N \times N$ block initial portion of the pattern is produced correctly with probability at least p in*

---

[2] Note that unlike the original proofreading transformation, the snake proofreading transformation does not result in a quarter plane tile system as it uses both strong and weak bonds.

[3] [4] also guarantees that the assembly is stable for a long time after it is complete, a concern we ignore in this paper. For fixed k, they also provide theorem 4.1, which guarantees reliable assembly of an $N \times N$ square in time $O(N^{1+8/k})$.

[4] Chen and Goel only prove their result for the case when the initial L seed assembly has arms that span exactly N blocks. We need to cover the case when an infinite L seed assembly is used. See Appendix A for a proof that their results can be extended to an infinite seed assembly.

time $O(Npoly(\log(N))$ by the $k \times k$ snaked proofreading tile system where $k = \theta(\log N)$, using the locking kTAM with appropriate G_{mc} and G_{se}.

To obtain this result, assembly conditions (G_{mc} and G_{se}) need be adjusted only slightly as N increases.[5]

The above construction requires increasing the scale of the produced pattern, even if only logarithmically in the size of the total desired size of the self-assembled pattern. Reif et al [17] pointed this out as a potential problem and proposed schemes for decreasing the effective error rate while preserving the scale of the pattern. However, they rely on certain specific properties of the original tile system, and do not provide a general construction that can be extended to arbitrary levels of error correction. Further, their constructions suffer from the same problem as the original Winfree and Bekbolatov proofreading system. In the next section we argue that the snaked proofreading construction can be adopted to achieve same-scale proofreading for sufficiently "simple" patterns.

3   Compact Proofreading Schemes for Simple Patterns

In this section we argue that a wide variety of sufficiently "simple" patterns can be produced with arbitrarily small effective error rates without increasing the scale of self-assembly, at the cost of slightly increasing the number of tile types and the time of self-assembly. Based on Reif et al's nomenclature [17], we call these proofreading schemes *compact* to indicate that the scale of the pattern is not allowed to change.

The following definition illustrates our goal:

Definition 1. *Let $p < 1$ be a constant (e.g., 0.99). A sequence of deterministic tile systems $\{\mathbf{T}_1, \mathbf{T}_2, \ldots\}$ is a **compact proofreading scheme** for pattern $\mathbf{P}$ if:*

(1: correctness) $\mathbf{T}_N$ *produces the full infinite pattern $\mathbf{P}$ under the aTAM.*
(2: conciseness) $\mathbf{T}_N$ *has poly$(\log N)$ tile types.*
(3: robustness) $\mathbf{T}_N$ *produces the correct $N \times N$ initial portion of pattern $\mathbf{P}$ (without scaling) with probability at least p in time $O(Npoly(\log N))$ in the locking kTAM for some G_{se} and G_{mc}.*

If you want to construct the initial $N \times N$ portion of pattern $\mathbf{P}$ with probability at least p in time $O(Npoly(\log N))$ you pick tile system $\mathbf{T}_N$ and the corresponding G_{se} and G_{mc}. The same tile system might be used for many N (i.e., the sequence of tile systems may have repetitions). The second condition indicates that we don't want this tile system to have too many tile types. For constructible quarter plane patterns, a constant number of tile types suffices to

---

[5] It is hard to say whether the snaked proofreading construction is asymptotically optimal. While the best possible assembly time in a model where concentrations are held constant with changing N is linear in N, we assume that G_{mc} and G_{se} are free to change as long as the relationship $f = r_2$ is maintained. Of course while decreasing G_{mc} and G_{se} speeds up the assembly process, the rate of errors is increased; thus, the optimal tradeoff is not obvious.

create the infinite pattern in the absence of errors. If the second condition is satisfied then the error correction itself is accomplished with a polylogarithmic number of additional tiles, which is comparable to the cost of error correction in other models studied in computer science. While one can imagine different versions of these conditions, the stated version gives the proofreading condition that can be obtained by adapting the snaked proofreading construction, as argued below. Finally, note that the tile systems $\{\mathbf{T}_1, \mathbf{T}_2, \ldots\}$ do not have to be quarter plane tile systems, and therefore our theorems will apply to a wide range of potential proofreading schemes.

For which patterns do there exist compact proofreading schemes? Given a pattern and a quarter plane tile system $\mathbf{T}$ producing it, consider any assembly of $\mathbf{T}$. For a given k, imagine splitting the assembly into $k \times k$ disjoint blocks starting at the origin. We'll use the term **block** to refer to aligned blocks, and **square** to refer to blocks without the restriction that they be aligned to integer multiples of k with respect to the origin. Each complete block contains k^2 tiles; two blocks at different locations are considered equivalent if they consist of the same arrangement of tile types. If there is some polynomial $Q(k)$ such that repeating this process for all assemblies and all k yields at most $Q(k)$ different (completed) block types, then we say that $\mathbf{T}$ **segments** into $poly(k)$ $k \times k$ block types.[6] Patterns produced by such tile systems are the "simple" patterns, for which, we will argue, there exist compact proofreading schemes; we term such patterns **robust** to indicate this.

On the other hand, there are patterns for which it is easy to see that no quarter tile system producing them segments into $poly(k)$ $k \times k$ block types. For example these include patterns which have $2^{\Omega(k)}$ different types of $k \times k$ squares of colors.[7] We'll prove negative results about such patterns, which we term **fragile** in the next section.[8]

---

[6] We use disjoint blocks aligned with the origin for simplicity in what follows. It is inessential that we define segmentation in terms of blocks rather than squares: A tile system segments into $poly(k)$ different $k \times k$ block types if and only if it produces assemblies that contain $poly(k)$ different types of non-aligned $k \times k$ squares. This is also true for other shapes than squares, as long as they have sufficient extent. See Appendix B for an example, the size-k diagonals.

[7] In what follows, we will consider both the number of blocks (or squares) in an assembly, in which case we mean blocks (or squares) of tile types, as well as the number of blocks (or squares) in a pattern, which which case we mean block (or squares) of colors. Since each tile type has a color, the latter is less than or equal to the former for patterns produced by quarter-plane tile systems.

[8] Analogous to the uncomputability of topological entropy for cellular automata [11], it is in general undecidable whether a tile set produces a robust or fragile pattern, due to the undecidability of the Halting Problem: a tile system that simulates a universal Turing machine may either produce a pattern that is eventually periodic (if the Turing machine halts), or else it may continue to produce ever more complicated subpatterns. The former patterns (that are eventually periodic) are formally robust, although only for very large k does this become apparent, while the latter patterns are fragile.

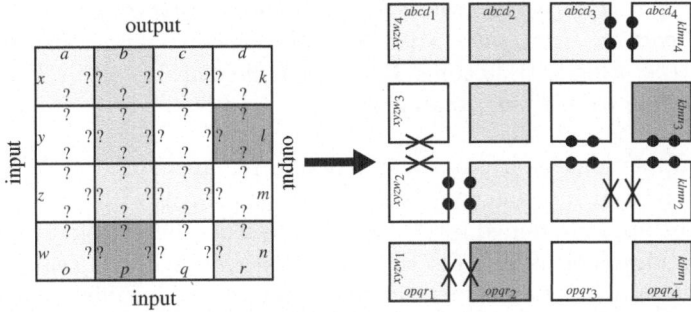

Fig. 4. Compact proofreading transformations using 4×4 blocks. Strength 2 bonds are indicated with 2 dots. Strength 0 bonds are indicated with a cross. Question marks indicate arbitrary bond types. All unlabeled (internal) bond types are unique (within the block and between blocks.) This construction is equivalent to "compressing" the $k \times k$ block on the left to a single tile and then applying the snaked proofreading construction, remembering to paint the resulting tiles with the original colors.

Definition 2. *A pattern* **P** *is called* **robust** *if it is constructible by a quarter plane tile system* **T** *that segments into poly(k) different $k \times k$ block types. A pattern* **P** *is called* **fragile** *if every quarter plane tile system segments into $2^{\Omega(k)}$ different $k \times k$ block types.*

The natural way to use Chen and Goel's construction to implement compact proofreading for robust patterns is as follows. For any k, for each of the $poly(k)$ $k \times k$ block types described above, create k^2 unique tile types with bond strengths according to the snaked proofreading blocks and colors according to the original pattern. The internal bond types are unique to each transformed $k \times k$ block type and do not depend upon the internal bond types in the original $k \times k$ block type. External bond types in the transformed block redundantly encode the full tuple of external bond types in the original block. (This transformation for a 4×4 block is illustrated in Fig. 4.) The L-shaped seed assembly must also be revised to use the new compound bond types. The above set of tile types together with this seed assembly yields a new tile system $\mathbf{T}(k)$.

It is easy to check that under aTAM $\mathbf{T}(k)$ correctly produces the pattern. At the corner between two existing blocks, only a tile that matches all the border tiles of both blocks, can attach. Any other internal tile must bind correctly since at least one side must match a bond type unique to the block. Since the original block assembled deterministically from its west and south sides, the transformed block also grows deterministically in the same direction. In fact, $\mathbf{T}(k)$ is locally deterministic [20], which makes a formal proof easy. Furthermore, for any particular choice of k, Chen and Goel's theorem 4.1 applies directly to our compact proofreading tile sets, but with multiplicative constants that increase with k. But we also claim the following, where $M = \lceil N/k \rceil$ is the size of our target assembly in units of blocks:

Lemma 1. *If a pattern* **P** *is robust then: For any constant $p < 1$, the $M \times M$ block initial portion of the pattern is produced correctly with probability at least p in time $O(Mpoly(\log M))$ by some* **T**(k) *(as defined above) where $k = \theta(\log M)$, using the locking kTAM with appropriate G_{mc} and G_{se}.*

Proof. Recall, as long as a particular location remains susceptible, insufficient attachments at that location constitute a Poisson process with rate $QO(e^{-3G_{se}})$. Here Q can be upper bounded by the total number of different blocks since that is the maximum number of different tile types that can be added as an insufficient attachment at any location. Thus, the maximum rate of insufficient attachments at any location is $q(G_{se}) = Q(k)O(e^{-3G_{se}})$, where $Q(k) = poly(k)$ since the pattern is robust.

The difference between the proof of Chen and Goel [4] and what we need is that Chen and Goel assumed that $Q(k)$ was a constant. Thus, whereas they were able to increase k without increasing the rate of insufficient attachments, q, we are not so fortunate. To remedy this situation, we must slow down growth slightly in order to sufficiently decreases the rate of insufficient attachments, but not so fast as to change the asymptotic form of the results.

Informally, note that Chen and Goel's bound on the probability of successfully completing the square within a certain time (scaled relative to f) depends only on the ratio q/f; the absolute time scale does not matter, nor does it matter whether q is the result of many or a few possible erroneous block types. Thus, we can slow down f by a polynomial in k without affecting the completion time asymptotics of $O(Mpoly(\log M))$, since $k = \Theta(\log M)$. Does q decrease enough? So long as it decreases faster relative to f, we can compensate for the polynomial increase in insufficient attachments. We will see that a factor of $Q(k)^2$ is sufficient.

Formally, assuming the maximum rate of insufficient attachments is any $\tilde{q}(G_{se}) = O(e^{-3G_{se}})$ independent of k, and the forward (=reverse) rate is any $\tilde{f}(G_{se}) = \Omega(e^{-2G_{se}})$, for any M, Chen and Goel give a value $\tilde{k}$ for k and $\tilde{G}_{se}$ for G_{se} such that with high probability the assembly completes correctly in time $t = O(Mpoly(\log M))$. We, of course, have $q(G_{se}) = O(Q(k)e^{-3G_{se}})$ and $f(G_{se}) = \Omega(e^{-2G_{se}})$. Now let us define $\tilde{q}(G_{se}) = q(G_{se} + \ln Q(k)) \cdot Q(k)^2$ and $\tilde{f}(G_{se}) = f(G_{se} + \ln Q(k)) \cdot Q(k)^2$. Observe that $\tilde{q}(G_{se}) = O(e^{-3G_{se}})$ and $\tilde{f}(G_{se}) = O(e^{-2G_{se}})$. This means that if the maximum rate of insufficient attachments and the forward rate were these $\tilde{q}$ and $\tilde{f}$, then Chen and Goel's proof gives values $\tilde{k}$ and $\tilde{G}_{se}$ such that with high probability the assembly completes correctly in time $t = O(Mpoly(\log M))$. But now note that if we set $G_{se} = \tilde{G}_{se} + \ln Q(\tilde{k})$ then the actual maximum rate of insufficient attachments and the forward rate are both exactly a factor of $Q(\tilde{k})^2$ slower than $\tilde{q}$ and $\tilde{f}$. Thus our system is simply overall slower by a factor of $Q(\tilde{k})^2$. This means that our system would finish correctly with the same high probability as achieved by Chen and Goel by time $O(tQ(\tilde{k})^2)$. But this is still $O(Mpoly(\log M))$ since $\tilde{k} = \theta(\log M)$ and $Q(\tilde{k}) = poly(\tilde{k})$. □

Theorem 2. *If a pattern* **P** *is robust then there exists a compact proofreading scheme for* **P***.*

Proof. Let us use the sequence $\{\mathbf{T}_N = \mathbf{T}(k)\}_N$ where k for each N is from lemma 1. Each of these tile systems can produce the whole pattern correctly under aTAM so the correctness condition of definition 1 is satisfied. Since $O(M \, poly(\log M)) = O(Npoly(\log N))$, lemma 1 implies that the sequence satisfies the robustness condition. Further, because $\mathbf{T}$ segments into $poly(k)$ different $k \times k$ block types and $k = \theta(\log M)$ implies $k = O(\log N)$, $\mathbf{T}_N = \mathbf{T}(k)$ has only $poly(k)k^2 = poly(\log N)$ tile types, satisfying the conciseness condition.   □

For some patterns, Chen and Goel's theorem can be applied directly (without requiring lemma 1). These include patterns whose quarter plane tile systems segment into a constant number of $k \times k$ block types. Furthermore, consider the Sierpinski pattern (Fig. 1(c)). The Sierpinski pattern is a fractal that has the following property: split the pattern into blocks of size $k \times k$ for any k that is a power of 2, starting at the origin. For any such k there are exactly 2 different types of blocks in the pattern. If you consider the assembly produced by the Sierpinski tile system in Fig. 1(d), there are exactly 4 different $k \times k$ blocks of tiles (the difference is due to the fact there are now two types of black and two types of white tiles.) We can let the sequence of tile systems for the compact proofreading scheme for the Sierpinski pattern consist only of $\mathbf{T}(k)$ for k that are a power of 2. Note that because of the restriction on k, we may have to use a block size larger than that which results from Chen and Goel's theorem. But since it does not have to be more than twice as large, definition 1 is still satisfied.

It would be interesting to identify constructible quarter plane patterns that have at least k^d different $k \times k$ block types for all k and for some constant $d \geq 1$.

4   A Lower Bound

In this section we will show that we cannot make compact proofreading schemes for fragile patterns using known methods.

First of all, note that although the definition of fragile patterns quantifies over all quarter plane tile systems, it can be very easy to prove that a pattern is fragile using the following lemma.

Lemma 2. *If a pattern $\mathbf{P}$ has $2^{\Omega(k)}$ different types of $k \times k$ squares of colors then it is fragile.*

Proof. If a pattern contains $2^{\Omega(k)}$ different types of $k \times k$ squares of colors, then any tile system producing it contains at least $2^{\Omega(k)}$ different types of $k \times k$ squares, and therefore comparably many block types.   □

The scheme described in the previous section does not work for quarter plane tile systems that segment into $2^{\Omega(k)}$ $k \times k$ block types (i.e., fragile patterns). This is because for $k = \theta(\log N)$, $\mathbf{T}(k)$ would then have $poly(N)$ tile types, violating the second condition (conciseness) of compact proofreading schemes (Definition 1).[9]

---

[9] Further, we believe Lemma 1 does not hold if the number of block types increases exponentially, rather than polynomially in k. This is an open question.

However, it is unclear whether other methods exist to make compact proof-reading schemes for patterns produced by such tile systems. While we cannot eliminate this possibility entirely, we can show that a variety of schemes will not work.

Existing attempts at making self-assembly robust through combinatorial means ([24, 4, 17]) are based on creating redundancy in the produced assembly. Specifically, knowing only a few tiles allows one to figure out a lot more of the surrounding tiles. Intuitively, this redundancy allows the tile system to "detect" when an incorrect tile has been incorporated and stall. We will argue that if a pattern is sufficiently complex, then only if there are many possible tile types can a few tiles uniquely determine a large portion of the pattern. Since the definition of compact proofreading schemes (Definition 1) limits the number of tiles types, we will be able to argue that for complex patterns there do not exist compact proofreading schemes that rely on this type of redundancy.

Definition 3. *An assembly A is (k,d)-**redundant** if there exists a decision procedure that, for any $k \times k$ (completed) square of tiles in A, querying at most d relative locations in the assembly for its tile type, can determine the types of all tiles in that square.*

The proofreading schemes of [24] and [4], using a block size $k \times k$, are $(k,3)$-redundant: even if the square is not aligned with the blocks, it is enough to ask for the types of the tiles in the upper-left, lower-left, and lower-right corners of the square. Because all tiles in a block are unique, and because the tile system is deterministic, these three tiles allow you to figure out all four blocks that the square may intersect. A proofreading construction that generalizes Reif et al's [17] 2-way and 3-way overlay tile sets to k-way overlays is shown in Appendix B to be $(k,3)$-redundant as well. This construction is not based on block transformations; the fact that its power is nonetheless limited by Theorem 3, below, illustrates the strength of our lower bound.

Lemma 3. *If a tile system $\mathbf{T}$ produces (k,d)-redundant assemblies in which more than 2^{ck} different types of (completed) $k \times k$ squares appear, then it must have at least $2^{ck/d}$ tile types.*

Proof. Let m be the number of tile types of $\mathbf{T}$. If an assembly produced by $\mathbf{T}$ is (k,d)-redundant, then it has no more than m^d types of squares of size $k \times k$ because the decision procedure's decision tree is of depth at most d and of fan-out at most m. But we assumed that $\mathbf{T}$ makes assemblies that have 2^{ck} different types of $k \times k$ squares. Thus, $m^d \geq 2^{ck}$, which can only happen if $m \geq 2^{ck/d}$. $\square$

Lemma 4 lets us limit the types of compact proofreading schemes that such complex patterns may have.

Theorem 3. *If a pattern is fragile then there does not exist a compact proofreading scheme $\{\mathbf{T}_1, \mathbf{T}_2, \ldots\}$ such that $\mathbf{T}_N$ produces assemblies that are $(\Omega(\log N), d)$-redundant (for any constant d).*

Proof. Any tile system producing this pattern makes $2^{\Omega(k)}$ different types of $k \times k$ (completed) squares of tiles. Suppose $\mathbf{T}_N$ produces assemblies which are $(c' \log N, d)$-redundant, for constants c', d. Take $k = c' \log N$ and note that for large k, $\mathbf{T}_N$ makes at least 2^{ck} $k \times k$ squares for some constant c. Apply Lemma 3 to conclude that $\mathbf{T}_N$ has at least $2^{ck/d} = N^{cc'/d}$ tile types, which violates the second condition of Definition 1. □

Even though both the Sierpinski pattern and the counter pattern (Fig. 1) are infinite binary patterns that can be constructed by very similar tile systems, they are very different with respect to error correction. We saw that the Sierpinski pattern has compact proofreading schemes. However, because the counter must count through every binary number, for any k there are 2^k rows that have different initial patterns of black and white squares. This implies that there are exponentially many (in k) different squares. By Theorem 3 this implies that the counter pattern does not have compact proofreading schemes that use $(\Omega(\log N), d)$-redundant assemblies. That is, no existing proofreading scheme can be adapted for making compact binary counters arbitrarily reliable.

This theorem suggests that in order to find universal compact proofreading schemes we must find a method of making self-assembly more error-robust without making it too redundant. However, we conjecture that there are inherent tradeoffs between robustness and conciseness (small number of tile types) raising the possibility that there do not exist compact proofreading schemes for patterns having an exponential number of $k \times k$ squares.

Acknowledgments

We thank Ho-Lin Chen, Ashish Goel, Paul Rothemund, Matthew Cook, and Nataša Jonoska for discussions that greatly contributed to this work. This work was supported by NSF NANO Grant No. 0432193.

References

1. L. M. Adleman, Q. Cheng, A. Goel, and M.-D. A. Huang. Running time and program size for self-assembled squares. In *ACM Symposium on Theory of Computing (STOC)*, pages 740–748, 2001.
2. G. Aggarwal, M. Goldwasser, M. Kao, and R. T. Schweller. Complexities for generalized models of self-assembly. In *Symposium on Discrete Algorithms (SODA)*, pages 880–889, 2004.
3. R. D. Barish, P. W. K. Rothemund, and E. Winfree. Two computational primitives for algorithmic self-assembly: Copying and counting. *NanoLetters*, to appear.
4. H.-L. Chen and A. Goel. Error free self-assembly using error prone tiles. In Ferretti et al. [7], pages 62–75.
5. J. Chen and J. Reif, editors. *DNA Computing 9*, volume LNCS 2943, Berlin Heidelberg, 2004. Springer-Verlag.
6. M. Cook, P. W. K. Rothemund, and E. Winfree. Self-assembled circuit patterns. In Chen and Reif [5], pages 91–107.

7. C. Ferretti, G. Mauri, and C. Zandron, editors. *DNA Computing 10*, volume LNCS 3384, Berlin Heidelberg, 2005. Springer-Verlag.

8. P. W. K. Rothemund. *Theory and Experiments in Algorithmic Self-Assembly*. PhD thesis, University of Southern California, Los Angeles, 2001.

9. P. W. K. Rothemund, N. Papakakis, and E. Winfree. Algorithmic self-assembly of DNA Sierpinski triangles. *PLoS Biology*, 2:e424, 2004.

10. P. W. K. Rothemund and E. Winfree. The program-size complexity of self-assembled squares. In *ACM Symposium on Theory of Computing (STOC)*, pages 459–468, 2000.

11. L. Hurd, J. Kari, and K. Culik. The topological entropy of cellular automata is uncomputable. *Ergodic Theory Dynamical Systems*, 12:255–265, 1992.

12. T. H. LaBean, H. Yan, J. Kopatsch, F. Liu, E. Winfree, J. H. Reif, and N. C. Seeman. Construction, analysis, ligation, and self-assembly of DNA triple crossover complexes. *Journal of the Americal Chemical Society*, 122:1848–1860, 2000.

13. M. G. Lagoudakis and T. H. LaBean. 2-D DNA self-assembly for satisfiability. In E. Winfree and D. K. Gifford, editors, *DNA Based Computers V*, volume 54 of *DIMACS*, pages 141–154, Providence, RI, 2000. American Mathematical Society.

14. C. Mao, T. H. LaBean, J. H. Reif, and N. C. Seeman. Logical computation using algorithmic self-assembly of DNA triple-crossover molecules. *Nature*, 407:493–496, 2000.

15. C. Mao, W. Sun, and N. C. Seeman. Designed two-dimensional DNA holliday junction arrays visualized by atomic force microscopy. *Journal of the Americal Chemical Society*, 121:5437–5443, 1999.

16. J. Reif. Local parallel biomolecular computing. In H. Rubin and D. H. Wood, editors, *DNA Based Computers III*, volume 48 of *DIMACS*, pages 217–254, Providence, RI, 1999. American Mathematical Society.

17. J. H. Reif, S. Sahu, and P. Yin. Compact error-resilient computational DNA tiling assemblies. In Ferretti et al. [7], pages 293–307.

18. R. Schulman and E. Winfree. Programmable control of nucleation for algorithmic self-assembly. In Ferretti et al. [7], pages 319–328.

19. R. Schulman and E. Winfree. Self-replication and evolution of DNA crystals. In M. S. Capcarrere, A. A. Freitas, P. J. Bentley, C. G. Johnson, and J. Timmis, editors, *Advances in Artificial Life: 8th European Conference (ECAL)*, volume LNCS 3630, pages 734–743. Springer-Verlag, 2005.

20. D. Soloveichik and E. Winfree. Complexity of self-assembled shapes, 2005. Extended abstract; preprint of the full paper is cs.CC/0412096 on arXiv.org.

21. E. Winfree. On the computational power of DNA annealing and ligation. In R. J. Lipton and E. B. Baum, editors, *DNA Based Computers*, volume 27 of *DIMACS*, pages 199–221, Providence, RI, 1996. American Mathematical Society.

22. E. Winfree. *Algorithmic Self-Assembly of DNA*. PhD thesis, California Institute of Technology, Pasadena, 1998.

23. E. Winfree. Simulations of computing by self-assembly. Technical Report CS-TR:1998.22, Caltech, 1998.

24. E. Winfree and R. Bekbolatov. Proofreading tile sets: Error-correction for algorithmic self-assembly. In Chen and Reif [5], pages 126–144.

25. E. Winfree, F. Liu, L. A. Wenzler, and N. C. Seeman. Design and self-assembly of two dimensional DNA crystals. *Nature*, 394:539–544, 1998.

26. E. Winfree, X. Yang, and N. C. Seeman. Universal computation via self-assembly of DNA: Some theory and experiments. In L. F. Landweber and E. B. Baum, editors, *DNA Based Computers II*, volume 44 of *DIMACS*, pages 191–213, Providence, RI, 1998. American Mathematical Society.

A   Extension of Chen and Goel's Theorem to Infinite Seed Boundary Assemblies

The following argument uses terms and concepts from [4].

First, suppose we desire to build an $(N + N^2) \times (N + N^2)$ block initial portion of the pattern starting with the L seed assembly having arms that are $N + N^2$ blocks long. The extra N^2 blocks will serve as a buffer region. Chen and Goel's [4] theorem 4.2 then gives us a $k = \theta(\log (N + N^2)) = \theta(\log N)$ and G_{se} s.t. with high probability no block error occurs in the $(N + N^2) \times (N + N^2)$ block region in time $O(N^2 poly(\log N))$ that it takes to finish it. Further, with high probability the initial $N \times N$ block portion of the pattern is completed in time $t_N = O(Npoly(\log N))$.

Now, let's suppose we use this k and G_{se} with an infinite L seed assembly, and we'll be interested in just the $N \times N$ block initial portion of the pattern. The only way the infinite seed assembly can affect us is if a block error outside the $(N + N^2) \times (N + N^2)$ block region propagates to the $N \times N$ initial region before it completes. For this to occur, at least N^2 tiles must be added sequentially, at least one per block through the buffer region, to propagate the error. The expected time for this to happen is N^2/f with standard deviation N/f (i.e., it is a gamma distribution with shape parameter N^2 and rate parameter f). However, the propagated error can only cause a problem if it reaches the $N \times N$ rectangle before time t_N. Since $t_N = O(Npoly(\log N))$, this becomes less and less likely as N increases by Chebyshev's inequality. Small N are handled by increasing k and G_{se} appropriately, which does not affect the asymptotic results. Thus we have a $k = \theta(\log N)$ and G_{se} such that with high probability (i.e., $\geq p$) the initial $N \times N$ block portion of the pattern is completed correctly in time $O(Npoly(\log N))$, even if we use an infinite L seed assembly.

B   An Overlay Proofreading Scheme

In this appendix we give an example showing that our lower bound on the complexity of same-scale proofreading schemes also applies to proofreading schemes that are not based on block transformations. Here, we consider a k-way overlay scheme (suggested by Paul Rothemund and Matt Cook) that generalizes the 2-way and 3-way overlay schemes introduced by Reif et al [17]. The construction is shown in Fig. 5.

Consider the assembly grown using some original tile set, as in Fig. 5a. When the shaded tile x was added, it attached to the tiles a and b to its west and to its south. Since we consider only deterministic quarter-plane tiles sets, the tile type at a particular location is a function of the tile types to its south and to its west, e.g., $x = f(a, b) = f_{ab}$. Therefore, it is possible to reconstruct the same pattern without keeping track of bond types, explicitly transmitting only information about tile types.

The 1-overlay tile set, derived from the original tile set, is a deterministic tile set for doing exactly that. As shown in Fig. 5b, for each triple of neighboring tiles a, b, and x that appears in the assembly produced by the original tile set (in the relative positions shown in (a)), create a new tile (x, x, b, a), colored the same as x, that "inputs" the original tile types of its west and its south neighbors, and "outputs" tile type x to both its north and its east neighbor. With an appropriately re-coded L-shaped boundary, the new tile set will produce exactly the same pattern as the original tile set: the output of the tile at location $\langle i, j \rangle$ in the 1-overlay assembly is the tile type at $\langle i, j \rangle$ in the original assembly. Supposing the original tile set T has $|T|$ tile types, the new tile set contains at most $|T|^2$ tile types, and possibly fewer if not all pairs of inputs a, b appear in the pattern.

Redundancy is achieved in a k-way overlay tile set by encoding not just one original tile, but k adjacent tiles along the diagonal growth front. Specifically, each tile in the k-way overlay assembly will output the k-tuple of original tile types that appear in the same location in the original assembly and locations to the east and south. For example, in Fig. 5c, the output of the tile at $\langle i, j \rangle$ in the 4-overlay assembly is the 4-tuple $abcd$ containing the tile types at locations $\langle i, j \rangle$, $\langle i + 1, j - 1 \rangle$, $\langle i + 2, j - 2 \rangle$, and $\langle i + 3, j - 3 \rangle$. Each new tile is colored according to the first tile type in its output tuple. The new tile set consists of all such tiles that appear in the k-overlay assembly[10,11]. The new tile set contains at most $|T|^{k+1}$ tiles, since there are at most $|T|^k$ input k-tuples, and the two inputs to a given tile will always agree at $k - 1$ indices. This is exponential in k, but for some patterns – e.g., robust patterns, as we will see – only a polynomial number of tile types will be necessary. Note that growth with the new tile set is still deterministic, since the tuple output by a tile is a function of the two input tuples.

In what sense is the k-overlay tile set guaranteed to be proofreading? Consider a growth site where a tile is about to be added. Unless the two input k-tuples agree at all $k - 1$ overlapping positions, there will be no tile that matches both inputs. Thus, every time that a tile is added without a mismatch, it provides a guarantee that $k - 1$ parallel computations are carrying the same information, locally. Note that the fact that site $\langle i, j \rangle$ in the original assembly contains tile type t is encoded in k locations in the k-overlay assembly. It is reasonable to conjecture that it is impossible for all k locations to have incorrect information, unless at least k insufficient attachments have occurred.

Unfortunately, like the original proofreading tile sets of [24] and the 2-way and 3-way overlay tile sets described in [17], the k-way overlay tile sets do not protect against facet nucleation errors, and therefore we do not expect error

---

[10] In addition, the L-shaped boundary must be properly re-coded to carry the boundary information in the form the new tiles require. This is easy to do if the pattern is consistent with a larger hypothetical assembly that extends k tiles beyond the quarter plane region, since then tuples on the boundary encode for tile types in this buffer zone. Otherwise a few extra tile types will be necessary, but as this does not change the nature of our arguments, we ignore this detail here.

[11] Note that the exact (minimal) set of such tiles is in general uncomputable, since the original tile set could be Turing-universal, and thus predicting whether a particular original tile appears in the assembly is equivalent to the Halting Problem. However, the new tile set is well-defined and in many cases can be easily computed.

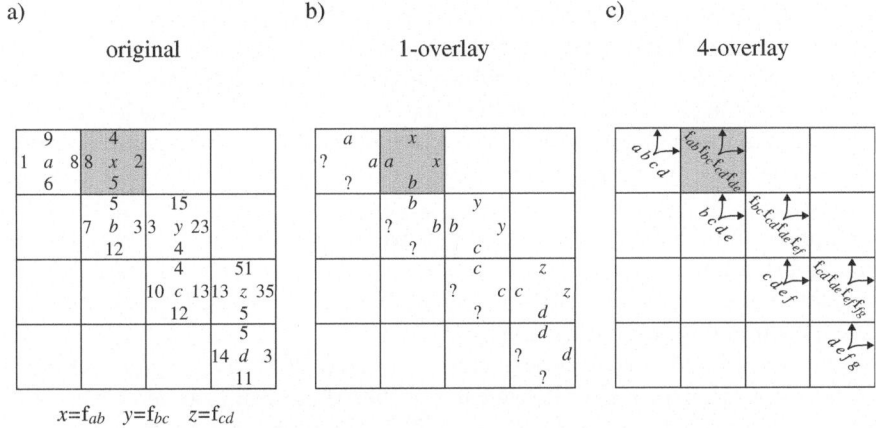

a)

a)   b)   c)

original   1-overlay   4-overlay

$x = f_{ab}$ $y = f_{bc}$ $z = f_{cd}$

Fig. 5. The construction for k-way overlay proofreading tile sets. **(a)** An original quarter-plane tile set T, containing $|T|$ tile types. Numbers indicate bond types. Letters name the tile types. For example, the tile $x = (4, 8, 5, 2)$. **(b)** The 1-overlay transformation of the original tile set. The question marks indicate that there may be several different new tile types that output a or b; **(c)** The 4-overlay transformation of the original tile set.

rates to decrease substantially with k. We do not see an obvious way to correct this deficiency.

Nonetheless, as a demonstration of the general applicability of our lower bound, we will show that *even if* the k-way overlay tile sets reduced errors sufficiently, for fragile patterns the k-way overlay tile sets will contain an exponential number of tile types and are thus infeasible, whereas for robust patterns the k-way overlay tile sets will contain a polynomial number of tile types and are thus feasible.

First we show that all k-overlay tile sets are $(k, 3)$-redundant, regardless of the original tile set. To determine all tile types in the $k \times k$ square with lower left coordinate $\langle i, j \rangle$, we need only know the tiles at $\langle i, j-1 \rangle$, $\langle i-k, j+k-1 \rangle$, and $\langle i+k-1, j-k \rangle$. The outputs of these tiles encodes for the entire diagonal from $\langle i-k, j+k-1 \rangle$ to $\langle i+2k-2, j-2k+1 \rangle$ in the original assembly. Deterministic growth from this diagonal results in a triangle of tiles with upper right corner at $\langle i+2k-1, j+k-1 \rangle$, in the original assembly. Thus all tile types are known for the input and output k-tuples of overlay tiles in the $k \times k$ square of interest.

Theorem 3 tells us that fragile patterns cannot have compact proofreading schemes that are $(\Omega(\log N), d)$-redundant for any constant d. Therefore, k-overlay tile sets can't work as compact proofreading schemes for fragile patterns; they must have an exponential number of tile types. This is what we wanted to show.

Alternatively, we could have directly bounded the number of tile types in k-overlay tile sets for fragile and robust patterns. For robust patterns, with $poly(k)$ $k \times k$ squares of tile types, clearly there are also $poly(k)$ size-k diagonals. Since each tile in the k-overlay tile set contains two inputs encoding size-k diagonals, there can be at most $poly(k)^2 = poly(k)$ tile types altogether. Thus, (although probably not satisfying the robustness criterion of Definition 1) k-overlay tile sets are at least concise for robust patterns. Conversely, concise k-overlay tile sets, having $poly(k)$ tile types by construction, have a comparable number of size-k diagonals in the original assembly. Consider now the original assembly. Since growth is deterministic, the diagonal determines the upper right half of a $k \times k$ square, and thus there are $poly(k)$ tops and $poly(k)$ sides; taking these as inputs to other squares, we see that there are $poly(k)^2 = poly(k)$ $k \times k$ squares. In this loose sense, k-overlay tile sets are neither more nor less concise than $k \times k$ snaked proofreading, for robust patterns.

On the other hand, for a fragile pattern, requiring $2^{\Omega(k)}$ $k \times k$ squares of tiles in any tile system that produces it, we can see that there will also be at least $2^{\Omega(k)}$ size-k diagonals of tiles. Specifically, if $S(k)$ is the number of such squares, and $D(k)$ is the number of such diagonals, then $S(k) \leq D(2k)$ because deterministic growth from a size-$2k$ diagonal results in the completion of a triangular region containing a $k \times k$ square. $S(k)$ being at least exponential therefore implies the same for $D(k)$. Conversely, a pattern generated by a tile system with $2^{\Omega(k)}$ size-k diagonals obviously also has at least that many $k \times k$ squares as well. Thus, our notions of fragile and robust patterns appears to be sufficiently general.

A Microfluidic Device for DNA Tile Self-assembly

Koutaro Somei[1], Shohei Kaneda[2], Teruo Fujii[2], and Satoshi Murata[1]

[1] Tokyo Institute of Technology, Yokohama, 226-8502 Japan
somei@mrt.dis.titech.ac.jp
murata@dis.titech.ac.jp
[2] The University of Tokyo, Tokyo, 153-8505 Japan

Abstract. This paper presents a microfluidic device specially designed for DNA tile self-assembly. The DNA tile is one of the most promising building blocks for complex nanostructure, which can be used as a molecular computer or a scaffold for functional molecular machineries. In order to build desired nanostructure, it is necessary to realize errorless self-assembly under thermal fluctuation. We propose a method to directly control environmental parameters of DNA self-assembly such as concentration of each monomer tile and temperature in the reaction chamber by using a microfluidic device. The proposed device is driven by a capillary pump and has an open reaction chamber which enables real-time observation by AFM. Results of preliminary experiments to evaluate performance of the device will be reported.

1 Introduction

The DNA Computing is one of the emerging nanotechnologies based on hybridization of DNA molecules. In the earliest DNA computer proposed by Adleman, computation is based on hybridization of linear DNA molecules [1]. Seeman and Winfree introduced "DNA tiles" which has four "sticky ends." It is proven that DNA tiles have much stronger computational power compared to linear DNA strands [2] [3] [9]. The results of experiments showed DNA tile's ability to self-assemble nanofabric, which has programmed periodic or aperiodic patterns corresponding to some computational processes of cellular-automata. Stimulated by their work, many researches related to DNA tile have been made. For instance, the idea of DNA tile is extended to self-assembling RNA tiles [7]. Self-folding octahedron composed of concatenation of DNA tile-like motif [8] will be a step to realize three-dimensional self-assembling nanoblocks. Using DNA molecules as building blocks for nanostructure is very advantageous because they specifically bind to other kinds of molecules such as proteins. Self-assembled DNA tiles can be used as a scaffold of functional molecular machinery.

Currently, suppression of assembly errors is the central problem of the DNA tile based nanotechnology, because in order to build desired structure, it is necessary to realize errorless self-assembly under thermal fluctuation. Numbers of error reduction methods have been proposed so far, but most of them consider

A. Carbone and N.A. Pierce (Eds.): DNA11, LNCS 3892, pp. 325–335, 2006.
© Springer-Verlag Berlin Heidelberg 2006

only the design of DNA tile sets [11]. They usually result in complicated tile sets, and thus, none of them are actually implemented yet.

Instead of complicating tile sets, we propose a new method to control environmental parameters of DNA self-assembly. Error rate of self-assembly is essentially dependent on parameters such as concentrations of each monomer tile and temperature in the reaction chamber. If we can directly control these parameters, it is not only possible to reduce errors, but also simplify the tile set.

Based on this idea, we propose a microfluidic device specially designed for DNA tile assembly. All the components of the microfluidic device are integrated on a single chip; a capillary pump and a stop valve to drive solution including DNA tiles, an open reaction chamber which enables direct AFM observation of self-assembly, and micro heater and sensor printed beneath the reaction chamber. It requires very small amount of DNA sample, and needs almost no external apparatus such as an array of syringe pumps. In the following sections, we will explain the details of the microfluidic device and some results of preliminary experiments.

2   Microfluidic Device for DNA Tile Self-assembly

In conventional methods, all the DNA tiles are mixed in a single test tube, annealed for self-assembly, and then the mixture is dropped on a mica surface for AFM observation. Since all kinds of tiles are assembled in one pot, DNA tile set must be very carefully designed such that each sticky end has an appropriate bonding specificity and strength to obtain desired structure. Actually, it is very difficult to keep concentrations of each monomer tile in one-pot self-assembly. Even in the best condition, resultant aggregates include more than 1% mismatch errors. In conventional methods, all the DNA tiles are mixed in a single test tube, annealed for self-assembly, and then the mixture is dropped on a mica surface for AFM observation. Since all kinds of tiles are assembled in one pot, DNA tile set must be very carefully designed such that each sticky end has an appropriate bonding specificity and strength to obtain desired structure. Actually, it is very difficult to keep concentrations of each monomer tile in one-pot self-assembly. Even in the best condition, resultant aggregates include more than 1% mismatch errors. Our purpose is to show a different way of DNA self-assembly. If the solution surrounding the aggregation can be replaced step by step, the self-assembly process is much easier to design and also rate of errors can be reduced (Fig. 1).

To illustrate the idea, we assume only three types of sticky ends here (represented by shapes of edges of each tile). If all kinds of tiles are mixed in one pot, you get a random aggregation (Fig.1 left). In contrast, in the "step by step self-assembly," you apply solutions one by one. Each solution contains no tile or only one kind of tile. Self-assembly is initiated by immobilized strands in the reaction chamber. After tiles aggregate, surplus unbound tiles can be flushed out by water, or, directly replaced by the next solution containing different tiles. By this process you will obtain an arranged layered lattice, in which each layer is made of single kind of tiles. Note that only three types of sticky ends are enough to make arbitrary sequence of layers. If you want the same thing by one-pot

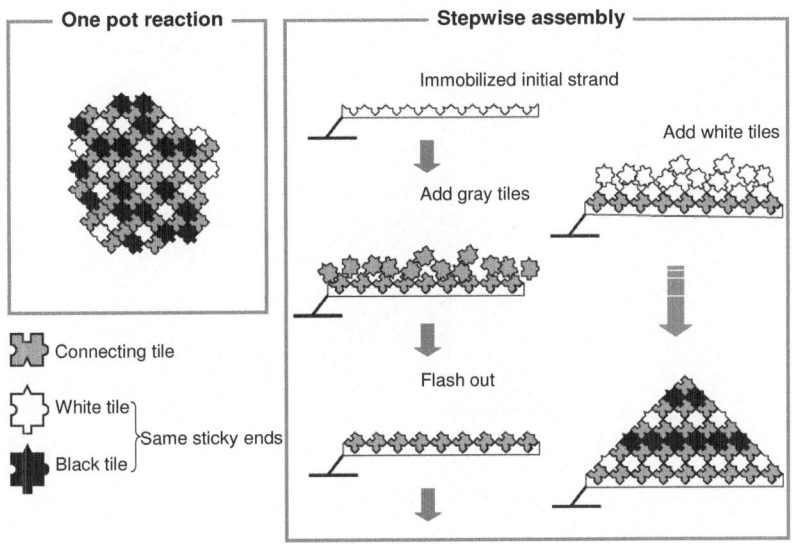

Fig. 1. One pot vs. Stepwise self-assembly

reaction, you need many different kinds of sticky ends. Different arrangements other than layered lattice are also possible depending on design of the tile set. A computational model of stepwise self-assembly based on the same idea was proposed by Reif [12].

In order to realize this, we have developed a microfluidic device. The proposed device has the following functions.

- Concentration control (replacement of solution)
- Temperature control
- Real-time observation of reaction chamber by AFM
- Requires very small amount of DNA sample
- Can be used as a platform for further experiments after self-assembly of DNA structure

3   Prototype of Microfluidic Device

Fig.2 shows the overall configuration of the first prototype. The device is one chip device made of PDMS (poly-dimethylsiloxane) on a glass substrate. Each chip has three identical sets of the fluidic channels with micro heater/ sensor made of ITO (Indium Tin Oxide). Each channel is composed of three parts; 1) a service port where DNA sample solution is applied by pipetting, 2) a reaction chamber where DNA tile self-assembly takes part under controlled temperature, and 3) a capillary pump which generates suction force to pull the next solution from the service port. The whole channel is covered by oil to prevent evaporation

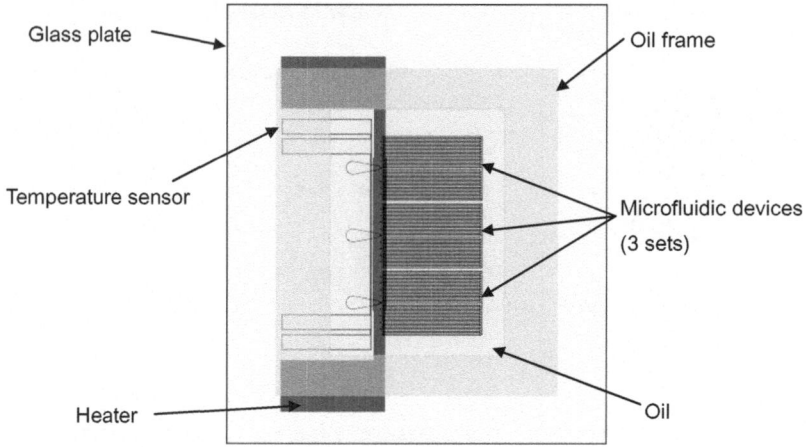

Fig. 2. Microfluidic device

of the solution. In the following subsections, each component will be explained in detail.

The PDMS fluidic chip was fabricated by soft lithography and the ITO sensor and heater were made by wet etching process [10]. The details of the fabrication methods are omitted in this paper.

3.1   Flow Control by Capillary Pump and Stop Valve

Capillary force is a phenomenon that we usually observe at the boundary of two different kinds of liquids (or liquid and gas) they do not mix. The capillary force is caused by surface tension, in other words, tendency to minimize free energy on the boundary surface.

We use a capillary pump, which is actually a very long channel engraved on a solid plate (PDMS in this case). Let us consider a flow in a channel. Capillary force Fc depends on the cross-sectional area of the channel A and the contact angle θ between the liquid and wall material.

$$F_c = \gamma A \cos\theta \tag{1}$$

where γ represents surface tension.

Driving force of a capillary pump is given in a form of pressure head,

$$P_c = \gamma \left(\frac{\cos\alpha_b + \cos\alpha_t}{d} + \frac{\cos\alpha_l + \cos\alpha_r}{w} \right) \tag{2}$$

where, d is depth and w is width of the channel, α_b, α_t, α_l, and α_r are contact angles at the bottom, upper, right, and left side of the channel section, respectively [4]. From equation (2), we can conclude that the thinner the channel, the more the suction force. Also, to get a large suction force, we need large θ or α, namely, the surfaces of the channel must be hydrophilic.

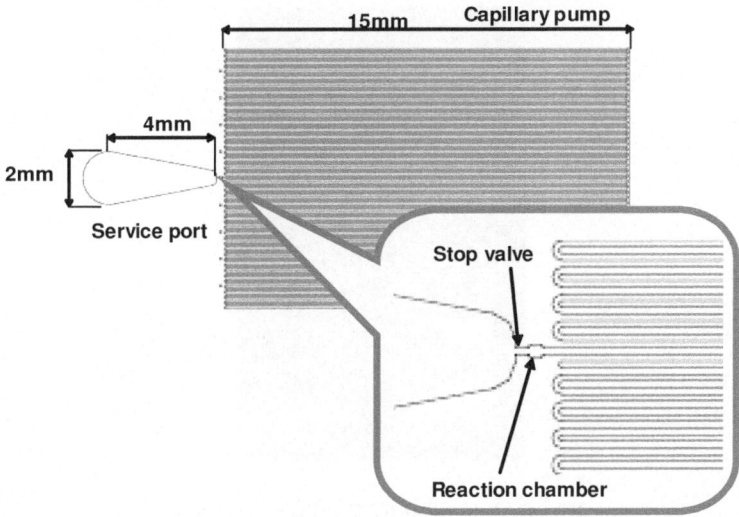

Fig. 3. Microfluidic channel

Fig.3 shows design of the overall channel. It is composed of a service port, a stop valve, a reaction chamber, and a capillary pump.

Fig.4 shows the profile of the capillary force along the channel. The channel is 1.5 m long in total, 50 μm in width, and 50 μm in depth. Tapered shape of service port is to introduce the droplet to the channel. A bottleneck between the service port and the reaction chamber is called a "stop valve." It has the same width with that of capillary pump.To begin with the self-assembly process, a droplet is injected to the service port by a pipette (Fig. 5 (1)). Through the

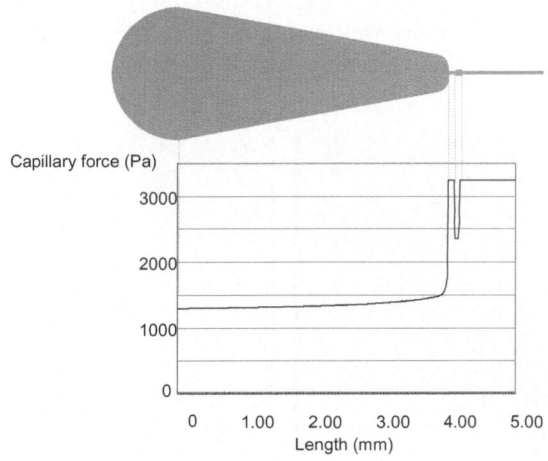

Fig. 4. profile of capillary force

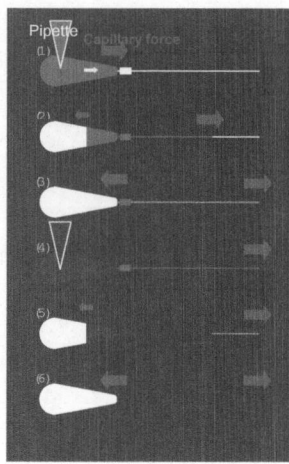

Fig. 5. Microfluidic Capillary System

stop valve, it goes to the reaction chamber (90 x 120 μm) and then sucked into the capillary pump (Fig. 5 (2)). In the reaction chamber, some seed strands are immobilized on the bottom, to initiates self-assembly process and also to hold resultant aggregations against flow. When the tail of the droplet reaches at the mouth of the stop valve (Fig.5 (3)), capillary force at the mouth balances with that of the capillary pump. As a result, first (blue) solution is kept in the reaction chamber. When the next (red) solution is injected (Fig.5 (4)), the boundary surface at the stop valve vanishes. Then the capillary pump restarts. Duration allowed to each step can be controlled by the interval of injection.

3.2 Experiment of Flow Control

Performance of proposed device was evaluated by experiments. Fig.6 shows microscope images of the reaction chamber. In the first experiment, water solution of

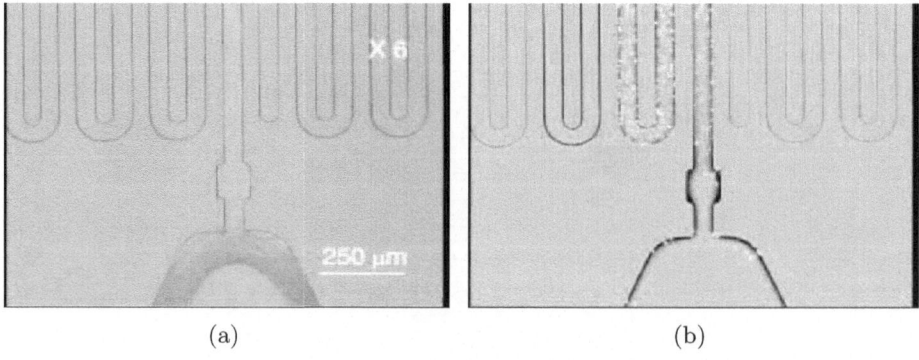

(a)   (b)

Fig. 6. Performance of stop valve

fluorescent beads was injected at the service port (Fig.6 (a)). 50 second later, the
tail of droplet reaches at the mouth of stop valve, and flow is stopped (Fig.6 (b)).

In the following experiment, dyed solutions with streptavedin (blue: 350nm,
green: 488nm, red: 546nm) were injected consecutively (green → water → red →
water → blue), and successful replacement of solutions is observed by fluorescent
microscope with multi-band-path filter.

3.3   Preventing Evaporation by Oil Coverage

During the above experiments, we found that evaporation of the solution makes
a serious problem. The solution dries out very quickly in micro scale, actually
almost all the solution in the service port is gone in about a minute. This is not
good for DNA tile assembly, because concentration of the tile must be kept at
desired level. Moreover, we want to observe the self-assembly process by AFM,
thus we cannot cover the channel with glass plate.

To solve the problem, the device was covered by oil. The whole channel including the service port and the capillary pump is covered by mineral oil for
PCR. It allows cantilever access for AFM observation and also doesn't effect
hybridization reaction. This method prevents evaporation perfectly.

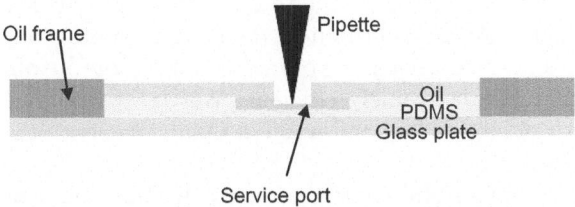

Fig. 7. Oil coverage

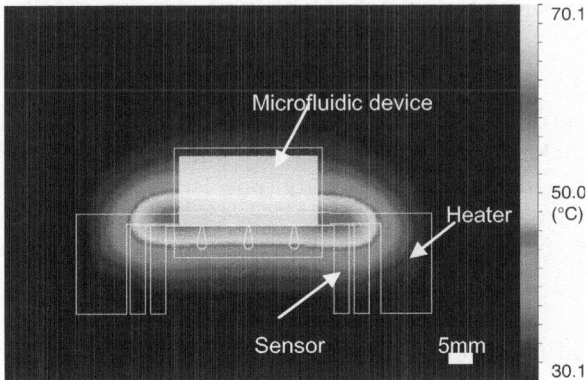

Fig. 8. Temperature distribution by ITO heater (24V)

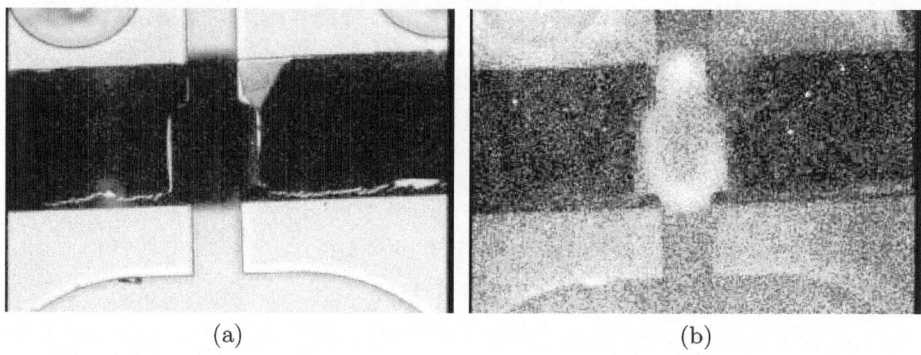

(a)   (b)

Fig. 9. Immobilized DNA at reaction chamber.(a) By using a rectangular mask 100 m in width, gold layer is patterned on PDMS surface (black horizontal band). DSP is immobilized on gold surface by Au-S bonding. (b) After DNA solution is supplied into the channel, fluorescence from Cy3 label on a complementary strand to the linker was detected.

3.4   Temperature Control

Another important parameter of DNA tile self-assembly is temperature. To obtain a perfect crystal of DNA tiles, the temperature must be kept slightly below the melting temperature [2]. Microfluidic device is advantageous to control temperature, because surface area per volume is very large in micro scale, which means heat exchange is extremely efficient. We use a combination of a heater and a sensor both made of ITO (Indium Tin Oxide) beneath the reaction chamber (Fig.8).

This figure also shows obtained temperature distribution measured by a thermography. More precise temperature measurement is possible by measuring resistance of sensor wires, but it is not used yet.

3.5   Immobilization of DNA Molecule at Reaction Chamber

Immobilization techniques are often used in biochemical experiments. Immobilization of DNA molecules is very important in practical use such as DNA microarrays [5] [6]. In our method, some initial strands have to be anchored in the reaction chamber. To realize this, DNA linker was immobilized on a gold surface using DSP (dithiobis-succinimidyl propionate: PIERCE, U.S.A.).

The gold surface was patterned by vapor deposition through stencil mask, thus we can specify the place to fix the initial strands (in this case the whole reaction chamber). Fig.9 shows immobilized DNA in the reaction chamber.

3.6   Microfluidic Channel Fabricated on Silicon

The microfluidic device described so far is made of PDMS. PDMS is widely used material for micro-fluidic devices because of its easy fabrication, but it is not the optimal material for our purpose for some reasons; 1) PDMS is hydrophobic in

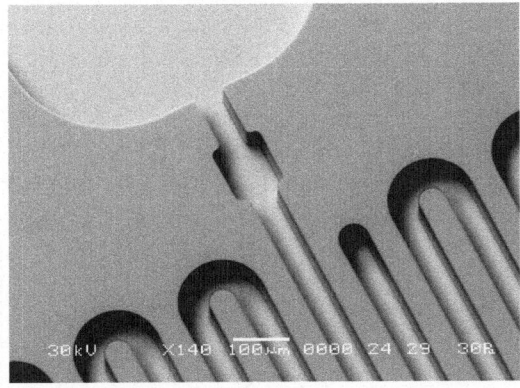

Fig. 10. Microfluidic channel fabricated on Silicon

nature, thus its surface must be changed to hydrophilic before experiments. We use oxygen plasma to change the surface, however its effect does not last long. 2) PDMS does not have enough solvent resistance. DMSO for DNA immobilization dissolves surface of fluidic channel. 3) PDMS is resin and its stiffness is not enough for AFM observation.

We are currently working on the second version of our microfluidic device made of silicon (Fig. 10). Most of the above problems will be solved by silicon based device, because the surface of silicon is spontaneously covered by silicon oxide, which is much smoother and more stiff than PDMS, and also hydrophilic and solvent-resistant. Gold deposition and DNA immobilization is much easier than PDMS.

4   Conclusions and Future Work

We propose a novel microfluidic device specially designed for DNA tile assembly in this paper. It has a capillary pump to drive water solution including DNA tiles, an open reaction chamber which enables AFM direct observation, and micro heater/sensor under the reaction chamber. All these components are patterned on a single chip made of PDMS. We have fabricated a prototype to evaluate its basic functionalities such as replacement of solution and temperature control.

Our goal is to achieve complex nano structure by the self-assembly, and the microfluidic device has a large potential in self-assembly of various kind of nano-particles such as DNA tiles and other motifs. In the complex self-assembly, the process should be stepwise, and also, the environment of the self-assembly must be precisely controlled. Fig. 11 illustrates such multi-stage self-assembler (Nano factory). In this figure, the self-assembly process has four stages. Each of them is controlled at the optimal temperature for each corresponding step (e.g. single tile assembly, lattice assembly etc.). Seed structure is immobilized at the reaction chamber (final stage) in which concentration of nano particles are continuously measured by fluorescents. Undesired spurious structures around the

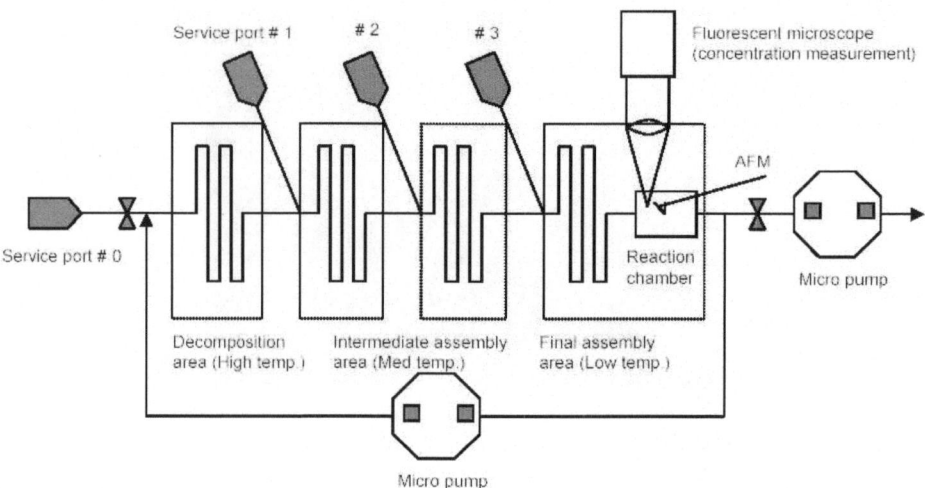

Fig. 11. Nano factory

seed are washed out by flow to give enough space and material around the seed. These surplus byproducts are fed back to the decomposition. This kind of controlled self-assembly will drastically improve the size and yield of errorless nano structure. It is also possible to produce desired nano structures on a patterned template in the reaction chamber for various applications.

Acknowledgement

This work was supported by Ministry of Education, Culture, Sports, and Science and Technology of Japan under Grant-in-Aid for Scientific Research on Priority Areas, No. 17059001, 2005.

References

1. Adleman, L.: Molecular Computation of Solutions to Combinatorial Problems, Science, Vol.266, pp.1021-1024, 1994.
2. Winfree, E.: Algorithmic Self-Assembly of DNA, Ph.D Thesis, California Institute of Technology, 1998.
3. Winfree, E., Yang, X. and Seeman, N. C.: Universal Computation via Self-assembly of DNA : Some Theory and Experiments, DNA based Computers 2, DIMACS Series in Discrete Mathematics and Theoretical Computer Science, Vol.44, pp. 191-213,1999.
4. Juncker, D., Schmid, H., Drechsler, U., Wolf, H., Wolf, M., Michel, B., de Rooij, N., and Delamarche, E.: Autonomous Microfluidic Cappilary System: Analytical Chemistry, Vol. 74, No. 24, pp.6139-6144, 2002.
5. Delamarche, E., Bernard, A., Schmid, H., Michel, B., Biebuyck, H.: Science, Vol. 276, pp.779-781, 1997.

6. Smith, E.A., Wanat, M.J., Cheng, Y., Barreira, S.V.P., Frutos, A.G., and Corn, R.M: Formation, Spectroscopic Characterization, and Application of Sulfhydryl-Terminated Alkanethiol Monolayers for the Chemical Attachment of DNA onto Gold Surfaces: Langmuir, Vol.17, pp.2502-2507, 2001.

7. Chworos, A., Severcan, I., Koyfman, A.Y., Weinkam, Y., Oroudjev, E., Hansma, H.G., and Jaeger, L.: Building Programmable Jigsaw Puzzles with RNA, Science Vol.306, pp.2068-2072, 2004.

8. Shih, W.M., Quispe, J.D., Joyce, G.F.: A 1.7-kilobase single-stranded DNA that folds into a nanoscale octahedron, Nature Vol.427, pp.618-21, 2004.

9. Winfree, E., Liu, F., Wenzler, L., Seeman, N.C.: Design and self-assembly of two-dimensional DNA crystals, Nature, Vol. 394, pp.539-544, 1998.

10. Yamamoto, T. Fujii, T. and Nojima, T: PDMS-glass hybrid microreactor array with embedded temperature control device. Application to cell-free protein synthesis: Lab on a Chip, Vol.2 (4), pp197 - 202

11. Several papers related to error suppression of DNA tiles are presented in: Preliminary Proceedings of Tenth International Meeting on DNA Computing (Edtors. C. Ferretti, et. al), Milan, 2004.

12. J.H. Reif: Local Parallel Biomolecular Computation: Proc. DNA-Based Computers, III: University of Pennsylvania, June 23-26, 1997.

Photo- and Thermoregulation of DNA Nanomachines

Keiichiro Takahashi[1], Satsuki Yaegashi[2],
Hiroyuki Asanuma[3], and Masami Hagiya[2,4]

[1]NovusGene Inc., 2-3 Kuboyama-cho, Hachioji-shi, Tokyo 192-8512, Japan
takahashi-k@novusgene.co.jp
[2]Japan Science and Technology Corporation (JST-CREST)
yaegashi@lyon.is.s.u-tokyo.ac.jp
[3]Department of Molecular Design and Engineering, Graduate School of Engineering,
Nagoya University, Chikusa, Nagoya 464-01, Japan
asanuma@mol.nagoya-u.ac.jp
[4]Department of Computer Science, Graduate School of Information Science and
Technology, University of Tokyo,
7-3-1 Hongo, Bunkyo-ku, Tokyo 113-0033, Japan
hagiya@is.s.u-tokyo.ac.jp

Abstract. We have been investigating DNA state machines, especially those based on the opening of hairpin molecules in which state transitions are realized as hairpin loops are opened by molecules called openers. This paper introduces photo- and thermoregulation of such hairpin-based DNA machines, in which the openers become active by sensing external signals in the form of light or heat. We conducted fluorescence experiments and show that photo- and thermoregulation is possible. In the experiments, the openers become active when they are irradiated by UV light or when they receive heat as external input. For photoregulation, we use azobenzene-bearing oligonucleotides developed by the third author.

1 Introduction

Implementing controllable molecular nanomachines made of DNA is one of the goals of DNA computing and DNA nanotechnology, and a variety of implementations of DNA machines have been reported [1, 2, 3, 4, 5]. One of the most typical methods for controlling such DNA machines is to use DNA strands that hybridize with target machines and drive their state transition [1, 7, 2]. DNA strands can also be used as catalysts for the formation of double helices in such machines [13, 6]. As another approach to control DNA machines, Mao et al. showed that the B-Z transition of DNA owing to a change in solution condition can switch the conformation of their DNA motor [3].

In this paper, we show that signals in the form of light or heat can be used as another means to control DNA machines. We have been investigating DNA state machines that are based on the opening of hairpin molecules, in which state transitions are realized as hairpin loops are opened by molecules called openers [8,9,10].

A. Carbone and N.A. Pierce (Eds.): DNA11, LNCS 3892, pp. 336–346, 2006.

Signals in the form of light or heat change the activity of the openers. The proposed reaction systems will be used as components of a larger molecular system, consisting of various sensors, computing elements, and actuators. We envision that such general-purpose molecular systems can be constructed from a network of DNA machines based on hairpins or other kinds of loop structure, as proposed by Seelig et al. [6] and the authors [12], the operations of which are driven by the formation of double helices and the dissociation of loops. The photo- and thermoregulation of hairpin-based machines introduced here can be incorporated into such systems and extend their range of application.

In this paper, we first briefly describe our previous experiments on hairpin dissociation using various kinds of opener. Then, we present experimental results of photo- and thermoregulation, and discuss our model of hairpin dissociation.

2   Previous Work

Before describing the reaction systems operated by light or heat, let us briefly introduce our previous work [11], which constitutes the basis of those systems. In these preliminary experiments, we measured the efficiency of various kinds of opener molecules using fluorescence, and introduced some techniques to achieve robust hairpin dissociation, as seen in the work of Yurke's group [13,14].

The main reaction system in our study is depicted in Figure 1. We call the hybridization site of the hairpin or opener the *lead section*. The overhang of the hairpin is the hairpin's lead section, and this hybridizes with the lead section of the opener. The substrand of the opener that invades the hairpin and replaces one of the stem strands via branch migration is called an *invasion section* .

In the study, we varied the length of the opener's lead section from 0 to 20 (20, 10, 7, and 0). The lead section with length 10 resulted in the most efficient kinetic rate (3.9×10^5 M^{-1}s^{-1}) in our experiment (0.05μM hairpin molecules and 0.05μM opener molecules in 1×SSC buffer at 25°C). Although a longer lead section causes a faster reaction in general [15], the kinetic rate for the 20-base lead section resulted in about 1.5×10^5 M^{-1}s^{-1} because the opener strand folds into a conformation more stably than the openers with a lead section 10 or 7 bases long (see Fig. 2(a)).

Then, we verified that openers with a mismatching lead section cannot open the hairpin, and that if the hairpin's lead section is covered with its complementary strand, no proper openers can open the hairpin (see Fig. 2(b)).

We also tested openers with a lead section (seven bases) that was complementary to the hairpin loop and with an invasion section that might invade the hairpin stem from top to bottom (in the direction opposite to that of ordinary openers) (see Fig. 2(b)). Unlike the so-called molecular beacon [16], the small loop of the hairpin inhibits the invasion of the openers. An extra random coil attached to the lead section strengthens this inhibition, as reported by Yurke et al. [13]

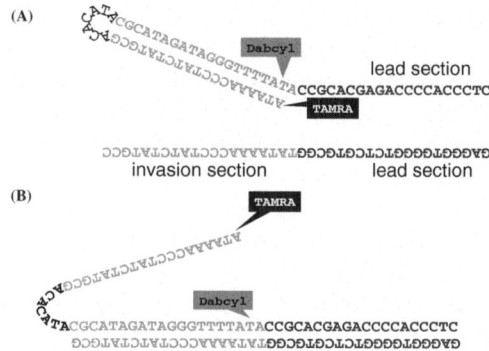

Fig. 1. [Hairpin Structure and Detection Scheme]: The hairpin molecule is labeled at the 5′-end with TAMRA and at the end of the stem with Dabcyl. (A) If TAMRA and Dabcyl are in close proximity, Dabcyl quenches the fluorescence of TAMRA. (B) The fluorescence intensity increases in proportion to the opening of the hairpin structure.

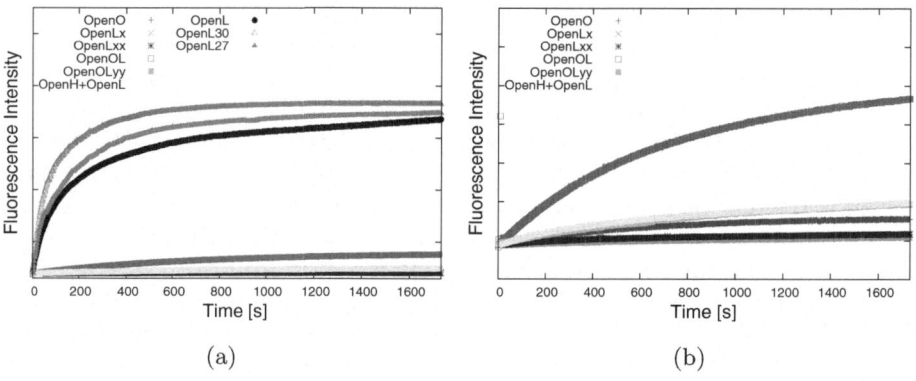

Fig. 2. [Experimental Results of the Previous Study]: (a) From the top curve, the efficiencies of the openers that have 10-, 7- and 20-base lead sections are each depicted. And the other curves show the efficiencies of the suppressed openers. (b) The fluorescence intensity of the suppressed openers (scaled up). The efficiency of the opener consisting of the invasion section and the lead section complementary to the hairpin loop is drawn in the pink line (top), and the opener with a random coil reached down to the light blue line (under the yellow curve). And we observed that the opener without any lead section cannot open the hairpin (the red curve), that the openers with a mismatching lead section cannot open the hairpin (the green and the blue curves), and that if the hairpin's lead section is covered with its complementary strand, the hairpin cannot be opened by any proper opener (the yellow curve).

3   Photoregulation

3.1   Photoregulation with Azobenzene

Using the isomerization of azobenzene residues in the side chain of an oligonucleotide (see Fig. 3(a)), hybridization between the oligomer and its complement

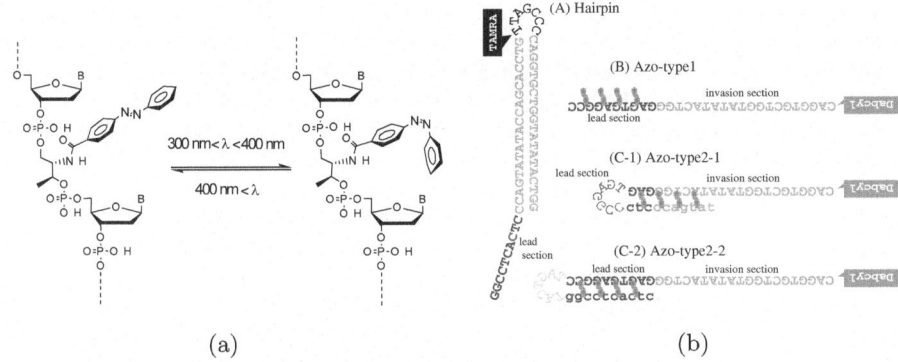

(a)   (b)

Fig. 3. [Photoregulation using Azobenzene]: (a) Isomerization scheme of an azobenzene in the side chain upon irradiation. (b) (A) A hairpin structure labeled at the top of the stem with TAMRA. (B) An opener for the hairpin structure labeled at the 5′-end with Dabcyl, which contains azobenzenes in its lead section. (C-1) and (C-2) Other openers with lead sections that form a stem region with their counterparts. These openers have supplementary bases allowing them to fold into a loop structure. The fluorescence intensity decreases as the reaction between the hairpin molecule and each opener molecule proceeds.

can be photoregulated [17,18]. When azobenzenes are isomerized from the *trans* form to the *cis* form by irradiating them with UV light (300 nm < λ < 400 nm), the melting temperature of the duplex is lowered considerably. Moreover, when the *cis*-form azobenzenes are irradiated with visible light (λ > 400 nm), the azobenzene residues are isomerized back to the *trans* form. Using these properties of azobenzenes, we show the feasibility of controlling hairpin opening with light.

Figure 3(b) depicts the reaction system used for photoregulation. Opener (B) has a 10-base lead section and openers (C-1) and (C-2) fold into a stem-loop in the *trans* condition. Even when opener (B) isomerized to the *cis* form is added to the solution containing hairpin molecules, the hairpin structure is retained because the lead section of the *cis* opener has a much lower melting temperature. Once the solution of dissolved molecules of the hairpin and its *cis* opener is irradiated with visible light, the reaction between the hairpin and the opener is permitted. By contrast, openers (C-1) and (C-2) will open the hairpin structure (A) when the solution is irradiated with UV light, because the stem-loop of the openers is dissociated and the lead section is exposed.

3.2   Materials and Methods

The oligomers intercalating azobenzenes, shown in Table 1, were synthesized by the third author, while *H-TAM* was synthesized by Sigma-Aldrich Japan, Genosys Division. Each opener strand has four azobenzenes in its lead section; these are located every two bases. The difference between *Azo-type2-1* and *Azo-type2-2* is that in *Azo-type2-1*, part of the lead section is exposed in the loop, while in *Azo-type2-2*, it is completely concealed in the stem.

Table 1. [Sequences of the Photo-Sensor Systems]: From top to bottom, these sequences correspond to structures (A), (B), (C-1), and (C-2) in Fig. 3(b). The sequence *H-TAM* folds into a hairpin stem with a sticky end. The base on the 5′-side of the closing pair is labeled with TAMRA. *Azo-type1, Azo-type2-1,* and *Azo-type2-2* are the hairpin openers intercalating azobenzenes, which are labeled with Dabcyl at the 5′-end. Each X represents an azobenzene residue.

H-TAM	: 5'-GGCCTCACTC-CCAGTATATACCAGCACCTG(-TAMRA)
	-TTAGCCC-CAGGTGCTGGTATATACTGG-3'
Azo-type1	: 5'-(Dabcyl-)CAGGTGCTGGTATATACTGG-GAXGTXGAXGGXCC-3'
Azo-type2-1	: 5'-(Dabcyl-)CAGGTGCTGGTATATACTGG-GAG
	-TGAGGCC-CTXC-CXCAXGTXAT-3'
Azo-type2-2	: 5'-(Dabcyl-)CAGGTGCTGGTATATACTGG-GAGTGAGGCC
	-TAGTCAT-GGXCCXTCXACXTC-3'

In an actual application, we would use light as input to the entire molecular system. In our experiments, however, instead of irradiating the entire system, which consists of both the opener and the hairpin machine, we irradiated only the openers in advance, to facilitate quantitative measurement. These experiments were conducted at 40°C, which is roughly the melting temperature of *trans* [18]. The openers *Azo-type1*, *Azo-type2-1*, and *Azo-type2-2* were irradiated for isomerization before being added to the solution. The temperature of the sample cell fixed in a HITACHI F-2500 spectrophotometer was maintained with a LAUDA RC6 thermostatic bath. In each measurement, the opener and hairpin machine were diluted to $0.05\mu M$ and $0.0225\mu M$ in $1\times$SSC buffer, respectively. The concentration of the hairpin molecules was planned to be half that of the openers ($0.025\mu M$), but an approximately 9% difference occurred after quantitative adjustments. We preprocessed the sample tube to dissolve the opener by heating it at 60 °C before irradiating it because we can attain a higher rate of azobenzene isomerization at a higher temperature. Subsequently, we exposed the tube to UV light through UV-D36C glass (a filter from Asahi Techno Glass that transmits UV and absorbs visible light) with a UVP B-100AP 100-W lamp (the original light filter was removed) for five minutes in order to effectively isomerize the azobenzenes from *trans* to *cis*. This treatment resulted in an isomerization rate of about 80% (roughly three of the four azobenzenes were isomerized), as observed using BECKMAN DU 650 spectroscopy (data not shown).

For the experiments with the *trans*-form openers, we irradiated the openers through an L-39 filter (a UV blocking filter from Asahi Techno Glass) with the 100-W lamp, because some *cis* azobenzenes might exist in the normal condition.

As the "half life" of *cis*-form azobenzenes is about twelve hours at 37°C and about one day at room temperature, we dropped each opener into the solution containing the *H-TAM* molecules immediately after irradiating it with UV light. Therefore, although our experiments were conducted at 40°C for thirty minutes, the temperature would not have affected the isomerization back to the *trans* form.

3.3   Experimental Results

Figure 4 shows the experimental result for the reaction between the hairpin machine *H-TAM* and the opener *Azo-type1*. The fluorescence intensity decreases as the hairpin machines are opened, since the TAMRA is quenched by the Dabcyl on the opener. Although we tethered fluorescent dyes at the bottom of the stem in the previous study while dyes are put at the top of the stem in this experiment, it seems that their position does not affect fluorescent properties judging from the kinetic rates (data will be shown later). The *trans* opener was expected to be far more efficient than the *cis* one for opening the hairpin structure. As the figure shows, however, in this experiment our expectation was rarely met. Assuming the reactions follow the second-order kinetics of ordinary duplex formation, we estimated that the rate constants are around 1.2×10^5 $M^{-1}s^{-1}$ for the *trans* openers and 8.1×10^4 $M^{-1}s^{-1}$ for the *cis* openers from fitting the fluorescence data. Therefore, even if the opener *Azo-type1* has a *cis* lead section, branch migration can proceed via partial hybridization between the lead sections.

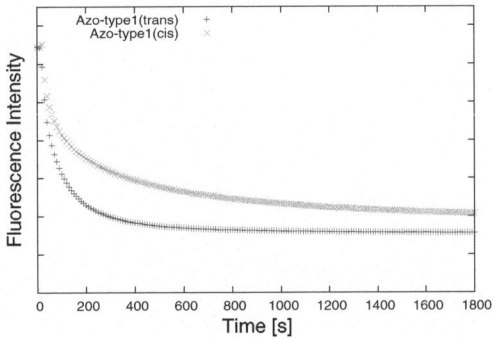

Fig. 4. [Comparison of the *trans* and *cis* forms of *Azo-type1*]: The upper line is the change in fluorescence intensity as the reaction between the *trans* opener *Azo-type1* and the hairpin structure *H-TAM* proceeds. The lower line is for the reaction between the *cis* opener *Azo-type1* and the hairpin structure.

Figure 5 depicts the fluorescence intensities of the reactions between the hairpin machine and the hairpin openers *Azo-type2-1* and *Azo-type2-2*. As the opener *Azo-type2-1* has a partially exposed lead section in its hairpin loop, this opener is more likely to open the hairpin machine in the *trans* condition as compared to *Azo-type2-2* (see Fig. 6(b)). In other words, the length of the lead section on the hairpin structure *H-TAM* is slightly shorter to avoid the interference between the lead sections [11]. By contrast, *Azo-type2-2*, which carries the *trans*-form azobenzenes, is more robust, as the lead section of the opener is completely concealed in the stem region. However, in exchange for this advantage, *Azo-type2-2* is less capable of breaking the hairpin machine (see Fig. 6(a)) because it is more stable. On the assumption that the reaction of the *Azo-type2* openers takes the second-order kinetics, we also tried to fit the curves and estimated the rates;

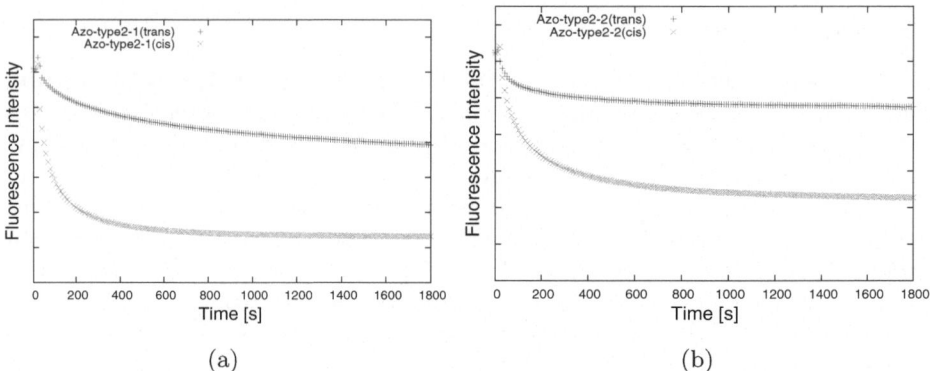

Fig. 5. [The Efficiencies of *Azo-type2-1* and *Azo-type2-2*]: (a) This figure shows the difference in the efficiency of *Azo-type2-1* between the *trans*(upper) and *cis*(lower) conditions. (b) This figure shows the difference in the efficiency of *Azo-type2-2* between the *trans*(upper) and *cis*(lower) conditions.

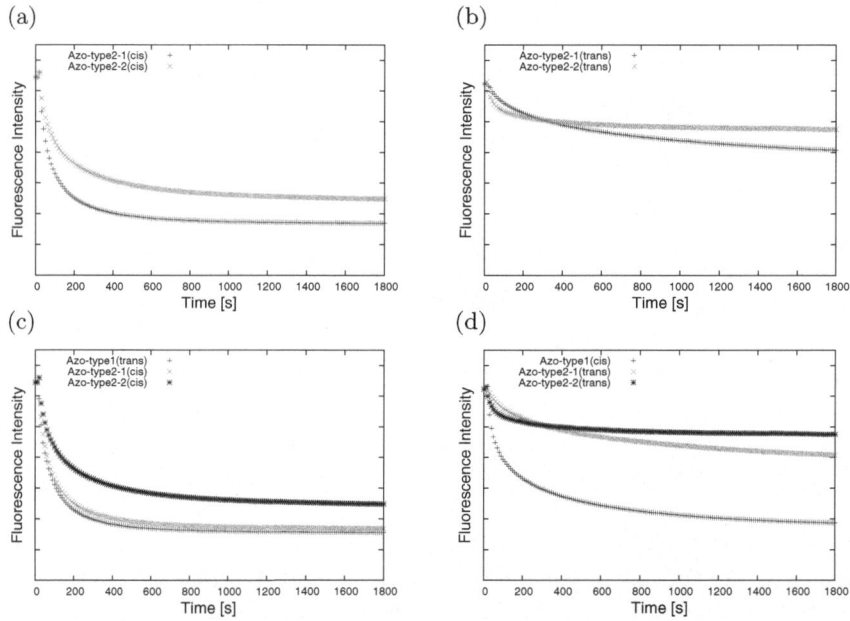

Fig. 6. Some Comparisons of the Photoregulation

1.2×10^5 $M^{-1}s^{-1}$ for the *cis* form *Azo-type2-1* and 8.3×10^4 $M^{-1}s^{-1}$ for the *cis* form *Azo-type2-2*. As for the *trans* form openers, we could not fit the curves well. If we borrow the coefficients obtained for the *cis* form other than the reaction rate and use the fluorescent intensity at 1800s, the reaction rate for the *trans* form is approximated by 5.1×10^4 $M^{-1}s^{-1}$ and 5.3×10^4 $M^{-1}s^{-1}$ for *Azo-type2-1* and *Azo-type2-2*, respectively.

Figure 6(c) compares the *switched-on* openers. *Azo-type1* with *trans* azobenzenes and *Azo-type2-1* with *cis* azobenzenes have equal ability, while the *cis* opener *Azo-type2-2* is slightly less efficient. Fig. 6(d) compares the *switched-off* openers. Unfortunately, *Azo-type1* works as if it were an *always switched-on* opener in contrast to other suppressed openers.

4 Thermoregulation

As another possible means of sensing environmental changes, this section introduces thermoregulation of hairpin opening.

4.1 Materials and Methods

We verified the dissociation of a hairpin machine as a thermo-sensor using the two openers *Th8* and *Th6* listed in Table 2. *Hairpin* is the same molecule that we used in our preliminary experiment. These oligomers were also synthesized by Sigma-Aldrich Japan, Genosys Division. The secondary structure of each sequence is shown in Fig. 8, where the hairpin loops of the openers might be closed by wobble pairs of (T, G).

Table 2. [Sequences of the Thermo-Sensor Systems]: *Hairpin* folds as the hairpin machine that is opened by the thermo-reactive opener strands *Th8* and *Th6*

Hairpin : TAMRA-5'-TATAAAACCCTATCTATGCG-ACACATA
-CGCATAGATAGGGTTTTAT(-Dabcyl-)A
CCGCACGAGACCCCACCCTC-3'
Th8　　: 5'-GTTTTATA-TCTCGTGCGG-TATAAAACCCTATCTATGCG-3'
Th6　　: 5'-TTTATA-TCTCGTGCGG-TATAAAACCCTATCTATGCG-3'

The openers have a lead section of length 10 enclosed in their hairpin loop. These openers exist as hairpins until they receive external input in the form of heat, i.e., until the temperature is raised (to 50°C in this system) so that their hairpin structure is dissociated. Therefore, the thermo-sensing system consisting of the hairpin machine and opener should measure the difference in temperature between 25°C (the room temperature) and 50°C. The hairpin machine and its opener were each diluted to $0.05\mu M$ in 1×SSC buffer, and each experiment was measured using a HITACHI spectrophotometer and a LAUDA thermostatic bath.

4.2 Experimental Results

Figure 7(a) plots the fluorescence intensity of TAMRA as a function of time during the reaction of *Hairpin* and *Th6*. The red line shows the efficiency of

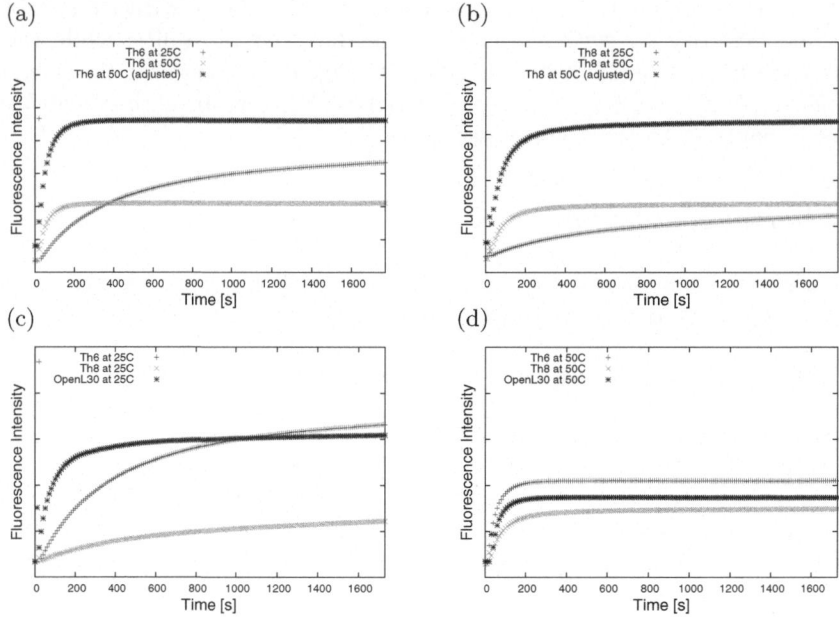

Fig. 7. Experimental Results of Thermoregulation

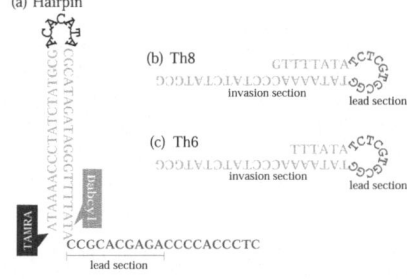

Fig. 8. [Thermo-Sensor Systems]: (a) The sequence of the hairpin structure is the same as the hairpin machine used in our preliminary experiments. The machine's lead section is a part of its overhang and the rest part functions as a hybridization inhibitor of the hairpin formed openers *Th8* and *Th6*. (b,c) The openers for the thermo-sensor machine. *Th8* has a stem of length 8 and *Th6* has a stem of length 6, and both have a lead section of length 10.

the hairpin-formed opener at 25 °C and the green line shows the efficiency of the single-stranded opener at 50 °C. As the fluorescence intensity of TAMRA is inversely proportional to the temperature, we adjusted the green curve to the fluorescence intensity at 25 °C and obtained the blue curve. In addition, the efficiency of opener *Th8* at 25 °C and 50 °C is shown in Figure 7(b).

Since these reactions do not follow the ordinal hybridization kinetics, we could not fit the curves. However, Figure 7(a) clearly shows that we cannot effectively control the dissociation of the haripin machine with *Th6*, while Figure 7(b) shows that *Th8* at 25°C does not have much ability to open the machine. Figure 7(c) and (d) depict the comparisons between the two openers at 25°C and 50°C, respectively. We also compare the two openers against the proper opener with the 10-base lead section used in our preliminary experiment (Fig. 2(a)). The adjustment of the fluorescence intensity in (a) and (b) is made according to the measurement of this opener. In exchange for the good controllability, *Th8* is totally inferior to *Th6* in terms of efficiency because it has a more stable stem.

5   Discussion

In this paper, we showed that photoregulation and thermoregulation of hairpin opening are possible in the same framework of hairpin opening. Note that there are a variety of methods for thermoregulation other than the method proposed in this paper, but the openers for thermoregulation in this study can be used in conjunction with other kinds of openers, including those for photoregulation and those containing aptamers as proposed by Dirks et al. [19].

In the photoregulation experiments, we succeeded in controlling the opening of the hairpin machine using UV light. However, control using visible light (as was expected for *Azo-type1*) remains a future goal. In the thermoregulation experiments, we succeeded in regulating the conformational change of the hairpin machine with openers that changed structure according to external input in the form of heat. These kinds of sensors will be used to regulate general-purpose molecular systems such as DNA logical circuits [12].

Compared with the functionally suppressed openers used in the preliminary experiments mentioned in the first section, we have not adequately inhibited the *switched-off* openers. In order to apply the current results effectively, we need to construct more robust sensor machines by suppressing the *switched-off* openers more strongly. As for the *switched-on* openers, on the other hand, we need to make hairpin opening more efficient. As suggested by an anonymous reviewer, in order to drive the equilibrium of the system towards the complete hairpin opening, we could add a few bases to the invading substrand, which complements the first few bases on the hairpin loop.

Acknowledgments

The authors would like to thank anonymous reviewers for valuable comments. The work reported here is also supported in part by Grand-in-Aid for Scientific Research on Priority Area No.14085202, Ministry of Education, Culture, Sports, Science and Technology, Japan.

References

1. B. Yurke *et al.* A DNA-fuelled molecular machine made of DNA. *Nature* **406**, 605–608 (2000)
2. F. C. Simmel *et al.* Using DNA to construct and power a nanoactuator. *Physical Review E* **63**, 041913 (2001)
3. C. Mao *et al.* A DNA nanomechanical device based on the B-Z transition. *Nature* **397**, 144–146 (1999)
4. Y. Benenson *et al.* Programmable and autonomous computing machine made of biomolecules. *Nature* **414**, 430–434 (2001)
5. M. Hagiya *et al.* Towards Parallel Evaluation and Learning of Boolean μ-Formulas with Molecules. *DNA Based Computers III, DIMACS Series in Discrete Mathematics and Theoretical Computer Science* **48**, 57–72 (1999)
6. G. Seelig *et al.* DNA Hybridization Catalysts and Catalyst Circuits *DNA10, Tenth International Meeting on DNA Based Computers, Preliminary Proceedings* 202–213 (2004)
7. H. Yan *et al.* A robust DNA mechanical device controlled by hybridization topology *Nature* **415**, 62–65 (2002)
8. H. Uejima *et al.* Secondary Structure Design of Multi-state DNA Machines Based on Sequential Structure Transitions. *Ninth International Meeting on DNA Based Computers, LNCS, Springer* **2943**, 74–85 (2004)
9. M. Kubota *et al.* Branching DNA Machines Based on Transitions of Hairpin Structures. *Proceedings of the 2003 Congress on Evolutionary Computation (CEC'03)*, 2542–2548 (2003)
10. A. Kameda *et al.* Conformational Addressing Using the Hairpin Structure of Single-Strand DNA. *Ninth International Meeting on DNA Based Computers, LNCS, Springer* **2568**, 219–223 (2004)
11. K. Takahashi *et al.* Preliminary Experiments on Hairpin Structure Dissociation for Constructing Robust DNA Machines. *Proceedings of the 2004 IEEE Conference on Cybernetics and Intelligent Systems (CIS'04)*, 285–290 (2004)
12. K. Takahashi *et al.* Chain Reaction Systems based on Loop Dissociation of DNA. *submitted*
13. Turberfield A. J *et al.* DNA fuel for free-running nanomachines. *Physical Review Letters* **90**, 118102 (2003)
14. B. Yurke *et al.* Using DNA to power nanostructures. *Genetic Programming and Evolvable Machines* **4**, 111–122 (2003).
15. L. E. Morrison *et al.* Sensitive fluorescence-based thermodynamic and kinetic measurements of DNA hybridization in solution. *Biochemistry* **32**, 3095–3104 (1993)
16. W. Tan *et al.* Molecular beacons for DNA biosensors with micrometer to submicrometer dimensions. *Analitical Biochem.* **283**, 56–63 (2000)
17. H. Asanuma *et al.* Photo-regulation of DNA function by azobenzene-tethered oligonucleotides. *Nucleic Acids Res. Supple.* **3**, 117–118 (2003)
18. H. Asanuma *et al.* Photoregulation of the Formation and Dissociation of a DNA Duplex by Using the *cis-trans* Isomerization of Azobenzene. *Angewandte Chemie International Edition*, **38**, 2393–2395 (1999)
19. R. M. Dirks *et al.* Triggered amplification by hybridization chain reaction. *PNAS* **101**, 15275–5278 (2004)

Chain Reaction Systems Based on Loop Dissociation of DNA

Keiichiro Takahashi[1], Satsuki Yaegashi[2],
Atsushi Kameda[2], and Masami Hagiya[2,3]

[1] NovusGene Inc., 2-3 Kuboyama-cho, Hachioji-shi, Tokyo 192-8512, Japan
takahashi-k@novusgene.co.jp
[2] Japan Science and Technology Corporation (JST-CREST)
yaegashi@lyon.is.s.u-tokyo.ac.jp, kameda@complex.eng.hokudai.ac.jp
[3] Department of Computer Science, Graduate School of Information Science and
Technology, University of Tokyo, 7-3-1 Hongo, Bunkyo-ku, Tokyo 113-0033, Japan
hagiya@is.s.u-tokyo.ac.jp

Abstract. In the field of DNA computing, more and more efforts are
made for constructing molecular machines made of DNA that work in
vitro or in vivo. States of some of those machines are represented by their
conformations, such as hairpin and bulge loops, and state transitions are
realized by conformational changes, in which such loops are opened. The
ultimate goal of this study is to implement not only independent molecular machines, but also networks of interacting machines, called *chain
reaction systems,* where a conformational change of one machine triggers
a conformational change of another machine in a cascaded manner. A
chain reaction system would result in a much larger computational power
than a single machine in the number of states and in the complexity of
computation. As a simple example, we propose a general-purpose molecular system consisting of logical gates and sensors. As a more complex
example, we present a new idea of constructing a DNA automaton by a
chain reaction system, which can have an arbitrary number of states.

1   Introduction

Seeman and Winfree's tile assembly model has made it possible to construct
large structures of DNA by programmed self-assembly of basic components. The
structures constructed so far include DNA nanotubes and planar patterns such
as Sierpinski triangle. Meanwhile, DNA nanomachines like the molecular tweez-
ers [11] have come under the spotlight of nanorobotics. Those nanomachines
realize finite state machines, where states are represented by their conformations, and conformational changes make state transitions.

The ultimate goal of this study is to construct general-purpose molecular
systems consisting of interacting molecular machines. In contrast to self-assembly
of DNA tiles, where static components only hybridize together, each component
of such a system is a DNA machine that changes its state through interactions
with other machines. In other words, we aim at constructing networks of DNA
machines interacting with one another, where a conformational change of one

A. Carbone and N.A. Pierce (Eds.): DNA11, LNCS 3892, pp. 347–358, 2006.

machine triggers a conformational change of another machine in a cascaded manner. We call such networks of machines *chain reaction systems* in this paper. By a chain reaction system, it would be possible to realize information processing at the molecular level. Recently, molecular systems involving a chain reaction have been investigated by several other groups. For example, Dirks et al. have already demonstrated self-assembly of two stable hairpin species, triggered by an initiator strand, which may contain a DNA aptamer and become active when ATP binds the aptamer and exposes a sticky end [19]. In the chain reaction, one hairpin opening mutually triggers another hairpin opening and long linear complexes are yielded.

In the next section, we briefly introduce the design method that we have suggested to construct chain reaction systems. And in the subsequent sections, we introduce two examples of chain reaction systems. The first one is the chain reaction system simulating AND-OR circuits, where an AND gate is represented by a DNA machine including two bulge loops and an OR gate is composed of two hairpin structure molecules. The thermo- and photo-regulated hairpins [18], can be used as sensors that produce inputs to such logical circuits. As the second example, we present DNA automata, where both the transition rules and states would be implemented by DNA machines.

2   Design Method of Chain Reaction Systems

There have been developed various methods for designing DNA sequences [4]. However, those methods are not enough for designing DNA machines and their networks as mentioned above. Developing a new and systematic design framework for constructing DNA machines is also a goal of our study [1, 2].

The framework for designing and implementing DNA machines advocated in this paper consists of the following three steps. We first pre-design candidate DNA sequences that fold into intended initial conformations. Secondly, we predict thermodynamic properties of the pre-designed sequences and select optimal ones. Finally, we actually verify behaviors of machines in laboratory experiments.

In the first step, we adopt the template method [5, 6] that can be used to systematically generate a set of DNA sequences in which different sequences are guaranteed to have a certain number of mismatches. Briefly, in the template method, a DNA library X can be derived as $X = \tau \cdot E := \{\tau \cdot w \mid w \in E\}$, where τ is a mismatch-guaranteed binary string called *template* and E is an error-correcting code. The $\cdot$ operator stands for a bit-wise product as $1 \cdot 1 = $G, $1 \cdot 0 = $C (or $1 \cdot 1 = $C, $1 \cdot 0 = $G), $0 \cdot 0 = $A and $0 \cdot 1 = $T (or $0 \cdot 0 = $T, $0 \cdot 1 = $A). However, since the original template method requires all sequences to have the same length, we have modified the method so that templates of different lengths can be mixed to generate a set of DNA sequences of different lengths. By concatenating DNA sequences of different lengths generated by the extended template method, we can obtain candidate sequences that fold into target structures.

As for the second step, many existing programs for thermodynamic analysis of DNA, such as the Vienna Package [7], can only handle a single DNA sequence.

Therefore, we have extended the Vienna Package to handle multiple sequences for both minimum free energies and partition functions [1, 2, 3]. Using the extended Vienna Package, we select optimal sequences from the candidates.

3   DNA Logical Circuits

Logical circuits consisting of wired logic elements are one of the simplest models of computation, and hence DNA-based simulation of logical circuits has been investigated by many researchers in DNA computing. For example, the first DNA-based simulation of logical circuits is reported by Ogihara and Ray [8], Amos et al. have proposed a DNA simulation of logical circuits using NAND gates in time proportional to the depth of the circuit [9], Carbone et al. have applied DNA tiles to the evaluation of circuits [10], and Seelig et al. have proposed the way of simulating circuits using hybridization catalysts [12].

Our chain reaction system realizing a DNA logical circuit consists of AND gate machines having two bulge loops and OR gate machines composed of two hairpin molecules. Figure 1 presents these logical gates. The AND gate contains regions that can hybridize with its inputs on the topside chain and its output as the downside chain. If the gate receives two input signals in the form of single-stranded molecules, it produces its output which will become an input to a successive gate. The dotted region on the left side of the second input prevents it from interacting with the first bulge loop before the first input breaks the stem region, constructing a robust kinetic wall that prevents the system from rapidly moving to the equilibrium [17].

Meanwhile, the OR gate consists of two hairpin machines that have the same hairpin loops, which work as a converter of the input. An input that has the lead section complementary to that of either hairpin machine opens the hairpin and exposes the lead section of the output. This means that the gate converts the lead section of the input to that of the output.

Inputs to those logical gates can be thermo- and photo-regulated hairpins, which we are now developing [18]. So far, we have verified that an input with a small hairpin that covers its lead section can be used as a thermo-sensor because the small hairpin is dissociated in high temperature. A small hairpin bearing azobenzenes in its stem can be used as a photo-sensor because UV light isomerizes the azobenzenes and dissociates the hairpin. In this way, we can construct a general-purpose molecular system consisting of logical gates and sensors, which combines various inputs and produces their boolean combination.

3.1   Design of an AND Gate

Hereafter, the term *unit* is used for a subsequence of a specified length on a given target sequence. When the given target sequence is scanned from 5′ to 3′ end, the pair of lengths of adjacent two units appearing on the sequence is called a *concatenation pattern,* denoted by $n * m$, where n is the length of the unit on the 5′-end side and m is the length of the unit on the 3′-end side. The set of

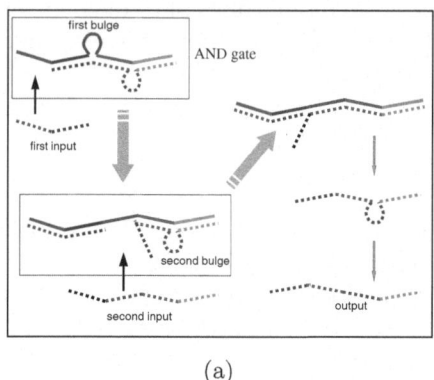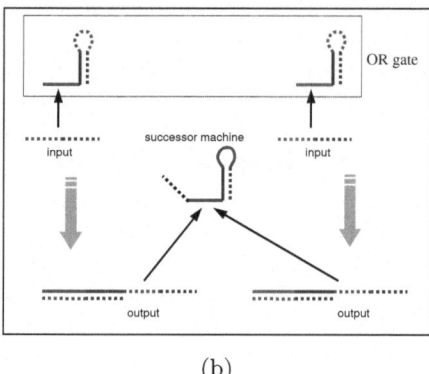

(a)   (b)

Fig. 1. [AND Gate and OR Gate]: The panel (a) shows the behavior of an AND gate. The AND gate is implemented by two strands that fold into a double-bulge structure. The panel (b) shows how an OR gate works. The OR gate consists of two hairpin molecules.

all concatenation patterns of a given chain reaction system is used as the only information to generate optimal template tuples.

We are currently conducting some preliminary experiments of an AND gate. The actual sequences for the AND gate and its inputs are shown in Table 1. These sequences are designed using the extended template method. We selected $10 * 15$, $15 * 10$ and $15 * 15$ as the concatenation patterns for the system and we chose the template tuple $(0001001011, 111011101000111)$ by the extended template method, taking their GC-content into account. While the template 00010010011 is used for the loop regions and overhanging sections, the template 111011101000111 is relevant to the stem regions. Since the stems of length 15 on the AND gate have to stabilize the two bulge loops of length 10, the stem regions

Table 1. [Sequences for AND Gate]: The sequences correspond to the sequences in Fig. 1(a). The upper two strands fold into a double bulge structure that functions as an AND gate when they hybridize together. For the sake of detecting behavior in the fluorescence experiment mentioned in the next subsection, the topside strand *TopB* is labeled at $3'$ end with the black hole quencher dye BHQ-1 and the downside strand *DownB* is labeled at $5'$ end with the reporter dye FAM. *FO* and *SO* are input molecules to the AND gate.

TopB	: 5'-AAACTTCACC-GGGTCCGTGAATGGG-ATACTCTACTGC
	-CGCACCCTGATACCG-CGCTGGCAGATTCCG-(BHQ-1)-3'
DownB	: 5'-(FAM-)CGGAATCTGCCAGCG-AATCTAGACC
	-CGGTATCAGGGTGCG-CCCATTCACCGGACCC-3'
FO	: 5'-CCCATTCACGGACCC-GGTGAAGTTT-3'
SO	: 5'-CGGAATCTGCCAGCG-CGGTATCAGGGTGCG
	-GCAGTAGTAT-CCGTGCGTCTTAGCG-3'

have a little extra GC-content of 67%. The actual DNA libraries of lengths 10 and 15 are obtained with BCH code, where the T_m values are about $55 \pm 1°C$ and about $25 \pm 1°C$ in our experimental conditions, respectively. Table 1 shows the sequences of the AND gate and the inputs.

The sequences *TopBulge* and *DownB* compose the AND gate, where *TopB* has regions that can hybridize with the two inputs, and *DownB* works as the output molecule after receiving the inputs. The sequence named *FO* is the first input and the sequence named *SO* is the second one.

3.2 Experimental Results of an AND Gate

By gel electrophoresis and fluorescence experiments, we have verified that the AND gate releases the output signal only when the two inputs exist.

Figure 2(a) shows the result of electrophoresis on 10% PAGE (non-denaturing gel). The boxes labeled with a lower-case letter on the gel classify the enclosed bands into the structures that we predict. The band on the cross of Lane 2 and the box (b) corresponds to the double-bulge structure of the AND gate. And the clear band on the cross of Lane 3 and the box (a) is the AND gate whose first bugle loop is opened by *FO*. Importantly, Lane 4 shows whether the output is produced only by the second input or not. In fact, the output molecules are rarely released from the AND gate, judging from the box (e) on which the output molecules are present. Through Lanes 5 to 7, the AND gate and the two inputs are applied to the gel. Each lane has almost the same appearance, but the order of mixing *FO* and *SO* is different in each lane. Lane 5 shows the result of the AND gate on receiving the two inputs at a time. In Lane 6, after hybridization between the AND gate and the first input, the second input is mixed. Conversely,

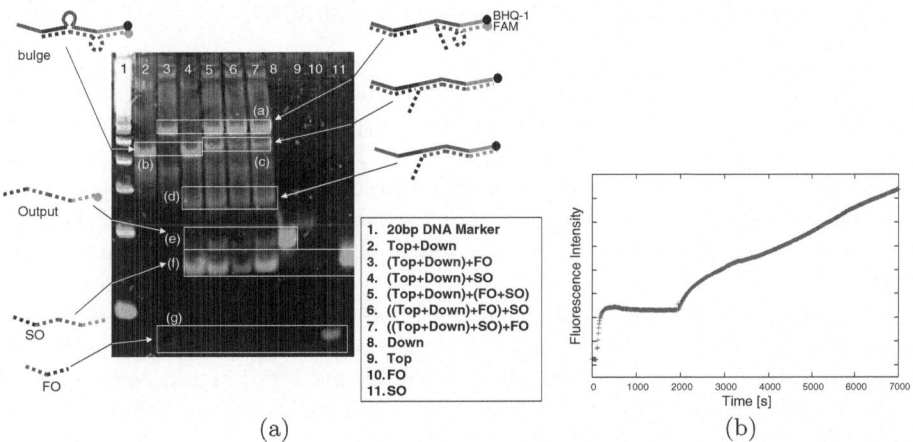

(a)   (b)

Fig. 2. [Experimental Results]: (a) The composition of each lane is given in the bottom right box, where each + means a hybridization process for thirty minutes. (b) The fluorescence intensity change from $0s$ to $2000s$ is by the reaction between the AND gate and the second input *SO*. *FO* is put at $2000s$, and all the molecules exist from $2000s$ to $7000s$.

after interaction between the AND gate and the second input, the first input is mixed in Lane 7. The important result is that the output is produced only when both of the inputs exist.

The electrophoresis result is in good agreement with the fluorescence experiment. In order to confirm that the output molecules are derived only if the AND gate receives the first and second input molecules, we mixed the AND gate with the second input before giving the first one (Fig. 2(b)). While the reaction between the AND gate and the second input reaches saturation relatively fast in a small change of fluorescence intensity, the output molecules are released in a larger quantity when the two inputs exist.

3.3   Discussion

We verified the function of the AND gate made of DNA molecules. However, the experiment unfortunately results in that most of the molecules of the AND gate hybridizing with the first input molecules are still intact even if the second input molecules exist as seen in the box (a) in Figure 2(a). The main reasons of this problem are as follows.

- There is a thermodynamic limitation in the system. Namely, since the system does not contain strands that the output molecules can hybridize with, the total energy change of the system is relatively small. Therefore, by preparing strands that are complementary to the final output molecules of a logical circuit, we should lower the energy of the system after the reaction.
- At the time of designing the sequences, our design framework did not take the structure of each sequence into consideration. Therefore, each sequence adopts a secondary structure that has many base pairs by itself. For example, the second input SO adopts the structure $((((..(((((.((((.........)))))))))).....))))$. $((......))$. (-8.86 kcal/mol) as the optimal one at 35°C.

The first problem does not arise in the approach using hybridization catalysts [12]. However, there is another problem concerning AND gates, which even the catalyst approach does not escape: the output will be attenuated if other AND gates that can receive the first signal exist. In our case, although the second input signal can be accepted by the AND gate only after the gate receives the first signal, the first input molecules can hybridize with any other AND gates that have the complementary lead section even if their second inputs do not exist.

A possible solution to this attenuation problem is to keep hybridization and dissociation between the AND gate and the first signal under certain equilibrium as shown in the left part of Figure 3(a). If the second input signal does not exist in the system, the equilibrium will be kept. And if the second input molecule exists in the system, the AND gate with its first bulge loop opened by the first input will take the second input, moving the equilibrium towards hybridization. Figure 3(b) depicts the example of the improved strategy. The AND gate situated in the middle can accept the first input, but the reaction is in the equilibrium as mentioned above since the second input to the gate does not exit. Therefore, only the intended reaction will gradually proceed.

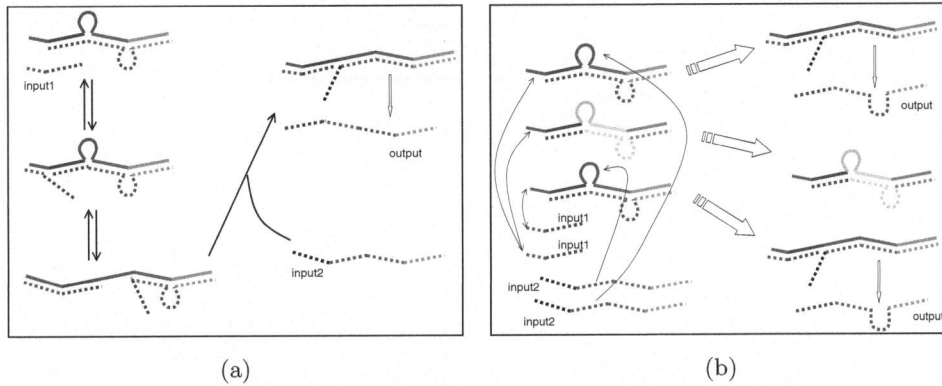

(a)   (b)

Fig. 3. [Improvement of the AND Gate]: (a) Keeping the three states on the left equilibrium, the decay of signals will be prevented. (b) This figure shows the improved AND gates.

4 DNA Automata

In this section, we propose another chain reaction system. A wide variety of ideas for implementing finite automata by DNA have been reported. Among them, Gao et al. have proposed to use double-stranded molecules including one bulge loop which encode transition rules and several kinds of enzymes [13, 14]. And Benenson et al. have actually implemented finite automata using more sophisticated encoding techniques [15]. And more recently, they have succeeded in analyzing mRNA levels of gene expression using their molecular computer *in vitro* [16].

Formally, a deterministic finite automaton (DFA) is a structure

$$M = (Q, \Sigma, \delta, s, F),$$

where Q is a finite set of *states*, Σ is a finite set called *input alphabet*, $\delta : Q \times \Sigma \to Q$ is the *transition function*, $s \in Q$ is the *initial state* and $F(\subset Q)$ is a set of *final states*.

Here we propose a new kind of deterministic finite automaton $(Q, \{0, 1\}, \delta, s, F)$ comprised of only DNA molecules, based on our chain reaction system. The automaton receives external input molecules one by one, which are manually put into the solution, and makes transitions by chain reaction. The design of our DNA automaton is flexible in that it can have arbitrarily many states.

To construct such a chain reaction system which simulates a DFA, three kinds of component molecules encoding the corresponding components of the DFA are required: state molecules, transition rule molecules and state activation molecules. Each state molecule uniquely encodes a state $q \in Q$ and also contains the hairpins corresponding to the acceptable input symbols 0 and 1 (Fig. 5(a)(1)). A transition rule consists of two modules of DNA molecules, where one is comprised of a bulge loop, an interior loop and a hairpin loop,

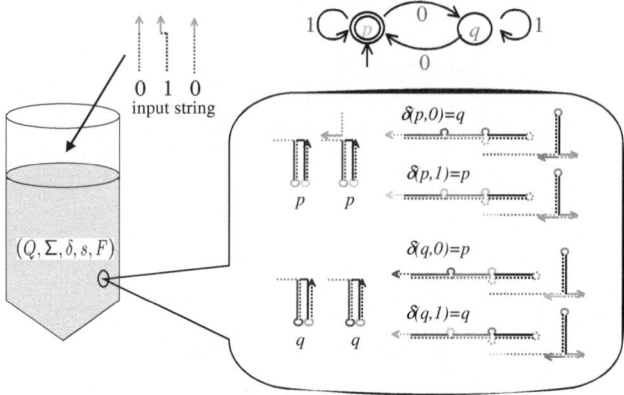

Fig. 4. [DNA Automaton]: The solution dissolving state molecules and transition rule molecules processes an input binary string

and the other consists of a hairpin loop and a 3-loop (Fig. 6(b)). The former module, triggered by the state molecule, hybridizes with the latter one, which then releases a state activation molecule at the completion of the transition. The state activation molecule identifies the next state and contains the hybridization site of input molecules (Fig. 5(a)(2)). Only state molecules which hybridize with the activation molecule represent the current state, and other state molecules remain inactive, that is, they cannot read any input symbols (Fig. 5(b)).

In order to explain how the chain reaction proceeds, let us show a simple DNA automaton (Fig. 4: for simplicity, blocker subsequences that hinder unwanted interference as found in [17] are omitted) which is found in the work of Benenson et al. [15]:

$$M = (Q, \Sigma, \delta, s, F),$$

where $Q = \{p, q\}$, $\Sigma = \{0, 1\}$, $s = p$, $F = \{p\}$ and $\delta : Q \times \Sigma \longrightarrow Q$ is specified by $\delta(p, 0) = q$, $\delta(p, 1) = p$, $\delta(q, 0) = p$ and $\delta(q, 1) = q$, as shown on the upper right panel of the figure.

At first, since the initial state p is the current state, the state activation molecule, which should be kept equimolar to the first input molecule, stochastically hybridizes with the state p molecule. If the molecule encoding the input symbol 0 is put into the solution, the initial state p takes in the symbol 0 via branch migration, opening the hairpin stem and exposing the region encoding the label $(p, 0)$ on the hairpin loop as in Figure 6(a). The region encoding $(p, 0)$ works as a specifier for the transition rule modules representing $\delta(p, 0) = q$. Note that the label $(p, 0)$ is hidden from the transition modules until the hairpin structure opens.

In the next step, the molecular interaction between the initial state p and the transition rule $\delta(p, 0) = q$ proceeds through the label $(p, 0)$ as in Figure 6(b). Figure 7 depicts the interaction of the two modules after the hybridization with the initial state. The first module hybridizing with the initial state uniquely communicates with the second module via the opened bulge structure, which

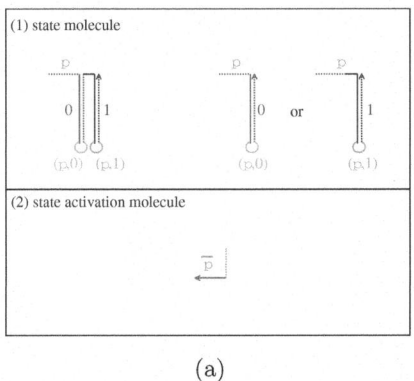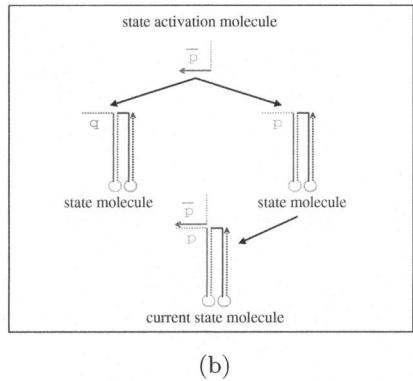

(a)   (b)

Fig. 5. [State Molecules and Current State Molecules]: (a) The panel (1) shows state molecules encoding state p. Each state molecule encodes the state in its overhang and the acceptable inputs 1 and 0 in its hairpin stem(s). The hairpin loops encode the elements of $Q \times \{0, 1\}$. If the transition rule applied to the state p is a partial function, the state molecule consists of one hairpin stem and overhang (the two molecules on the right hand side). And the panel (2) describes a state activation molecule, which contains the subsequence $\bar{p}$ complementary to the state p. And the molecule also has the hybridization site of input molecules on the rest of it. The state activation molecule is released from a transition rule molecule (Fig. 6(b)) if the state transition succeeds. (b) A state molecule that an state activation molecule hybridizes with works as the current state and receives an input molecule.

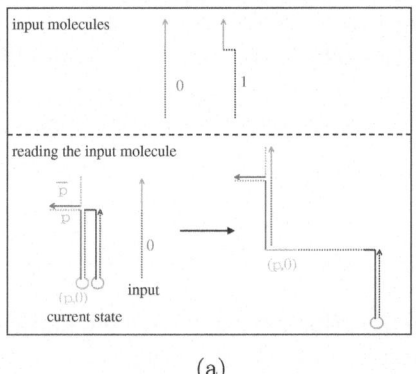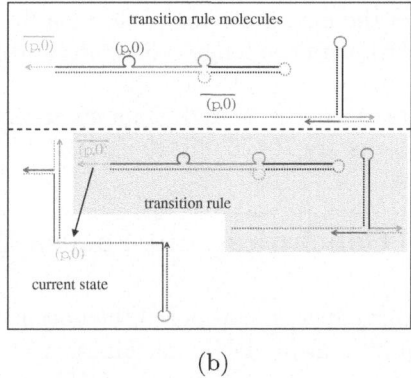

(a)   (b)

Fig. 6. [Input Molecules and Transition Rule Molecules]: (a) Input molecules are shown in the upper box. The red region encodes the input symbol 0 and the blue region encodes the input symbol 1. The ocher region is common to both the input molecules, which is complementary to the counterpart of state activation molecules. The bottom box shows the current state reading input 0. (b) The pair of two molecules in the upper box represents the transition rule $\delta(p, 0) = p$. The transition rule is only applied to the current state through the hybridization of $(p, 0)$ and $\overline{(p, 0)}$.

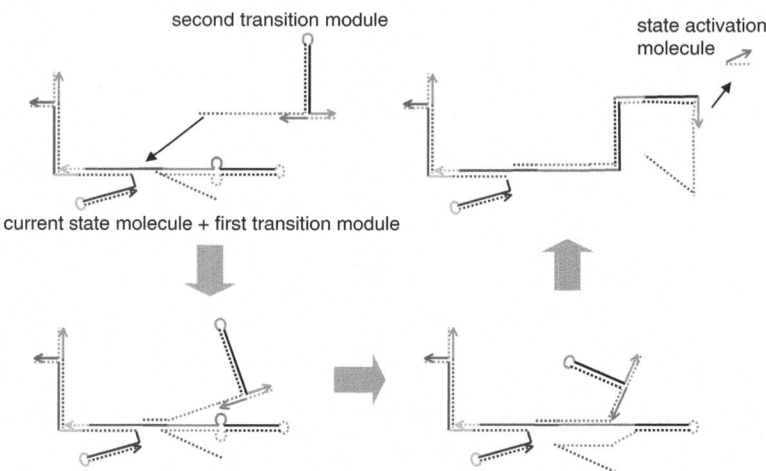

Fig. 7. [State Transition]: The transition rule releases a state activation molecule for the next state

encodes the label $\delta(p, 0)$ and which is complementary to the end part of the overhang on the second module (Fig. 5(b)).

At the end of the transition reaction, the state activation molecule for the next state q is released. And then, the activation molecule will also stochastically hybridize with the state q molecule and represent the current state. At this point, the next input molecule can be mixed. For the detection of acceptance, we can label the final state molecule with fluorescent dye.

Although the lengths of stems and loops (or sticky ends) have to be carefully chosen, we expect that the DNA automaton can be stably constructed with units of lengths 10 and 25 for stems and loops, respectively, where concatenation patterns are 10*25, 25*10 and 25*25.

5   Conclusion and Future Work

In this paper, we proposed the notion of chain reaction systems based on interaction of multiple DNA machines. And as examples of chain reaction systems, we proposed and explained the DNA logical circuit and the DNA automaton. The former system would realize a multi-sensor system and the latter is expected to have more states than existing DNA automata.

We also showed the results of preliminary experiment for the AND gate. Implementation of an actual multi-sensor system consisting of multiple AND gates, OR gates and sensor machines remains as future work. We are currently refining AND gates so that they efficiently receive the pair of input signal molecules and release the output molecule. And also, the actual design of the DNA automaton is future work.

In order to construct a robust chain reaction system, it is essential to establish techniques to keep the current state of each machine stable unless it is given an input signal that takes part in the cascading chain of transitions, and it is also important to make each cascading reaction rapid. For the first problem, we proposed the combination of

1. preventing loop opening by blocker substrands,
2. mismatches systematically guaranteed by the (extended) template method, and
3. optimal selection of sequences by the extended Vienna Package.

We expect that this combination of techniques is effective in constructing large chain reaction systems, although we are currently dealing with a single gate.

For the second problem, however, the current gate is very slow because of the inefficient reaction between SO and the gate. According to Figure 2(b) from $0s$ to $2000s$ and Lane 4 of Figure 2(a), we can observe that a certain amount of SO rapidly completes a reaction with the gate without FO (the cross of Lane 4 and the box(d) in Figure 2(a)), while the rest of SO remains independent of the gate as intended (the cross of Lane 4 and the box(f) in Fig. 2(a)). And some of the latter SO seems allowed to push away the output from the gate quite rapidly as soon as FO completely hybridizes with the gate (from $2000s$ to $4000s$ in Figure 2(b)). The efficiency of hybridization between FO and the gate is indicated in the boxes (a) and (g) of Figure 2(a). After $4000s$ in Figure 2(b), the remaining SO seems allowed to react with the gate very slowly, due to some kinetic trap in which SO is considered to form a certain structure with the gate and FO. As mentioned in Subsection 3.3, we consider that in order to avoid unwanted kinetic traps and facilitate the reaction we should enhance the driving force of the system and use small structure openers as well as select proper lengths of stems and loops.

Acknowledgments

The authors would like to thank anonymous reviewers for valuable comments. The work reported here is also supported in part by Grand-in-Aid for Scientific Research on Priority Area No.14085202, Ministry of Education, Culture, Sports, Science and Technology, Japan. And the authors deeply thank Professor Satoshi Kobayashi (University of Electro-Communications) for designing DNA libraries that we used in our early experiments and advices on the extended template method.

References

1. H. Uejima *et al.* Secondary Structure Design of Multi-state DNA Machines Based on Sequential Structure Transitions. *Ninth International Meeting on DNA Based Computers, LNCS, Springer* **2943**, 74–85 (2004)
2. M. Kubota *et al.* Branching DNA Machines Based on Transitions of Hairpin Structures. *Proceedings of the 2003 Congress on Evolutionary Computation (CEC'03)*, 2542–2548 (2003)

3. K. Takahashi *et al.* On Computation of Minimum Free Energy and Partition Function of Multiple Nucleic Acid Sequences. *FIT2004, Forum on Information Science and Technology*, 91–92 (2004)

4. R. M. Dirks *et al.* Paradigms for computational nucleic acid design. *Nucleic Acids Res.*, **32**, 1392–1403 (2004)

5. M. Arita *et al.* DNA Sequence Design Using Templates. *New Generation Computing*, **20**, 263–277 (2002)

6. S. Kobayashi *et al.* On Template Method for DNA Sequence Design. *DNA Based Computers (DNA8), LNCS, Springer*, **2568**, 205–214 (2003)

7. I. L. Hofacker. Vienna RNA secondary structure server. *Nucleic Acids Res.* **31**, 3429–3431 (2003)

8. M. Ogihara *et al.* Simulating Boolean circuits on a DNA computer. *Algorithmica*, **25**, 239–250 (1999)

9. M. Amos *et al.* DNA simulation of Boolean circuits. *Proc. of the Third Annual Conference on Genetic Programming*, Morgan Kaufmann, 679–683 (1998)

10. A. Carbone *et al.* Circuits and programmable self-assembling DNA structures. *PNAS*, **99**, 12577–12582 (2002)

11. B. Yurke *et al.* A DNA-fuelled molecular machine made of DNA. *Nature*, **406**, 605–608 (2000)

12. G. Seelig *et al.* DNA Hybridization Catalysts and Catalyst Circuits *DNA10, Tenth International Meeting on DNA Based Computers, Preliminary Proceedings*, 202–213 (2004)

13. Y. Gao *et al.* DNA implementation of nondeterminism. *DNA Based Computers III, DIMACS Series in Discrete Mathematics and Theoretical Computer Science* **48**, 137–148 (1999)

14. M. Garzon *et al.* In vitro Implementation of Finite-State Machines. *Proc. Workshop on Implementing Automata (WIA '97)*, 56–74 (1998)

15. Y. Benenson *et al.* Programmable and autonomous computing machine made of biomolecules. *Nature* **414**, 430–434 (2001)

16. Y. Benenson *et al.* An autonomous molecular computer for logical control of gene expression. *Nature* **429**, 423–429 (2004)

17. K. Takahashi *et al.* Preliminary Experiments on Hairpin Structure Dissociation for Constructing Robust DNA Machines. *Proceedings of the 2004 IEEE Conference on Cybernetics and Intelligent Systems (CIS'04)*, 285–290 (2004)

18. K. Takahashi *et al.* Photo- and Thermo-Regulation of DNA Nanomachines. *submitted*

19. R. M. Dirks *et al.* Triggered amplification by hybridization chain reaction. *PNAS* **101**, 15275–5278 (2004)

A Local Search Based Barrier Height Estimation Algorithm for DNA Molecular Transitions*

Tsutomu Takeda**, Hirotaka Ono***,
Kunihiko Sadakane, and Masafumi Yamashita

Dept. of Electrical Engineering and Computer Science, Kyushu University,
6-10-1 Hakozaki, Higashi-ku, Fukuoka, Fukuoka 812-8581, Japan
takeda@tcslab.csce.kyushu-u.ac.jp, {ono, sada, mak}@csce.kyushu-u.ac.jp

Abstract. An accurate estimation of the barrier height between two given secondary structures of DNA molecules is known to be a fundamental and difficult problem. In 1998 Morgan and Higgs proposed a heuristic algorithm based on the shortest path between the two structures, and in DNA 10, Kubota and Hagiya did an exact algorithm based on the flooding. The former runs in a practical time for sufficiently large length n of molecule and would always show a good performance if the barrier always appeared near the shortest path. The only but crucial drawback of the latter on the other hand is that it cannot run for a large n; we found an instance of even length $n = 46$ for which the run did not complete because of the memory. In this paper we formulate it as an optimization problem, and then propose a new heuristic algorithm based on the local search strategy. We use the Morgan and Higgs' heuristics as the engine to find a locally optimal solution, and based on the local search paradigm, we repeat this local search starting from the solution of the previous local search, with the hope that this sequence of improvements will eventually reach the optimum solution. We also discuss some techniques to improve the performance. We demonstrate that for size about 200, our algorithm runs in 5 seconds, and for many of the cases (13 cases out of 16) in which the Kubota and Hagiya's algorithm can complete, our algorithm exactly answers the optimum values.

1 Introduction

Molecular/DNA computing is a computing paradigm that utilizes molecular reactions as state transitions; computation is implemented by conformational change or structural formation of DNA molecules [3]. In this sense, how to control reactions of DNA molecules is a fundamental issue in DNA computing, and understanding how DNA molecular structures change is indispensable to design DNA molecules.

---

* This work was supported by the Scientific Grant-in-Aid by the Ministry of Education, Science, Sports and Culture of Japan.
** The first author is now working on Fujitsu Limited.
*** Corresponding author.

A. Carbone and N.A. Pierce (Eds.): DNA11, LNCS 3892, pp. 359–370, 2006.

The energy barrier is one of the clearest barometers of the efficiency / inefficiency of transitions between two structures; the higher (resp., lower) the energy barrier of the transition between the two structures is, the slower (resp., faster) the corresponding reaction (transition) would be. In fact, several DNA computing machines based on the heights of the energy barriers have been proposed. For example, Uejima et al. [9] designed the "multi-state DNA machine" which sequentially changes the formation of repeating open DNA hairpin structures by extending the fuel DNA model [12]. The computation utilizes the differences of the heights of the energy barriers among several structures.

However, it is not easy to find the energy barrier between two given structures, because the barrier is defined by all the intermediate structures of all the transition paths, the number of which is exponential of the length of the DNA sequence [2]. For the problem, Morgan and Higgs proposed a heuristic algorithm (Morgan and Higgs' heuristics, MH heuristics, for short) based on the shortest path between the two structures in 1998 [7]. MH heuristics runs in a practical time even for sufficiently large length n of DNA sequence since MH heuristics assumes that the barrier would appear near the shortest path and searches areas near the shortest path in a greedy manner. However, if the path of the barrier are quite far from the shortest path, the performance is of course not necessarily good. To overcome this, Kubota and Hagiya proposed an exact algorithm, called *the minimum basin algorithm* (MB algorithm, for short), in DNA 10 [5]. The idea of the algorithm is based on the flooding algorithm [8] that captures the landscape of all the secondary structures, though the algorithm does not search the whole landscape but only the landscape near the given structures. They implemented the algorithm and applied it to a DNA sequence with length 154, and the run time is just a few seconds. However, the algorithm is sensitive not only to the length of the sequence but also to the landscape itself, e.g., the number of stable structures; we found an instance of a DNA sequence with length $n = 46$ and structures for which the run halts without outputting the result after a few days' computation, because of the memory shortage.

In this paper, we formulate the problem as a combinatorial optimization problem, and apply a popular metaheuristics of the local search [10]. The generic local search is as follows: Start from an initial solution and repeat replacing with a better solution in the the *neighborhood* of the current solution [1]. Since the strategy is quite natural and simple, many researchers propose local search-based algorithms for many combinatorial optimization problems, but the "performances" are not always good; the design of a good local search algorithm is not trivial. For example, if we adopt a very large neighborhood, the algorithm might find a very good solution, but the algorithm would require a very large amount of run time. (If the neighborhood is whole the solution space, the "local search" algorithm will find the optimal solution, but useless.) Contrarily, if we adopt a very small neighborhood, each neighborhood search will finish very quickly, but it would be difficult to find a better solution. (If the neighborhood is just the current solution itself, the search will immediately finish, but also useless.) These observations imply that it is important to adopt an adequate neighborhood and an efficient neighborhood search in order to design practically good local search algorithms.

Taking these into consideration, we propose a local search algorithm for solving this problem. The key idea of our local search is to consider that the solution space is the set of (s, f)-paths (transitions) where s is the initial structure and f is the final structure. In this solution space, an (s, f)-path is considered a collection of several sub-paths each of which is a "shortest path" from a structure to another structure. This device enables to control the size of the solution spaces; an adequate size of the neighborhood is defined. Another point is that we best utilize MH heuristics for an efficient neighborhood search algorithm: By using the property that paths found by MH heuristics in sequential searches are similar, we present an efficient neighborhood search.

We then conduct computational experiments. We apply our algorithm to 24 instances with length 46, 100, 154 and 194. Some of them are real DNA sequences, i.e., they were used in bio-experiments, and the others are randomly generated. For comparison, we also apply MB algorithm, which solved 16 instances (i.e., found the optimal energy barrier), though for the rest 8 instances it could not find the solutions because of the memory problem and the run times are not practical even for some instances of the solved 16 ones; it takes a few days to find the solution. On the other hand, our algorithm found the solutions for all of the above 24 instances in a few seconds, and the quality of the solutions is good; actually in total 17 instances are guaranteed to be optimally solved [1]. These experimental results show that our algorithm has a sufficiently good performance for the practical estimation of the energy barrier.

The remainder of the paper is organized as follows: In Section 2, we formulate the problem of estimating the energy barrier and present the idea of MH heuristics. In Section 3, we propose the local search algorithm for the problem, and Section 4 discusses some techniques for accelerating the neighborhood search. In Section 5, we report the experimental results, and then Section 6 concludes this paper.

2 Preliminaries

2.1 The Transition Path and the Barrier Height

Let us consider a DNA molecule X that undergoes a change from one conformation s_0 (initial structure) to another s_f (final structure). For a molecular or a sequence X, $S(X)$ denotes the set of all the secondary structures formed by X. A structural transition path p from s_0 to s_f is a series of structures

$$p = \langle s_0, s_1, ..., s_f \rangle,$$

---

[1] MB algorithm guaranteed the optimality of the 13 instances among the 17, and the rest 4 are guaranteed by the estimation of the lower bound of the energy barrier, although we omit the detail due to the space limitation.

where s_i conformationally changes into s_{i+1} by either of "addition of one base pair" or "removal of one base pair". We define 1 transition step as the transition from s_i to s_{i+1} . A transition path whose step size is the shortest is called a *direct path*. We denote the set of all the structural transition paths from s_0 to s_f by $P(s_0, s_f)$.

The free energy of a structure s is denoted by $E(s)$. The top energy, or simply the energy, of a transition path $p = \langle s_0, s_1, ..., s_f \rangle$ is defined by

$$E_{top}(p) = \max\{E(s_i) \mid s_i \in p = \langle s_0, s_1, ..., s_f \rangle\}.$$

The energy barrier $E_{bar}(s_0, s_f)$ of the structural transitions between s_0 and s_f is defined by

$$E_{bar}(s_0, s_f) = \min\{E_{top}(p) \mid p \in P(s_0, s_f)\}.$$

The energy barrier estimation is a problem of finding a path p that minimizes $E_{top}(p)$ for given a sequence X, an initial structure s_0, and a final structure s_f. As mentioned in Section 1, the size of $P(s_0, s_f)$ is an exponential of the length of X. This means that estimating $E_{bar}(s_0, s_f)$ is non-trivial.

2.2   MH Heuristics

In this subsection, we present a key idea of MH heuristics, since our local search algorithm searches the solution spaces defined by MH heuristics; The local search repeatedly invokes MH heuristics.

MH heuristics searches for a direct path whose intermediate structures have as many base pairs as possible. The heuristics consists of 2 steps: One is *remove step* in which a base pair in s_0 that does not exist in s_f (*incompatible pair*, say) is removed, and the other is *add step* in which a base pair in s_f that does not exist in s_0 is added. (See Fig. 1.)

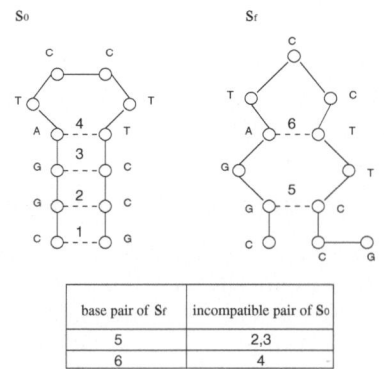

base pair of S_f	incompatible pair of S_0
5	2,3
6	4

Fig. 1. An example of incompatible pairs

Morgan and Higgs' Heuristics [7]

1. Find the pair in s_f which has the least number of incompatible pairs in s_0.[2]
2. Remove the incompatible pair found in 1. from s_0, and add a pair in s_f to s_0 if possible[2]. This creates a new structure s'.
3. Repeat this procedure with s' instead of s_0, until the intermediate structure is transformed into the final structure s_f.

The policy of MH heuristics is based on the hypothesis that if the structure has the more base pairs then it is more stable (i.e., the free energy is lower).

2.3   The Local Search

The local search is one of the most popular metaheuristics approaches, and many other metaheuristics are considered to be based on the local search algorithm. Also, it is reported to be quite practical for many combinatorial optimization problems [1, 10].

The local search starts from an initial solution x and replaces x with a better solution x' on condition that evaluate function $g(x') < g(x)$ in its *neighborhood* $N(x)$, where $N(x)$ is a set of solutions obtainable by slight perturbations. It repeats replacing until no better solution is found in $N(x)$. The solution found eventually is called *locally optimal* (Fig.2).

As mentioned in Introduction, the performance of a local search algorithm highly depends on the neighborhood used and the neighborhood search (algorithm). Thus the points of this paper are how adequate the proposed neighborhood is (Section 3) and how elaborate the algorithm searches the neighborhood (Section 4).

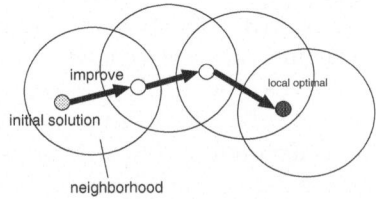

Fig. 2. Local Search

3   A Local Search for Barrier Height Estimation

3.1   Formulation: Barrier Height Estimation Problem

In this section, we propose a local search algorithm solving the barrier height estimation problem. First, we formulate the problem as a combinatorial

---

[2] If there are several pairs with an equal number, choose one that reduces the energy maximally or increases the energy minimally one by one.

optimization problem. Here, we consider a structural transition path p a solution itself, and use $E_{top}(p)$ as evaluate function. The problem is formally described as follows:

Problem: Barrier Height Estimation (BHE)

Input : DNA sequence X, initial structure s_0 and final structure s_f.
Goal : Find a $p \in P(s_0, s_f)$ that gives $E_{bar}(s_0, s_f)$, i.e.,
$$E_{top}(p) = \max\{E_{top}(q) \mid q \in P(s_0, s_f)\}$$

3.2 Neighborhood Solutions

Under the above formulation, we give a basic idea of our local search algorithm as follows.

Algorithm Local Search for BHE:

Step 1. Generate a transition path p_0 from s_0 to s_f. Set $p := p_0$.
Step 2. Search for a transition path $p' \in N(p)$ such that $E_{top}(p') <$
    $E_{top}(p)$. If such p' exists, set $p := p'$ and return to step2.
Step 3. Output p and stop.

In Step 1., we need to find a transition path p_0 as an initial solution. Here, we find a path by using MH heuristics. We then consider how we define the "neigborhood paths" $N(p)$ in Step 2. If we use sufficiently large $N(p)$, then the solution found might be optimal. However, it is not practical because very large $N(p)$ requires quite a large amount of time to search the neighborhood. Thus we consider to define a suitable neighborhood to be searched. For this purpose, we again use MH heuristics.

For two structures s_i and s_j, we define $MH(s_i, s_j)$ as the path from s_i to s_j which is the solution of MH heuristics with initial structure s_i and final structure s_j. Also we define the *replacement* of the transition path $p = \langle s_0, \ldots, s_i, s_{i+1}, \ldots, s_{j-1}, s_j, \ldots, s_f \rangle$ with $p' = \langle s_i, s'_{i+1}, \ldots, s'_{j-1}, s_j \rangle$, as $p'' = \langle s_0, \ldots, s_i, s'_{i+1}, \ldots, s'_{j-1}, s_j, \ldots, s_f \rangle$, and we simply write this $p'' = \langle p[0, i-1], p', p[j+1, f] \rangle$, where $p[a, b]$ is denoted by $\langle s_a, \ldots, s_b \rangle$ part of p.

Now we consider to generate a new transition path from the current transition path p. Since the new transition path p' is better to have a lower top energy, the new path should not contain structure $top(p)$, where $top(p)$ denotes the structure s_h satisfying $E(s_h) = E_{top}(p)$. Namely, a new path p' should be a path avoiding $top(p)$. To obtain a path having this property, we utilize MH heuristics and define *k–back neighborhood* as

$$N(p)_k = \bigcup_{h-k+1 \leq i \leq h} \{\langle p[0, i-1], MH(s'_i, s_f) \rangle \mid s'_i \neq s_i.\},$$

where $s_h = top(p)$. Note that s'_i is a structural transformed from s_{i-1} by 1 step (Fig. 3). By using this, we define the following three neighborhood searches:

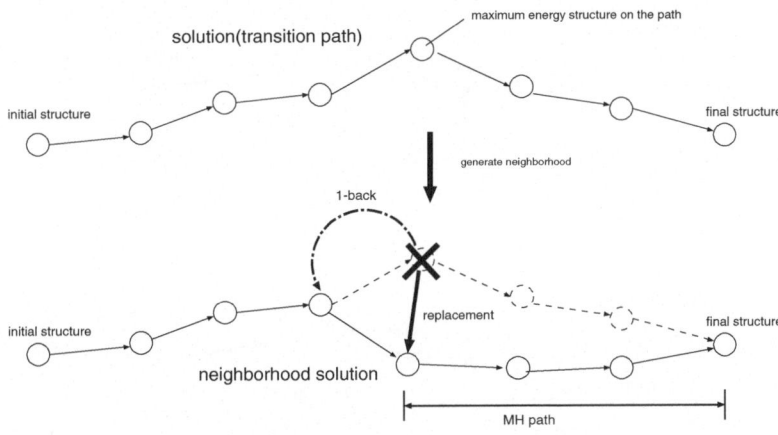

Fig. 3. Generate the Neighborhood Solution

- **1–back neighborhood search**
  Its search space is 1–back neighborhood, and it repeats the search until no better solution is found.
- **Full–back neighborhood search**
  Its search space is Full–back neighborhood $(N(p)_{Full} = \{p'(p, s_i) \mid 1 \leq i \leq h)\}$, and it repeats the search until no better solution is found.
- **VD–back neighborhood search**
  First, the search space is set 1–back neighborhood. If no better solution is found in neighborhood, then it is enlarged to be 2–back, 3–back, one after another. When a better solution is found, the search space is set 1–back neighborhood again (This scheme is called *variable depth search* [11]).

VD–back neighborhood search

1. $k := 1$.
2. If no better solution is found in k–back neighborhood, go to 3. Otherwise go to 1.
3. If the neighborhood is not able to be enlarged, stop. Otherwise $k := k + 1$ and go to 2.

Note that we employ MH heuristics for each neigborhood search.

4 Accelerating the Neighborhood Search

In this section, we discuss the possibility of the speeding up the proposed local search and the techniques. In the local search, the current solution and the previous solution have similar solution structures in general. This means that the current and the previous solutions (transition paths) may have some common partial paths in our local search. Here, by using this property, we consider to accelerate our local search. Figure 4 provides an idea of the acceleration. The upper two figures show how the original local search searches the paths. The

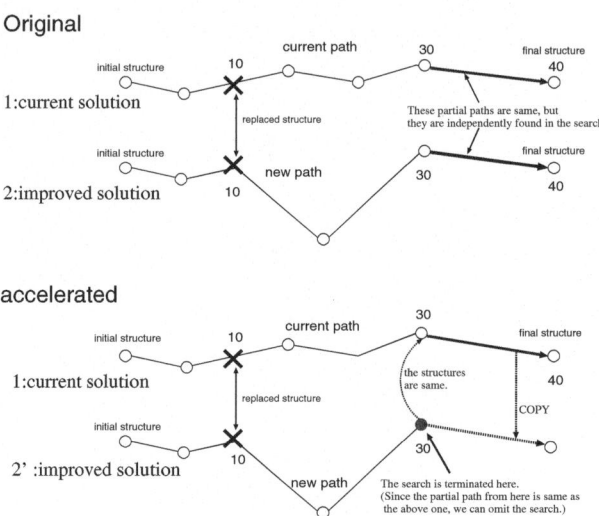

Fig. 4. Same Path Omission

numbers (e.g., 10, 30, and 40) mean step numbers in the structural transition. Suppose that Path 1 is a current solution and Path 2 is the next (improved) solution. Path 2 is obtained by replacing node 10 and then applying MH heuristics for structural transition between step 10 to the final structure.

In the example, the partial path from step 30 to 40 in path 1 and the one of path 2 have the same structures (i.e., the partial paths are common). Actually, this kind of phenomenon can be frequently seen. This is because of the property of MH heuristics, which is the engine of our local search. The paths found by MH heuristics have the following property: For two transition paths $p_a = \langle s_a, \ldots, s_f \rangle$ and $p_b = \langle s_b, \ldots, s_f \rangle$ by MH heuristics, if there exists a structure s such that $s \in p_a$ and $s \in p_b$ (i.e., $p_a = \langle s_a, \ldots, s, \ldots, s_f \rangle$ and $p_b = \langle s_b \ldots, s \ldots, s_f \rangle$), then

$$p_a = \langle s_a, \ldots, s'_a, MH(s, s_f) \rangle \text{ and } p_b = \langle s_b, \ldots, s'_b, MH(s, s_f) \rangle$$

hold.

From the observation, we consider the device of accelerating the local search by omitting the search of the partial paths as mentioned above. (See the lower two figures of Fig. 4.)

How to Omit the Partial Path Search

Here, we consider how the above omission is performed. For the omission, first we need to find a path that has a same partial path with the previous solution (structural transition), that is, we need to check the correspondence between each structure in the previous solution and the structure which MH heuristics is now constructing (expanding). The easiest way of checking the correspondence between two structures is to match structures sequentially, though it is time-consuming. Thus, we elaborate the method of checking the correspondence efficiently.

The point is to utilize the property of MH heuristics again. MH heuristics constructs the transition path in the order of adding and removing base pairs which is defined by the numbers of incompatible base pairs. Instead of matching the structures sequentially, we just compare the numbers of incompatible base pairs, which greatly reduces the comparison time. To describe the idea, we introduce an *incompatible-table* as follows: Let m denote the number of base pairs in the final structure s_f, and associate the number $1, ..., m$ to the base pairs. Given a structure s, we define incompatible-table $IT(s, s_f)$ of s, by the number t_i of incompatible base pairs of the base pair i, for $i = 1, \ldots, m$. That is,

$$IT(s, s_f) = (t_1(s), t_2(s), \ldots, t_m(s)).$$

Note that the m is much smaller than n, the length of DNA molecule. Roughly speaking, for two given structures s and s', we can judge if $s = s'$ by comparing $IT(s, s_f)$ and $IT(s', s_f)$ instead of s and s' themselves [3].

5   Computational Experiments

To evaluate the performance of the proposed local search, we conducted computational experiments. In the experiments, we used a version of Vienna package modified by Uejima et al. [9] to calculate the free energy of DNA structures. The Vienna RNA Package is developed by Vienna university and distributed[4] [4].

Table 1 shows the instances we used. "instance" and "sequence length" columns represent the IDs of instances and their length. For example, 46(1–5) and 46 mean that we have instances of ID 46(1), 46(2), ..., 46(5) and their length is 46, respectively. "initial structure" and "final structure" represent how the initial/final structure of the instance is given: "random" means that the pairs of the structure are randomly chosen, "real" means real DNA structure used in [6], and "mes" means the structure with the minimum free energy. In the initial structures of 100(6), 154(8), "min(x)+min(y)" means the combination of two minimum free energy structures with length x and y.

We then implemented the local search algorithm by 1–back and Full–back or VD–back neighborhood search. We also run MB algorithm for comparison. The programs are run on Pentium 4 CPU 2.4GHz with 512MB memory.

Table 2 shows the experimental results by 1–back, Full–back and the VD–back neighborhood search after generating an initial solution by MH heuristics for the instances in Table 1. Also the results by MB algorithm are shown. The time column shows the run time (second). The column E_{top}(kcal/mol) shows the maximum energy on the path found by the algorithm and the column initial shows that the value of E_{top} of the initial solution (i.e., the value by MH heuristics). The hyphens in the results of MB algorithm mean that the result could not be obtained because of the memory shortage. The energy value in boldfaced

---

[3] More precisely, there are several additional conditions to affirm that they are exactly same structures, though they can be quickly checked anyway.

[4] http://www.tbi.univie.ac.at/ivo/RNA

Table 1. Instances

instance	sequence length	initial structure	final structure
46(1–5)	46	random	mes
46(6–7)	46	random	random
100(1–5)	100	random	mes
100(6)	100	min(50)+min(50)	mes
154(1–5)	154	random	mes
154(6–7)	154	real	mes
154(8)	154	min(77)+min(77)	mes
194(1)	194	real	mes
194(2–3)	194	random	random

Table 2. Run Time(s), the Value of $E_{top}(p)$(kcal/mol) and Lower Bound

instance	initial $E_{top}(p)$	time	1–back $E_{top}(p)$	time	Full–back $E_{top}(p)$	time	VD–back $E_{top}(p)$	minimum basin time	$E_{top}(p)$	lower bound
46(1)	33.02	0.08	28.60	0.08	28.60	0.07	28.60	0.11	28.60	28.60
46(2)	21.27	0.08	18.35	0.08	18.35	0.09	18.35	0.14	18.35	18.35
46(3)	28.81	0.07	27.24	0.08	27.24	0.08	27.24	0.14	27.24	27.24
46(4)	34.41	0.08	31.33	0.08	31.33	0.08	31.33	0.13	31.33	31.33
46(5)	18.36	0.09	14.30	0.09	14.30	0.09	14.30	0.14	14.30	14.30
46(6)	33.91	0.07	28.60	0.09	28.60	0.08	28.60	-	-	28.60
46(7)	30.92	0.08	21.71	0.09	19.56	0.08	21.71	-	-	18.35
100(1)	70.89	0.12	63.32	0.13	63.32	0.11	63.32	0.50	63.32	63.32
100(2)	59.24	0.12	52.09	0.14	52.09	0.12	52.09	3.22	52.09	52.09
100(3)	50.20	0.11	44.50	0.11	44.50	0.10	44.50	4.74	44.50	44.50
100(4)	58.62	0.11	52.55	0.10	52.55	0.10	52.55	0.74	52.55	52.55
100(5)	68.23	0.11	64.36	0.11	64.36	0.11	64.36	0.43	64.36	64.36
100(6)	3.32	0.17	-1.34	0.18	-1.34	0.18	-1.34	0.17	-1.34	-1.95
154(1)	73.80	0.28	69.39	0.28	63.39	0.28	69.39	1.27	69.39	69.39
154(2)	80.23	0.87	73.85	0.87	73.85	0.87	73.85	-	-	73.85
154(3)	87.06	0.85	80.89	1.03	80.89	0.84	80.89	-	-	80.89
154(4)	68.92	0.48	65.18	0.48	65.18	0.48	65.18	-	-	65.18
154(5)	87.55	0.44	75.82	0.44	75.82	0.44	75.82	11.85	75.82	75.82
154(6)	-43.49	0.66	-46.27	0.65	-46.27	0.65	-46.27	1.94	-48.41	-48.78
154(7)	-43.43	1.46	-46.71	2.43	-46.71	2.03	-46.71	262.85	-48.92	-51.49
154(8)	-35.13	1.08	-36.01	2.36	-36.01	2.06	-36.01	133.84	-38.34	-42.03
194(1)	-47.48	3.61	-52.73	3.62	-52.73	3.61	-52.73	-	-	-53.70
194(2)	109.01	0.22	102.43	3.59	102.23	3.28	102.23	-	-	93.78
194(3)	102.53	0.71	95.79	6.78	95.79	1.85	95.79	-	-	90.43

type means that the corresponding result is optimal, i.e., the true energy barrier height is found. The last column "lower bound" shows lower bounds of the energy barrier obtained by another kind of approximation algorithms.[5] Even in the case where the optimal value cannot be found, such as 154 (2), (3), (4), if the solution value coincides with the "lower bound", the solution is confirmed to be optimal.

---

[5] The energy barrier can be found also by solving the dual problem of the barrier height estimation, which implies that approximate solutions of the dual problem give lower bounds on the solutions of the (primal) barrier height estimation, though we omit the detailed explanation. The values of the last column "lower bound" were obtained by calculating the minimum energy value of a cut-set between the initial and the final structures approximately.

Comparison with MB Algorithm

Although MB algorithm found the solutions for many small instances (e.g., 46(1)–46(5) and 100(4)–100(6)) in very short time, there are several instances of which MB could not find the solution. On the other hand, all the variations of the proposed local search find the solution for most cases in less than 1 second, and at most 7 seconds. Also the run time of the proposed algorithm is not so influenced by the growth of the instance length, while the one of MB algorithm is not. As for the solution quality, our algorithm found the optimal solutions for 17 instances among 24 instances, while the number of the instances optimally solved by MB algorithm is 16. Even in the cases where our algorithm could not find the optimal solution, the differences from the optimal values are quite small.

Difference by Neighborhood Search

The differences of both the run times and the solution qualities between 1–back, Full–back and VD–back neighborhood search are small for the short length instances. For the larger instances, also the solution qualities have almost no differences, but Full–back is slightly better than 1–back. As for the run times, Full–back tend to be a little larger than 1–back and VD–back, but also the differences are small.

The readers might feel strange, because the differences of the neighborhood sizes would cause the differences of (at least) the run times. However, it is not necessarily true because of the following reasons: One Full–back neighborhood search may tend to require larger run time than one 1–back, but the number of improvements of Full-back neighborhood search may tend to be much less than the ones of 1–back search. Of course, the total amount of run time depends on not only the size of the neighborhood but also the number of the improvements. Also, Full–back neighborhood search may be not so different from 1–back neighborhood search, when the maximum energy structure is very near to the initial structure.

Anyway, there might exist a small trade-off between the run time and the solution quality, though we need more experimental studies to confirm the relationship.

6   Conclusion

In this paper, we proposed a local-search based heuristic algorithm for barrier height estimation problem, which uses well-known Morgan Higgs' heuristics algorithm as the engine. Experimental results for synthetic/real DNA sequences show that the proposed algorithm can optimally find solutions for many instances some of which cannot be solved by the existing exact algorithm, the minimum basin algorithm. Also the run time is fast enough and it completes the calculation within a few seconds for most cases. The authors believe that the proposed algorithm is practically useful for the sequence design in molecular/DNA computing fields.

Acknowledgements

The authors would like to thank Professor Masami Hagiya and Mr. Mitsuhiro Kubota of University of Tokyo, who suggested the problem and gave the source code of the minimum basin algorithm [5], and the anonymous reviewers for their helpful comments which improved the presentation of this paper.

References

1. E. H. L. Aarts, J. K. Lenstra, eds., *"Local Search in Combinatorial Optimization"*, Chichester England: Wiley, 1997.
2. J. Cupal, C. Flamm, P.F. Stadler *"Density of States, Metastable States, and Saddle Points Exploring the Energy Landscape of an RNA Molecule"*, ISMB 1997, pp. 88-91, 1997.
3. M. Hagiya *"Towards Molecular Programming"*, Modelling in Molecular Biology (G. Ciobanu, G. Rozenberg, Eds.), Natural Computing Series, Springer, pp. 125-140, 2004.
4. I.L. Hofacker, W. Fontana, P.F. Stadler, S. Bonhoeffer, M. tacker, P. Schuster *"Fast Folding and Comparison of RNA Secondary Structures"*, Monatsh. Chem. 125, pp. 167-188, 1994.
5. M. Kubota, M. Hagiya, *"Minimum Basin Algorithm: An Effective Analysis Technique for DNA Energy Landscapes"*, 10th International Workshop on DNA Computing, DNA10, LNCS 3384, pp. 202-214, 2005.
6. M. Kubota, K. Ohtake, K. Komiya, K. Sakamoto, M. Hagiya, *"Branching DNA Machines Based on Transitions of Hairpin Structures"*, CEC2003, pp. 2542-2548, 2003.
7. S. R. Morgan, P. G. Higgs, *"Barrier heights between ground states in a model of RNA secondary structure"*, J.Phys.A: Math. Gen. 31, pp. 3153-3170, 1998.
8. P. F. Stadler, C. Flamm, *"Barrier Trees on Poset-Valued Landscapes"*, J. Gen. Prog. Evol. Machines 4, pp. 7-20, 2003.
9. H. Uejima, M. Hagiya *"Secondary Structure Design of Multi-state DNA Machine Based on Sequential Structure Transitions"*, DNA9, Springer LNCS 2943 pp. 74-85, 2004.
10. M. Yagiura, T. Ibaraki, *"On Metaheuristic Algorithms for Combinatorial Optimization Problems"*, Systems and Computers in Japan 32(3), pp. 33-55, 2001.
11. M. Yagiura, T. Yamaguchi, T. Ibaraki, *"A Variable Depth Search Algorithm for the Generalized Assignment Problem"*, in: S. Voss, S. Martello, I.H. Osman and C. Roucairol, eds., Meta-Heuristics: Advances and Trends in Local Search Paradigms for Optimization, Kluwer Acad. Publ., pp. 459-471, 1999.
12. B. Yurke, A. J. Turberfield, A. P. Mills, Jr., F. C. Simmel and J. L . Neumann, *"A DNA-fuelled molecular machine made of DNA"*, Nature 406, pp. 605-608, 2000.

Specificity of Hybridization Between DNA Sequences Based on Free Energy

Fumiaki Tanaka[1], Atsushi Kameda[2],
Masahito Yamamoto[2,3], and Azuma Ohuchi[2,3]

[1] Graduate School of Engineering, Hokkaido University,
North 13, West 8, Kita-ku, Sapporo 060-8628, Japan
[2] CREST, Japan Science and Technology Corporation,
4-1-8, Honmachi, Kawaguchi, Saitama 332-0012, Japan
[3] Graduate School of Information Science and Technology, Hokkaido University,
North 14, West 9, Kita-ku, Sapporo 060-0814, Japan
{fumiaki, kameda, masahito, ohuchi}@dna-comp.org

Abstract. We investigated the specificity of hybridization based on a minimum free energy (ΔG_{min}) through gel electrophoresis analysis. The analysis, using 94 pairs of sequences with length 20, showed that sequences that hybridize each other can be separated using the constraint $\Delta G_{min} \leq -14.0$, but cannot be separated using the number of base pairs (BP) in the range from 9 to 18. This demonstrates that the ΔG_{min} is superior to the BP in terms of the capability to separate specific from non-specific sequences. Furthermore, the comparison between sequence design based on ΔG_{min} and that based on the BP, done through a computer simulation, showed that the former outperformed the latter in terms of the number of sequences designed successfully as well as the ratio of successfully designed sequences to the total number of sequences checked.

1 Introduction

Sequence design is an essential step towards success in various applications of DNA computing, including DNA-based computation [1][2] and nano-fabrication [3][4]. Many efforts have been made to design a set of sequences that hybridize only with their complementaries based on the Hamming distance (i.e., the number of base pairs, BP) [5][6][7] or the minimum free energy (ΔG_{min}) [8]. Although many algorithms have been proposed for sequences whose BP or ΔG_{min} values exceed a threshold for satisfactory hybridization specificity, the threshold itself is still unknown. Furthermore, it is not known whether sequence design with the appropriate threshold is best based on the BP or on ΔG_{min}.

We have investigated the specificity of hybridization by analyzing 94 pairs of sequences with length 20 using gel electrophoresis based on the BP or ΔG_{min}. Based on this experiment, we estimated the thresholds that the BP and ΔG_{min} must reach to enable satisfactory hybridization specificity under regulated conditions such as where two sequences hybridize each other with a 1:1 concentration ratio. We then compared the number of sequences that can be designed

A. Carbone and N.A. Pierce (Eds.): DNA11, LNCS 3892, pp. 371–379, 2006.
© Springer-Verlag Berlin Heidelberg 2006

based on the BP with that based on ΔG_{min} with these thresholds. Furthermore, through the gel electrophoresis analysis, we found that sequences containing sub-sequence "GGGGG" formed an unintended structure, which appeared to be the four-stranded G4-DNA structure [9]. To confirm that this structure was formed by interaction between GGGGG and GGGGG, we analyzed mutated sequences obtained by changing the sequences with GGGGG.

2   Materials and Methods

Two sequences were hybridized with each other and then analyzed through gel electrophoresis to investigate the hybridization specificity. Because gel electrophoresis does not require an enzyme reaction (e.g., kination and PCR), we can investigate the hybridization specificity while avoiding the influence of extra experimental steps. We checked whether two sequences, A and B, hybridized each other as follows. Sequences A, B, and $A + B$ with a 1:1 concentration ratio were electrophoresed through a 10% polyacrylamide gel. If the band in the lane for $A + B$ corresponded to neither the band in the lane for A nor that for B, we assumed A and B hybridized each other (Figure 1). Thus, if any extra bands in the lane for $A + B$ were observed by eye, we classified the outcome as "hybridize"; otherwise, we classified it as "non-hybridize". However, the double strand between A and B could break down into two single strands during the gel electrophoresis, so we had to take into account that we could not distinguish these from the sequences that did not hybridize with each other. Although this simple model only focuses on the hybridization between two sequences without any competitive sequences, the sequences found to hybridize in the experiment are likely to be harmful even under other conditions. Therefore, it is better to avoid such sequences to avoid blocking a specific hybridization.

The BP between sequences A with length n and B with length m is defined as

$$BP := \min(n, m) - \min_{-m < k < n} H(A, \sigma^k(\overline{B})), \tag{1}$$

where $H(*, *)$ denotes the Hamming distance, σ^k denotes the right (left) bit shift in the case of $k > 0$ $(k < 0)$, k denotes the number of the shift, and $\overline{B}$ denotes the reverse complementary of B. Note that the BP is equivalent to the H-measure proposed by Garzon *et al.* [10] in the case of $n = m$.

We calculated ΔG_{min} between two sequences using the extended algorithm for the ΔG_{min} calculation of a single strand [11]. The calculation was done as reported previously [8].

We analyzed 94 pairs of sequences with length 20 having various values of ΔG_{min} for each BP in a range from 9 to 18. The 94 pairs of sequences were chosen as follows. First we randomly generated 100,000 pairs of sequences for each BP through a computer simulation where T_M was in the range $69.58 \leq T_M \leq 72.58$ and the ΔG_{min} between each sequence and itself was greater than or equal to a threshold, -3.0, so that the sequence would not form secondary structures by itself. The T_M values of 69.58 and 72.58 were, respectively, $T_M^{ave} -$ 1.5 and $T_M^{ave} + 1.5$, where T_M^{ave} is the average calculated from 10,000 randomly generated sequences with length 20. The frequency distribution curves in Figure 2 show that the number of sequences with a particular BP varies with ΔG_{min}. We then chose pairs of sequences that would contain the maximum and minimum ΔG_{min} value for each BP. When the BP was 12, for example, the selected sequences included those with $\Delta G_{min} = -0.54$ and those with $\Delta G_{min} = -21.24$, respectively the maximum and minimum from 100,000 pairs of sequences.

Oligonucleotides were supplied by Hokkaido System Science and were synthesized using column purification. All oligonucleotides were dissolved in a buffer containing 1 M NaCl, 10 mM Na_2HPO_4, and 1 mM Na_2EDTA with a pH of 7.0. The oligonucleotide concentrations (Ct) of each sample were determined from the difference in absorbance at 260 nm and that at 320 nm using extinction coefficients calculated from dinucleoside monophosphates and nucleotides [12]. The oligonucleotides were hybridized by increasing the temperature to 90 °C for 10 min and lowering the temperature to 20 °C at heating rates of 0.08 and 0.02 °C/s, respectively. It took about 14 and 58 minutes, respectively, to go from 20 °C to 90 °C and from 90 °C to 20 °C: this is almost the typical protocol for the thermodynamic analysis [13]. All gel electrophoresis profiles were obtained using a 10% polyacrylamide gel in a 1×TAE buffer at 200 V for 35 min. We used 2 μl samples at a concentration of 1 μM. Bands in the gels were dyed using SYBR Gold nucleic acid gel stain for 20 min.

3 Experimental Results

3.1 Specificity of Hybridization Based on BP Versus ΔG_{min}

Figure 1 shows an example where the BP was 14 with length 20 and the sequences used in this example. A pair of sequences with $\Delta G_{min} = -18.64$ or $\Delta G_{min} = -16.41$ formed double strands resulting in a new band, while that with $\Delta G_{min} = -5.39$ or $\Delta G_{min} = -4.49$ remained two single strands with no extra band. Similar experiments were iterated using 94 pairs of sequences containing the above sequences where the BP was in the range from 9 to 18 with length 20.

The results are shown in Figure 2. All the pairs of sequences that hybridized with each other can be separated from the other pairs by the constraint $\Delta G_{min} \leq -14.0$, but these two groups could not be separated using the BP in the range from 9 to 18. Table 1 shows the number of sequences from 100,000 sequences where $\Delta G_{min} \leq -14.0$ for each BP. The BP had to be less than 13 to guarantee that the number of sequence pairs where $\Delta G_{min} \leq -14.0$ would be lower than 5% of the total. These results demonstrate that ΔG_{min} is superior to the BP in terms of the capability to separate specific from non-specific sequences. However, there seemed to be some pairs of sequences that did not hybridize with each other even though $\Delta G_{min} \leq -14.0$ (e.g., the pair of sequences where $\Delta G_{min} = -15.1$

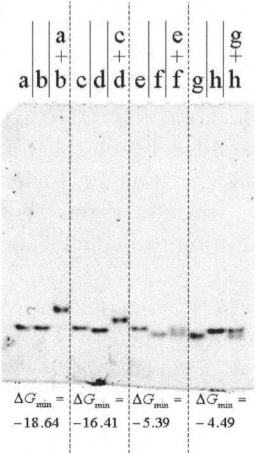

Name	Sequence
a	CACAGTCCCGATTTAGCCAG
b	ACTCAACTGGCTAAATCGGG
c	GAGTGCTTGGGGTCAATTTG
d	ATGACCCAAAGCACTCCTTG
e	ACCTCCCCGTTTATTAAGCA
f	TGATTGAGAAAGCGAGAGGT
g	CATTGTGCGGGATTACAAGC
h	GCGTGTAGTGACCCAAAATG

Fig. 1. LEFT: An example of experimental results from the gel electrophoresis. Four sets of sequences, whose BP was 14, were analyzed. The lanes in each set correspond (from left to right) to a sequence A, a sequence B, and sequences $A + B$. RIGHT: Sequences used in the left figure are listed in the direction 5' to 3' from left to right. The letters correspond to lanes from the gel electrophoresis in the left figure.

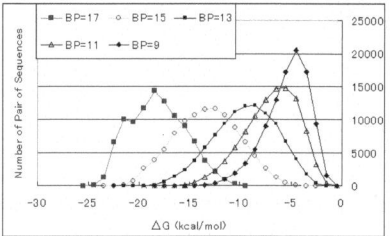
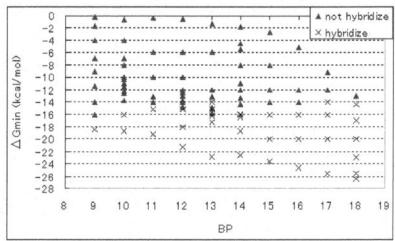

Fig. 2. LEFT: The frequency distribution curve of 100,000 sequences with length 20 for each odd-numbered BP from 9 to 18. RIGHT: Specificity of hybridization based on BP versus ΔG_{min} from the gel electrophoresis analysis using 94 pairs of sequences.

Table 1. Number of sequences out of 100,000 sequences where $\Delta G_{min} \leq -14.0$ for each BP in the range from 9 to 18

BP	9	10	11	12	13	14	15	16	17	18
Number of Sequences	46	179	757	2,934	8,544	20,333	41,587	64,716	92,754	99,944

and the BP was 13). This was probably due to the prediction error for ΔG_{min} and the limit of separability with gel electrophoresis.

Through the above experiment, we found that five single oligonucleotides resulted in unexpected bands on gels with slow mobility. All of these sequences contained sub-sequence "GGGGG", while the others did not. We believe the sequences containing GGGGG may have formed the four-stranded G4-DNA structures [9].

3.2 Sequences Forming Four-Stranded G4-DNA Structures

Sen *et al.* discovered that guanine-rich sequences form four-stranded structures, called G4-DNA, that are linked by Hoogsteen-bonded guanine quartets [9]. In particular, they observed that sequences containing GGGGG formed G4-DNA. In addition, the characteristic feature of G4-DNA is its slow electrophoretic mobility, which is consistent with our results. Thus, we think that the unexpected bands, which were observed through the experiment in previous subsection, were due to interaction between GGGGG and GGGGG.

To confirm this, we analyzed five sets of sequences; each set consisted of a sequence with GGGGG, its complementary with CCCCC, and two mutated sequences. The two mutated sequences were generated as follows. One contained GGGG rather than GGGGG, while the other did not contain base G except for GGGGG (Figure 3). For example, in set 'a' in Figure 3, AAGGGGTTCTATGGT-GTATT and AGGGGGTTCTATACTCTATT were, respectively, the sequence containing GGGG and the sequence containing no Gs except for GGGGG, where the underlined base is the base changed from the sequence AGGGGGTTC-TATGGTGTATT.

Figure 3 shows that sequences with GGGGG formed a structure with slow electrophoretic mobility regardless of the presence of other Gs, while the sequence

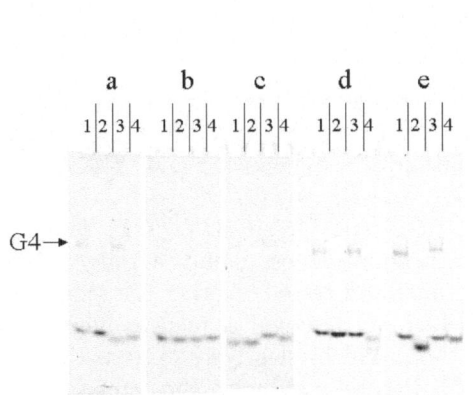

Set	No.	Sequence
	1	AGGGGGTTCTATGGTGTATT
a	2	AAGGGGTTCTATGGTGTATT
	3	AGGGGGTTCTATACTCTATT
	4	AATACACCATAGAACCCCCT
	1	AAAGTTCTCAAAAGAGGGGG
b	2	AAAGTTCTCAAAAGAGGGGC
	3	AAACTTCTCAAAACAGGGGG
	4	CCCCCTCTTTTGAGAACTTT
	1	TCTTGTTATCTCGTAGGGGG
c	2	TCTTGTTATCTCGTAAGGGG
	3	TCTTATTATCTCATAGGGGG
	4	CCCCCTACGAGATAACAAGA
	1	CTCTTGTGGGGGTGTATTTT
d	2	CTCTTGTAGGGGTGTATTTT
	3	CTCTTATGGGGGTATATTTT
	4	AAAATACACCCCCACAAGAG
	1	CCTTAACATTCTAGGGGGGT
e	2	CCTTAACATTCTAGGGGAAT
	3	CCTTAACATTCTACGGGGGT
	4	ACCCCCCTAGAATGTTAAGG

Fig. 3. LEFT: Five sets of sequences consisting of a sequence with GGGGG and mutated sequences were analyzed. In each set, 1: sequence with GGGGG; 2: sequence with GGGG; 3: sequence without G except for GGGGG; 4: complementary of sequence 1 with CCCCC. RIGHT: Sequences used in the left figure are listed in the direction 5' to 3' from left to right. The letters and numbers correspond to lanes for the gel electrophoresis in the left figure. Sequence GGGGG is shown in boldface. The underlined bases show mutated bases from the sequence with GGGGG.

with GGGG and the sequence with CCCCC did not form such a structure. This indicates that the structures of the unexpected bands were formed by interaction between GGGGG and GGGGG.

The structures of the unexpected bands, which we believe are G4-DNA, must compete with the specific hybridization and will be intermediate to the unintended structures. Therefore, we conclude that sequences with GGGGG should be avoided when designing specific sequences.

3.3   Sequence Design Based on ΔG_{min} Versus That Based on BP

To evaluate sequence design based on ΔG_{min}, we compared it with sequence design based on the BP in terms of the number of sequences successfully designed within 10 hours. We designed sequences with length 20 such that $69.58 \leq T_M \leq 72.58$ and $\Delta G_{min} > \Delta G^*_{min}$ ($BP < BP^*$) in the combinations described below, where ΔG^*_{min} and BP^* are the thresholds. We set $\Delta G^*_{min} = -14.0$ and $BP^* = 11$, 12, or 13 based on Figure 2 and Table 1; for $BP = 11$, 12, or 13, there were, respectively, 757, 2,934, or 8,544 pairs of sequences (out of the 100,000 pairs of sequences) where $\Delta G_{min} \leq -14.0$. In the case that n sequences were to be designed, the combinations to be considered for the ΔG_{min} (BP) calculation were as follows.

1. $< U_i, U_j U_k > (0 \leq i, j, k \leq n - 1)$
2. $< U_i, U_j V_k > (0 \leq i, j, k \leq n - 1), i \neq k$
3. $< U_i, V_j U_k > (0 \leq i, j, k \leq n - 1), i \neq j$
4. $< U_i, V_j V_k > (0 \leq i, j, k \leq n - 1), (i \neq j) \wedge (i \neq k),$

where U_i is the $i - th$ sequence, V_j is the complementary of U_j, $X_j X_k$ ($X_j \in \{U_j, V_j\}$, $X_k \in \{U_k, V_k\}$) is the concatenation of sequences X_j and X_k in that order, and $< U_i, X_j X_k >$ is the combination of sequences U_i and $X_j X_k$. For example, if $U_i = CCCCC$, $U_j = AGAGA$, and $U_k = TCTCT$, $< U_i, U_j U_k >$ means the combination of sequences $CCCCC$ and $AGAGATCTCT$. The algorithm for both sequence designs was a random generate-and-test algorithm that generated a candidate sequence randomly and tested whether the sequence satisfied the constraints. Furthermore, when we designed sequences based on ΔG_{min}, we used the ΔG_{gre} filtering method, which effectively excluded inappropriate sequences where $\Delta G_{min} \leq \Delta G^*_{min}$, thereby reducing the computation time. Finally, the sequence design based on ΔG_{min} checked the candidate sequence with the T_M, ΔG_{gre}, and ΔG_{min} filters in that order, while that based on BP checked the candidate sequence with T_M and BP filters in that order (see reference [8] for details). The computational experiments were performed using Turbolinux Workstation 7.0 on a computer with a Pentium 4 2.26-GHz CPU and 256 MB of memory. The experiments were iterated five times with a different seed for the random generator. The results are shown in Table 2. The number of sequences successfully designed based on ΔG_{min} exceeded that based on the BP. This shows that sequence design based on ΔG_{min} is more effective than that based on the BP when designing specific sequences that hybridize with only the complementary.

Table 2. Number of sequences successfully designed within 10 hours based on ΔG_{min} versus BP. The experiments were iterated five times with a different seed for the random generator. In the column $\Delta G_{min} > -14.0$, the numbers in parentheses correspond to the design strategy without ΔG_{gre} filtering. Using ΔG_{gre} filtering enables the design of more sequences. Sequence design based on ΔG_{min} outperformed that based on the BP even without ΔG_{gre} filtering.

Trial	$\Delta G_{min} > -14.0$	$BP < 11$	$BP < 12$	$BP < 13$
1	106 (92)	11	27	64
2	106 (90)	11	27	65
3	96 (87)	13	27	60
4	103 (87)	12	24	62
5	104 (87)	10	25	62
Average	103 (88.6)	11.4	26	62.6
Standard Deviation	4.1 (2.3)	1.1	1.4	1.9

3.4 Comparison Between the Solution Space Based on ΔG_{min} and That Based on BP

Above we demonstrated that more sequences can be successfully designed based on ΔG_{min} than based on the BP. However, the number of sequences that can be designed also depends on the sequence design algorithm. Thus, one might think that sequence design based on the BP with a more sophisticated algorithm might outperform that based on ΔG_{min}. It is difficult to prove that any and all algorithms based on ΔG_{min} are superior to those based on the BP. Instead, we investigated the ratio of successfully designed sequences to the total number of sequences checked because this ratio corresponds to the size of the solution space that can be designed under predefined constraints. Table 3 shows that the ratio of sequences successfully designed based on ΔG_{min} was far larger than that based on the BP (e.g., $2.8\% \gg 1.9 \cdot 10^{-4}\%$). This means that the solution space that can be designed based on ΔG_{min} is undoubtedly larger than that based

Table 3. Ratio of successfully designed sequences to total number of sequences checked. The experiments were iterated five times with a different seed for the random generator. In the column $\Delta G_{min} > -14.0$, the numbers in parentheses correspond to the design strategy without ΔG_{gre} filtering.

Trial	$\Delta G_{min} > -14.0$	$BP < 11$	$BP < 12$	$BP < 13$
1	3.0 (5.1) %	$4.0 \cdot 10^{-6}$ %	$2.3 \cdot 10^{-5}$ %	$2.0 \cdot 10^{-4}$ %
2	2.8 (4.9) %	$1.6 \cdot 10^{-6}$ %	$1.2 \cdot 10^{-5}$ %	$2.2 \cdot 10^{-4}$ %
3	2.4 (4.5) %	$5.9 \cdot 10^{-6}$ %	$1.4 \cdot 10^{-5}$ %	$1.9 \cdot 10^{-4}$ %
4	3.2 (4.2) %	$1.8 \cdot 10^{-6}$ %	$2.3 \cdot 10^{-5}$ %	$1.7 \cdot 10^{-4}$ %
5	2.6 (5.1) %	$1.7 \cdot 10^{-6}$ %	$1.1 \cdot 10^{-5}$ %	$1.9 \cdot 10^{-4}$ %
Average	2.8 (4.8) %	$3.0 \cdot 10^{-6}$ %	$1.7 \cdot 10^{-5}$ %	$1.9 \cdot 10^{-4}$ %
Standard Deviation	0.3 (0.4) %	$1.9 \cdot 10^{-6}$ %	$0.6 \cdot 10^{-5}$ %	$0.2 \cdot 10^{-4}$ %

on the BP. Therefore, although the time complexity of BP is less than that of ΔG_{min}, the number of sequences that can be designed based on ΔG_{min} is greater than that for the BP (Table 2). These results demonstrate the rationality of sequence design based on ΔG_{min}.

4 Conclusions

We conclude that using ΔG_{min} is preferable to using the BP to separate specific hybridization from non-specific hybridization. With an appropriate threshold, sequence design using ΔG_{min} outperformed that using the BP in terms of the number of sequences that could be successfully designed. Comparison of the ratio of successfully designed sequences to the total number of sequences checked showed that the superiority of ΔG_{min} over the BP probably does not depend on the algorithm used for the sequence design.

In addition, our analysis of sequences with GGGGG and their mutated sequences suggested that the sequences with GGGGG formed G4-DNA. Thus, sequences with GGGGG should be avoided when designing specific sequences.

References

1. D Faulhammer, AR Cukras, RJ Lipton, and LF Landweber. Molecular computation: RNA solutions to chess problems. *Proc Natl Acad Sci U S A*, 97(4):1385–9, Feb 2000.
2. Ravinderjit S Braich, Nickolas Chelyapov, Cliff Johnson, Paul W K Rothemund, and Leonard Adleman. Solution of a 20-variable 3-SAT problem on a DNA computer. *Science*, 296(5567):499–502, Apr 2002.
3. Hao Yan, Xiaoping Zhang, Zhiyong Shen, and Nadrian C Seeman. A robust DNA mechanical device controlled by hybridization topology. *Nature*, 415(6867):62–5, Jan 2002.
4. William M Shih, Joel D Quispe, and Gerald F Joyce. A 1.7-kilobase single-stranded DNA that folds into a nanoscale octahedron. *Nature*, 427(6975):618–21, Feb 2004.
5. M. Arita and S. Kobayashi. DNA sequence design using templates. *New Generation Computing*, 20:263–277, 2002.
6. D. Tulpan, H. Hoos, and A. Condon. Stochastic local search algorithms for DNA word design. *Proceeding of 8th International Workshop on DNA-Based Computers, LNCS*, 2568:229–241, 2002.
7. Satoshi Kashiwamura, Atsushi Kameda, Masahito Yamamoto, and Azuma Ohuchi. Two-step search for DNA sequence design. *IEICE TRANSACTIONS on Fundamentals of Electronics, Communications and Computer Sciences Special Section on Papers Slected from 2003 International Technical Conference on Circuts/Systems, Computer and Communications (ITC-CSCC 2003)*, E87-A(6):1446–1453, 2004.
8. Fumiaki Tanaka, Atsushi Kameda, Masahito Yamamoto, and Azuma Ohuchi. Design of nucleic acid sequences for DNA computing based on a thermodynamic approach. *Nucleic Acids Res*, 33(3):903–11, 2005.
9. D Sen and W Gilbert. Formation of parallel four-stranded complexes by guanine-rich motifs in DNA and its implications for meiosis. *Nature*, 334(6180):364–6, Jul 1988.

10. M. Garzon, R. Deaton, P. Neather, D. R. Franceschetti, and R. C. Murphy. A new metric for DNA computing. In *Poceedings of 2nd Annual Genetic Programming Conference*, volume GP-97, pages 472–8, 1997.

11. M Zuker and P Stiegler. Optimal computer folding of large RNA sequences using thermodynamics and auxiliary information. *Nucleic Acids Res*, 9(1):133–48, Jan 1981.

12. DM Gray, SH Hung, and KH Johnson. Absorption and circular dichroism spectroscopy of nucleic acid duplexes and triplexes. *Methods Enzymol*, 246:19–34, 1995.

13. Fumiaki Tanaka, Atsushi Kameda, Masahito Yamamoto, and Azuma Ohuchi. Thermodynamic parameters based on a nearest-neighbor model for DNA sequences with a single-bulge loop. *Biochemistry*, 43(22):7143–50, Jun 2004.

A Poor Man's Microfluidic DNA Computer

Danny van Noort

Biointelligence Lab., School of Computer Science and Engineering,
Seoul National University, San 56-1, Sinlim-dong, Gwanak-gu,
Seoul 151-742, Korea
danny@bi.snu.ac.kr

Abstract. Not everyone has the knowledge or facility to do microfabrication. This paper will show that with a few simple off-the-shelf items a microfluidic DNA computer can be build. It also shows hybridisation and annealing working in such a system. Furthermore, there are some tentative results of a small computing problem. Finally, in the conclusion, it gives some suggestion of ways of automation.

Keywords: microfluidics, DNA hybridisation and denaturing, finemachining, Perspex®, Delrin®, automation.

1 Introduction

Compared to test-tubes, microfluidics has certain advantages. Some of these advantages are the small amount of solution, and therefore DNA and capture probes, high reaction speeds, because of the fast diffusion rate in small scale structures and the possibility for easy automation [1]. However, there are also some disadvantages, the most important being the lack of expertise and/or facilities. Unfortunately, to manufacture microfluidics one needs sophisticated equipment and cleanroom, as well as an operator or technician. These facilities are not always readily available. While test-tubes are cheap and can be easily implemented in DNA computing, automating the process can be expensive (requiring, for example, a pipetting robot [2]), but they don't have the advantages of microsystems.

The alternative is to utilize fine-mechanics, micro-bore tubing and connectors. The latter is readily available off-the-shelf (for example, Upchurch Scientific, Oak Harbor, WA, USA), as are the filters, nuts, frits and ferrules which are necessary for connection and bead retention. A standard fine-mechanics workshop can drill channels in plastic down to a diameter of 250 μm. The most preferred materials are Perspex® (poly(methyl methacrylate)) and Delrin® (Polyoxymethylene) for their inertness to most chemicals. Tubes and channels can be filled with capture probes, immobilised to beads, to perform hybridisation and therefor selection [3]. This paper will show two design possibilities and the ease of connectivity.

2 Experimental Set-Up

Performing Boolean logic in fluidics is straight forward [4, 5]. In principle, one only needs a concatenation of selection probes, whether it is in gels or microreactors filled

A. Carbone and N.A. Pierce (Eds.): DNA11, LNCS 3892, pp. 380–386, 2006.
© Springer-Verlag Berlin Heidelberg 2006

with beads. To perform a selection in the poor man's DNA computer, selection modules can be made from short sections of micro-bore tubing, which are then filled with functionalised beads. The inner dimension of the tubing is truly microfluidic, for example, an 1/32" outer diameter PEEK tubing can be obtained with a 25 μm bore. However, in transparent tubing, TEFLON has the smallest bore of 150 μm. These tube sections are then capped with frits containing inline filters (Fig. 1; Upchurch Scientific, Oak Harbor, WA, USA) to retain the beads in the tube. An OR function is two selection modules in parallel, while an AND is two selection modules in sequential flow, similar to the logic proposed earlier by van Noort et al. [5].

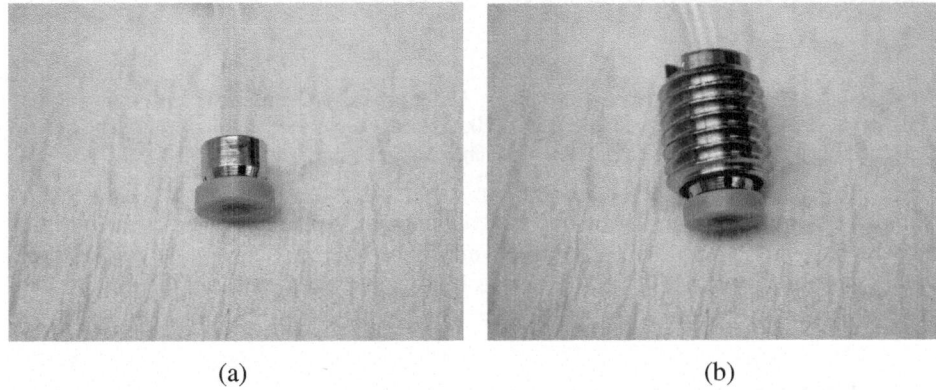

(a)   (b)

Fig. 1. (a) A ferrule with a frit. The plastic part of the ferrule has a filter, i.e. a frit, embedded in it. This can be used to retain beads in tube sections. (b) A nut falls over the ferrule. By tightening the nut, the metal part of the ferrule pushes down on the tube, effectively immobilising the tube.

There are two ways to set up a simple microfluidic system. The first is to buy all the items off-shelf. For instances, an OR function can be made by using two T-junctions, while an AND function is made by connecting two tube sections with a union. Actually, for an AND function, the beads representing the variables can be mixed in one tube section, i.e. selection module. The whole system is then connected to a syringe pump (PHD 2000; Harvard Apparatus, Holliston, MA, USA).

However, a fluidic biocomputer like this, while very flexible and very easy to build up, will be difficult to work with under a microscope. Therefore, I have opted for fine-machined holders. Again, there are two possibilities. Two holders can be used to support tube sections in between (Fig. 2a). These holders were made from Perspex® or Delrin® while the tube sections were capped with filters and connected to the holders with short nuts. The second possibility is to make the holder from Perspex® and fill the fine-machined channel between the connectors with beads (Fig. 2b). In both cases TEFLON tubing (Upchurch Scientific, Oak Harbor, WA, USA) with a 150 μm inner diameter were used to make the connections. Both type of holders were manufactured by the in-house fine-mechanics workshop. And both holder types could be easily placed under the microscope. The advantage of the first holder set is that the selection modules with beads can be easily exchanged, making it a very flexible system, while the advantages of the second system is its compactness.

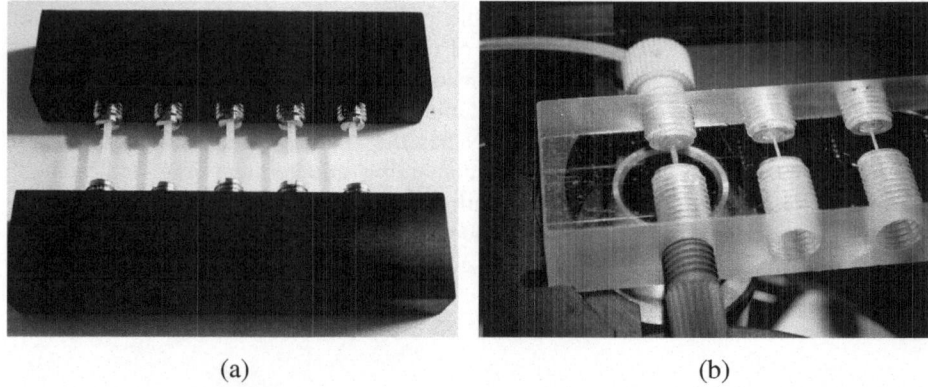

(a)   (b)

Fig. 2. (a) A set of 2 Delrin holders with short Teflon tubes. There are frits with filters at both sides of the tubes. The tubes diameter is 150 μm. (b) A Perspex holder where the micromachined channel is used to retain the beads. The channel's diameter is 250 μm.

Furthermore, the materials have both advantages and disadvantages. Perspex is transparent, while Delrin is easier to fine-machine. The disadvantage is that Perspex, for reasons unknown, can form hair cracks that widen when cleaning the system with acetone, while Delrin is black.

3 Hybridisation and Denaturing

This section will show that hybridisation and denaturing is possible with this setup. Streptavidin functionalised Sepharose beads with a diameter of 34 μm were used (Amersham Bioscience Corp., Piscataway, NJ, USA) as a support for capture probes. Biotinylated single $d(T)_{25}$ strands (Integrated DNA Technologies, Coralville, IA, USA) were immobilised to these beads in a PBS buffer (pH 7.4). Hereafter, they were directly injected into the selection module, capped with 2 μm pore sized filters. 10 μl, 1 μM of $d(A)_{25}$ was loaded into the system, after which the tubing was connected to the syringe pump. The flow rate was 5 μl/min. The hybridisation was visualised by using the intercalater JOJOTM-1 (ex. 529, em. 545; Molecular Probes, Eugene, OR, USA), which is only activated when a double stranded DNA is present.

The holder was placed on an inverted fluorescence microscope (Zeiss, Jena, Germany) and images were taken with a high-resolution peltier cooled monochrome CCD camera (AxioCam, Zeiss, Jena, Germany) at a 1 sec. interval. The images were processed by a programme written in Mathematica® (Wolfram Research Inc.). Figure 3a shows the hybridisation curve under the previous mentioned conditions. The flattening out of the curve was caused by signal saturation.

Figure 3b shows the denaturation curve. Denaturation was achieved by flowing a solution of NaOH at pH 12 through the selection module. The flow rate was again 5 μl/min. The noise in the signal is due to the movement of the beads, as flow had been stopped, NaOH injected and the flow re-started. From the figures it can be seen that while hybridisation occurs in 30 sec, denaturing is a 2 times faster process.

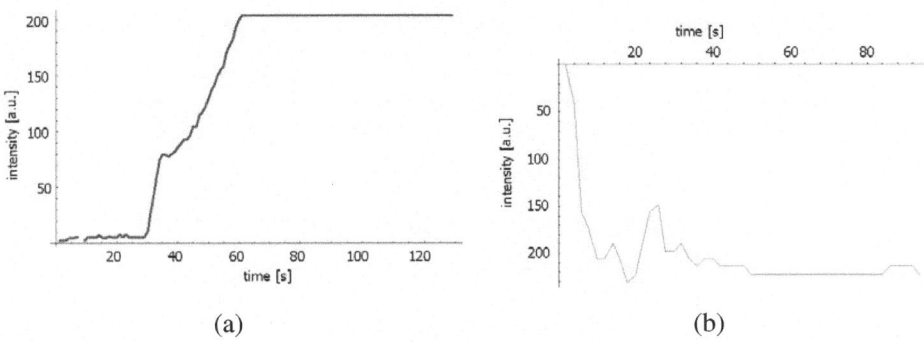

(a)   (b)

Fig. 3. (a) Intensity versus time of the hybridisation of $d(T)_{25}$ functionalised beads with 10 μl of 1 μM single stranded $d(A)_{25}$. (b) Intensity versus time of the denaturing process with NaOH at a pH of 12.

4 Tentative Boolean Problem

A first attempt was made to solve a small 3-bit Boolean problem in the system described above (Fig. 4). A subset of the DNA library reported by Faulhammer et al. [6] was used (Table 1), as it was proven to be a reliable set. The problem was given by:

$$A1 \wedge B0 \wedge C1$$

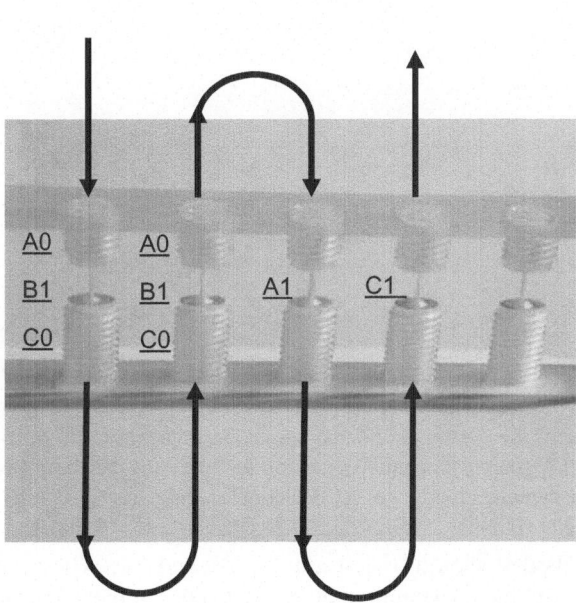

Fig. 4. The connections and placement of the beads with complimentary sequences to solve the simple Boolean equation: $A1 \wedge B0 \wedge C1$

Table 1. The DNA library used in the Boolean problem. A0 mean string A with bit value 0, while <u>A0</u> is the complementary of A0.

A0	CTCTTACTCAATTCT	<u>A0</u>	AGAATTGAGTAAGAG
A1	TCCTCACATTACTTA	<u>A1</u>	TAAGTAATGTGAGGA
B0	CATATCAACATCTTA	<u>B0</u>	TAAGAGGTTGATATG
B1	ACTTCCTTTATATCC	<u>B1</u>	GGATATAAAGGAAGT
C0	ATCCTCCACTTCACA	<u>C0</u>	TGTGAAGTGGAGGAT
C1	TTATAACAAACATCC	<u>C1</u>	GGATGTTTGTTATAA

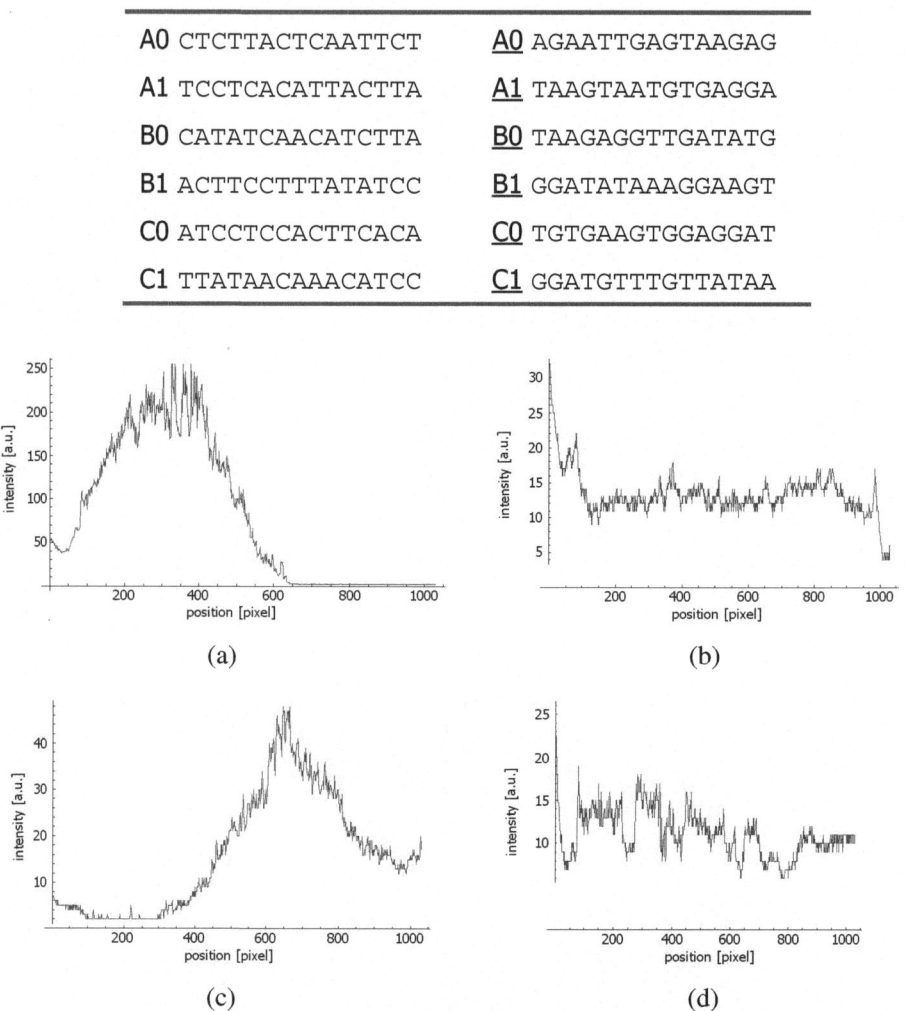

Fig. 5. Intensity versus the position in the modules. The data was taken from over the length of the channel. (a) <u>A0</u>, <u>B1</u> and <u>C0</u> negative selection module. (b) the control of the first module, containing the same capture strands. (c) <u>A1</u> detection module. (d) <u>C1</u> detection module.

The only strands that should be left are A1, B0 and C1. The problem was solved by negative selection, i.e. everything that is not wanted in the solution is hybridised to the capture probes while all others are passed on to the next selection module. This means in order to solve the problem, A0, B1 and C0 should be retained, i.e. captured, by the complimentary sequences <u>A0</u>, <u>B1</u> and <u>C0</u>.

When an AND operation is performed in selection modules, all the variables of that AND can be actually located in only one module, instead of concatenating the single variables. Here, the beads with <u>A0</u>, <u>B1</u> and <u>C0</u> were injected into the firstmodule. As a control for the effectiveness of negative selection, a second selection module was made identically to the first one. After the selection, two of the remaining three strands were detected by positive selection. For this purpose, modules were filled with complimentary strands <u>A1</u> and <u>C1</u>.

The success of the experiment was mixed. The first modules showed a typical hybridisation behaviour, while the second module, the check to see whether negative in the first selection module was successful, showed no significant signal, as expected (Fig. 5a, b). The third module, a detection module capturing A1, showed a weaker signal than the first module, which was expected as well (Fig. 5c). The first and the second module had 3 times more binding capacity compared to the third one, as the third module only had one capture probe, compared to three in the first two. The intensity is collected in a 2 dimensional space, so the intensity of the third module should be 9 times lower than the first. The results show that this was indeed measured. However, the fourth module containing the C1 probe didn't show any signal, while it should have shown the same signal as the A1-module (Fig. 5d).

A reason why the selection procedure didn't work properly is because the operating temperature was at room temperature and not at the melting temperature, at around 40° C. This can cause unpredicted hybridisations.

5   Conclusion

The poor man's microfluidic computer is very promising for groups with limited facilities, and it is not just limited to DNA computing. Off-the-shelf components and small fine-machined holders are needed to make a extremely flexible setup. Any problem size can be handle with these components, while it lends it self for full automation. By using computer controlled selection ports (e.g. Valco Instruments Co. Inc., Houston, TX, USA) and software packages like LabVIEW (National Instruments), selection modules can be programmed, making the computer programmable. Using a standard fluorescence microscope, together with some image processing software, is all what is needed to check the selections. Detections modules can also be incorporated so that there is no need for any gel-electrophoresis.

Acknowledgement

The author wish to acknowledge the support from DARPA award F30602-01-2-0560 to Laura F. Landweber and NSF award 0121405 to Lydia L. Sohn and L. F. L.; the support from the Molecular Evolutionary Computing (MEC) project of the Korean Ministry of Commerce, Industry and Energy, and the National Research Laboratory (NRL) Program from the Korean Ministry of Science and Technology. And thanks to LiChin Wong at Princeton University for the help in preparing the DNA samples.

References

[1] van Noort, D., Tang Z.-L. and Landweber, L. F. (2004) Fully controllable microfluidics for molecular computers. JALA 9, 5 October 2004.

[2] Suyama, A. (2002) Programmable DNA computer with application to mathematical and biological problems. Preliminary Proceedings, Eighth International Meeting on DNA Based Computers, June 10-13, 2002, Japan, 91.

[3] van Noort, D. and Landweber, L. F. (2004) Towards a re-programmable DNA computer. LNCS 2943, 190 – 197.

[4] Adleman, L., 1994. Molecular computation of solutions to combinatorial problems. Science 266, 1021-1024.

[5] van Noort, D., Wagler, P. and McCaskill, J. S. (2002) The role of microreactors in molecular computing. Smart Mater. Struct. 11, 756-760.

[6] Faulhammer, D., Cukras, A. R., Lipton, R. J. & Landweber, L. F. (2000) Molecular computation: RNA solutions to chess problems. PNAS 97, 1385-1389.

Two Proteins for the Price of One: The Design of Maximally Compressed Coding Sequences*

Bei Wang[1], Dimitris Papamichail[2], Steffen Mueller[3], and Steven Skiena[2]

[1] Dept. of Computer Science, Duke University, Durham, NC 27708
`beiwang@cs.duke.edu`
[2] Dept. of Computer Science, State University of New York, Stony Brook, NY 11794
`{dimitris, skiena}@cs.sunysb.edu`
[3] Dept. of Microbiology, State University of New York, Stony Brook, NY 11794
`smueller@ms.cc.sunysb.edu`

Abstract. The emerging field of *synthetic biology* moves beyond conventional genetic manipulation to construct novel life forms which do not originate in nature. We explore the problem of designing the provably shortest genomic sequence to encode a given set of genes by exploiting alternate reading frames. We present an algorithm for designing the shortest DNA sequence simultaneously encoding two given amino acid sequences. We show that the coding sequence of naturally occurring pairs of overlapping genes approach maximum compression. We also investigate the impact of alternate coding matrices on overlapping sequence design. Finally, we discuss an interesting application for overlapping gene design, namely the interleaving of an antibiotic resistance gene into a target gene inserted into a virus or plasmid for amplification.

1 Introduction

The emerging field of *synthetic biology* moves beyond conventional genetic manipulation to construct novel life forms which do not originate in nature. The synthesis of poliovirus from off-to-shelf components [1] attracted worldwide attention when announced in July 2002. Subsequently, the bacteriophage PhiX174 was synthesized using different techniques in only three weeks [2], and Kodumal, et al. [3] recently set a new record for the longest synthesized sequence, at 31.7 kilobases. The ethics and risks associated with synthetic biology continue to be debated [4], but the pace of developments is quickening. Indeed, Tian, et al. [5] have just proposed a method for DNA synthesis based on microarrays and multiplex PCR that promises a substantial reduction in cost.

Once you can synthesize an existing genome from scratch, you can do the same for new and better designs as well. In this paper, we explore an interesting problem in genome design, namely designing the provably shortest genomic sequence to encode a given set of genes, by exploiting alternate reading frames and the redundancy of the genetic code. Theoretically, up to six proteins can be

---

* This research was partially supported by NSF grants EIA-0325123 and DBI-0444815.

A. Carbone and N.A. Pierce (Eds.): DNA11, LNCS 3892, pp. 387–398, 2006.
© Springer-Verlag Berlin Heidelberg 2006

encoded on the same genomic sequence using three alternate reading frames on both strands. Indeed, long gene overlaps occur frequently in nature.

Our contributions in this paper are:

- *Finding Shortest Encodings for Given Protein Pairs* – We present an algorithm for designing the shortest DNA sequence simultaneously encoding two given amino acid sequences. Our algorithm runs in worst-case quadratic time, but we provide an expected-case analysis explaining its observed linear running time when employing the standard DNA triplet code.

- *Comparing Natural and Synthetic Coding-Pair Sequences* – We compare the overlapping gene designs constructed by our algorithm to those occurring in natural viral sequences. We show that the coding sequence of naturally occurring pairs of overlapping genes in general approach maximum compression, meaning that it is impossible to design overlapping shorter coding sequences for them which save more than 1-2% over independent genes. This counterintuitive result has natural explanation in terms of the evolutionary mechanics of overlapping gene sequences.

  Further, we show interesting differences between the preferred phase (reading frame), strand, and orientation of natural and optimized overlapping sequences.

- *Impact of Alternate Coding Matrices on Overlapping Sequence Design* – Protein designs are not immutable; indeed, certain pairs of amino acids share such similar physical/chemical properties they can be fairly freely substituted without altering protein function. This freedom can be exploited to design substantially shorter encodings for a given pair of proteins.

  We investigate the impact of increasingly permissive amino acid substitution matrices (derived from the well-known PAM250 matrix) on the potential for constructing tight encodings. Extremely tight encodings are often possible while largely preserving the hydrophobicity of the associated residues. Further, the encodings designed under each of these matrices shows interesting differences between the preferred phase (reading frame), strand, and orientation.

- *Biotechnology Applications of Nested Encodings* – We propose an interesting application for overlapping gene design, namely the interleaving of an antibiotic resistance gene into a target gene inserted into a virus or plasmid for amplification. Selective pressures tend to quickly remove such target genes as disadvantageous to the host. However, coupling such a target with a resistance gene provides a means to select for individuals *containing* the arbitrarily selected target gene.

  To demonstrate the feasibility of this technique, we apply our algorithm to encode each of five important antibiotic resistance genes within the body of the Hepatitis C virus. In fact, we demonstrate there are many possible places to encode each resistance gene within the virus, assuming a sufficiently (but not excessively) permissive codon replacement matrix.

These sequence design problems naturally arise in our project, currently underway, to design and synthesize weakened viral strains to serve as candidate vaccines [6]. This work also follows our previous efforts to design encoding sequences for proteins which minimize or maximize RNA secondary structure [7] and avoid restriction sites [8].

This paper is organized as follows. In Section 2, we survey the literature concerning why gene overlaps occur in nature and how they evolve. We present our algorithm for constructing optimal encodings in Section 3, with associated analysis, and compare our synthesized designs with wildtype viral encodings. In Section 4, we study the impact of alternate codon substitution matrices on the size and parity of minimal pairwise gene encodings. Finally, in Section 5, we present our results on encoding antibiotic resistance genes within viral coding sequences.

2   Overlapping Genes in Nature

Overlapping genes are adjacent genes whose coding regions are at least partly overlapping. They occur most frequently in prokaryotes, bacteriophages, animal viruses and mitochondria, but are seen in higher organisms as well. Gene overlapping presumably results from evolutionary pressure to minimize genome size and maximize encoding capacity [9]. For viruses, this is manifested in two ways; first when genome size substantially affects the speed of replication, and second when an upper bound on the genome size is imposed by packaging.

Overlapping genes are common for viruses with prokaryotic hosts because they must be able to replicate sufficiently fast to keep up with their host cells [10]. As an example, many bacteriophages have compact genomes which maximize coding information into the minimum genome size [10].

In term of evolutionary pressure to minimize genome size, packaging size pressure (the packaging size of the virus particle as the amount of nucleic acid which can be incorporated into the virion) sets the genome size upper bound for viruses with eukaryotic hosts [10].

Overlap between genes is very common in genomes mutating at high rates, such as bacteria and mitochondria, but especially viruses. Although a mutation in an overlapping region can impair more than one protein and would be naturally selected against, there are several reasons overlapping genes can benefit an organism:

- By reducing the size of the genome, without affecting the number of genes encoded.
- By generating new (or sometimes more complex) proteins without increasing the size of the genome.
- By coordinating the expression levels of functionally related genes.
- By coordinating the expression levels of genes, where the expression of one gene requires the deactivation of the other.

The first two functions are supported by the theory of "overprinting", which attempts to describe the origin of new genes from an existing genome with minimal mutational change [11]. Size reduction is considered important under the assumption that replication rate is inversely related to genome length, since it has an obvious effect in increased rates of replication and minimization of mutation load [12].

Overlapping reading frames can serve to expedite efficient translation. Overlaps can bring translation machinery close to both overlapping genes, which can co-ordinate or co-regulate their expression [12]. In other cases an overlap can bring the termination site of one gene into the same region as the translation initiation site for the next gene [13].

The rate of evolution can be expected to be slower in overlapping genes [14]. Since point mutations in overlapping regions can affect two genes simultaneously, a mutant variant produced with a mutation in an overlapping region will have a lower growth rate and in most cases cannot compete with the wild type variant [15].

Although high mutation rates and selection towards a compacted genome would indicate that overlapping genes should occur mostly in viral and cellular prokaryotic genomes and mitochondria, recent studies show that mammalian genomes have relatively frequent occurrences of overlapping genes too. The observed 774 overlapping genes in the human and 542 overlapping genes in the mouse genome [16] do not compare favorably with the 806 overlapping gene pairs in the genome of E.Coli [9], since the latter genomes is three orders of magnitude smaller. Nevertheless, the same mechanisms of evolution, like rearrangements or loss of parts and utilization of neighboring gene signals, provide explanation for the origin of these overlaps.

Overlapping genes offer an efficient way to study how coding and control sequences have evolved. With direct comparison of the overlapping genes for related species, one can determine how the overlaps evolved and under which conditions, like neighboring gene distance (for example, in closely related bacterial species it has been observed that most of the overlapping genes were generated or degraded in gene pairs that have a short intergenic region [17]). By comparing gene overlaps that are not conserved between related species, the mutational changes that caused the diversion can often be identified. In other cases further species sequencing are necessary to decipher the evolutionary mechanisms and tendencies (see [9] [16] and [17]).

In bacterial species it has been observed that the total number of overlapping genes depends on the genome size or the total number of genes, which could imply that the rates for the accumulation and degradation of overlapping genes are universal among bacterial species [17].

Overlapping gene regions can also provide information for evolution patterns among classes of organisms and seem to converge with ribosomal RNA phylogenetic methods' results [18]. For certain bacterial species, the extent of conservation of unidirectional overlaps correlates with the evolutionary distances between pairs of species [9]. Gene overlaps have even been correlated with certain human disease genes; further genomic rearrangements are likely to occur within overlapping regions, possibly as a consequence of anomalous sequence features prevalent in these regions [19].

3   Finding Maximally Compressed Gene-Pair Encodings

Our algorithm for constructing the maximally compressed encoding for a given pair of amino acid sequences P_1 and P_2 can be most succinctly described via a dynamic program. We consider the canonical case where the encoding of P_1 starts to the left (5' end) of P_2 as shown in Figure 1; the reverse case follows by simply relabeling the proteins. We present only the algorithm for the case of same-strand encodings; the case of alternate strand encodings follows analogously.

Let P_1 contain n residues and P_2 m residues, respectively. Let o_1, o_2, and o_3 denote possible DNA sequences of 0 to 3 bases in length. There are two general cases:

- We say that $C[i, j, o_1, top]$ is *realizable* iff there exists a pair of sequences o_2, o_3 such that $o_1 o_2$ codes for residue $P_1[i]$, $o_2 o_3$ codes for residue $P_2[j]$, and $C[i + 1, j, o_3, bottom]$ is realizable or $i = n$.
- We say that $C[i, j, o_1, bottom]$ is *realizable* iff there exists a pair of sequences o_2, o_3 such that $o_1 o_2$ codes for residue $P_2[j]$, $o_2 o_3$ codes for residue $P_1[i]$, and $C[i, j + 1, o_3, top]$ is realizable or $j = m$.

An exception occurs only when the residues are aligned, where only one case is needed, in which we advance both indices i and j and we check for reaching both ends of the proteins.

The basis cases for the canonical labeling assert that an overlap is attainable ($C[n, j, o_1, top]$ or $C[i, m, o_1, bottom]$ is realizable) iff $C[j, 1, o_1, top]$ is realizable for some $1 < j < n$.

Since there are only a constant number of possible short strings o_1, o_2, and o_3, it takes constant time to evaluate a given value of $C[i, j, o, b]$ given the solution of all smaller cases. With $\Theta(mn)$ values to evaluate, the algorithm runs in worst case $\Theta(mn)$ time.

By ceasing evaluation once no realizable values remain, the longest overlap can be computed in $O((n + m)l)$, where l is the length of the longest overlap between the protein sequences. Below, we argue that l should in general be of constant length on non-degenerate substitution matrices; this states that on average this algorithm should run in linear time on such matrices.

We say that two overlapping proteins are *in-phase* if the overlap length is congruent to 0 mod 3, i.e. they align along codon boundaries. Non-trivial

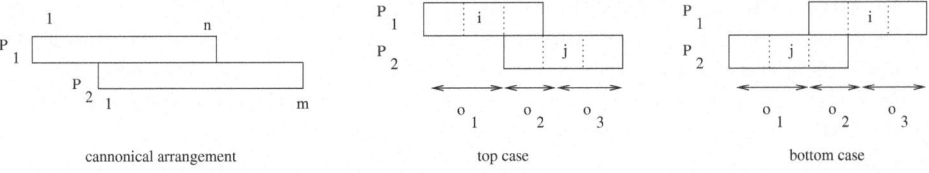

Fig. 1. Notation for the gene encoding algorithm: the canonical encoding (left), with the top (center) and bottom (right) overhang cases

in-phase, same-strand overlapping designs are in principle forbidden by the fact that proteins must end with stop codons. However, we consider an abstraction of this case to simplify the analysis.

Here we consider the expected length of the maximal overlap as a function of the *residue equivalence probability*, defined as the probability that two randomly selected amino acids have an equivalent codon between them. This residue equivalence probability p is a function both of the codon substitution matrix and the distribution of amino acids in the proteins.

Assuming independence of the protein sequences, the expected length of the longest left-right overlap $E(O)$ of two random sequences P_1 and P_2 is given by

$$E_1(O) = \sum_{l=0}^{\infty} l p^l \prod_{i=l+1}^{\infty} (1 - p^i) \tag{1}$$

For the case of two-sided overlaps (i.e. either P_1 or P_2 may occur on the left side of the alignment),

$$E_2(O) = \sum_{l=0}^{\infty} l(2p^l - p^{2l}) \prod_{i=l+1}^{\infty} (1 - (2p^i - p^{2i})) \tag{2}$$

The above analysis demonstrates that the expected maximum overlap length remains quite small until the residue equivalence probability approaches 1. This suggests that two arbitrary proteins are unlikely to permit substantially compressed in-phase encodings except under a forgiving (degenerate) coding matrix.

Still, all is not lost. Our analysis of both wildtype and synthetic overlaps demonstrates that out-of-phase encodings are likely to be substantially longer than in-phase encodings. This phenomenon appears to be difficult to analyze in general because it strongly depends upon the properties of the codon equivalence matrix.

Each amino acid is encoded by a minimum of one and a maximum of six different codons. In total, 61 of the 64 codons encode 20 amino acids while the other three are stop codons, a termination point for protein-synthesizing machinery. Thus there is an approximate 1-to-3 correspondence between amino acids and their codon encodings. It is this redundancy that offers the flexibility in amino acid sequence encoding.

To study the extent of gene overlapping in viruses, we analyzed all 1058 completely sequenced viral genomes available in Genbank as of February 22, 2004. After excluding 273 genomes containing a single annotated gene (and hence not a candidate for gene overlapping) and 108 genomes with sequence ambiguity or obvious annotation errors, we were left with 677 viruses of interest.

In total, these viruses contained 3,232 pairs of overlapping genes, 2,407 of which had overlaps of length greater than four bases.[1]

Figure 2 presents the frequency distribution of gene overlaps by length, the tail of which demonstrates that long overlaps occur with surprisingly high frequency.

---

[1] Overlaps of less than four bases are not particularly interesting, since the possible overlaps are restricted to the start and stop codons possessed by every gene.

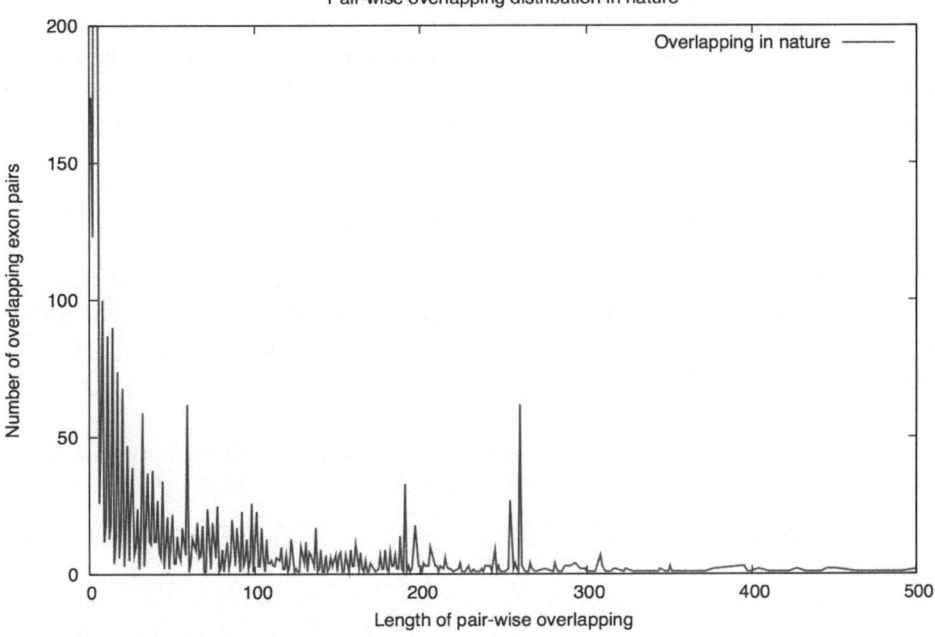

Fig. 2. Length distribution of pairwise-overlapping genes in viral genomes

Table 1 partitions these overlaps into disjoint cases, distinguished by whether the genes occur on the same strand, or are head-to-head or tail-to-tail on opposing strands. Same strand overlaps dominate in the sample. Table 1 also partitions these overlaps by the length mod 3. In-phase overlaps are understandably rare (any stop codon breaks both same strand sequences), but there is also a clear preference for 2 mod 3 parity over 1 mod 3.

Table 1. Parities of natural gene overlaps, ties discarded. All 3232 gene pairs (left). The 2407 gene pairs with overlap > 4.

Pattern	All overlaps, parity mod 3				Length > 4, parity mod 3			
	0	1	2	All	0	1	2	All
SAME	0.0%	23.1%	39.9%	63.0%	0.0%	12.9%	53.3%	66.2%
HH	4.4%	1.9%	4.8%	21.1%	5.9%	4.9%	6.5%	17.3%
TT	3.0%	1.9%	11.0%	15.9%	4.0%	2.6%	9.9%	16.5%
Total	7.4%	36.9%	55.7%	100%	9.9%	20.4%	69.7%	100%

Using our gene pair encoder, we attempted to find more compressed representations of the wildtype gene pairs. In general we failed badly, with the vast majority of cases having zero or insignificant improvement (recall that approximately one third of all natural overlaps were of length 4 or less). In no case were we able to increase the overlap length of such an overlapping gene pair by more than 20 bases.

The lesson here is that gene overlaps occur because the proteins evolved together – significant potential overlaps are extremely unlikely to arise in unrelated sequence pairs because the genetic code does not provide sufficient flexibility. Figure 3 presents the results of optimally encoding 135,869 pairs of unrelated proteins. In no case were we able to reduce the length of an overlapping gene pair by more than 30 bases.

More interesting is the breakdown of our optimized encodings by strand and parity, reported in Table 2. The optimized encodings show sharply different preferences than the wildtype encodings. Functional demands likely constrict the choice of same strand encodings, although it is less obvious why there is such dramatic difference in head-to-head and tail-to-tail preferences. The difference in preferred parity is largely explained by the change in strand encoding distribution.

Table 2. Longest optimized overlap using codon matrix, ties discarded. All 135,869 overlapping gene pairs (left), the 14,925 overlapping gene pairs of length > 4 (right).

Pattern	All overlaps, parity mod 3					Length > 4, parity mod 3			
	0	1	2	All		0	1	2	All
SAME	0.01%	31.92%	1.47%	33.40%		0.05%	3.04%	13.16%	16.25%
HH	5.09%	35.51%	2.31%	42.91%		45.55%	7.77%	21.07%	74.39%
TT	0.08%	0.91%	22.70%	23.69%		0.62%	8.28%	0.46%	9.36%
Total	5.18%	68.34%	26.48%	100.00%		46.22%	19.09%	34.69%	100.00%

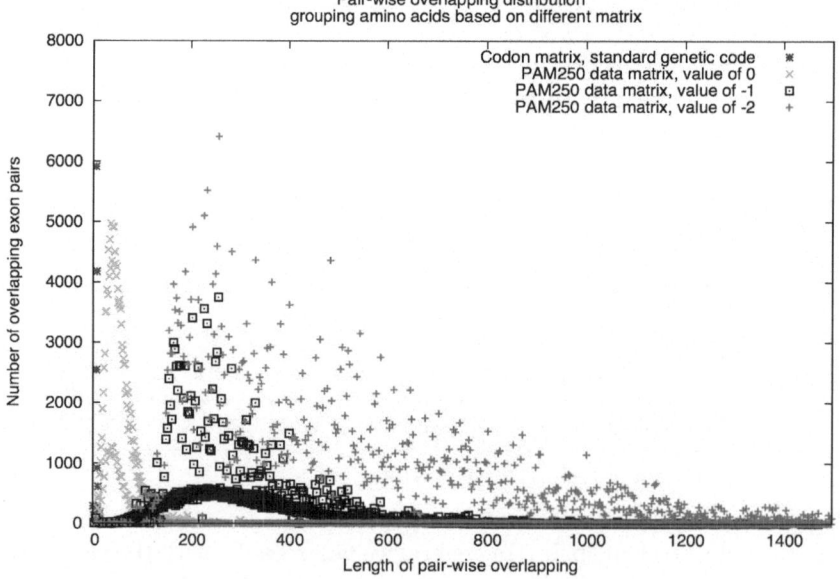

Fig. 3. Distribution of maximum overlaps under four different codon substitution matrices

4   Experiments in Synthetic Gene Encoding

Recent studies [20] have demonstrated that the genetic code maximizes the likelihood that a gene mutation will not harm and may even improve the protein. In general, the code is resilient to random mutations leading to significant changes of the affected amino-acid properties, so that a misread codon often codes for the same amino acid or one with similar biochemical properties. Furthermore, simulations by Gilis et al. [21] have shown that taking the amino-acid frequency into account further increases the resilience of the code compared to random codings. It is also known that proteins with a limited number of point mutations which lead to non-synonymous substitutions fold in similar ways, in a degree that homology database search can detect function similarity in proteins differing in up to 50% of their amino-acid compositions [22].

Based on these results, we decided to further investigate the pairwise gene overlapping possibilities using non-synonymous amino-acid substitution matrices, which increase the combinatorial possibilities of compressed overlapping representations at the cost of minor changes in the residues in the underlying proteins.

Our substitution matrices are derived from the well-known PAM 250 amino acid substitution scoring matrix. The value of each entry describes the reward or penalty in replacing an instance of the first amino acid with the second in aligned sequences. Positive values contribute favorably to an alignment, and negative values unfavorably. We may derive a permissive codon equivalence matrix from PAM 250 as a function of a threshold t by permitting replacement of amino acid x with y if the score is $\geq t$. By decreasing t, we can define a sequence of increasingly permissive substitution matrices for our experiments.

Clearly other substitution matrices are possible (e.g. Levitt's hydrophobicity scoring matrix [23]), and perhaps even preferable. Our primary interest is establishing the flexibility for compressed sequences as a function of more tolerant substitution matrices.

The results of our overlapping experiments with the use of the alternate substitution matrices are shown in Figure 3. One can observe the significant increase in both the number and frequency of long overlaps with increasing length as the matrices become more permissive. In particular, almost arbitrarily long overlaps appear possible under $t \geq -3$ substitution.

5   Hiding Short Genes in Long Genes

Here we report on proof-of-concept simulations of two related biotechnology applications for carefully designed overlapping of synthetic gene sequences:

- *Plasmid incorporation into mammalian cells* – A common technique for incorporating target gene expression into mammalian cell involved plasmid incorporation and mammalian cell transfection. Initially, the plasmid containing the target gene is propagated in bacteria. The naked plasmid DNA is extracted and then introduced into the mammalian cell by transfection.

Typically the target gene is paired with an antibiotic resistance gene, so as to create a marker for selection in the eukaryotic cell. All cells not expressing this marker can be eliminated with the corresponding antibiotic drug (ex. geneticin or G418), to isolate cells expressing the target protein. Sometimes, however only one gene is expressed, such as when the cell fails to incorporate the entire plasmid. Because the plasmid is linearized to be incorporated in a chromosome, the cut may also occur in the target gene location.

By overlapping the target and marker genes, we reduce the probability that either the cut will eliminate the target gene but the not the marker, as well as the probability that the two genes will be separated.

– *Foreign gene incorporation into viruses* – RNA viruses are very prone to recombination, so an added sequence has a high probability to be deleted. Since RNA viruses are streamlined to perform a limited number of specific tasks, the addition of a gene slows down the virus processes, merely by extending slightly its length. Since the foreign gene is undesirable, its deletion will result in a faster produced replicon that will eventually outgrow the engineered virus we implanted.

Interleaving the target gene into a gene that the virus needs can prohibit its deletion through reversion.

Positive indications in the direction of gene overlap engineering are the recent results of [20], which show that the amino acid code minimizes the effects of mutations and maximizes the likelihood that a gene mutation will improve the resulting protein. Additionally, methods of local sampling (see [24] and [25]) can help us simulate the behavior of slightly altered proteins in respect to folding and docking, so that we can test the codon substitution effects without lab experimentation.

To evaluate the potential for such synthetic overlap encodings, we attempted to find maximal encodings of five important antibiotic genes (whose length ranges from 375 to 1026 nucleotides) within the coding region of the Hepatitis C virus (HCV). Consistent with the results of the previous section, only trivial overlaps can be obtained using synonymous substitutions.

However, multiple complete encodings are possible under $t \geq -2$ and $t \geq -3$ substitution for each of the five antibiotic resistance genes, as reported in Table 3. There is a strong bias for alternate strand encodings, although all five

Table 3. Number of fully-enclosed $t \geq -2$ and $t \geq -3$ encodings of antibiotic resistance genes within the Hepatitis C virus, same strand (SS) and alternate strand (AS)

Gene	Accession	Length	$t \geq -2$ encodings		$t \geq -3$ encodings	
			SS	AA	SS	AS
Hygromycin	X03615	1026	0	1	4	1
Neomycin	M55520	795	0	1	2	3
Puromycin	X92429	600	0	11	16	25
Blasticidin	AYI96214	423	56	250	217	442
Zeocin	A31902	375	35	132	163	175

antibiotic resistance genes offer same strand encodings for $t \geq -3$. In fact, the preferable target for the inserted gene encoding (and promoter region) in the virus application is the minus strand, so this bias appears fortunate.

Based on these results, we are pursing more rigorous designs for intended synthesis and implementation.

Acknowledgments

We thank Eckard Wimmer for his interest and support. We also thank Chen Zhao, Huei-Chi Chen, and Rahul Sinha for discussions and contributions to this research.

References

1. J. Cello, A. Paul, and E. Wimmer. Chemical synthesis of poliovirus cDNA: Generation of infectious virus in the absence of natural template. *Science*, 297:1016–1018, 2002.

2. H. Smith, C. Hutchison, C. Pfannkoch, and J. C. Venter. Generating a synthetic genome by whole genome assembly: phix174 bacteriophage from synthetic oligonucleotides. *Proc. Nat. Acad. Sci.*, 100:15440–15445, 2003.

3. S. Kodumal, K. Pael, R. Reid, H. Menzella, M. Welch, and D. Santi. Total synthesis of long DNA sequences: Synthesis of a contiguous 32-kb polyketide synthase gene cluster. *Proc. Nat. Acad. Sci.*, 44:15573–15578, 2004.

4. P. Ball. Starting from scratch. *Nature*, 431:624–626, 2004.

5. J. Tian, H. gong, N. Sheng, Z. Zhou, E. Gulari, X. Gao, and G. Church. Accurate multiplex gene synthesis from programmable DNA microchips. *Nature*, 432:1050–1054, 2004.

6. S. Skiena and E. Wimmer. Gene design for vaccines and theraputic phages. NSF ITR Award 0325123, 2003.

7. B. Cohen and S. Skiena. Natural selection and algorithmic design of mrna. *J. Computational Biology*, 10:419–432, 2003.

8. S. Skiena. Designing better phages. *Bioinformatics*, 17:253–261, 2001.

9. Y. Fukuda, T. Washio, and M. Tomita. Evolution of overlapping genes: Comparative genomics of mycoplasma genitalium and mycoplasma pneumoniae. The Ninth Workshop on Genome Informatics, 1998.

10. Cann A.J. *Principles of Molecular Virology*. Academic Press, 1993.

11. P. Keese and A. Gibbs. Origins of genes: "big bang" or continuous creation? *Proc. Natl. Acad. Sci.*, 89:9489–9493, 1992.

12. D. C. Krakauer. Evolutionary principles of genomic compression. *Comments on Theor. Biol.*, 2002.

13. D. Oppenheim and C. Yahofsky. Translational coupling during expression of the tryptophan operon of e. coli. *Genetics*, 95:785–795, 1980.

14. T. Miyata and T. Yasunaga. Evolution of overlapping genes. *Nature*, 272:532–535, 1978.

15. D. C. Krakauer. Stability and evolution of overlapping genes. *Evolution*, 54(3):731–739, 2000.

16. V. Veeramachaneni, W. Makalowski, M. Galdzicki, R. Sood, and I. Makalowska. Mammalian overlapping genes: The comparative method. *Genome Research*, 14:280–286, 2004.

17. Y. Fukuda, Y. Nakayama Y, and M. Tomita. On dynamics of overlapping genes in bacterial genomes. *Gene*, 323:181–7, 2003.

18. I. Rogozin, A. Spiridonov, A. Sorokin, Y. Wolf, J. King, R. Tatusov, and E. Koonin. Purifying and directional selection in overlapping prokaryotic genes. *Trends Genet.*, 18(5):228–232, 2002.

19. S. Karlin, C. Chen, A. Gentles, and M. Cleary. Associations between human disease genes and overlapping gene groups and multiple amino acid runs. *Proc. Natl. Acad. Sci.*, 99(26):17008–13, 2002.

20. S. Freeland and L. Hurst. Evolution encoded. *Sci Am.*, 290(4):84–91, 2004.

21. D. Gilis, S. Massar, N.J. Cerf, and M. Rooman. Optimality of the genetic code with respect to protein stability and amino-acid frequencies. *Genome Biol.*, 2(11), 2001.

22. M.A. Marti-Renom, A.C. Stuart, A. Fiser, R. Sanchez, F. Melo, and A. Sali. Comparative protein structure modeling of genes and genomes. *Annu Rev Biophys Biomol Struct.*, 29:291–325, 2000.

23. M. Levitt. A simplified representation of protein conformations for rapid simulation of protein folding. *J. Mol. Biol.*, 104:59–107, 1976.

24. R. Elber and M. Karplus. Enhanced sampling in molecular dynamics: Use of the time-dependent hartree approximation for a simulation of carbon monoxide diffusion through myoglobin. *J. Am. Chem. Soc.*, 112:9161–9175, 1990.

25. V. Hornak and C. Simmerling. Generation of accurate protein loop conformations through low-barrier molecular dynamics. *Proteins*, 51:577–590, 2003.

Design of Autonomous DNA Cellular Automata*

Peng Yin[1], Sudheer Sahu[1], Andrew J. Turberfield[2], and John H. Reif[1]

[1] Department of Computer Science, Duke University,
Box 90129, Durham, NC 27708-0129, USA
{py, sudheer, reif}@cs.duke.edu
[2] University of Oxford, Department of Physics, Clarendon Laboratory,
Parks Road, Oxford OX 1 3PU, UK
a.turberfield@physics.ox.ac.uk

Abstract. Recent experimental progress in DNA lattice construction, DNA robotics, and DNA computing provides the basis for designing DNA cellular computing devices, *i.e.* autonomous nano-mechanical DNA computing devices embedded in DNA lattices. Once assembled, DNA cellular computing devices can serve as reusable, compact computing devices that perform (universal) computation, and programmable robotics devices that demonstrate complex motion. As a prototype of such devices, we recently reported the design of an Autonomous DNA Turing Machine, which is capable of universal sequential computation, and universal translational motion, *i.e.* the motion of the head of a single tape universal mechanical Turing machine. In this paper, we describe the design of an Autonomous DNA Cellular Automaton (ADCA), which can perform *parallel* universal computation by mimicking a one-dimensional (1D) universal cellular automaton. In the computation process, this device, embedded in a 1D DNA lattice, also demonstrates well coordinated parallel motion. The key technical innovation here is a molecular mechanism that synchronizes pipelined "molecular reaction waves" along a 1D track, and in doing so, realizes parallel computation. We first describe the design of ADCA on an abstract level, and then present detailed DNA sequence level implementation using commercially available protein enzymes. We also discuss how to extend the 1D design to 2D.

1 Introduction

DNA has recently been used, with great success, to fabricate nanoscale lattices and tubes [3, 8, 11, 14, 15, 16, 18, 20, 21, 22, 29, 31, 32], to construct nanomechanical devices [2, 4, 9, 10, 24, 25, 27, 23, 28, 33, 34, 36, 37], and to perform computation [1, 5, 6, 7, 13, 17, 19, 26]. The progress in these three fields together provides the basis for the next step forward: designing and constructing autonomous DNA computing devices embedded in well defined DNA lattices, which are capable of (universal) computation. We call such devices *DNA cellular computing devices*.

Once assembled, DNA cellular computing devices can serve as reusable, compact computing devices that perform (universal) computation, and programmable robotics

---

* The work is supported by NSF ITR Grants EIA-0086015 and CCR-0326157, NSF QuBIC Grants EIA-0218376 and EIA-0218359, and DARPA/AFSOR Contract F30602-01-2-0561.

A. Carbone and N.A. Pierce (Eds.): DNA11, LNCS 3892, pp. 399–416, 2006.

devices that demonstrate complex motion. First, DNA cellular computing devices represent a step forward beyond the prior molecular computing schemes, such as algorithmic self-assembly of DNA tiles, which performs a one-time computation during the assembly. In contrast, DNA cellular computing devices, once assembled, can perform multi-round computations. The output, embedded in the DNA lattices, is preserved and can serve as input for future computations. In addition, DNA cellular computing devices are more compact than the tiling scheme. A 1D Autonomous DNA Turing machine holds equivalent computational power as a 2D tiling lattice. Second, DNA cellular computing devices can demonstrate sophisticated programmable motion, for example, universal translational motion, which we define as the motion of the head of a single tape universal mechanical Turing machine. As such, DNA cellular computing devices may promise interesting applications in nanorobotics and nano-computation, as well as, nano-fabrication, nano-sensing, and nano-actuated electronics.

As one prototype of the DNA cellular computing devices, we previously reported the design of an autonomous unidirectional DNA mechanical device capable of universal sequential computation, termed as Autonomous DNA Turing Machine [35]. Here, we extend our previous work and obtain the design of an Autonomous DNA Cellular Automaton (ADCA). By mimicking a 1D universal cellular automaton, the ADCA can perform *parallel* universal computation, and in the process, demonstrate well coordinated parallel motion. The parallel computation and motion is realized using a novel molecular mechanism that synchronizes pipelined "molecular reaction waves" along a 1D track.

A cellular automaton is a set of "colored" cells on a grid of specified shape that evolve through discrete time steps according to a set of transition rules based on the colors of neighboring cells [30]. If the lattice is a one (resp. two) dimensional lattice, the cellular automaton is called a one (resp. two) dimensional cellular automaton. Figure 1 (a) shows the cells of an example one-dimensional cellular automaton. Each cell of this automaton can have one of two states, or equivalently two colors, WHITE and BLACK. The evolving process of a cellular automaton is specified by the transition rules. Figure 1 (b) illustrates one example rule set for the cellular automaton shown in Figure 1 (a). The rule set consists of 8 transition rules (numbered (1) - (8) in the figure). For example,

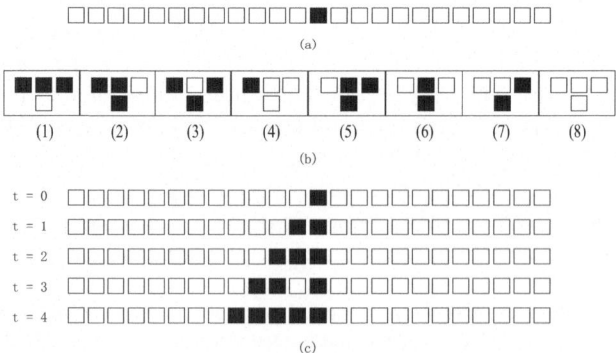

Fig. 1. A universal cellular automaton with two colors: Rule 110. The figure is adapted from [30].

according to rule (1), if the current cell and both of its neighbors have color BLACK, at next time step, the middle cell will change to color WHITE. This rule is denoted as BLACK, BLACK, BLACK → WHITE. Applying the rules in Figure 1 (b) to the initial configuration in Figure 1 (a), we obtain the transition history table depicted in Figure 1 (c). Cellular automaton can hold universal computing power. The cellular automaton depicted in Figure 1 is one such universal cellular automaton, known as rule 110, as described in [30]. This cellular automaton can be simulated by the ADCA described in this paper.

The rest of the paper is organized as follows. In Sect. 2, we describe the structural design and operational process of our ADCA. In Sect. 3, we give a detailed molecular implementation of the design presented in Sect. 2. We then briefly describe how to extend the design of ADCA to two-dimensions in Sect. 4 and close in Sect. 5.

2   Design

2.1   Structure

The ADCA operates in a solution system. Figure 2 illustrates an example abstract cellular automaton in the top panel, and the structure of the corresponding ADCA in the bottom panel. The ADCA is composed of four parts: a rigid *symbol track*, a linear array of *dangling DNA molecules* tethered to the symbol track, a set of *floating DNA molecules*, and a group of floating *protein enzymes*.

– **Symbol track.** The symbol track provides a rigid structural platform on which the dangling-molecules are tethered. It can be implemented, for example, as a rigid addressable DNA lattice, such as the barcode DNA lattice reported in [31].

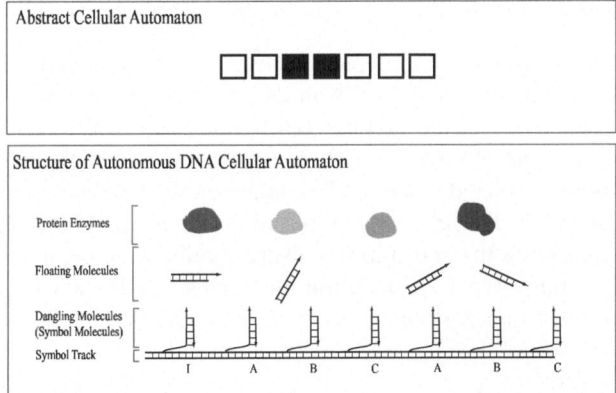

Fig. 2. Top panel: the initial configuration of an abstract cellular automaton. Bottom Panel: schematic drawing of the structure of an ADCA corresponding to the abstract cellular automaton in the top panel. The backbones of DNA strands are depicted as directed line segments. The short bars represent base pairing between DNA strands.

- **Dangling DNA molecules.** The array of dangling-molecules, also called symbol-molecules, tethered to the symbol track represent the array of cells (symbols) in the cellular automaton (and hence the name symbol-molecule). A dangling-molecule is a duplex DNA fragment, with one end tethered to the symbol track via a flexible single strand DNA fragment and the other end possessing a single strand DNA extension (the sticky end). Due to the flexibility of the single strand DNA linkage, a dangling-molecule moves rather freely around its joint on the symbol track. We require that the only possible interactions between two dangling-molecules are those between two immediate neighbors. This requirement can be ensured by properly spacing the dangling-molecules along the rigid track.
- **Floating DNA molecules.** In addition to the array of dangling-molecules, the system contains *floating-molecules*. A floating-molecule is a free floating (unattached to the symbol track) duplex DNA segment with a single strand overhang at one end (sticky end). A floating-molecule floats freely in the solution and thus can interact with another floating-molecule or a dangling-molecule provided that they possess complementary sticky ends. There are two kinds of floating-molecules: the rule-molecules and the assisting-molecules. The rule-molecules collectively specify the computational rules and are the programmable part of the ADCA, while the assisting-molecules assist in carrying out the operations of the ADCA, which we describe in detail in Sect. 2.2 and Sect. 2.3.
- **Protein enzymes.** The system also contains floating DNA ligase and three types of DNA endonucleases. The enzymes perform ligations and cleavages on the DNA molecules to effect the designed structural changes and hence the information processing. The cleavage patterns of the endonucleases are described in Figure 3.

2.2   Structural Changes

Figure 4 illustrates the structural changes during the operation of ADCA.

Initial configuration. Figure 4 (a) depicts an example abstract cellular automaton in its top panel, and a corresponding ADCA in its bottom panel. For simplicity and clarity, the floating enzymes and the floating DNA molecules in the ADCA are omitted from the figure; the symbol track, as well as the duplex and sticky end portions of a dangling-molecule, is depicted as a thick line segment; the flexible hinge of a dangling-molecule as a thin curve. The leftmost symbol-molecule is a special *initiator* dangling-molecule, I, representing the cell 0 in the abstract cellular automaton (see Figure 4). To the right of I, three types of dangling-molecules, A, B, and C, are positioned evenly along the track in a periodic order such that cells $3i + 1$, $3i + 2$, and $3i + 3$,

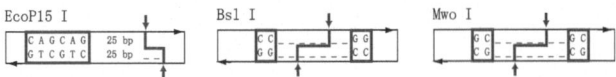

Fig. 3. Three endonucleases used in the molecular implementation of the ADCA. The recognition site of an enzyme is bounded by a box and the cleavage site indicated with a pair of arrows. The symbol "−" indicates the position of a base that does not affect endonuclease recognition.

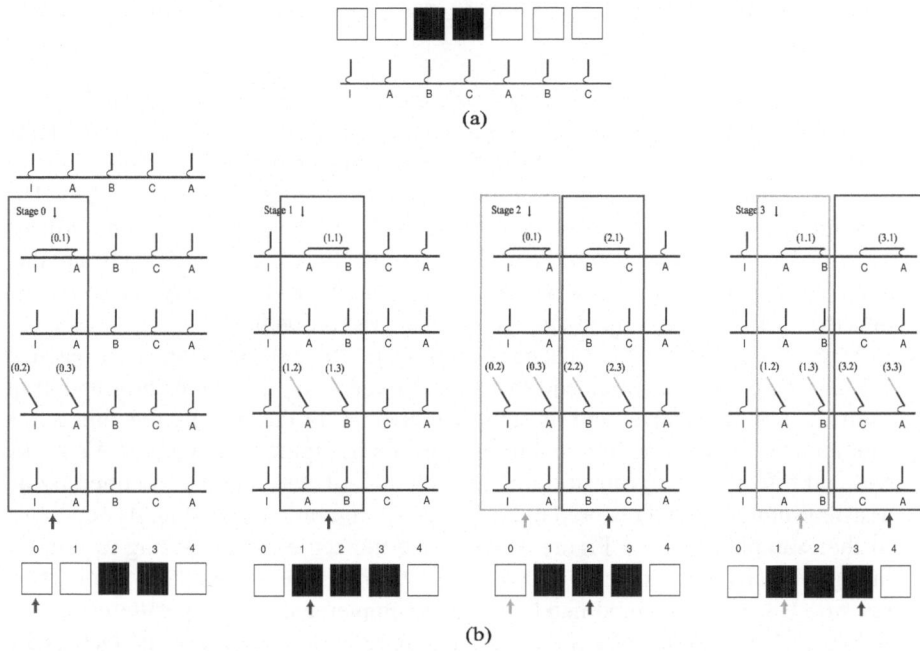

Fig. 4. Structural changes during the operation of an ADCA. In (b), red(dark) and green(grey) boxes indicate two pipelined.

where i is a non-negative integer, in the abstract cellular automaton are represented in the ADCA by symbol-molecules A, B, and C, respectively. The symbol-molecules differ in their *default sticky ends*, *i.e.* the sticky ends they possess in their respective initial configurations before the reaction starts. As we shall see later, this is essential for the synchronization of the operation of the ADCA. The color of each cell in the abstract cellular automaton is encoded in a corresponding symbol-molecule in the ADCA.

Pipelined reaction waves. Figure 4 (b) illustrates structural changes. During the operation of the ADCA, the initiator molecule I sends out a "reaction wave" that travels down the track from left to right. A critical novel property of the ADCA is that multiple reaction waves can travel down the track in a "pipelined" fashion. However, we have carefully engineered a "synchronization" mechanism so that a reaction wave that starts at a later stage can never overtake one that starts earlier. This ensures the synchronization of the state changes of the ADCA, and hence its correct operation.

Figure 4 (b) illustrates two consecutive reaction waves, respectively indicated with red (dark) and green (grey) boxes and arrows. Next, we focus on the first reaction wave (indicated by dark boxes and arrows), and describe the structural changes of ADCA.

In Stage 0, the reaction wave starts at the initiator I at position 0, then travels sequentially to A in Stage 1, B in Stage 2, and C in Stage 3. The reaction wave finishes one full cycle in Stages 1, 2, and 3, and thus goes on inductively down the track.

In Stage i, where $i = 0, 1, 2$, and 3, three types of reactions occur, namely reactions $i.0$, $i.1$. and $i.2$.

- In Stage 0, I has a complementary sticky end to its right neighbor A and is thus ligated to A, and the ligation product is subsequently cleaved by an endonuclease (Reaction 0.1). Next, I is "modified" by an assisting-molecule, depicted as a pink (grey) line segment, and restored to its default configuration (Reaction 0.2). The "modification" will be implemented as ligation and cleavage events and will be described in detail in Sect. 3. In a parallel reaction 0.3, A is also modified by another assisting-molecule such that A will possess a complementary sticky end to B, and thus the reaction wave is ready to enter Stage 1 (Reaction 0.3).
- In Stage 1, similar structural changes occur as in Stage 0. However, after reaction 1.1, A will possess a sticky end that encodes the state, *i.e.* color, information of itself, its left neighbor I, and its right neighbor B. In the ensuing reaction 1.2, a rule-molecule corresponding to a transition rule in Figure 1 recognizes A's sticky end and effects a state transition of molecule A. A will then be modified by an assisting-molecule and restored to its default configuration, encoding its new state. In the example shown in Figure 4 (b), a rule-molecule corresponding to rule (7) in Figure 1 changes the color encoded in A from WHITE to BLACK. In the parallel reaction 1.3, B will be modified to posses a complementary sticky end to C.
- In Stages 2 and 3, reactions of the same nature as in Stage 1 will occur. Details are omitted for brevity.

2.3 Information Flow

Notation. We next describe the information flow during the operation of the ADCA. For ease of exposition, we first introduce some notation. An information encoding DNA molecule is denoted as $X^a[y]^b$, where X is its duplex portion, $[y]$ is its sticky end portion, and a and b respectively represent the state information encoded in X and $[y]$. This is illustrated in Figure 5. As shown in the figure, there are two ways to encode information a in the duplex X. In Figure 5 (a), a is encoded as a unique DNA sequence GTA; in Figure 5 (b), a is encoded as the number of base pairs (L bp in the Figure) between an endonuclease recognition site and the sticky end of DNA molecule. The sequence of the sticky end $[y]$, in this case CGC, encodes the state information b. Furthermore, we

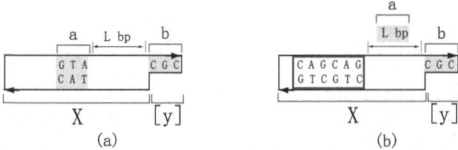

Fig. 5. Encoding state information in DNA molecules. The backbones of DNA strands are depicted as directed line segments. X and $[y]$ represent the duplex portion and the sticky end of the DNA molecule, respectively. The bases shaded in blue are used to encode state information, a or b. "L bp" indicates L DNA base pairs, where L is a non negative integer. In Figure (b), the number L is used to encode state information a and is thus shaded in blue. The red (dark) box indicates the recognition site for an endonuclease, in this case, EcoP15 I.

use $[\bar{y}]$ to denote the complementary sticky end of $[y]$. Finally, we describe how to represent ligation and cleavage events. The ligation of two molecules $X^a[y]^b$ and $[\bar{y}]^c Z^d$ is described by the equation

$$X^a[y]^b + [\bar{y}]^c Z^d \rightarrow XY.$$

Suppose XY incorporates an endonuclease recognition site and is cut into $X^{a'}[u]^{b'}$ and $[\bar{u}]^{c'} Z^{d'}$. This is represented as

$$XY \rightarrow X^{a'}[u]^{b'} + [\bar{u}]^{c'} Z^{d'}.$$

We can combine the above two equations and obtain,

$$X^a[y]^b + [\bar{y}]^c Z^d \rightarrow X^{a'}[u]^{b'} + [\bar{u}]^{c'} Z^{d'}.$$

Initial configuration. Figure 6 shows the ADCA in its default configuration before the reaction starts. As mentioned in Sect. 2.2, molecules A, B, and C possess different default sticky ends: $[\bar{u}]$, $[\bar{v}]$, and $[\bar{w}]$. Note that the state information a, b, and c are encoded in the duplex portions of A, B, and C, *not* their sticky ends. This is essential to ensure that repeated reactions between neighboring symbol-molecules can occur for multiple rounds, as described below.

Information flow. Figure 7 illustrates the information flow. We follow the framework of four-stage structural changes presented in Sect. 2.2 and enumerate the involved reactions below.

1. *Reaction 0.1.* Initiator molecule $I^i[u]$ and its immediate right neighbor $[\bar{u}]A^a$ share complementary sticky ends u and $[\bar{u}]$, and result in reaction,

$$I^i[u] + [\bar{u}]A^a \rightarrow I[t]^{ia} + [\bar{t}]^{ia}A.$$

Note that the sticky end $[\bar{t}]$ of product A encodes both the state information i from reactant I and the state information a from reactant A.

2. *Reaction 0.2.* The rule-molecule $[\bar{t}]^{ia}R$ restores $I[t]^{ia}$ to its default configuration in reaction,

$$I[t]^{ia} + [\bar{t}]^{ia}R \rightarrow I^i[u] + [\bar{u}]R.$$

3. *Reaction 0.3.* Molecule $[\bar{t}]^{ia}A$ is modified by assisting-molecule $T^{ia}[t]^{ia}$ in reaction

$$T^{ia}[t]^{ia} + [\bar{t}]^{ia}A \rightarrow T[\bar{v}] + [v]A^{ia}.$$

Now A is transformed to $[v]A^{ia}$. This essentially transduces the state information ia initially encoded in the sticky end of A to its duplex portion. Hence we term the assisting-molecule $T^{ia}[t]^{ia}$ as a *transducer-molecule*. The above reaction also modifies A's sticky end to $[v]$, which is complementary to the default sticky end of A's immediate right neighbor B. This makes A ready to interact with B.

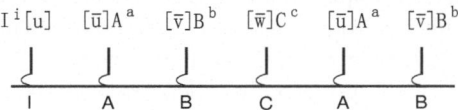

Fig. 6. Initial configuration

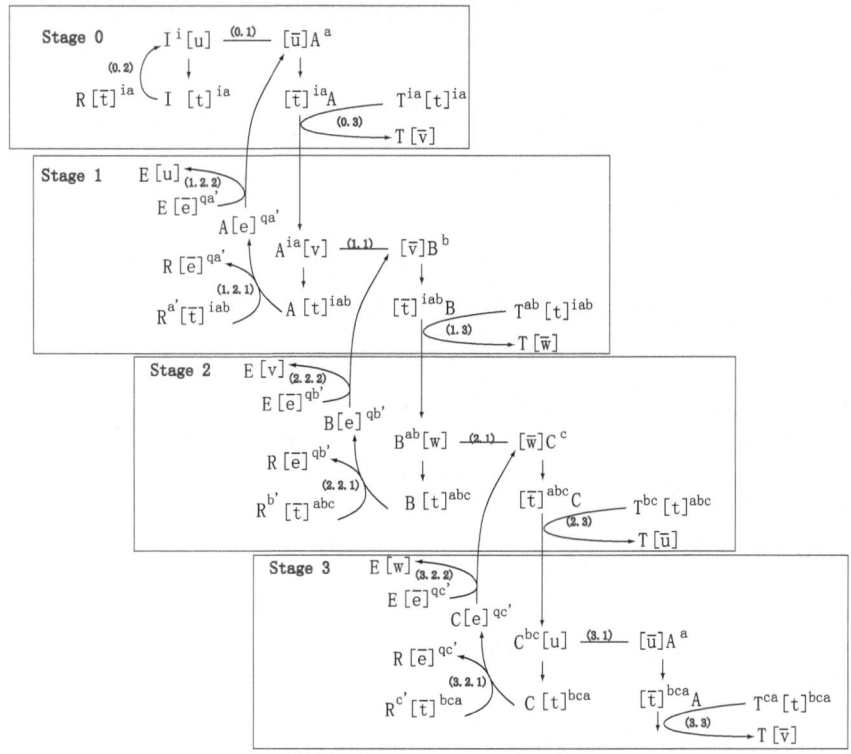

Fig. 7. Information flow

4. *Reaction 1.1.* Molecule $A^{ia}[v]$ interacts with its right neighbor $[\bar{v}]B^b$ in reaction,

$$A^{ia}[v] + [\bar{v}]B^b \rightarrow A[t]^{iab} + [\bar{t}]^{iab}B.$$

Now the sticky end of the product A encodes state information iab, *i.e.* the current state of A's left neighbor, the current state of A, and the current state of A's right neighbor. This suffices to specify a transition rule shown in Figure 1 and results in Reaction 1.2 below.

5. *Reaction 1.2.* Reaction 1.2 has two steps. In step 1.2.1, $A[t]^{iab}$ interacts with a rule-molecule $[\bar{t}]^{iab}R^{a'}$ in reaction,

$$A[t]^{iab} + [\bar{t}]^{iab}R^{a'} \rightarrow A[e]^{qa'} + [\bar{e}]^{qa'}R.$$

This effects a state transition of molecule A, as specified by the rule $iab \rightarrow a'$. However, to enable A to repeatedly perform computation, we need to further restore A to its default configuration, *i.e.* a configuration with a default sticky end $[\bar{u}]$ and encoding its new state a' in its duplex portion. This task is carried out by another kind of assisting-molecule called *extension-molecule* E. However, as a floating molecule, E needs to not only recognize A's current state but also distinguish A from the other two types of symbol-molecules, B and C. As such, we require the sticky end of A encodes not only its state information but also its type information q, where $q \in \{q_A, q_B, q_C\}$. Hence, in the above equation, the product A possesses a sticky end $[e]$ encoding both its type information q and its new state a' (a' not shown in Figure 7 (b)). This molecule $A[e]^{qa'}$ is then modified by extension-molecule $E[\bar{e}]^{qa'}$ in reaction 1.2.2,

$$A[e]^{qa'} + [\bar{e}]^{qa'} E \rightarrow A^{a'}[\bar{u}] + [u]E.$$

6. *Reaction 1.3.* Molecule $[\bar{t}]^{iab} B$ is modified by transducer-molecule $T^{ab}[t]^{iab}$ in reaction,

$$T^{ab}[t]^{iab} + [\bar{t}]^{iab} B \rightarrow T[\bar{w}] + [w] B^{ab}.$$

Note that now B encodes state ab in its duplex portion (state i is not kept since it is not required for effecting B's transition), and possesses sticky end $[w]$, which is complementary to the default sticky end of C.

7. *Other reactions.* Similar to reactions 1.1, 1.2, and 1.3.

3 Molecular Implementation

3.1 Step-by-Step Implementation

To demonstrate the practicality of our design, we next give a detailed description of the molecular implementation of the ADCA. The complete DNA molecule set will be described in Sect. 3.2.

1. *Reaction 0.1.* Figure 8 depicts an example molecular implementation of reaction 0.1,

$$I^i[u] + [\bar{u}] A^a \rightarrow I[t]^{ia} + [\bar{t}]^{ia} A.$$

For simplicity, only the end sequences of dangling-molecule A are depicted; for full sequences, see Figure 15 (a) in Appendix. Panels (a) and (b) respectively illustrate cases when $a = \texttt{WHITE}$ and $a = \texttt{BLACK}$. In molecule $[\bar{u}] A^a$, the state information $a \in \{\texttt{WHITE}, \texttt{BLACK}\}$, is encoded by the presence or absence of a DNA base pair between the sticky end $[\bar{u}]$ (sequence $\texttt{TA}$) and the half recognition site for endonuclease Bsl I (sequence $\texttt{GG/CC}$) in the duplex portion. This is further indicated in Figure 8 by the shaded blue (grey) region. In the case $a = \texttt{WHITE}$, the cleavage of ligation product IA by Bsl I produces a sticky end sequence $\texttt{GGT}$ for molecule I, and $\texttt{CCA}$ for molecule A. Both these unique sticky end sequences encode state information ia. When $a = \texttt{BLACK}$, a different pair of sticky ends $[t]/[\bar{t}]$ are generated ($\texttt{CGG/GCC}$).

Molecule A contains a pair of unnatural bases, *i.e.* synthetic bases other than the natural bases $\texttt{A}$, $\texttt{C}$, $\texttt{G}$, and $\texttt{T}$. They are required because the four-letter $\texttt{ACGT}$ natural vocabulary does not provide sufficient encoding space for our construction. For a survey on experimental synthesis of unnatural bases, see [12].

2. *Reaction 0.2.* Figure 9 depicts an example molecular implementation of reaction 0.2,

$$I[t]^{ia} + [\bar{t}]^{ia} R \rightarrow I^i[u] + [\bar{u}] R.$$

The endonuclease involved is EcoP15 I. This restores I to its default configuration with sticky end sequence $[u] = \texttt{AT}$.

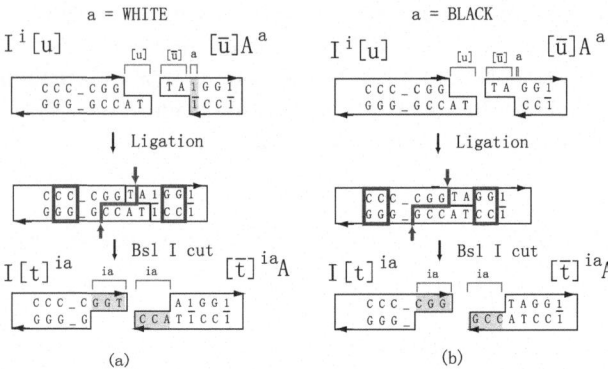

Fig. 8. Example molecular implementation of reaction 0.1. The red (dark) box and red (dark) arrows respectively indicate the recognition and cleavage sites for endonuclease Bsl I. The encoded state information is indicated with blue (grey) region. Base pair $1/\bar{1}$ is a pair of unnatural bases.

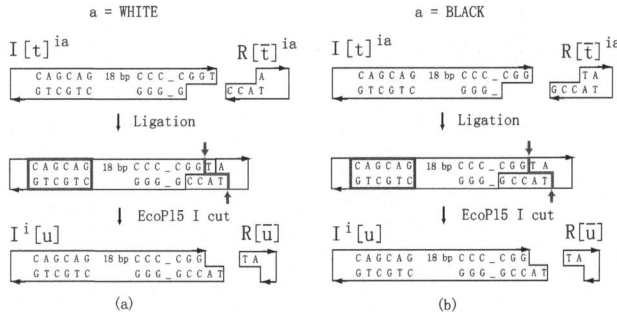

Fig. 9. Example molecular implementation of reaction 0.2

3. *Reaction 0.3.* Figure 10 depicts an example molecular implementation of reaction 0.3,

$$T^{ia}[t]^{ia} + [\bar{t}]^{ia}A \rightarrow T[\bar{v}] + [v]A^{ia}.$$

Here, we only illustrate the case $a = \mathtt{WHITE}$, and omit the similar case $a = \mathtt{BLACK}$ for brevity. Note that the state information ia initially encoded in the sticky end $[\bar{t}]$ (sequence CCA) of A is now encoded in the blue (grey) region in its duplex portion (sequence CGA/GCT).

4. *Reaction 1.1.* Figure 11 depicts an example molecular implementation of reaction 1.1,

$$A^{ia}[v] + [\bar{v}]B^b \rightarrow A[t]^{iab} + [\bar{t}]^{iab}B.$$

Again, we only illustrate the case $a = \mathtt{WHITE}$ and $b = \mathtt{WHITE}$ for brevity. This reaction is similar to reaction 0.1. Both the sticky ends $[t]$ and $[\bar{t}]$ encode state information iab.

Two technical points warrant explanation in this reaction. First, the bases labeled with red (dark) circles contain phosphorothioate bond and are hence resistant to enzyme cleavage. This modification of the bases is required to prevent the unwanted

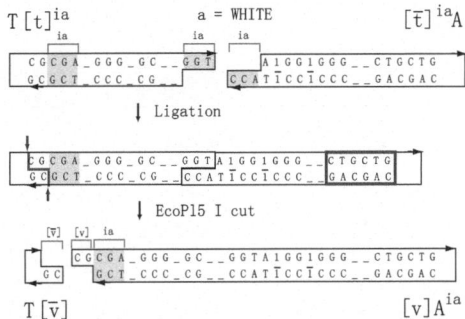

Fig. 10. Example molecular implementation of reaction 0.3

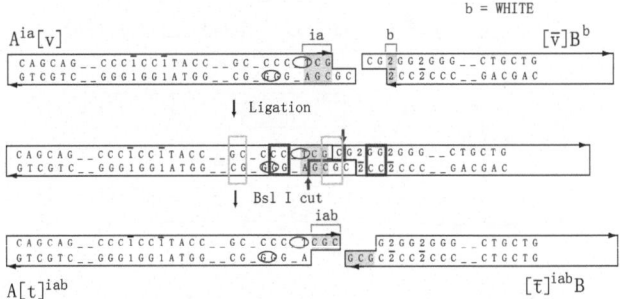

Fig. 11. Example molecular implementation of reaction 1.1. The red (dark) box and red (dark) arrows respectively indicate the recognition and cleavage sites for endonuclease Bsl I. The bases labeled with red (dark) circles contain phosphorothioate bond and are hence resistant to enzyme cleavage. The light blue (grey) box indicates the Mwo I recognition site.

cleavage of the DNA duplex by Mwo I, whose recognition sites are indicated with blue (grey) boxes in the figure. This trick will be used again in reaction 1.2.2. Second, we assume here that the cleavage by endonuclease Bsl I will occur, but the cleavage by EcoP15 I will not occur (note that both molecule A and molecule B contain EcoP15 I recognition site, *i.e.* CAGCAG/CTGCTG). This assumption is based on the fact that Bsl I, a Type II endonuclease, can act far more efficiently than EcoP15 I, a Type III endonuclease.

5. *Reaction 1.2.1.* Figure 12 (a) depicts an example molecular implementation of reaction 1.2.1,

$$A^q[t]^{iab} + [\bar{t}]^{iab} R^{a'} \rightarrow A[e]^{qa'} + [\bar{e}]^{qa'} R.$$

Recall that this reaction effects a state transition for molecule A, as specified by the rule $iab \rightarrow a'$. Here, the rule molecule incorporates a spacer region, the length of which (L bp) encodes the new state information a'. In particular, when $L = 2$, $a' = $ BLACK; when $L = 6$, $a' = $ WHITE.

When $a' = $ WHITE (the EcoP15 I cleavage step not shown in Figure 12 (a)), molecule A is restored to the desired target configuration $[\bar{u}]A^{a'}$, with the default

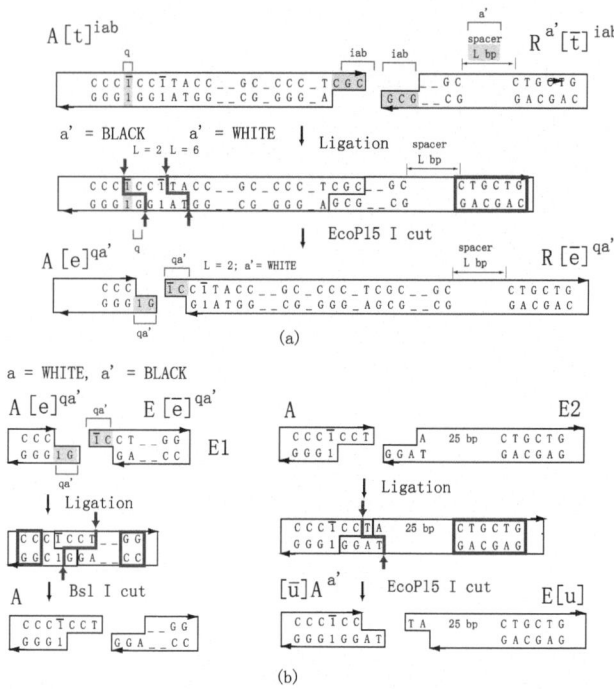

Fig. 12. Example molecular implementation of reaction 1.2. Panel (a): reaction 1.2.1. Panels (b): reaction 1.2.2.

sticky end $[\bar{u}] = $ AT. In this case, the next step, reaction 1.2.2, is not required. The reaction can thus be rewritten as,

$$A^q[t]^{iab} + [\bar{t}]^{iab} R^{a'} \to A^{a'}[\bar{u}] + [u]R.$$

However, when $a' = $ BLACK (the EcoP15 I cleavage step shown in the figure), molecule A is modified to $A[e]^{qa'}$, with a unique sticky end $[e] = $ 1G that encodes both A's type information $q = q_A$ and A's new state a' (Recall that $q \in \{q_A, q_B, q_C\}$ encodes type information, in the case illustrated here, $q = q_A$. This information is initially encoded in the blue (grey) duplex portion of A, in the form of sequence $1/\bar{1}$). Then the reaction proceeds to the next step, reaction 1.2.2, which will finish the state transition for A.

In our molecular implementation, in a transition $xyz \to y'$, the values of y, z, and z' cooperatively determine the spacer length L, which in turn decides the result of the transition. For detail, see Figure 15 (c) in Appendix.

6. *Reaction 1.2.2.* Figures 12 (b) depicts the case $a = $ WHITE, $a' = $ BLACK, which follows from the case illustrated in Figures 12 (a).

$$A[e]^{qa'} + [\bar{e}]^{qa'} E \to A^{a'}[\bar{u}] + [u]E.$$

Here, extension-molecule $E[\bar{e}]^{qa'}$ restores A to its default configuration with sticky end $[\bar{u}] = $ AT. However, now A encodes new state a' in its duplex portion.

7. *Other reactions.* Similar to the above reactions, and hence omitted for brevity.

3.2 Complete Molecule Set

The complete set of DNA molecules constituting ADCA are described in Figure 15 and Figure 16 in the Appendix. The dangling-molecules (Figure 15 (a)) and the floating rule-molecules (Figure 15 (c)) are the programmable parts of the ADCA: the selections of dangling-molecules and rule-molecules respectively determine the initial configuration and the transition rules of the ADCA. Note that all the sticky ends of rule-molecules are unique. In contrast, the floating assisting-molecules, *i.e.* transducer-molecules (Figure 16) and extension-molecules (Figure 15 (b)), only assist in the proper operation of the ADCA and are non-programmable.

3.3 Futile Reactions

Besides the main reactions described above, there exist *reversible futile reactions* in the system. These futile reactions are carefully engineered such that they will not block the main reactions. Futile reactions, however, can also be used to maintain a dynamic balance among the floating molecules, and, in doing so, ensure the proper operation of the devices. Figure 13 shows an example where futile reactions are used to maintain a balance between R and T molecules that have complementary sticky ends. For a more detailed discussion of futile reactions, see [35].

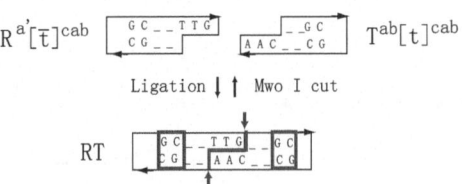

Fig. 13. An example futile reaction

3.4 Computer Simulation

Fully debugging the above molecular implementation requires meticulous inspection of every step of the ADCA operation, which can become exceedingly tedious. We thus developed a computer simulator and used it to test and debug the ADCA. The simulator takes as input the DNA sequences which specifies an ADCA instance, simulates the operations of ADCA, and gives graphical output. For detail, see http://pengyin.org/paper/dnaCA/.

4   Two-Dimensional ADCA

To illustrate the operational principle of 2D ADCA, we first present an abstract view of the 1D ADCA in Figure 14 (a) and (b). Figure 14 (a) illustrates a reaction wave of the 1D ADCA: the reaction wave starts at initiator I and travels sequentially down the one-dimensional track. Figure 14 (b) examines one individual dangling-molecule X, where $X = A, B, C$. Assume w.l.o.g., $X = B$. As shown in Figure 14 (b), B in a 1D ADCA undergoes the following four phases in one full reaction cycle.

1. *Phase 1.* B has a sticky end that is complementary to its left neighbor A (indicated by a solid square to the left of B in Figure 14 (b)). This is before reaction 1.1 as depicted in Figure 7. In this phase, B encodes in its duplex portion its own state information denoted by C (C for center).
2. *Phase 2.* In reaction 1.1, B interacts with its left (*i.e.* west) neighbor, and enters phase 2. Now B encodes in its sticky end both the state information of itself, C, and the state information of its west neighbor, W. This sticky end is complementary to a floating transducer-molecule, indicated by a circle around B.
3. *Phase 3.* After reaction 1.3, B now encodes state information CW in its duplex portion, and possesses a sticky end complementary to its right (*i.e.* east) neighbor (indicated by a solid square to the right of B).
4. *Phase 4.* In reaction 2.1, B interacts with its east neighbor, and enters phase 4. Now B encodes in its sticky end the state information of itself C, its west neighbor W, and its east neighbor E. This sticky end thus encodes all the state information required to effect a state transition for B, and is recognized by a floating rule-molecule (indicated by a thick circle). And this state transition restores B to its default configuration, finishing a full cycle.

With the above understanding of the 1D ADCA, we can extend it to 2D as follows. First, we take care of reaction waves by positioning two arrays of initiators as shown

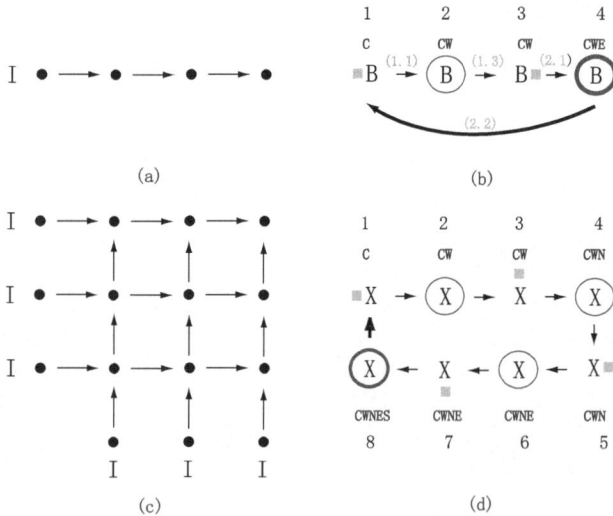

Fig. 14. (a) (b) Operational overview of 1D ADCA. (c) (d) Operational overview of two-dimensional ADCA. In panels (b) and (d), black numbers indicate the phases of X; blue (grey) numbers indicate reactions corresponding to reactions depicted in Figure 7; red (dark) letters indicate the state information carried by X.

in Figure 14 (c). Each initiator can send out a reaction wave that travels either horizontally or vertically. Next, we take care of the information flow and synchronization, by again examining one single molecule X. As shown in Figure 14 (d), we engineer the system such that molecule X undergoes 8 phases. During these 8 phases, X sequentially interacts with its west (W), north (N), east (E), and south (S) neighbors to garner the state information from each of them. As such, upon entering phase 8, X carries in its sticky end the state information CWNES. This state information is sufficient to effect a state transition for X. As in the 1D case, X will undergo a state transition and re-enters phase 1, completing a full circle.

5 Conclusion

In this paper, we present the theoretical design and molecular implementation of 1D ADCA and describe how to extend it to 2D.

Open question: can we simplify the current complex design?

References

1. L. Adleman. Molecular computation of solutions to combinatorial problems. *Science*, 266:1021–1024, 1994.
2. P. Alberti and J.L. Mergny. DNA duplex-quadruplex exchange as the basis for a nanomolecular machine. *Proc. Natl. Acad. Sci. USA*, 100:1569–1573, 2003.
3. R. Barish, P.W.K. Rothemund, and E. Winfree. Algorithmic self-assembly of a binary counter using DNA tiles. 2005. In preparation.
4. J. Bath, S.J. Green, and A.J. Turberfield. A free-running DNA motor powered by a nicking enzyme. *Angew. Chem. Intl. Ed.*, 44:4358–4361, 2005.
5. Y. Benenson, R. Adar, T. Paz-Elizur, Z. Livneh, and E. Shapiro. DNA molecule provides a computing machine with both data and fuel. *Proc. Natl. Acad. Sci. USA*, 100:2191–2196, 2003.
6. Y. Benenson, B. Gil, U. Ben-Dor, R. Adar, and E. Shapiro. An autonomous molecular computer for logical control of gene expression. *Nature*, 429:423–429, 2004.
7. Y. Benenson, T. Paz-Elizur, R. Adar, E. Keinan, Z. Livneh, and E. Shapiro. Programmable and autonomous computing machine made of biomolecules. *Nature*, 414:430–434, 2001.
8. N. Chelyapov, Y. Brun, M. Gopalkrishnan, D. Reishus, B. Shaw, and L. Adleman. DNA triangles and self-assembled hexagonal tilings. *J. Am. Chem. Soc.*, 126:13924–13925, 2004.
9. Y. Chen, M. Wang, and C. Mao. An autonomous DNA nanomotor powered by a DNA enzyme. *Angew. Chem. Int. Ed.*, 43:3554–3557, 2004.
10. L. Feng, S.H. Park, J.H. Reif, and H. Yan. A two-state DNA lattice switched by DNA nanoactuator. *Angew. Chem. Int. Ed.*, 42:4342–4346, 2003.
11. Y. He, Y. Chen, H. Liu, A.E. Ribbe, and C. Mao. Self-assembly of hexagonal DNA two-dimensional (2D) arrays. *J. Am. Chem. Soc.*, 127:12202–12203, 2005.
12. A.A. Henry and F.E. Romesberg. Beyond A, C, G, and T: augmenting nature's alphabet. *Curr. Opin. Chem. Biol.*, 7:727–733, 2003.

13. J. Kuramochi and Y. Sakakibara. Intensive in vitro experiments of implementing and executing finite automata in test tube. In *Proc. 11th International Meeting on DNA Computing*, pages 59–67, 2005.

14. T.H. LaBean, H. Yan, J. Kopatsch, F. Liu, E. Winfree, J.H. Reif, and N.C. Seeman. The construction, analysis, ligation and self-assembly of DNA triple crossover complexes. *J. Am. Chem. Soc.*, 122:1848–1860, 2000.

15. D. Liu, M. Wang, Z. Deng, R. Walulu, and C. Mao. Tensegrity: Construction of rigid DNA triangles with flexible four-arm dna junctions. *J. Am. Chem. Soc.*, 126:2324–2325, 2004.

16. J. Malo, J.C. Mitchell, C. Venien-Bryan, J.R. Harris, H. Wille, D.J. Sherratt, and A.J. Turberfield. Engineering a 2D protein-DNA crystal. *Angew. Chem. Intl. Ed.*, 44:3057–3061, 2005.

17. C. Mao, T.H. LaBean, J.H. Reif, and N.C. Seeman. Logical computation using algorithmic self-assembly of DNA triple-crossover molecules. *Nature*, 407:493–496, 2000.

18. J.C. Mitchell, J.R. Harris, J. Malo J, J. Bath, and A.J. Turberfield. Self-assembly of chiral DNA nanotubes. *J. Am. Chem. Soc.*, 126:16342–16343, 2004.

19. P.W.K. Rothemund. A DNA and restriction enzyme implementation of Turing machines. In R. J. Lipton and E.B. Baum, editors, *DNA Based Computers: Proceedings of the DIMACS Workshop, April 4, 1995, Princeton University*, volume 27, pages 75 – 119, Providence, Rhode Island, 1996. American Mathematical Society.

20. P.W.K. Rothemund. Generation of arbitrary nanoscale shapes and patterns by scaffolded DNA origami. 2005.

21. P.W.K. Rothemund, A. Ekani-Nkodo, N. Papadakis, A. Kumar, D.K. Fygenson, and E. Winfree. Design and characterization of programmable DNA nanotubes. *J. Am. Chem. Soc.*, 126:16344–16353, 2004.

22. P.W.K. Rothemund, N. Papadakis, and E. Winfree. Algorithmic self-assembly of DNA sierpinski triangles. *PLoS Biology 2 (12)*, 2:e424, 2004.

23. N.C. Seeman. From genes to machines: DNA nanomechanical devices. *Trends in Biochemical Sciences*, 30:119–125, 2005.

24. W.B. Sherman and N.C. Seeman. A precisely controlled DNA biped walking device. *Nano Lett.*, 4:1203–1207, 2004.

25. J.S. Shin and N.A. Pierce. A synthetic DNA walker for molecular transport. *J. Am. Chem. Soc.*, 126:10834–10835, 2004.

26. M.N. Stojanovic, S. Semova, D. Kolpashchikov, J. Macdonald, C. Morgan, and D. Stefanovic. Deoxyribozyme-based ligase logic gates and their initial circuits. *J. Am. Chem. Soc.*, 127:6914–6915, 2005.

27. Y. Tian, Y. He, Y. Chen, P. Yin, and C. Mao. Molecular devices - a DNAzyme that walks processively and autonomously along a one-dimensional track. *Angew. Chem. Intl. Ed.*, 44:4355–4358, 2005.

28. A.J. Turberfield, J.C. Mitchell, B. Yurke, Jr. A.P. Mills, M.I. Blakey, and F.C. Simmel. DNA fuel for free-running nanomachines. *Phys. Rev. Lett.*, 90:118102, 2003.

29. E. Winfree, F. Liu, L.A. Wenzler, and N.C. Seeman. Design and self-assembly of two-dimensional DNA crystals. *Nature*, 394(6693):539–544, 1998.

30. S. Wolfram. *A new kind of science*. Wolfram Media, Inc., Champaign, IL, 2002.

31. H. Yan, T.H. LaBean, L. Feng, and J.H. Reif. Directed nucleation assembly of DNA tile complexes for barcode patterned DNA lattices. *Proc. Natl. Acad. Sci. USA*, 100(14):8103–8108, 2003.

32. H. Yan, S.H. Park, G. Finkelstein, J.H. Reif, and T.H. LaBean. DNA-templated self-assembly of protein arrays and highly conductive nanowires. *Science*, 301(5641):1882–1884, 2003.

33. H. Yan, X. Zhang, Z. Shen, and N.C. Seeman. A robust DNA mechanical device controlled by hybridization topology. *Nature*, 415:62–65, 2002.

34. P. Yin, A.J. Turberfield, and J.H. Reif. Designs of autonomous unidirectional walking DNA devices. In *Proc. 10th International Meeting on DNA Computing*, pages 119–130, 2004.

35. P. Yin, A.J. Turberfield, S. Sahu, and J.H. Reif. Design of an autonomous DNA nanomechanical device capable of universal computation and universal translational motion. In *Proc. 10th International Meeting on DNA Computing*, pages 344–356, 2004.

36. P. Yin, H. Yan, X.G. Daniell, A.J. Turberfield, and J.H. Reif. A unidirectional DNA walker moving autonomously along a linear track. *Angew. Chem. Int. Ed.*, 43:4906–4911, 2004.

37. B. Yurke, A.J. Turberfield, Jr. A.P. Mills, F.C. Simmel, and J.L. Neumann. A DNA-fuelled molecular machine made of DNA. *Nature*, 406:605–608, 2000.

Appendix: Complete Molecule Set

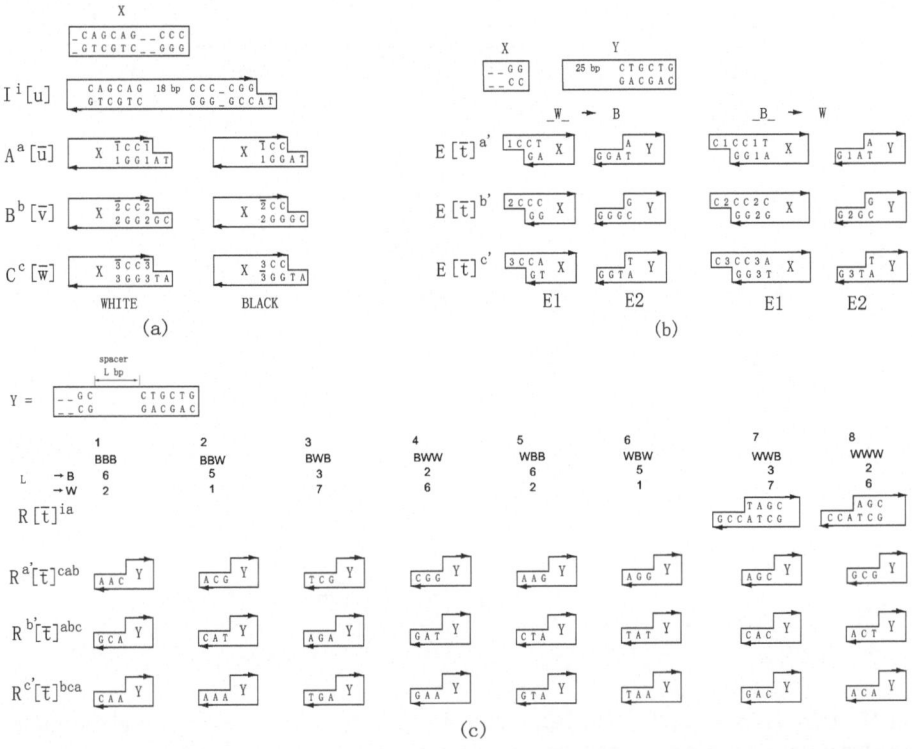

Fig. 15. (a)Dangling-molecules. Left and right panels respectively depict the cases when the encoded information $a/b/c =$ WHITE and BLACK. (b) Extension-molecules. Left and right panels respectively depict the cases when the transition is x, WHITE, $z \rightarrow$ BLACK and x, BLACK, $z \rightarrow$ WHITE, where $x, z \in \{$BLACK, WHITE$\}$. (c) Rule-molecules. The eight columns (1-8) correspond to the eight possible configurations of xyz in the rule $xyz \rightarrow y'$, where $x, y, z, y' \in \{$BLACK, WHITE$\}$. The symbol BBW stands for the configuration $xyz = \{$BLACK, BLACK, WHITE$\}$. The value of L is determined cooperatively by y, z, and z'.

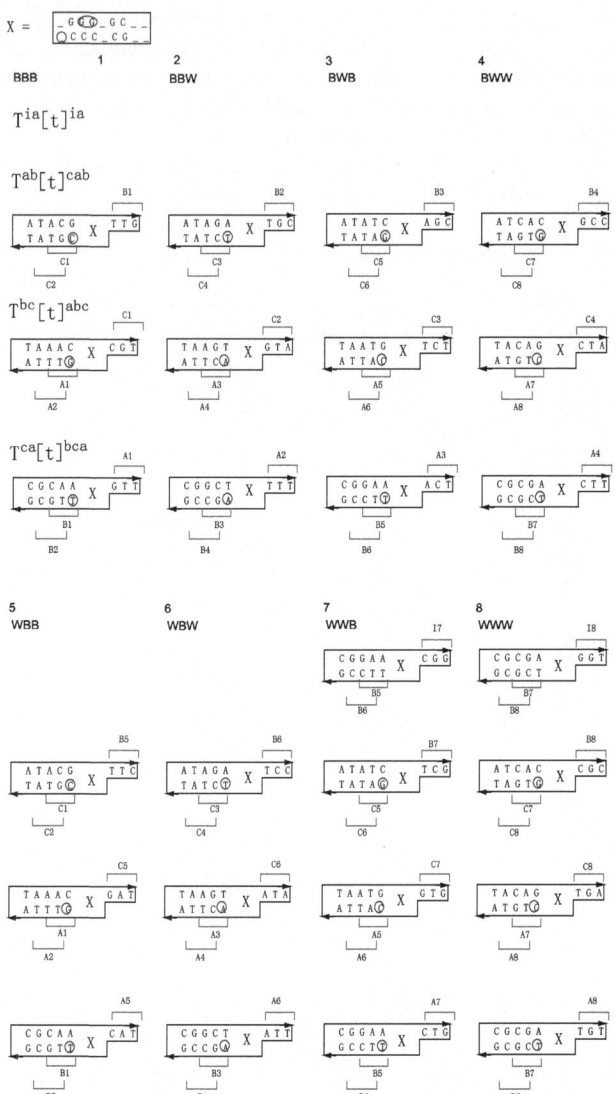

Fig. 16. Transducer-molecules. The bases labeled with red (dark) circles contain phosphorothioate bond and are hence resistant to enzyme cleavage (see reaction 1.1, Figure 11).

Use of DNA Nanodevices in Modulating the Mechanical Properties of Polyacrylamide Gels

Bernard Yurke[1], David C. Lin[2], and Noshir A. Langrana[2]

[1] Bell Laboratories, 600 Mountain Ave., Murray Hill, NJ 07974
yurke@lucent.com
[2] Department of Mechanical and Aerospace Engineering, Rutgers University,
Piscataway, NJ 08854
dclin@eden.rutgers.edu, langrana@rutgers.edu

Abstract. Here we show that bulk materials can be given new properties through the incorporation of DNA-based nanodevices. In particular, by employing simple nanodevices as crosslinks in polyacrylamide gels we have made the mechanical properties of these gels responsive to the presence of particular DNA strands. Two examples will be focused on here. One consists of a polymer system that can be switched between a sol and a gel state though the application of DNA strands that either form crosslinks or remove crosslinks. The other consists of a hydrogel whose crosslinks incorporate a motor domain. The stiffness of this hydrogel can be altered through the application of fuel strands, which stiffen and lengthen the crosslinks, or through the application of removal strands which remove the fuel strands form the motor domain. Such DNA-responsive gels may find applications in biomedical technology ranging from drug delivery to tissue engineering.

1 Introduction

A variety of DNA-based nanomachines have been devised [1]-[14]. Many of these employ fuel and removal strands or set and reset strands to advance the machine through its various states. It is likely that such machines will have application in bio-medical fields, chemistry, and nanotechnology. Here, by way of examples of how polyacrylamide gels can be given new functionality by the incorporation of DNA-based nanodevices, we demonstrate that DNA-based nanodivices may have application in materials science as well. Two examples will be focused on. One consists of a polymer system that can be converted from a sol to a gel or from a gel to a sol through the application of set and reset DNA strands [15, 16]. In contrast to the more usual thermoset gels, these gels do not require cycling of the material across a sol-gel transition by varying the temperature. Nor do they require a change of the chemical composition of the buffer, apart from the addition of DNA strands. The other example consists of a hydrogel whose stiffness can be modulated as a function of time through the application of DNA strands [17]. The crosslinks of this gel posses a motor domain to which fuel strands can bind. The binding of fuel strands to the motor domain lengthens and stiffens the DNA-crosslinks and thereby stiffens the gel. The gel can

A. Carbone and N.A. Pierce (Eds.): DNA11, LNCS 3892, pp. 417–426, 2006.

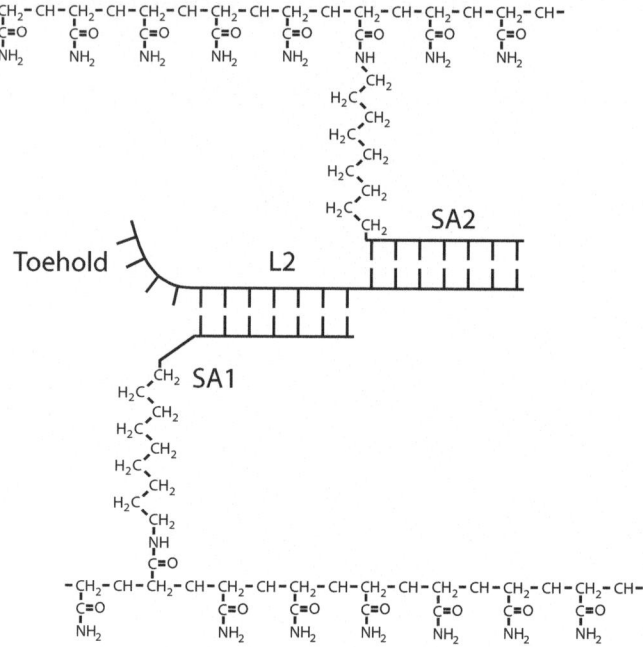

Fig. 1. A DNA-crosslinked gel

be returned to a more relaxed state through the application of removal strands that remove the fuel strands though toe-hold mediated strand exchange. The ability to change the stiffness of a gel as a function of time and in a biologically compatible way makes these gels an attractive candidate for tissue engineering applications, particularly since the shape, movement, proliferation rate, and differentiation of tissue cells are influenced by the mechanical properties of their environment.

2   DNA Induced Sol-Gel Transitions

The use of DNA as a crosslinking agent for a hydrogel was first reported by Nagahara and Mathsuda [18]. In their studies poly(N,N-dimethylacrylamide-co-N-acryloxyloxysuccinimide) was reacted with 5'-amino-modified 10-mer oligonucleotides consisting of all adenine bases (oligo A) or thymine bases (oligo T) to form polymer chains with short DNA side branches. Crosslinking was achieved by two different methods. One method consisted of mixing polymer with oligo A side branches with polymer having oligo T side branches. Hybridization between the oligo A and oligo T side branches formed a crosslinked network. The second method consisted of mixing 20-mer adenine strands with polymer having oligo T side branches to form a crosslinked network.

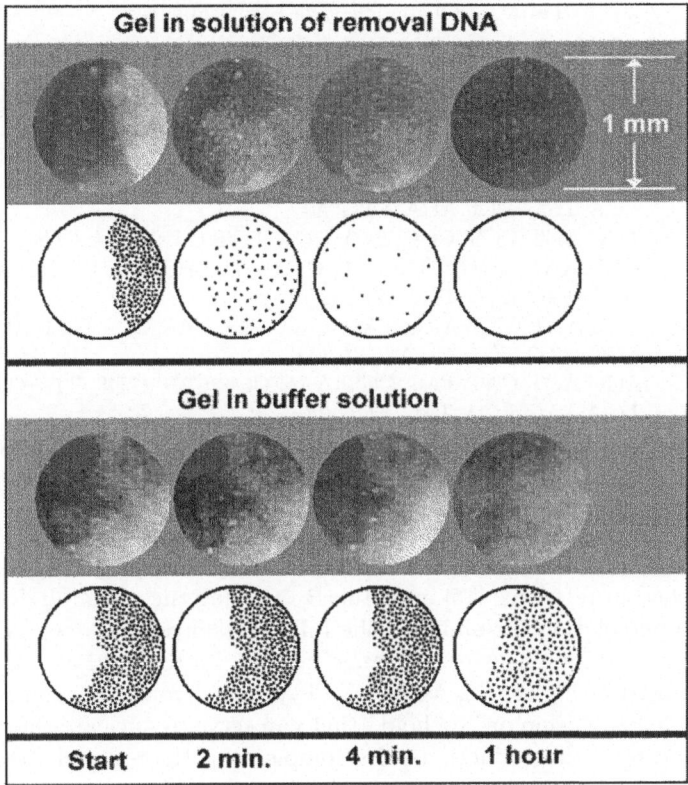

Fig. 2. Dissociation of a DNA crosslinked gel through the application of a removal strand

The method we have used to establish DNA crosslinks in a polyacrylamide gel is most similar to the second of the methods of Nagahara and Mathsuda. The structure of the crosslink is illustrated in Fig. 1. Polyacrylamide having 20-mer oligo sidebranches labeled SA1 was prepared by copolymerizing acrylamide with Acrydite modified SA1 oligomers. Polyacrylamide having a 20-mer oligo sidebranches labeled SA2 was similarly prepared. The methods are given in the Methods section and the sequences for the oligomers are given in Tabel 1. The resulting polymer solutions consisted of viscous fluids which when mixed together remained fluid. Upon the addition of the oligomer L2, which has a region that is complementary to SA1 and a region that is complementary to SA2, and mixing, hybridization of L2 with the SA1 and SA2 sidebranches, establishes crosslinks resulting in rapid gelation of the mixture. The gel sets on the time scale of a few seconds. The stiffness of the gel is a function of the amount of L2 added, that is, the number of crosslinks formed. Our study of the stiffness of this gel as a function of crosslink density along with our studies of the viscocity as a function

Table 1. Oligonucleotide base sequences

Strand	Sequence (5' to 3')
SA1	ACG GAG GTG TAT GCA ATG TC
SA2	CAT GCT TAG GGA CGA CTG GA
L2	ACT AAT CCT CAG ATC CAG CTA AGT AGG TGT GTG CGA TAC TTT ACA TTG AT
L3	TCC AGT CGT CCC TAA GCA TGT GTT CGA CGG TAC AAG AAG AGG GTT ACG CTA ATG AGT GCT GAC ATT GCA TAC ACC TCC GT
R1	ATC AAT GTA AAG TAT CGC ACA CAC CTA CTT AGC TGG ATC TGA GGA TTA GT
F1	AGC ACT CAT TAG CGT AAC CCT CTT CTT GTA CCG TCG AAC AGA TAG AGC TG
CF1	CAG CTC TAT CTG TTC GAC GGT ACA AGA AGA GGG TTA CGC TAA TGA GTG CT

of temperature when the gel is heated above the DNA melting temperature have been published in reference [15] and [16]. By varying the crosslink density from close to the percolation threshold to the fully crosslinked state we were able to vary the Young's modulus over the range from 39 Pa to 9.0 kPa.

As illustrated in Fig. 1, L2, when fully hybridized with SA1 and SA2, has a single stranded extension or "toehold" that can serve to nucleate the removal of the L2 strand by the introduction of its complement, the removal strand R1. To demonstrate that the DNA-hydrogel can be dissociated by the application of the R1 strand, we prepared a gel in which L2 was added at 33% the amount required for full crosslinking. This amount of L2 is well above that needed to achieve the percolation threshold of 23.6%. The gel incorporated fluorescent beads (Flourosbrite Carboxy YG latex microspheres; Polysciences, Inc.) to enhance visibility. A small piece of this gel having a volume of ∼0.01 mm^3 was placed in each of two wells, one containing 20 μl of TE buffer and the other containing a like volume of 1 mM R1 stock solution. The photographs of figure two show a sequence of images of the two gels. The gel in the presence of the removal strand dissociates in about 4 min. In contrast, the gel in buffer solution exhibited some swelling but remained intact over a period of 24 hr after which the experiment was terminated. We have thus shown that DNA-crosslinked gels can be formed and dissociated without having to heat or cool through a transition temperature, or without having to change the buffer composition, apart from adding the appropriate DNA strands to the buffer solution.

3   Crosslinks with Motor Domains

In this section we consider the case when the crosslinking strand L2 of Fig. 1 is replaced by a crosslinking strand L3 which, when fully hybridized with SA1 and SA2 has a central region that remains single stranded as shown in Fig. 3.

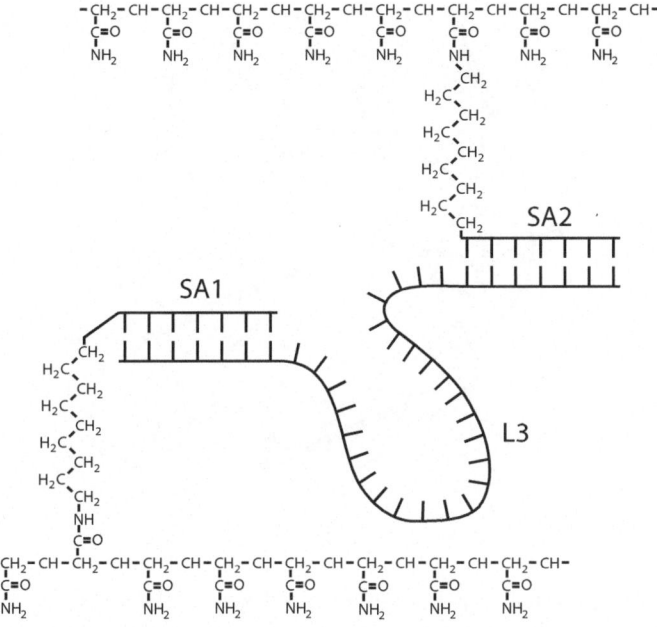

Fig. 3. A DNA-crosslinked gel in which the crosslike possesses a motor domain

This central region functions as a motor domain to which the fuel strand F1 of Table 1 can bind. As shown in Fig. 4(b), when the F1 strand binds with the motor domain M the DNA crosslink is transformed from a floppy random coil structure to a relatively rigid strand of duplex DNA. In the process the two ends of the crosslink will be pushed apart. The maximum force that can develop on the ends of the crosslinks can be estimated as the ratio of the average free energy change in forming a base pair divided by the distance the ends move apart as a base pair is formed. The force estimated from such a calculation [17] is 17 pN. For long DNA crosslinks this force will not be realized due to Euler buckling. For the case of the 60 bp crosslink used in our investigations one expects Euler buckling to occur once 5 pN of force has been developed [17]. This force sets the scale for the maximum outward force that the DNA crosslinks are expected to exert on the polyacrylamide network. As the ends of the crosslinks are pushed apart, the polyacrylamide strands will be stretched from a random coil configuration to a more linear configuration as illustrated by the change in the acrylamide strand configurations in going from (a) to (b) in Fig. 4. The opposing force that the crosslink must do work against as the ends of the crosslinks are pushed apart is thus the entropic force developed by the stretched polyacrylamide strands. One thus expects that after the formation of duplex crosslinks, the crosslinks will be under compression while the polyacrylamide strands will be under tension. A structure possessing some members that are under compression while others are under tension is said to posses prestress. Prestress is generally expected to stiffen the gel.

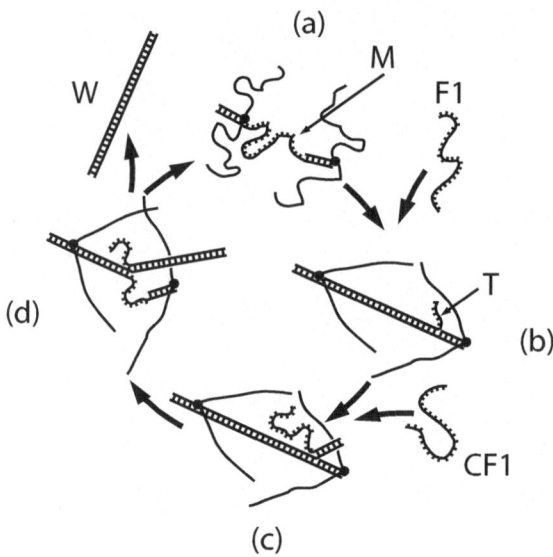

Fig. 4. Cycle of operation of a DNA-crosslinked gel in which the crosslinks posses a motor domain

As illustrated in Fig. 4 by the transitions from (b) to (c) to (d) to (a), the fuel strands can be removed from the gel network through the addition of the complement CF1 of the fuel strand. When the fuel strand F1 is fully hybridized with the motor domain M of the crosslink, a single stranded region remains, as shown in (b). This single stranded region or toehold serves as an attachment site for the removal strand CF1, as shown in (c). This initiates a strand displacement process that proceeds via the random walk of a three strand junction, as shown in (d). Once the branch point reaches the far side of the motor domain F1 and CF1 become fully hybridized and form the waste product W. In the process the gel is restored to its relaxed configuration (a).

To test whether a DNA-crosslinked gel possessing motor domains would stiffen upon application of fuel strands and return to its relaxed configuration upon application of the removal strand the apparatus depicted schematically in Fig. 5 was constructed. The apparatus provides a means of introducing fuel or removal strands into the DNA crosslinked gel through electrophoresis and a means for measuring the gel compliance by applying a known magnetic force on a magnetic inclusion embedded in the gel. The apparatus consisted of a funnel whose upper portion served as a buffer reservoir in which the negative electrode for gel electrophoresis was immersed. The neck of the funnel had a spheroidal chamber in which the DNA-crosslinked gel and the magnetic inclusion resided. A bis-crosslinked 20% polyacrylamide gel plug at the bottom of the funnel prevented loss of DNA-crosslinked gel from the sample chamber. The bottom of the funnel was immersed in buffer reservoir into which the positive electrode for electrophoresis was immersed. The magnetic inclusion consisted of a

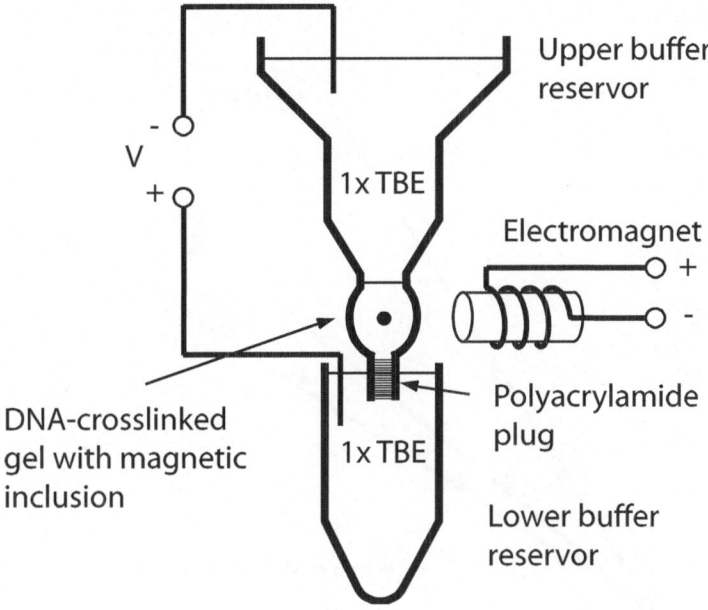

Fig. 5. Schematic of apparatus for measuring gel stiffness

spherical steal bead 0.79 mm in diameter and force was applied to the bead via an electromagnet. Details on the calibration procedure for the force is given in reference [17]. In addition to 1× TBE buffer the reservoirs contained a small amount of universal pH indicator solution (pH 2-10, Science Kit, Tonawanda, NY) to monitor the buffer condition. The buffer was replaced by fresh buffer when the indicator solution signaled that the buffer had become exhausted. TAMRA or TET dye labeled fuel strands were used so that the migration of the fuel strands into the gel by electrophoresis could be followed. The use of unlabeled removal strands allowed the progress of the clearing of the motor strands of the fuel strands to also be followed.

For the experiment reported here 12 μl each of SA1 and SA2 were added to the sample chamber, followed by enough $L3$ solution to produce 50% crosslinking. The magnetic inclusion was added and the sample was heated above the melting temperature and stirred to insure uniformity of composition. The sample was removed from heat and manipulated to maintain the magnetic inclusion at the center of the cell as the sample cooled and the gel set. Figure 6 shows data from one of our runs. The vertical axis is the magnitude F of the force applied on the magnetic inclusion and the horizontal axis is the resulting displacement δ. Three sets of data points are shown. The diamonds correspond to the measurements performed on in the gel before fuel or removal strands were added. The triangles correspond to the measurements performed once the fuel strands had fully penetrated the gel. The triangles correspond to the measurements after the removal strands had cleared the gel of fuel strands. Each of these data sets are closely

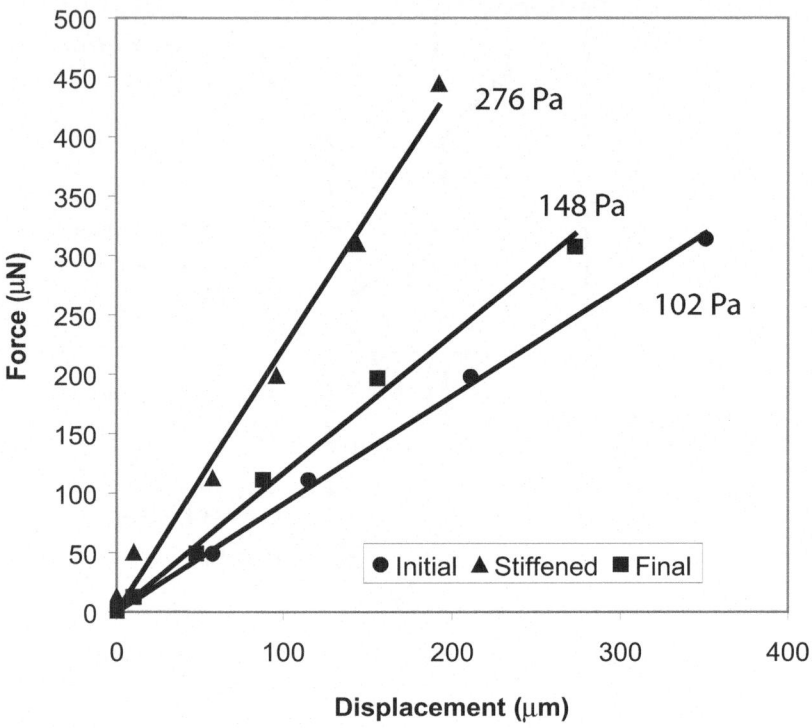

Fig. 6. Gel compliance data

dispersed about a straight line indicating that the measurements were performed in a linear regime. The Young's modulus E of the gel can be determined form the slope of force vs displacement curve. To perform this conversion we used the formula

$$E = \frac{F}{\delta} \left[\frac{4n^5 - 5n^4 - 5n^3 + 5n^2 + 5n - 4}{8\pi R_0(n^5 + n^4 + n^3 + n^2 + n)} \right], \tag{1}$$

where R_0 is the radius of the spherical inclusion and n is the ratio R_1/R_0 where R_1 is the distance between the center of the inclusion and the sample chamber wall. This equation is strictly true for a spherical inclusion in a spherical sample chamber [17, 19]. However, via validation experiments and numerical simulations [17], we have shown that it gives accurate results for our sample chamber geometry and when n is taken as the shortest distance between the center of the inclusion and the sample chamber wall for an approximately centered inclusion. As indicated in Fig. 6, the initial elastic modulus of the gel was measured to be 102 Pa. Upon addition of the fuel strands the gel stiffened to a value of 276 Pa. After the fuel strands had been removed the elastic modulus dropped to a value of 148 Pa. Thus a stiffening of more than a factor of 2.5 was observed. The incomplete return of the gel to its initial stiffness may be due to residual fuel strands not cleared by the removal strand or may be due to gel swelling.

4 Conclusion

We have shown that, by incorporating DNA-crosslinks into a hydrogel, the gel can be given functionality that would be difficult to obtain by other means. Gels can be constructed whose sol-gel transitions are not tied to a thermodynamic transition. Also, gels can be constructed whose stiffness as a function of time can be controlled by the application of DNA. For tissue cells, it is known [20, 21] that the mechanical properties of the extracellular matrix influences cell morphology, cell movement, and cell differentiation. DNA-crosslinked gels, whose mechanical properties can be changed in a relatively benign way through the application of DNA strands may thus find application in tissue engineering and organ growth.

5 Methods

The oligomers, including Acrydite modified oligomers were purchased, purified, from Integrated DNA Technologies (Coralville IA). The manufactures numbers for quantity of DNA were relied upon in preparing the stock solutions at 3 mM concentration in TE buffer and were not independently checked. Polymer with SA1 side branches was produced by first preparing a solution consisting of 10% acrylamide solution (consisting of 40% acrylamide in H_2O, by weight), 60% SA1 stock solution, 20% H_2O, and 10% 10× TBE buffer. Quantities ($\sim$ 100 μl) of this solution were degassed by bubbling dry nitrogen through the solution for five minutes. Polymerization was initiated by adding 1 μl of a freshly prepared solution of $\sim$20% ammonium persulfate and catalyzed by the addition of 1 μl TEMED. Nitrogen was bubbled through the mixture for another five minutes. Polymer with SA2 side branches was similarly prepared.

References

1. C. Mao, W. Sun, Z. Shen, and N. C. Seeman, "A nanomechanical device based on the B-Z transition of DNA," Nature **397**, 144 (1999).
2. B. Yurke, A.J. Turberfield, A.P. Mills, Jr., F.C. Simmel, and J.L. Neumann, "A DNA fuelled molecular machine made of DNA," Nature **406**, 605 (2000).
3. F.C. Simmel and B. Yurke, "Using DNA to construct and power a nanoactuator," Phys. Rev. E **63**, 041913 (2001).
4. F.C. Simmel and B. Yurke, "A DNA-based molecular device switchable between three distinct mechanical states," Appl. Phys. Lett. **80**, 883 (2002).
5. H. Yan, X. Zhang, Z. Shen, and N. C. Seeman, "A robust DNA mechanical device controlled by hybridization topology," Nature **415**, 62 (2002).
6. B. Yurke and A.P. Mills, Jr., "Using DNA to power nanostructures," Genet. Program. Evol. Mach. **4**, 111 (2003).
7. A.J. Turberfield, J.C. Mitchell, B. Yurke, A.P. Mills, Jr., M.I. Blakey, and F.C. Simmel, "DNA fuel for free-running nanomachines," Phys. Rev. Lett. **90**, 118102 (2003).
8. L. Feng, S.H. Park, J.H. Reif, and H. Yan, "A two-state DNA lattice switched by DNA nanoactuator," Angew. Chem. Int. Ed. **42**, 4342 (2003).

9. J.J. Li and W. Tan, "A single DNA molecule nanomotor," Nano Lett. **2**, 315 (2002).
10. P. Alberti and J.L. Mergny, "DNA duplex-quadruplex exchange as the basis for a nanomolecular machine," Proc. Natl. Acad. Sci. U. S. A. **100**, 1569 (2003).
11. W. U. Dittmer, A. Reuter, and F. C. Simmel, "A DNA-based machine that can cyclically bind and release thrombin," Angew. Chem. Int. Ed. **43**, 3549 (2004).
12. S. P. Liao and N. C. Seeman, "Translation of DNA signals into polymer assembly instructions," Science **306**, 2072 (2004).
13. W. B. Sherman and N. C. Seeman, "A precisely controlled DNA biped walking device," Nano Lett. **4**, 1203 (2004).
14. J. S. Shin and N. A. Pierce, "A synthetic DNA walker for molecular transport," J. Am. Chem. Soc. **126**, 10834 (2004).
15. D. C. Lin, B. Yurke, and N. A. Langrana, "Mechanical properties of a reversible, DNA-crosslinked polyacrylamide hydrogel," J. Biomech. Eng. **126**, 104 (2004).
16. D.C. Lin, B. Yurke, and N.A. Langrana, "Use of rigid spherical inclusions in Young's moduli determination: application to DNA-crosslinked gels," J. Biomech. Eng. **127**, 571 (2005).
17. D. C. Lin, B. Yurke, and N. A. Langrana, "Inducing reversible stiffness changes in DNA-crosslinked gels," J. Mater. Res. **20**, 1456 (2005).
18. S. Nagahara and T. Matsuda, "Hydrogel formation via hybridization of oligonucleotides derivatized in water-soluble vinyl polymers," Polym. Gels Networks **4**, 111 (1996).
19. D.C. Lin, N.A. Langrana, and B. Yurke, "Force-displacement relationships for spherical inclusions in finite elastic media," J. Appl. Phys. **97**, 043510 (2005).
20. E. J. Semler, and P. V. Moghe, "Engineering hepatocyte functional fate through growth factor dynamics: the role of cell morphologic priming," Biotechnol. Bioeng. **75**, 510 (2001).
21. E. J. Semler, C. S. Ranucci, and P. V. Moghe, "Mechanochemical manipulation of hepatocyte aggregation can selectively induce or repress liver-specific function," Biotechnol. Bioeng. **69** 359 (2000).

Molecular Learning of wDNF Formulae

Byoung-Tak Zhang and Ha-Young Jang

Biointelligence Laboratory, Seoul National University, Seoul 151-742, Korea
{btzhang, hyjang}@bi.snu.ac.kr
http://bi.snu.ac.kr/

Abstract. We introduce a class of generalized DNF formulae called wDNF or weighted disjunctive normal form, and present a molecular algorithm that learns a wDNF formula from training examples. Realized in DNA molecules, the wDNF machines have a natural probabilistic semantics, allowing for their application beyond the pure Boolean logical structure of the standard DNF to real-life problems with uncertainty. The potential of the molecular wDNF machines is evaluated on real-life genomics data in simulation. Our empirical results suggest the possibility of building error-resilient molecular computers that are able to learn from data, potentially from wet DNA data.

1 Introduction

Disjunctive normal form (DNF) is a disjunction of conjunctions of Boolean variables, such as (x_1 AND x_2) OR (x_1 AND $\bar{x}_3$) OR (x_2 AND x_3) where x_i represent attributes or binary-valued variables and $\bar{x}_i$ are their negations. The conjunctions in the form of (x_1 AND x_2) are called terms. DNF offers an interesting structure for representing knowledge in a logical form. For example, any Boolean function can be represented by a finite set of terms. Although previous research shows that the k-term DNF, i.e. DNF having k terms at most, is learnable with attribute noise if the noise rate is known exactly [10, 8], the pure Boolean logical nature of DNF restricts its application [11].

Here we introduce a generalized form of DNF that is more resilient to noisy and/or incomplete data thus applicable beyond the pure logical problems. This weighted DNF or wDNF formula extends DNF twofold. On the conjunction level the attributes can appear multiple times, e.g. $x_1 x_1 = x_1^2$ as well as x_1. This allows for higher-order attributes, enhancing the expressive power of DNF (We hurry to mention that logically x_1 AND $x_1 = x_1$, but this is true only when the variable does not contain noise). On the disjunction level the terms are permitted to appear multiple times. Thus the entire formula is a "disjunctive ensemble of conjunctions of higher-order terms". The number of copies of the terms represents the weight of voting in decision making, hence the "weighted DNF".

We show that the wDNF formulae can be learned from training examples using DNA computing, resulting in molecular wDNF machines. The probabilistic nature of the computation performed by the molecular wDNF machines is discussed along with its robustness against uncertainty arising from both internal (e.g., molecular reaction) and external (e.g., data) sources. The general

A. Carbone and N.A. Pierce (Eds.): DNA11, LNCS 3892, pp. 427–437, 2006.

setting for learning the wDNF machines is similar to the genetic programming (GP) framework where the programs for digital computers are evolved using the principle of natural selection [5, 14, 7]. While GP evolves tree-structured expressions, here we evolve the wDNF expressions encoded in DNA molecules. Starting from a random combinatorial library of wDNF expressions our "molecular programming" (MP) method evolves a wDNF formula that best fits to the learning sample. The potential of the molecular wDNF machines is evaluated on a DNA-based diagnosis problem. The terms in wDNF in this particular case represent the conjunctive rules of DNA markers for diagnosing a leukemia. Our simulation studies demonstrate a robust and highly competitive performance of the molecular wDNF machines on this real-life data.

The paper is organized as follows. Section 2 presents a formal description of the wDNF form. Section 3 presents the molecular algorithm for learning a wDNF formula. Section 4 explains the probabilistic nature of the molecular wDNF machines based on statistical mechanics of DNA hybridization reactions. Section 5 shows the simulation results on the diagnosis problem. We also evaluate the robustness of wDNF formulae by analyzing wDNF formulae containing small portions of the complete combinatorial terms. Section 6 draws conclusions.

2   Weighted Disjunctive Normal Form (wDNF)

Let x_i denote an attribute or a Boolean variable, i.e. $x_i \in \{0, 1\}$. A literal consists of a variable x_i or its negation $\bar{x}_i$. The former is called a positive literal and the latter a negative literal. For notational simplicity, the negative literal can be considered as a new positive literal by renaming it as $x_j = \bar{x}_i$. We shall adopt this convention in the following, unless otherwise noted. More generally, we consider the powers of literals and denote a literal of degree r by x_i^r, where r is an integer.

Then, a term is defined as a conjunction of the (positive) literals of degree one:

$$C_i = (x_{i_1}, x_{i_2}, \cdots, x_{i_k}, \cdots, x_{i_{n_i}}) = x_{i_1} x_{i_2} \cdots x_{i_k} \cdots x_{i_{n_i}}, \qquad (1)$$

where $x_{i_k} \in \{x_1, x_2, ..., x_n\}$. For example, $C_i = (x_{i_1}, x_{i_2}, x_{i_3}) = x_1 x_4 x_5$ represents a term consisting of three literals x_1, x_4 and x_5. In general, the number of variables n_i in a term C_i may vary. A disjunctive normal form, DNF, on n literals is defined as the disjunction of the terms:

$$DNF = \{C_1, C_2, \cdots, C_j, \cdots, C_N\} = C_1 + C_2 + \cdots + C_j + \cdots + C_N, \qquad (2)$$

where C_j is a term of an arbitrary number of literals out of $x_1, ..., x_n$. A k-term DNF formula is a DNF formula with maximum k terms. For instance, $\{x_1 x_2 x_3, x_4 x_5, x_1 x_3 x_5\}$ is an example of a 3-term DNF formula on five literals x_1, x_2, x_3, x_4, x_5.

The weighted DNF (wDNF) generalizes the DNF in two ways. First, at the conjunction level the terms can be of higher degree r. Second, at the disjunction

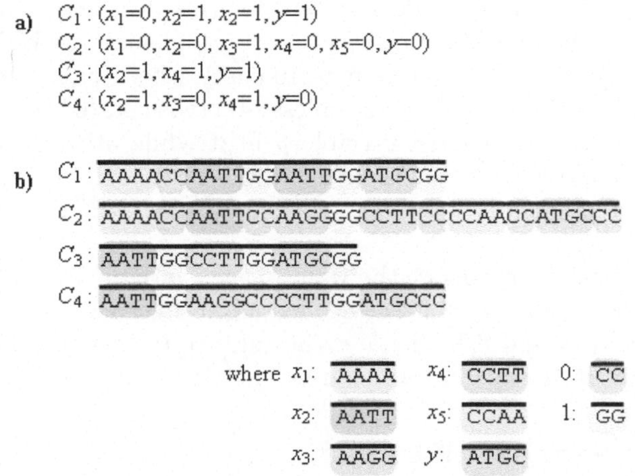

a) $C_1 : (x_1=0, x_2=1, x_2=1, y=1)$
    $C_2 : (x_1=0, x_2=0, x_3=1, x_4=0, x_5=0, y=0)$
    $C_3 : (x_2=1, x_4=1, y=1)$
    $C_4 : (x_2=1, x_3=0, x_4=1, y=0)$

b) $C_1 :$ AAAACCAATTGGAATTGGATGCGG

    $C_2 :$ AAAACCAATTCCAAGGGGCCTTCCCCAACCATGCCC

    $C_3 :$ AATTGGCCTTGGATGCGG

    $C_4 :$ AATTGGAAGGCCCCTTGGATGCCC

where x_1: AAAA x_4: CCTT   0: CC

      x_2: AATT x_5: CCAA   1: GG

      x_3: AAGG y: ATGC

Fig. 1. A wDNF formula in two different representations: (a) a collection of terms, (b) a library of DNA molecules corresponding to (a). The DNA code shown is illustration-purposes only.

level a term can appear w copies. Note that in Eqn. (1) all the literals in DNF are of maximum degree 1. In wDNF the term is generalized to contain literals of arbitrary degree. A term of degree r is defined then a conjunction of literals of the form:

$$C_i = (x_{i_1}^{r_{i_1}}, x_{i_2}^{r_{i_2}}, \cdots, x_{i_k}^{r_{i_k}}, \cdots, x_{i_{n_i}}^{r_{i_{n_i}}}) = x_{i_1}^{r_{i_1}} x_{i_2}^{r_{i_2}} \cdots x_{i_k}^{r_{i_k}} \cdots x_{i_{n_i}}^{r_{i_{n_i}}} \qquad (3)$$

where $x_{i_j} \in \{x_1, x_2, ..., x_n\}$ and $r_{i_j} \leq r$, $j = 1, ..., n_i$ for fixed r. For example, $C_i = (x_{i_1}^2, x_{i_2}^3, x_{i_3}^1) = x_{i_1}^2 x_{i_2}^3 x_{i_3}^1 = x_{i_1} x_{i_1} x_{i_2} x_{i_2} x_{i_2} x_{i_3}$ represents a term of degree 3. The generalized term is satisfied only if every occurrence of the literal is bound to a "sample" value. There are the designated instantiated "samples" of each literal.

Using the terms of degree r on n literals, a wDNF formula is defined as:

$$wDNF = \{w_i C_1, w_2 C_2, \cdots, w_j C_j, \cdots, w_N C_N\}$$
$$= w_1 C_1 + w_2 C_2 + \cdots + w_j C_j + \cdots + w_N C_N, \qquad (4)$$

where $w_j C_j$ means w_j copies of the term C_j. The coefficient w_j is interpreted to represent the "weight" or strength of the term. Thus, the number of variables matter in the generalized terms and the wDNF formulae.

To be more concrete, consider a wDNF formula for DNA-based diagnosis of disease shown in Figure 1(a). In this case the wDNF consists of four terms $C_1, ..., C_4$. A term is said to be "instantiated" if the values of the variables are bound to specific values. The instantiated term $C_1 = (x_1 = 0, x_2 = 1, x_2 = 1, y = 1)$, for example, encodes a diagnosis rule, where $y = 1$ indicates the label for disease. The meaning is that a DNA sample is decided positive ($y = 1$) if it contains two of the DNA marker 2 ($x_2 = 1$, $x_2 = 1$) and does not contain the

DNA marker 1 (other variables do not care for this decision). This procedure can be implemented by hybridization reaction of complementary DNA molecules. For example, bead separation can be used to check whether the required values are contained or not. In the following, unless otherwise stated, we shall assume every term in wDNF has a label variable y in it while other x-variables may appear or not. This does not lose the generality of the method since y-variable can be incorporated as an extra x-variable, but it makes the presentation more readable.

As shown in Figure 1 we encode the value of each variable as a DNA oligomer. For example, if we assume $x_1 = 0$ be encoded as a 6-mer like 'AAAACC', where 'AAAA' represents x_1 and 'CC' denotes the value 0. In this encoding scheme, a term consisting of 10 literals in total can be encoded as a 60-mer DNA.

3   Learning a wDNF Formula

In this section we describe the molecular algorithm for learning the wDNF formulae. The theoretical backgrounds of this procedure is given in the next section.

The goal is to learn a wDNF formula that best fits to a data set. We assume the training set D of K labeled DNA samples be given in the form

$$D = \{(\mathbf{x}_i, y_i)\}_{i=1}^{K}$$
$$\mathbf{x}_i = (x_{i_1}, x_{i_2}, ..., x_{i_n}) \in \{0, 1\}^n$$
$$y_i \in \{0, 1\},$$

where $\mathbf{x}_i$ is the sample data and y_i is the associated label. In the DNA-marker-based diagnosis problem, a training example $(10101, 1)$ means the sample is diagnosed positive $(y = 1)$ if it contains the DNA markers numbered 1, 3, and 5 $(x_1 = 1, x_3 = 1, x_5 = 1)$ and does not contain the rest $(x_2 = 0, x_4 = 0)$. Figure 1 shows an example in DNA encoding.

To learn the formula we initialize a library of DNA molecules representing random combinatorial wDNF terms as shown in Figure 2. Given a query pattern $\mathbf{x}_q$ we extract from the library all the molecules (terms) that match the query. The extraction can be implemented using hybridization reaction in the same way to check which markers exist. The idea is to chop the query sequence into subsequences for individual variables. These chopped query sequences hybridize with the wDNF formula in the library. Only the fully double-stranded sequences are then separated (by selecting out the single-stranded sequences by beads).

These molecules will have class labels from which we decide the majority label as the class of the query pattern. To perform the matching between $\mathbf{x}_i$ and $\mathbf{x}_q$ for $i = 1, ..., N$ in parallel, we present multiple copies (up to the number of the library size) of it. That is, we generate a collection

$$Q = \{\Delta c(x_1), \Delta c(x_2), ..., \Delta c(x_n), \Delta c(y)\}, \tag{5}$$

where $\Delta c(\cdot)$ denotes copies made by PCR. The class decision is made by comparing the number of elements in class 1, N_1, with that in class 0, N_0:

$$y^* = \arg\max_{y}\{N_y\}, \tag{6}$$

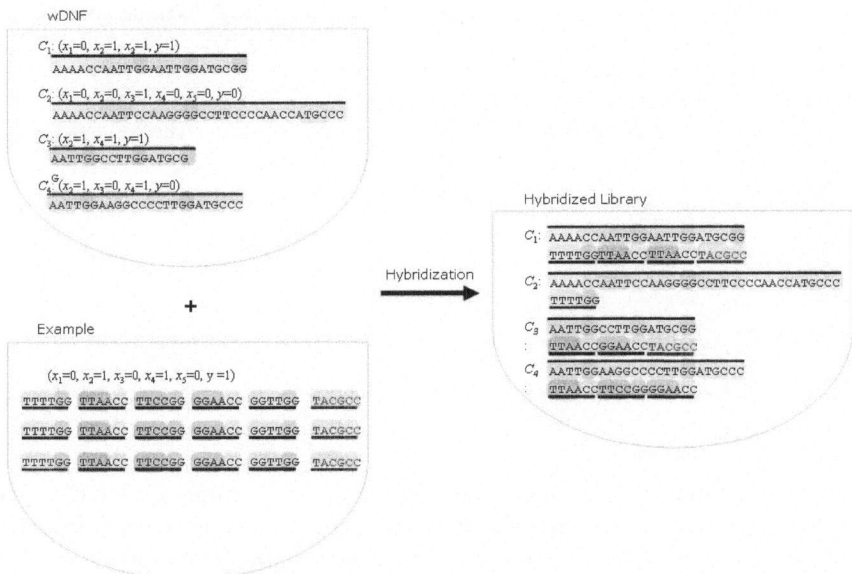

Fig. 2. Illustration of the decision-making procedure using the population of DNA-encoded terms. The query sample is chopped and provided in multiple copies to hybridize in parallel with the terms in the library.

where y takes 0 or 1. The next section discusses a theoretical background for this rule. For learning, we prepare two collections, M_+ and M_-, consisting of library elements that correctly (or incorrectly) classifies the query sample as follows:

- $M_+ = \{(\mathbf{u}_i, v_i) | \mathbf{u}_i \text{ consists of } x_i \text{ for } i = 1, ..., n \text{ and } v_i = y\}$
- $M_- = \{(\mathbf{u}_i, v_i) | \mathbf{u}_i \text{ consists of } x_i \text{ for } i = 1, ..., n \text{ and } v_i \neq y\}$.

Now, we describe how the library is revised to learn from newly observed data. The basic protocol is similar to that described in [13]. The difference is that the DNA molecules are now describing the generalized terms rather than simple examples and the whole test tube represents a wDNF formula. As a new training example $(\mathbf{x}, y)$ is given, we extract from the library the terms whose x-part matching with $\mathbf{x}$. The class y^* of $\mathbf{x}$ is determined by the classification procedure described above. Then, the matching terms (library patterns) are modified in their frequency depending on their contribution to the correct or incorrect classification of $\mathbf{x}$. If the label v of the library pattern $(\mathbf{u}, v)$ matching $\mathbf{x}$ is correct, i.e. $v = y$, it is reproduced:

$$L_t \leftarrow L_t + \Delta c(M_+). \tag{7}$$

If the label v is incorrect, i.e. $v \neq y$, the matching library pattern is removed from the library:

$$L_t \leftarrow L_t - \Delta c(M_-). \tag{8}$$

- 1. Let the library $L_0 = \{(\mathbf{u}_i, v_i)\}$ contain the initial wDNF formula. Let $t = 0$.
- 2. Let $t \leftarrow t + 1$.
- 3. Get a training example $(\mathbf{x}, y) = (x_1, x_2, ..., x_n, y)$.
- 4. Let $Q = \{\Delta c(x_1), \Delta c(x_2), ..., \Delta c(x_n), \Delta c(y)\}$.
- 5. Classify $\mathbf{x}$ using L_t as described in the text and construct the following:
  - $M_+ = \{(\mathbf{u}_i, v_i) | \mathbf{u}_i$ consists of x_i for $i = 1, ..., n$ and $v_i = y\}$.
  - $M_- = \{(\mathbf{u}_i, v_i) | \mathbf{u}_i$ consists of x_i for $i = 1, ..., n$ and $v_i \neq y\}$.
- 6. Update the library L as follows:
  - $L_t \leftarrow L_{t-1} + Q$.
  - $L_t \leftarrow L_t + \Delta c(M_+)$.
    Optionally, $L_t \leftarrow L_t - \Delta c(M_-)$.
- 7. Go to Step 2 if not terminated.

Fig. 3. The molecular programming (MP) procedure for learning a wDNF formula from examples

Figure 3 summarizes the molecular algorithm for learning a wDNF formula. Note that the library represents a kind of associative memory learned from data. In contrast to other molecular computation models of associated memory [2, 3, 6] proposed so far, the wDNF models contain higher-order patterns. An explicit probabilistic semantics underlying wDNF is also distinguished from other related work.

4   The Molecular wDNF Machine as a Probabilistic Computer

We consider the hybridization reaction between two single-stranded DNA molecules $\mathbf{x}_i$ and $\mathbf{x}_q$. Without loss of generality we consider $\mathbf{x}_i$ as the ith element (a term in wDNF) in the library and $\mathbf{x}_q$ as a query data. The probability of the ith term being retrieved by the query pattern is then expressed as Boltzmann distribution

$$P(\mathbf{x}_i|\mathbf{x}_q) = \frac{\exp\left(-\Delta G(\mathbf{x}_i|\mathbf{x}_q)/k_B T\right)}{\sum_j \exp\left(-\Delta G(\mathbf{x}_j|\mathbf{x}_q)/k_B T\right)}, \tag{9}$$

where j runs over the possible states of hybridization at the absolute temperature T [12]. ΔG is the Gibbs free energy change for the hybridization reaction and k_B is the Boltzmann constant [9].

A direct computation of this probability is difficult. However, we can approximate this by a Monte Carlo method performed "*in vitro*". To do this, we duplicate the molecules, both $\mathbf{x}_i$ and $\mathbf{x}_q$, let them hybridize, and count the double-stranded DNA at a fixed temperature T below the melting temperature T_m. The estimated value is obtained by averaging the values over the sample of size $|S|$:

$$P(\mathbf{x}_i|\mathbf{x}_q) \approx \frac{1}{|S|} \sum_{i=1}^{|S|} \rho(\mathbf{x}_i, \mathbf{x}_q), \tag{10}$$

where $\rho(\mathbf{x}_i, \mathbf{x}_q) = 1$ if i and q form a double-strand, and $\rho(\mathbf{x}_i, \mathbf{x}_q) = 0$ otherwise. The approximation can be made arbitrarily accurate by increasing the number of copies of the molecules.

By generalizing the above idea of "molecular" Monte Carlo simulation into the collection L of terms, $\mathbf{x}_i$, and a collection Q of a excessive number of the query pattern, $\mathbf{x}_q$, we can compute the probability distribution over the term patterns matching with a query pattern by

$$P_L(X|\mathbf{x}_q) \approx \frac{1}{|L|} \sum_{i=1}^{|L|} P(\mathbf{x}_i|\mathbf{x}_q), \tag{11}$$

where we assume that an excessive number of query molecules are put into the test tube so that all the terms have a fair chance of hybridizing with a query.

We now consider the library representing a k-wDNF, the wDNF formula with terms consisting only of k variables of degree 1. The ith molecule representing a term with k variables can be considered as a point estimator $f_i^{(k)}(X_1, X_2, ..., X_n, Y)$ of the probability distribution $P_L(X, Y)$. The whole library can then be thought of as a table representing the empirical distribution of the patterns

$$P_L(X, Y) \approx \frac{1}{|L|} \sum_{i=1}^{|L|} f_i^{(k)}(X_1, X_2, ..., X_n, Y), \tag{12}$$

$$f_i^{(k)}(X_1, X_2, ..., X_n, Y) = \frac{\exp\left(-\Delta G(X_1, X_2, ..., X_n|\mathbf{x}_q)/k_B T\right)}{\sum_j \exp\left(-\Delta G(X_1, X_2, ..., X_n|\mathbf{x}_q)/k_B T\right)}, \tag{13}$$

where ΔG is the Gibbs free energy change for the hybridization reaction and k_B is the Boltzmann constant.

Given the statistical physical interpretation of DNA hybridization and the wDNF representation as an empirical probability distribution, the learning process can be formulated in a probabilistic framework. The objective of learning is to find a wDNF or the library L that best predicts the output label y given input variables $\mathbf{x}$ for all possible training data $(\mathbf{x}, y)$ in the problem space (X, Y). The L can be found iteratively by starting with an initial L_0 and updating it as new sample $\mathbf{x}$ is observed:

$$L^* = \arg\max_L P(L|\mathbf{x}, y) = \arg\max_L \frac{P(\mathbf{x}, y|L)P(L)}{P(\mathbf{x}, y)}$$
$$= \arg\max_L P(\mathbf{x}, y|L)P(L) = \arg\max_L P_L(\mathbf{x}, y)P(L), \tag{14}$$

where we used the Bayes rule $P(A|B) = \frac{P(B|A)P(A)}{P(B)}$.

Once an L is given, the best class label for a query pattern can be determined by computing the probability of each class conditional on the input pattern $\mathbf{x}$, and then determining the class whose conditional probability is the highest, i.e.

$$y^* = \arg\max_{Y \in \{0,1\}} P_L(Y|\mathbf{x}) = \arg\max_{Y \in \{0,1\}} \frac{P_L(Y, \mathbf{x})}{P(\mathbf{x})}, \tag{15}$$

where we used the relation $P(A, B) = P(A|B)P(B)$ and Y represents the candidate classes.

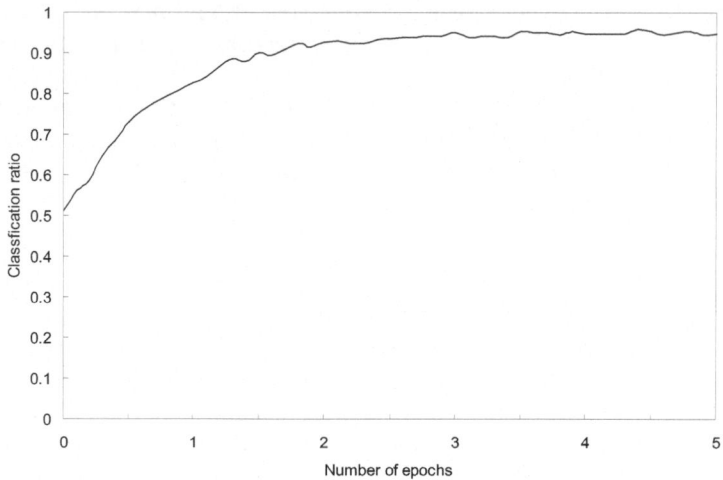

Fig. 4. Learning curve of the complete wDNF library. Shown are the average values for 12 runs. Parameters used are the learning rate $= 10^{-2}$ and $\beta = 20$.

5   Simulation Results and Discussion

We evaluated this method on a real-life medical diagnosis problem in simulation. Gene expression data are collected from microarray experiments for AML/ALL leukemia [4]. The microarray data are preprocessed and 10 genes were selected out of 12600 genes. The training set consists of 120 examples each consisting of 10 genes plus the associated leukemia class. A 6-fold cross-validation is used for testing the performance. That is, the whole data set of 120 examples is partitioned into 6 subsets and a total of six learning trials are executed, where each trial used a subset of 20 examples for test and the remaining 100 examples for training. The library was initialized to contain each and every term of wDNF on the 10 variables. These include $(x_1 = 0, y = 0), (x_1 = 0, y = 1), (x_1 = 1, y = 0), (x_1 = 1, y = 1), (x_1 = 0, x_2 = 0, y = 0), (x_1 = 0, x_2 = 0, y = 1), (x_1 = 1, x_2 = 0, y = 0), \ldots$ Thus, the total number of the different library elements is

$$N = \sum_{k=1}^{10} {}_{10}C_k \cdot 2^k \cdot 2 = 118,096,$$

where ${}_{10}C_k$ denotes the number of cases choosing k variables out of 10. For the simulation of in vitro computation of the wDNF formula, we used the library size of $118,096 \times 10^6$, i.e. the initial library was generated by copying each element 10^6 times. Thus, the library consists of multiple copies of the same terms and we evolved the distributions of the terms through the molecular programming procedure.

For decision making, we used a sigmoid squashing function:

$$f(x) = \frac{1}{1 + \exp(-\beta x)} \tag{16}$$

where β is a constant which reflects the level of noise and sets the decision boundary. As mentioned in the previous section, we count the number of each

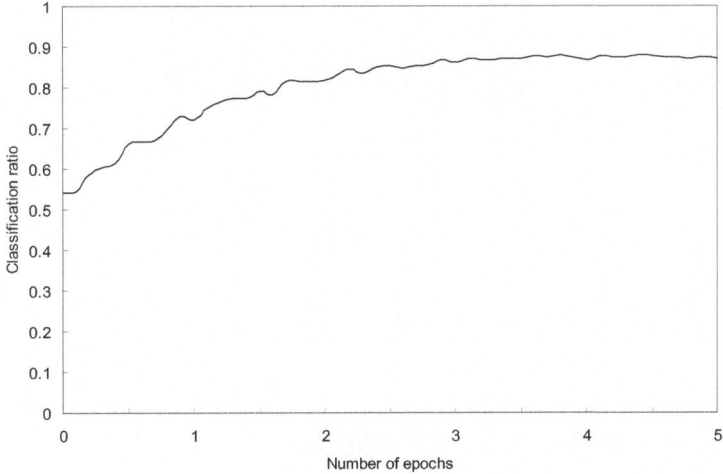

Fig. 5. Learning curve of the partial wDNF libraries (average over 12 runs each) containing 10 % of the complete wDNF. Same parameters as in Figure 4.

term which answers positive or negative. Then, the proportion of the positives and the negatives is calculated. This result is the input to the sigmoid function. We make a decision probabilistically based on the output of the sigmoid function.

Figure 4 shows the evolution of the performance as learning proceeds. We presented the positive training example and the negative example alternatingly. It should be mentioned that one generation consists of the presentation of one positive and one negative example. The performance was measured at every generation, i.e. each time a pair of new training examples was observed. One sweep through the training set constitutes an *epoch* which is equivalent to 60 generations in this experiment. The best performance of approximately 95% correct classification on the test data set was obtained in 2 epochs.

It is observed that the total number of the different library elements grows exponentially as we allow higher-order terms in the wDNF formulae. That is, if the dimension of the query is high, the total number of the different library elements grows very rapidly. Considering this, it is interesting to know how the total number of library elements affects the performance of wDNF formulae. In order to see how much the performance of wDNF formulae dependends on the complexity of structures, we ran simulations with partial libraries which are generated by eliminating some terms from the complete library. The results are shown in Figure 5. As expected, the wDNF formulae with partial terms perform less than the complete wDNF formulae. However, the results of the partial libraries are still robust. In particular, in the extreme case of the partial library consisting of only 10 % of the full combinatorial terms achieved approximately 90% in absolute accuracy. Figure 6 compares the performances of different partial

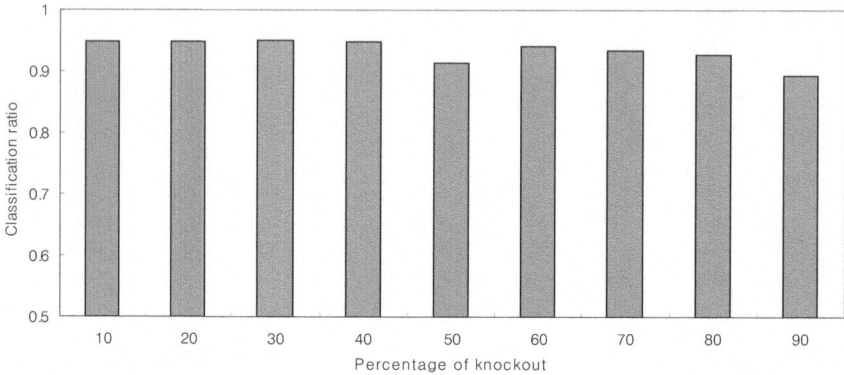

Fig. 6. The performance of partial libraries in which some portion of libraries are eliminated after 5 epochs. From left to right 10%, 20%, ..., 90% of whole libraries are eliminated.

libraries made by knocking out 10 % to 90 %. The performance was measured in 5 epochs. These results clearly show that the full library is not absolutely necessary to solve this real-life problem using wDNF formulae, suggesting the potential for robust decision making in vitro experiments.

6   Conclusion

We introduced the weighted disjunctive normal form (wDNF) as a scheme for representing probability distributions and presented a method for learning a wDNF formula from examples. The learning approach is distinguished from other DNA computing tasks in that the computational result here is a program or machine that can be reused for solving multiple instances of the problem. As the genetic programming provides an automatic programming method for digital computers, the molecular programming provides a method for automatic programming of molecular computers, in our case a wDNF machine.

The results on the leukemia diagnosis problem show that effective solution is possible using the wDNF learning. In particular, our simulation results were competitive to existing state-of-the-art machine learning algorithms. This is somewhat surprising considering the fact that the terms are random conjunctive combinations of Boolean variables. Our analysis suggests that even though the individual terms are simple, their collection as a whole, i.e. wDNF, has a weighted, ensemble representation with redundancy that leads to error-resilient decision making.

Our results on DNA-based diagnosis also suggest a potential use of the molecular learning method for automatically deriving decision rules from wet DNA data. Recently, Benenson et al. [1] demonstrate the possibility of in vitro or in vivo diagnosis. Here the decision rules for diagnosis are hard-coded by the designer. The wDNF learning approach may provide a further step forward into this direction of research by providing a potential means for automatically constructing the robust decision rules from raw data.

Acknowledgements

This research was supported by the Molecular Evolutionary Computing (MEC) Project of MICE and by the National Research Laboratory (NRL) Program of MOST.

References

1. Benenson, Y., Gil, B., Ben-Dor, U., Adar, R., Shapiro, E. "An autonomous molecular computer for logical control of gene expression," *Nature*, 429, 423-429, 2004.
2. Baum, E. B., "Building an associative memory vastly larger than the brain," *Science*, 268:583-585, 1995.
3. Chen, J. Deaton, R. and Wang, Y.-Z., "A DNA-based memory with in vitro learning and associative recall," DNA9, *Lecture Notes in Computer Science* 2943: 145-156, 2004.
4. Cheok, M.*v* et al., "Treatment-specific changes in gene expression discriminate in vivo drug response in human leukemia cells," *Nature Genetics*, 34:85-90, 2003.
5. Koza, J. R., *Genetic Programming: On the Programming of Computers by Means of Natural Selection*, MIT Press, Cambridge, MA, USA, 1992.
6. Reif, J.H., LaBean, T.H., Pirrung, M., Rana, V.S., Guo, B., Kingsford, C., Wickham, G.S., "Experimental construction of very large scale DNA databases with associative search capability," DNA7, *Lecture Notes in Computer Science* 2340: 231-247, 2002.
7. Rose, J. A., Deaton, R. J., Hagiya, M., Suyama, A., "A DNA-based in vitro genetic program", *Journal of Biological Physics*, 28:493-498, 2002.
8. Sakakibara, Y., "Solving computational learning problems with Boolean formulae on DNA computers," DNA6, *Lecture Notes in Computer Science*, 2052:220-230, 2001.
9. SantaLucia, J. and Hicks, D., "The thermodynamics of DNA structural motifs," *Annu. Rev. Biophys. Biomol. Struct.*, 33:415-440, 2004.
10. Shackelford, G. and Volper, D., "Learning k-DNF with noise in the attributes," *COLT '88: Proc. First Annual Workshop on Computational Learning Theory*, 97-103, 1988.
11. Valiant, L., "Robust logics", *Proc. ACM Symposium on the Theory of Computing (STOC 99)*, pp. 642-651, 1999.
12. Wartel, R.M. and Benight, A.S., "Thermal denaturation of DNA molecules: A comparison of theory with experiments," *Physics Reports*, 126(2):67-107, 1985.
13. Zhang, B.-T. and Jang, H.-Y., "A Bayesian algorithm for in vitro molecular evolution of pattern classifiers," *Proc. of 10th Int. Meeting on DNA Computing*, Milan, Italy, pp. 294-303, 2004.
14. Zhang, B.-T. and Mühlenbein, H., "Balancing accuracy and parsimony in genetic programming," *Evolutionary Computation*, 3(1):17-38, 1995.

Author Index

Lecture Notes in Computer Science

For information about Vols. 1–3891

please contact your bookseller or Springer

Vol. 3944: J. Quiñonero-Candela, I. Dagan, B. Magnini, F. d'Alché-Buc (Eds.), Machine Learning Challenges. XIII, 462 pages. 2006. (Sublibrary LNAI).

Vol. 3943: N. Guelfi, A. Savidis (Eds.), Rapid Integration of Software Engineering Techniques. X, 289 pages. 2006.

Vol. 3942: Z. Pan, R. Aylett, H. Diener, X. Jin, S. Göbel, L. Li (Eds.), Technologies for E-Learning and Digital Entertainment. XXV, 1396 pages. 2006.

Vol. 3939: C. Priami, L. Cardelli, S. Emmott (Eds.), Transactions on Computational Systems Biology IV. VII, 141 pages. 2006. (Sublibrary LNBI).

Vol. 3936: M. Lalmas, A. MacFarlane, S. Rüger, A. Tombros, T. Tsikrika, A. Yavlinsky (Eds.), Advances in Information Retrieval. XIX, 584 pages. 2006.

Vol. 3935: D. Won, S. Kim (Eds.), Information Security and Cryptology - ICISC 2005. XIV, 458 pages. 2006.

Vol. 3934: J.A. Clark, R.F. Paige, F.A. C. Polack, P.J. Brooke (Eds.), Security in Pervasive Computing. X, 243 pages. 2006.

Vol. 3933: F. Bonchi, J.-F. Boulicaut (Eds.), Knowledge Discovery in Inductive Databases. VIII, 251 pages. 2006.

Vol. 3931: B. Apolloni, M. Marinaro, G. Nicosia, R. Tagliaferri (Eds.), Neural Nets. XIII, 370 pages. 2006.

Vol. 3930: D.S. Yeung, Z.-Q. Liu, X.-Z. Wang, H. Yan (Eds.), Advances in Machine Learning and Cybernetics. XXI, 1110 pages. 2006. (Sublibrary LNAI).

Vol. 3929: W. MacCaull, M. Winter, I. Düntsch (Eds.), Relational Methods in Computer Science. VIII, 263 pages. 2006.

Vol. 3928: J. Domingo-Ferrer, J. Posegga, D. Schreckling (Eds.), Smart Card Research and Advanced Applications. XI, 359 pages. 2006.

Vol. 3927: J. Hespanha, A. Tiwari (Eds.), Hybrid Systems: Computation and Control. XII, 584 pages. 2006.

Vol. 3925: A. Valmari (Ed.), Model Checking Software. X, 307 pages. 2006.

Vol. 3924: P. Sestoft (Ed.), Programming Languages and Systems. XII, 343 pages. 2006.

Vol. 3923: A. Mycroft, A. Zeller (Eds.), Compiler Construction. XIII, 277 pages. 2006.

Vol. 3922: L. Baresi, R. Heckel (Eds.), Fundamental Approaches to Software Engineering. XIII, 427 pages. 2006.

Vol. 3921: L. Aceto, A. Ingólfsdóttir (Eds.), Foundations of Software Science and Computation Structures. XV, 447 pages. 2006.

Vol. 3920: H. Hermanns, J. Palsberg (Eds.), Tools and Algorithms for the Construction and Analysis of Systems. XIV, 506 pages. 2006.

Vol. 3918: W.K. Ng, M. Kitsuregawa, J. Li, K. Chang (Eds.), Advances in Knowledge Discovery and Data Mining. XXIV, 879 pages. 2006. (Sublibrary LNAI).

Vol. 3917: H. Chen, F.Y. Wang, C.C. Yang, D. Zeng, M. Chau, K. Chang (Eds.), Intelligence and Security Informatics. XII, 186 pages. 2006.

Vol. 3916: J. Li, Q. Yang, A.-H. Tan (Eds.), Data Mining for Biomedical Applications. VIII, 155 pages. 2006. (Sublibrary LNBI).

Vol. 3915: R. Nayak, M.J. Zaki (Eds.), Knowledge Discovery from XML Documents. VIII, 105 pages. 2006.

Vol. 3914: A. Garcia, R. Choren, C. Lucena, P. Giorgini, T. Holvoet, A. Romanovsky (Eds.), Software Engineering for Multi-Agent Systems IV. XIV, 255 pages. 2006.

Vol. 3910: S.A. Brueckner, G.D.M. Serugendo, D. Hales, F. Zambonelli (Eds.), Engineering Self-Organising Systems. XII, 245 pages. 2006. (Sublibrary LNAI).

Vol. 3909: A. Apostolico, C. Guerra, S. Istrail, P. Pevzner, M. Waterman (Eds.), Research in Computational Molecular Biology. XVII, 612 pages. 2006. (Sublibrary LNBI).

Vol. 3908: A. Bui, M. Bui, T. Böhme, H. Unger (Eds.), Innovative Internet Community Systems. VIII, 207 pages. 2006.

Vol. 3907: F. Rothlauf, J. Branke, S. Cagnoni, E. Costa, C. Cotta, R. Drechsler, E. Lutton, P. Machado, J.H. Moore, J. Romero, G.D. Smith, G. Squillero, H. Takagi (Eds.), Applications of Evolutionary Computing. XXIV, 813 pages. 2006.

Vol. 3906: J. Gottlieb, G.R. Raidl (Eds.), Evolutionary Computation in Combinatorial Optimization. XI, 293 pages. 2006.

Vol. 3905: P. Collet, M. Tomassini, M. Ebner, S. Gustafson, A. Ekárt (Eds.), Genetic Programming. XI, 361 pages. 2006.

Vol. 3904: M. Baldoni, U. Endriss, A. Omicini, P. Torroni (Eds.), Declarative Agent Languages and Technologies III. XII, 245 pages. 2006. (Sublibrary LNAI).

Vol. 3903: K. Chen, R. Deng, X. Lai, J. Zhou (Eds.), Information Security Practice and Experience. XIV, 392 pages. 2006.

Vol. 3902: R. Kronland-Martinet, T. Voinier, S. Ystad (Eds.), Computer Music Modeling and Retrieval. XI, 275 pages. 2006.

Vol. 3901: P.M. Hill (Ed.), Logic Based Program Synthesis and Transformation. X, 179 pages. 2006.

Vol. 3900: F. Toni, P. Torroni (Eds.), Computational Logic in Multi-Agent Systems. XVII, 427 pages. 2006. (Sublibrary LNAI).

Vol. 3899: S. Frintrop, VOCUS: A Visual Attention System for Object Detection and Goal-Directed Search. XIV, 216 pages. 2006. (Sublibrary LNAI).

Vol. 3898: K. Tuyls, P.J. 't Hoen, K. Verbeeck, S. Sen (Eds.), Learning and Adaption in Multi-Agent Systems. X, 217 pages. 2006. (Sublibrary LNAI).

Vol. 3897: B. Preneel, S. Tavares (Eds.), Selected Areas in Cryptography. XI, 371 pages. 2006.

Vol. 3896: Y. Ioannidis, M.H. Scholl, J.W. Schmidt, F. Matthes, M. Hatzopoulos, K. Boehm, A. Kemper, T. Grust, C. Boehm (Eds.), Advances in Database Technology - EDBT 2006. XIV, 1208 pages. 2006.

Vol. 3895: O. Goldreich, A.L. Rosenberg, A.L. Selman (Eds.), Theoretical Computer Science. XII, 399 pages. 2006.

Vol. 3894: W. Grass, B. Sick, K. Waldschmidt (Eds.), Architecture of Computing Systems - ARCS 2006. XII, 496 pages. 2006.

Vol. 3893: L. Atzori, D.D. Giusto, R. Leonardi, F. Pereira (Eds.), Visual Content Processing and Representation. IX, 224 pages. 2006.

Vol. 3892: A. Carbone, N.A. Pierce (Eds.), DNA Computing. XI, 440 pages. 2006.